CRITICAL INFRASTRUCTURE PROTECTION IN HOMELAND SECURITY

CRITICAL INFRASTRUCTURE PROTECTION IN HOMELAND SECURITY

DEFENDING A NETWORKED NATION

Ted G. Lewis
Naval Postgraduate School
Monterey, California

WILEY-INTERSCIENCE
A JOHN WILEY & SONS, INC., PUBLICATION

Published by John Wiley & Sons, Inc., Hoboken, New Jersey.
Published simultaneously in Canada.

For general information on our other products and services or for technical support, please contact our Customer Care Department within the United States at (800) 762-2974, outside the United States at (317) 572-3993 or fax (317) 572-4002.

Wiley also publishes its books in a variety of electronic formats. Some content that appears in print may not be available in electronic formats. For more information about Wiley products, visit our web site at www.wiley.com.

Library of Congress Cataloging-in-Publication Data:

Lewis, T. G. (Theodore Gyle), 1941–
Critical infrastructure protection in homeland security: defending a networked nation/Ted G. Lewis.
p. cm.
"Published simultaneously in Canada."
ISBN-13: 978-0-471-78628-3
ISBN-10: 0-471-78628-4
1. Computer networks—Security measures—United States. 2. Computer security—United States—Planning. 3. Terrorism—United States—Prevention. 4. Terrorism—Government policy—United States. 5. Civil defense—United States. I. Title.

QA76. 9. A25L5 2006
005.8'0973—dc22

2005026306

Printed in the United States of America

10 9 8 7 6 5 4 3 2 1

CONTENTS

PREFACE

This book explains why the various infrastructure sectors have evolved into today's critical infrastructures and then proposes several quantitative procedures for evaluating their vulnerability and establishing optimal policies for reducing these vulnerabilities. It is the first *scientific* study of the new field called *critical infrastructure protection*. By "scientific" I mean that I have tried to explain why infrastructure sectors as diverse as water supply systems and the Internet have surprisingly similar structures. Even more important, I propose a rigorous approach for studying these sectors that I believe is general enough to analyze fundamental sectors such as water, power, energy, telecommunications, and the Internet.

The reader will learn how to apply quantitative vulnerability analysis to a variety of infrastructure sectors and then be able to decide the best way to allocate limited funding in such as way as to minimize overall risk. As far as I know, this is the first time anyone has tried to formalize and quantify the field of critical infrastructure protection.

I have attempted to establish the foundations of a new discipline made necessary by the al-Qaeda attacks on the United States on September 11, 2001 (9/11). Before 9/11, the security of infrastructure was taken for granted. It was unthinkable for anyone to purposely destroy power plants, cut off water supplies, disable voice and data communications, deny access to information-processing computers, and render the Internet useless. Consequently, these systems were optimized for profit, efficient operation, and low cost. Security was sacrificed for economy. This public policy—in operation for more than a century—has left nearly all infrastructure systems vulnerable to manmade destruction.

The question addressed by this book is, "what should be protected, and how?" This question is nontrivial because of the enormous size and complexity of infrastructure in the United States. The solution is made even more challenging by the entangled regulatory and system interdependencies of the various infrastructure sectors. The answer is to allocate the nation's scarce resources to the most critical components of each infrastructure—the so-called *critical nodes*. In short, the best way to protect infrastructure is to identify and protect (harden) the critical nodes of each sector. But what parts of these vast structures are "critical?" This question is key. I claim that the optimal policy for critical infrastructure protection is to identify and protect a handful of critical assets throughout the United States. For example, perhaps less than 100 essential servers maintain the World Wide Web. There are perhaps fewer than a dozen critical nodes in the nation's energy supply chain, and maybe as few as 1000 key links in the major power grids that all other sectors depend on so heavily.

Chapter 1 surveys the national strategy and recommends five principles as guides for how to approach the protection of infrastructures. Although critical infrastructure protection is a massive problem, it turns out that a handful of principles can be applied to solve this problem, or at least to start the journey that will lead to a solution. Chapter 1 also analyzes the national strategy and points out several gaps between the ideal approach and the reality.

In Chapter 2, I briefly review the history of infrastructure protection—from the 1962 Cuban Missile Crisis to the formation of the U.S. Department of Homeland Security in 2003. This historical account of how the United States became aware of, and concerned for, infrastructures sets the stage for subsequent chapters. However, it does not offer any solutions to organizational and structural problems that plague government. I leave this challenge to another author.

Chapter 3 surveys some challenges to protecting the nation's infrastructures and key assets. This necessity is preliminary so that the reader can put the challenges into perspective and understand why I have narrowed the study of infrastructures down to a much smaller subset than actually exists. I have attempted to carve out a small enough territory that it can be adequately covered in a single book.

In Chapters 4–6, I establish the theory needed to master critical infrastructure protection as a scientific, formal discipline. I begin by claiming that all infrastructures of interest can be represented as a network of connected components called *nodes*. These nodes are connected by *links*. By simply counting the number of links at each node, I can identify the critical nodes. In most cases, there are one or two critical nodes, which reduces the problem of protection by several orders of magnitude. In this way, the concept of an infrastructure as a network is established and used to reduce the complexity of size—a challenge we need to surmount because of the vastness of each critical infrastructure. Without network theory, the problem is too large—we can never protect every mile of railroad, every power line, and every telephone pole. Only through network analysis can we reduce the problem of critical infrastructure protection to a workable (and solvable) problem!

Chapters 5 and 6 describe a method of vulnerability analysis and risk assessment based on network theory and the reliability engineer's *fault tree* technology. In these two chapters, I present a five-step vulnerability and risk assessment process that uses estimates of the cost and probability of an attack to compute an investment strategy aimed at reducing risk in the most effective way. Chapter 5 is focused on modeling the infrastructure as a fault tree, and Chapter 6 is focused on computing the best way to allocate the risk reduction budget.

The first step in the process described in Chapters 5 and 6 is to model the infrastructure as a network, find the critical nodes of this network, and then represent the structure of the critical node as a fault tree. The fault tree is converted into an event tree that enumerates all possible vulnerabilities and combinations of vulnerabilities. The fault and event trees identify the single- and multiple-combination events as well as the most-probable events that may occur because of threats.

Chapter 6 describes a variety of risk assessment algorithms. The idea is to allocate limited resources (money) in an optimal fashion such that overall risk is minimized. But how is vulnerability and risk defined? And what is the objective of resource allocation? Is it to reduce risk, eliminate all vulnerabilities, or simply prevent the worst thing from happening? As it turns out, we must decide which strategy is best, from among several competing strategies.

This method of assessment—called model-based vulnerability analysis (MBVA) for obvious reasons—is based on sound principles of logic, probability, and cost minimization. MBVA provides the policy maker with a scientific answer to the questions, "what is worthwhile protecting, and by how much?" MBVA is the only known method of vulnerability analysis and risk assessment that combines asset identification with quantitative analysis to reach a policy decision. It tells the decision maker how much money to spend on protecting the most critical components of the infrastructure.

Chapters 7–14 simply apply the MBVA technique to level 1 infrastructures: water, power and energy, information (telecommunications, Internet, Web, and cyber-security), and the monitoring and management networks that control them (SCADA). Power and energy are treated separately because of their size and complexity. The information sector is discussed in several chapters because it is a large and important topic. Unfortunately, the remaining eight major sectors defined by the national strategy are not covered in this volume because of their shear size and complexity.

There are several companions to this book: a website at http://www.CHDS.us, a CD containing audio and video lectures and articles, and the software described in this book. Both of these companions contribute more depth to the subject. The electronic media (Web and disk) contain executable programs for demonstrating concepts and reducing the mathematical labor during vulnerability analysis. Program FTplus.html",4>FTplus.html (FT.jar",4>FT.jar on the desktop), for example, performs the optimal resource allocation calculations described in Chapter 6. The RSA program calculates a public key and encrypts text automatically, thus providing the reader with hands-on tools for studying encryption. Other programs perform simulations, such as POWERGRAPH, which shows how a scale-free network emerges from a random network—the basis for today's critical infrastructure architectures. A program called TERMITES reinforces one of the most important concepts of this book: how and why critical nodes are formed. TERMITES illustrates clustering—the concentration of assets around one or more critical nodes, which then become the most vulnerable components of all. A novel program called NetworkAnalysis.html (a.k.a. NA.jar on the desktop) uses complex adaptive system algorithms to allocate resources to networks that tend to fail by cascading, such as the power grid. This program computes the best way to protect an infrastructure by allocating resources optimally to critical components of the infrastructure.

FTplus.html and *NetworkAnalysis.html* run from within any standard browser whether on a Microsoft Windows or Apple Macintosh computer. *NA.jar* and *FT.jar* run as stand-alone desktop applications and allow you to save your work as a local file. The source code is available as well, so you can modify it and do your own research.

In addition, the website and disk contain several audio tracks of the materials covered in the book. The audio tracks may be downloaded into a computer and then into an MP-3 player for mobile learning. For example, the 2003 National Strategy for the Protection of Critical Infrastructure and Key Assets is available as a collection of several audio tracks. For history and political science students, the foundational presidential directives (PDD-39, PDD-63, and HSPD-7) have been similarly transcribed into an audio book and are available online and on the disk.

This book is one component of blended learning—the combination of text, audio/video disk, and Web page. Specifically, several electronic lectures have been produced for online and CD viewing. All you need is a browser and either access to http://www.CHDS.us or a copy of the companion CD. The audio/video streaming lectures are tuned to this book, the website, and other content, such as the software for simulating various sectors and demonstrating vital concepts. In this way, the self-taught learner or classroom instructor can elect to learn by reading, listening, looking, or through participation in a traditional classroom setting.

I began developing the ideas for this book in the fall of 2002 and published a draft textbook in 2003. The material was class-tested in 2004, revised, and republished in 2004–2005. This book, the website, and the associated electronic media have been used extensively to teach a course labeled, CS 3660 Critical Infrastructure: Vulnerability and Analysis—one of a dozen courses given to military and civilian students enrolled in the Master of Arts in Security Studies, Homeland Defense, and Security curriculum at the Naval Postgraduate School, Monterey, CA. It is appropriate for upper division undergraduate and first-year graduate students majoring in national security, computing, and policy subjects where both policy and technical decisions are analyzed. Although it has been thoroughly class tested, it is still now without flaws. I take responsibility for any errors, inconsistencies, and exaggerations that may still exist in this edition.

I would like to thank my students for their insights, feedback, and contributions to this work. They have taught me well! Additionally, I would like to thank Steve McNally of Bellevue University, Hilda Blanco, University of Washington, and her students, for giving feedback on early drafts of this book. Joe Weiss was invaluable as a careful reader and critic of the SCADA chapter. Rudy Darken made many important contributions to the ideas and delivery methods used over 2 years of class testing.

TED G. LEWIS

December 2005

ABOUT THE AUTHOR

Ted G. Lewis has a distinguished 35-year career as a computer scientist, author, businessman, and scholar. His undergraduate degree is in Mathematics from Oregon State University (1966), and his graduate degrees were awarded in 1970 (M.S. Computer Science) and 1971 (Ph.D. Computer Science) from Washington State University. Since 1971 he has participated in several significant "firsts": In the late 1970s, he wrote the first personal computer book (*How to Profit From Your Personal Computer*); in 2002, he co-created, with Paul Stockton, the first graduate degree program in Homeland Security. In between, Lewis helped create the first Internet Car while serving as President and CEO of DaimlerChrysler Research and Technology, North America. During his technical career, he invented several important algorithms in software engineering (horizontal–vertical algorithm for deadlock detection); parallel processing (static scheduling of parallel programs on arbitrary architectures); and the model-based vulnerability analysis method of critical infrastructure risk assessment. And now he has written the first textbook to establish the study of critical infrastructure protection as a formal, scientific discipline. He has over 100 publications, including over 30 books. In 1997–1999, his books, *Friction-Free Economy* and *Microsoft Rising* documented the technical and economic forces that shaped the Internet Bubble, and in March 2000, he predicted its precipitous fall (*IEEE Computer Magazine*, March 2000). He is perhaps best known to the members of the IEEE Computer Society for a series of provocative articles appearing in the Binary Critic column of *IEEE Computer* from 1995 to 2000. His management experience began in 1988 as Technical Director of the Oregon Advanced Computing Institute. During 1993–1997, he was chairman of the Computer Science Department at the Naval Postgraduate School, and during 1999–2000, he was CEO of DaimlerChrysler Research and Technology NA in Palo Alto, CA. From 2001 to 2002, he was Senior Vice President of Digital Development for the Eastman Kodak Company. Currently, he is professor of Computer Science and Academic Associate of the Homeland Defense and Security curriculum at the Naval Postgraduate School.

"With the much-awaited publication of *Critical Infrastructure Protection in Homeland Security*, Professor Ted Lewis has provided Homeland Security specialists, law enforcement personnel, emergency managers, critical infrastructure experts and those whose day-to-day duties involve infrastructure security and protection with a timely, relevant and invaluable resource for defending the very essence of the American homeland. Lucidly written, perceptively analyzed and exhaustively researched, *Critical Infrastructure Protection in Homeland Security* is a work that reflects the concerns of our time while providing a viable blueprint for protecting our shared technological heritage. For those interested in critical infrastructure protection, Homeland Security and national defense, there is no better one-stop resource than this book. Read it, and you'll never be in the dark again when it comes to critical infrastructure protection."

David Longshore, New York City Homeland Security Liaison

"Professor Lewis's definitive textbook on critical infrastructure protection is a fascinating study of one of the challenges facing the nation in combating terrorism. In clear and concise language he establishes the foundation for his theory that critical infrastructure sectors are networks of critical nodes and links. Through network analysis, he identifies the most critical components of water systems, telecommunication systems, power grids, energy supply chains, and cyber systems such as the Internet and the World Wide Web. This is a must-read for anyone who wants to understand how to protect the nation's most-valuable physical assets."

Richard Elster, Provost, Naval Postgraduate School

CHAPTER 1

Strategy

What is the motivation for studying critical infrastructure protection? What are the central issues that need to be addressed to create a meaningful strategy for dealing with threats against infrastructure? Moreover, what is the national strategy? This chapter introduces the reader to the national strategy for the protection of critical infrastructure; identifies the roles and responsibilities of the federal, state, and local governments; lays out the approach being taken by the Department of Defense (DOD), and the newly created Department of Homeland Security (DHS), and then postulates five strategic principles of critical infrastructure.

This chapter makes the following claims and arguments:

1. Protection of critical infrastructure such as water, power, energy, and telecommunications is vital because of the impact such destruction would have on casualties, the economy, the psychology, and the pride of the nation.
2. Homeland defense and homeland security differ: The DOD is responsible for defense; and federal, state, and local governments are responsible for security. This division of responsibility defines roles and responsibilities of each, but their intersection—say along the borders and coastline—remains blurred at this time.
3. Federalism dictates a division of labor across federal, state, and local jurisdictions. The DHS is responsible for cross-sector roles such as standardization, research, and education. Delegation of major responsibility for critical infrastructure protection to state and local government is suboptimal, because nearly all critical infrastructure spans counties, states, and regions of the country.
4. When it comes to critical infrastructure protection, we will learn that it takes a network to fight a network. The United States and its collaborating cities, states, and regional partners are organized as a hierarchical bureaucracy. In other words, the nation is not network-centric. On the other hand, the organizational architecture of terrorist organizations is a network, which means they can flex and react quickly—much more quickly than hierarchical

Critical Infrastructure Protection in Homeland Security: Defending a Networked Nation, edited by Ted G. Lewis

organizations. By understanding network architectures in general, we will be able to apply network theory to both organizational and physical structures.

5. Network-structured critical infrastructure sectors can be protected most effectively by identifying and securing the hubs, not the spokes of each infrastructure sector, Hence, the best strategy for infrastructure protection will be based on identification and protection of these hubs. It is called *critical node analysis*, because hubs will be shown to be the critical nodes.[1]
6. The optimal strategy for critical infrastructure protection will follow the familiar 80–20% rule: 80% of our resources should be spent on 20% of the country. Although this national strategy is most effective for preventing attacks on infrastructure, it may not be politically feasible, because it does not distribute funding everywhere—only where it can be used in the most effective manner.
7. In addition to an uneven distribution of funding to prevent failures in critical infrastructure, we must learn to think dual-purpose. Most critical infrastructure is in the hands of corporations whose first responsibility is to shareholders, and not to homeland security. Therefore, one way to coax the necessary investment from profit-making private sector corporations is to couple investments in security with productivity and efficiency enhancements.
8. Critical infrastructure is too vast and complex to protect it all. The attacker has the luxury of attacking anytime, anywhere, using any weapon. As defenders we have the duty to protect everything, all the time, with infinite funding. Assuming we have less-than-infinite resources, our only alternative is to think asymmetric. Asymmetric thinking means thinking of new ways to protect our vast and complex infrastructure from attack. It means we must be clever.
9. Perhaps the biggest claim made by the author in this chapter is that critical infrastructure responsibility has to reside at the federal level because intelligence gathering and analysis is controlled by federal organizations, most infrastructure is controlled by interstate commerce laws at the federal level and therefore not local, and local communities are ill-prepared to wage war on global terrorism.

DEFINING CRITICAL INFRASTRUCTURE

The phrase, "critical infrastructure protection," did not appear in print until 1997, but the concept of infrastructure security has been evolving ever since the 1962 Cuban Missile Crisis, when President Kennedy and Premier Khrushchev had difficulty communicating with one another because of inadequate telecommunication technology. Therefore, telecommunications is the first sector to be considered "critical." But it would take decades for the United States to become aware of the importance of other sectors—an evolution described in more detail in Chapter 2.

[1]Critical nodes will be synonymous with hubs, in this book.

The "Marsh Report" (1997) and the subsequent executive order EO-13010 (1998) provided the first definition of infrastructure as "a network of independent, mostly privately-owned, man-made systems that function collaboratively and synergistically to produce and distribute a continuous flow of essential goods and services." A *critical infrastructure* is, "an infrastructure so vital that its incapacity or destruction would have a debilitating impact on our defense and national security."[2] Critical infrastructure could also have become known as *vital infrastructure*, according to this early definition. An infrastructure is considered critical because it is *vital* to national security, but the Marsh Report does not provide a concise definition of the term "vital." Indeed, one primary challenge of critical infrastructure will be the determination of which assets should be protected and which ones should not. It is primarily an issue of *resource allocation*, or the process of committing dollars, people, equipment, and legal assets to the prevention of attacks on various sectors. Resource allocation goes beyond target hardening and encompasses the formulation of a strategy to protect vital assets.

Today's definition of critical infrastructure includes 11 sectors and 5 key assets. This definition has grown out of an earlier definition that included only five sectors and is likely to expand even further over the next decade. According to the national strategy, critical infrastructure and key assets encompass the following sectors:

1. Agriculture and food
2. Water
3. Public health
4. Emergency services
5. Defense industrial base
6. Telecommunications
7. Energy
8. Transportation
9. Banking and finance
10. Chemicals and hazardous materials
11. Postal and shipping

The "key assets" are as follows:

1. National monuments and icons (Statue of Liberty)
2. Nuclear power plants
3. Dams
4. Government facilities (offices and governmental departments)
5. Commercial key assets (major skyscrapers)

[2]Critical Foundations: Protecting America's Infrastructures," The Report of the President's Commission on Critical Infrastructure Protection, October 1997.

Critical infrastructure protection is defined as the strategies, policies, and preparedness needed to protect, prevent, and when necessary, respond to attacks on these sectors and key assets. This definition, and how it evolved out of the Cuban Missile Crisis, is explored in greater detail in Chapter 2.

THE IMPORTANCE OF STRATEGY

The definition of critical infrastructure is evolving, and so is the strategy for protecting the various sectors that make up critical infrastructure. What is the national, state, city, and local level strategy for protection of these vital systems and services? Are the strategies adequate? Do they lead to a safer and more secure infrastructure? Answering these and related questions is the aim of this book. To reach the answer, however, it is necessary to understand the regulatory, technical, and dynamic structure of each sector. This task is daunting because of the enormous size and complexity of each individual sector. Furthermore, we know that each sector is interdependent with others, which complicates the problem even more. It is why we need an overarching strategy. Hence, the first step in the study of critical infrastructure protection is to establish a framework that will set a course for successful policies at the national, state, and local levels. We need to understand what we mean by *strategy*.

Johnson and Scholes[3] define strategy as follows:

> Strategy is the direction and scope of an organization over the long-term: which achieves advantage for the organization through its configuration of resources within a challenging environment, to meet the needs of markets and to fulfill stakeholder expectations.

Although this definition is aimed at business managers, it serves us well because the problems of critical infrastructure protection, like the problems of industry, are organizational, technical, and resource allocation problems. Similarly, strategy is a long-term plan, and we know that the war on terrorism is a long-term war. There will be no quick fixes to the problem of critical infrastructure protection. Additionally, the elements of infrastructure size and complexity exist in a challenging environment—the challenge posed by terrorism. Finally, the American public has high expectations for success on the part of federal, state, and local governments in their efforts to protect the most valuable and essential components of modern life—our water, power, energy, telecommunications, Internet, and other infrastructure sectors. Any strategy for securing these assets must be perfect; the citizens of the United States will not tolerate near-perfection.

Strategy is important because it provides a roadmap for solving complex problems involving organizations, technologies, and resource allocation within a

[3]G. Johnson and K. Scholes, *Exploring Corporate Strategy: Text and Cases*. Englewood Cliffs, NJ: Prentice-Hall, 2001.

challenging environment. This description accurately describes the situation with respect to homeland security. But, the question is, "what is a winning strategy for the United States?" "What is the current national strategy, and is it adequate?

Actually, there are several strategies in play at this time. First, there is a difference between homeland defense and homeland security. The DOD has responsibility for *homeland defense*, which the author defines as all defense activities outside of the country, or when called upon and under civilian control, within the country.[4] Nonetheless, the strategy of the DOD is instructive because it provides an example that illustrates the division of roles and responsibilities between civilian and military organizations. It also illustrates one of the most vexing organizational problems in the war on global terrorism—that of establishing who is in control and of what they are controlling.

DOD Strategy: Layered Defense

The homeland defense strategy articulated by the DOD is called a *layered defense*. The term "layered" means different approaches are applied to different layers—regions—of the world, depending on the geopolitical environment. The goal of this strategy is to suppress and deter threats from as far away as possible, first and foremost, and then suppress and deter the threat layer by layer, as the threat approaches our borders.

The first layer starts with other countries (Afghanistan and Iraq). The next layer is on the High Seas (Navy), within a 300-mile buffer for commercial shipping, then at the 12-mile zone with the Coast Guard, and finally at the border, with the cooperation of U.S. Customs. If called in by civilian authorities, the DOD's role is to provide assistance within the border through a newly formed Northern Command.

Layered defense is a well-known ancient strategy. It was used by Rome to hold the barbarians at bay. The strategy is as effective today as it was then, as long as the threat is the same. Unfortunately, the threat of chemical, biological, radiological, nuclear, and asymmetric explosive attacks is radically different today than 2000 years ago. Asymmetric conflict brings the battle home, inside the borders. Hence, the layered strategy of DOD is only part of the answer. A domestic strategy is needed, because the global war on terrorism is not restricted to the other side of the border.

DHS Strategy: Preparedness, Response, and Research

The DHS was legally created in 2002 and physically implemented in 2003 to address the problem of security within our borders. The roles and responsibilities of the DHS

[4]At the time of this writing, homeland defense was undergoing revision. In an Army document dated April 1999, the term was described as follows: "There is currently no definition of homeland defense. The proposed definition shows the Army's mission to protect our territory, population, and critical infrastructure by; deterring/defending against foreign and domestic threats, supporting civil authorities for crisis and consequence management, and helping to ensure the continuance [of] critical national assets." http://www.fas.org/spp/starwars/program/homeland/executive-overview.htm.

are still evolving, but at a minimum, its responsibility is to protect the citizens of the United States through bolstering of:

- Intelligence and warning
- Border and transportation security
- Domestic counter-terrorism
- Critical infrastructures and key assets
- Defending against catastrophic terrorism
- Emergency preparedness and response

Indeed, the terrorist attack of September 11, 2001 (9/11) was an attack on banking and finance, using the transportation sector. Therefore, two critical infrastructure sectors have already been attacked or involved in a major attack.

The devastation of 9/11 demonstrates that attacks on infrastructure can result in massive casualties, sizeable economic, political, and psychological damage, not to mention damage to the American psyche. These are collectively called "attacks on the American Way of Life" and because of the potential to disrupt an entire society, critical infrastructure protection must be one pillar of the homeland security strategy.

The importance of infrastructure protection has steadily risen over the past 40 years, as described in Chapter 2. The importance of a national strategy for the protection of critical infrastructure and key assets has yet to mature. The components of a national strategy are in the process of evolving—starting with the Stafford Act of 1988, which established the Federal Emergency Management Agency (FEMA) for coping with natural disasters; to the Nunn-Lugar-Domenici Act of 1999, which provided for defense against weapons of mass destruction (WMDs); and finally, to the DHS (HSPD-7) declaration by President Bush in late 2003. Therefore, by 2004, the outlines of a national strategy were known, but not fully implemented.

In the next section, we dissect the national strategy for homeland security as it pertains to the protection of infrastructure. The final section of this chapter will go one step further and describe a set of strategic principles for guiding future policies.

HOMELAND SECURITY AND CRITICAL INFRASTRUCTURE

What are the roles and responsibilities of the federal government? What are the responsibilities of state and local governments? These questions were addressed (if not fully answered) by the 2003 National Strategy, which is summarized here as follows:

1. Take stock of our most critical facilities, systems, and functions and monitor their preparedness across sectors and governmental jurisdictions. The first step in any vulnerability analysis is to take inventory—find out what you have, and how it works. This process was started in 2003 and continues today. State and local governments are required to perform an analysis of critical infrastructure every 3 years. But this requirement is problematic, because at this stage, it is

not clear how to categorize the components of each sector, how to place a value on each component, and therefore, how to estimate the associated risk. Taking stock is made difficult because there is no clear-cut metric for what is critical.

2. Assure that federal, state, local, and private entities work together to protect critical facilities, systems, and functions that face an imminent threat and/or whose loss would have significant, national-level consequences. To be used effectively, information must be shared. Before 9/11 little information sharing among federal, state, and city jurisdictions took place. Failure in the intelligence system before 9/11 enabled the al-Qaeda to succeed on 9/11. Information sharing is essential, but difficult, because of decades of separation among the various intelligence collection and dissemination agencies. It will take decades to fix, but we have much less time than this to correct the problem.
3. Create and implement comprehensive, multitiered protection policies and programs. Exactly what these policies are, or should be, remains a question. "Multitiered" means "across jurisdictions and federalists lines." This strategy is currently at cross-purposes with most jurisdictions and federalist lines. For several reasons, the structure of government is stove-piped; that is, most agencies operate in relative isolation from one another. These "seams" are where the organizational vulnerabilities exist, and hence they are targets of opportunity for terrorists. Once again, this strategy is laudable but extremely difficult to implement.
4. Explore potential options for enablers and incentives to encourage public- and private-sector entities to devise solutions to their unique protection impediments. The national strategy realized that most critical infrastructure is in the hands of corporations, not governments. How does a state or city government get these corporations to cooperate and share information that, when aggregated across competitive corporations, gives a revealing picture of an impending attack? Take the Internet as an example. Each corporate entity is left to its own devices when it comes to abolishing computer viruses and defending against cyber-attacks, and yet the solution to the problem requires sector-wide remedies. It is unlikely that we will defeat cyber-attacks through the actions of one Internet company at a time.
5. Develop protection standards, guidelines, and protocols across sectors and jurisdictions. Again, this goal is laudable, but what these standards are, and who should follow them, is left to the imagination. One main problem addressed by this book is that of vulnerability analysis. What should be the standard method of vulnerability analysis and risk assessment? Currently, there is no standard; hence, it is impossible to compare vulnerabilities across jurisdictions or sectors. And yet the DHS requires vulnerability analysis.
6. Facilitate the exchange of critical infrastructure and key asset protection best practices and vulnerability assessment methodologies. In other words, share knowledge of what works and what does not. Once again, the goal is commendable, but its implementation is made difficult by the competitive

nature of corporations, the lack of techniques, and the lack of motivation for the private sector to invest in security. From a practical standpoint, it is better for a power utility, telecommunications company, or transportation company to buy insurance against a low-probability, high-cost incident than it is to invest in target hardening or redundant systems. This strategy is too idealistic. What is needed, instead, is a strategy that creates incentives for companies to do the right thing.

7. Conduct demonstration projects and pilot programs. This objective is a corollary to the previous one, which is to discover what works and what does not, through the processes of demonstrations and pilots. The question this objective raises is, "what projects and pilot programs should we invest in?" The challenge currently facing state and local jurisdictions when it comes to performing vulnerability and risk analysis is a clear-cut candidate for a pilot program, but when it comes to more complex initiatives, such as information-sharing, multijurisdictional cooperation, and cross-sector implementation, this strategy begins to show its weaknesses.
8. Seed the development and transfer of advanced technologies while taking advantage of private-sector expertise and competencies. It is the traditional role of research and development. It is the stated goal of the homeland security strategy to use advanced technology to defeat terrorists. All highly technological societies are vulnerable to failures in technology. Therefore, it is a kind of "fight fire with fire" strategy, whereby we can leverage our excellence in technology to fend off attacks to our highly technical infrastructure.
9. Promote national-level critical infrastructure and key asset protection education and awareness, and improve its ability to work with state and local responders and service providers through partnership. Education and training is an essential ingredient of technology transfer and improved readiness. In particular, policy makers must know enough about each infrastructure sector to make good policy. Because nearly all infrastructures are highly specialized, both in terms of regulatory and technical architectures, this task is not easy. The perfect policy maker must understand a variety of subjects ranging from politics to electrical and chemical engineering. A good example of this challenge is the World Wide Web. Most cyber-attacks are rooted in software flaws—flaws that cannot be legislated away. Rather, the policy maker of the future must understand how the World Wide Web works before he or she can make good policy decisions about the telecommunications and information sector. Currently, few policy makers can do this.

The national strategy also recognizes the importance of state and local participation. Indeed, terrorist attacks are local attacks. Therefore, terrorism becomes the burden of state and local jurisdictions. The 2003 National Strategy lists three major objectives. State and local governments should:

1. Organize and conduct protection and continuity of operations planning, and elevate awareness and understanding of threats and vulnerabilities to critical

facilities, systems, and functions. In other words, prepare to respond to attacks. The problem with this policy is that state and local first responders do not enjoy the intelligence information needed to gain an "understanding of threats" needed to effectively prepare, and they lack the information and knowledge of complex infrastructure components such as telecommunication hotels, energy pipeline dynamics, and supervisory control and data acquisition (SCADA) control systems to be of much assistance to the companies that own and operate "critical facilities, systems, and functions." State and local emergency management organizations are squeezed between federal and corporate organizations.

2. Identify and promote effective sector-specific, risk-management policies and protection practices and methodologies. An ounce of prevention is worth a pound of response. The best policy is to prevent successful attacks in the first place. Unfortunately, there is never enough time, people, and money to prevent all attacks. Even more significant, there currently is no widely accepted risk-management strategy for state and local government agencies to apply. This book is designed to fill this knowledge gap, but in the end, this remains a research problem. In the meantime, the techniques described in subsequent chapters can be applied at the federal, state, and local levels.
3. Expand voluntary, protection-related information sharing among private entities within sectors, as well as between government and private entities. The national strategy calls for the formation of Information Sharing and Analysis Centers (ISACs) in all sectors. ISAC organizations may be extensions of the federal government, as in the case of the telecommunications ISAC, which is managed by the National Communications System within the DHS, or completely industry run as in the case of the water-ISAC, which is a nonprofit arm of the water sector's professional society. In any case, information sharing and cooperation across competitive organizations is key but extremely difficult. What are the motivations for competitors to share information that may result in loss of competitive advantage? Why increase the cost of doing business if the probability of an attack is extremely low and the damages that could be done by a successful attack are extremely high? Insuring the components of a sector against a high-cost, low-probability incident is better, in the mind of a corporate chief executive, than making a major capital investment in time, people, and equipment to harden the components.

In addition, the 2003 National Strategy also addresses what is called "cross-sector" issues. These all-encompassing tasks need to be carried out by someone and they benefit everyone:

1. Planning and Resource Allocation. The new DHS will provide strategies and guidelines for state and local jurisdictions and then provide (partial) funding to realize them.
2. Information Sharing and Indications and Warnings. DHS will aggregate and disseminate information and warnings to the states and cities.

3. Personnel Surety, Building Human Capital, and Awareness. This function provides for education and training.
4. Technology and Research and Development. The science and technology division of DHS will fund technology transfer and research activities to fight terrorism.
5. Modeling, Simulation, and Analysis. The national strategy is somewhat unique in its emphasis on modeling and simulation. DHS believes this technology is especially useful for solving some of the problems it faces. Indeed, we will apply simple modeling and simulation techniques in this book, especially in the analysis of vulnerabilities. Furthermore, several simulation programs for computing optimal resource allocation and other programs for modeling vulnerability and risk will be described and used in subsequent chapters.[5]

One of the most fundamental assumptions made regarding the national strategy is that critical infrastructure protection is the responsibility of state and local governments. This is like claiming that the attack on Pearl Harbor in 1941 was Hawaii's problem, not a national problem. In fact, there are three major reasons to claim that critical infrastructure is a federal responsibility: centralization of intelligence at the federal level, interstate commerce laws, and state and local capabilities—or lack thereof—to ward off attackers. The following brief arguments support an alternative view: Critical infrastructure protection is too big for state and local governments to handle on their own.

Intelligence collection and analysis is concentrated at the federal level—the CIA, DIA, FBI, and DHS are federal organizations that have historically had difficulty sharing information with states and cities. This cultural bias will perhaps never change, but if it does, it may take to long to modify. The "trickle down" theory of information sharing may be too slow to counter the network-centric speed of terrorists.

Most critical infrastructure is governed by interstate commerce legislation. Telecommunications, gas and oil pipelines, electrical power, and transportation are subject to interstate commerce regulation and standardization. The national strategy provides for local planning, prevention, and response, but infrastructures are, for the most part, national assets. Because of their interstate connectivity, state and local strategies will always be inadequate. For example, the 2003 Eastern Power Grid Blackout that affected 50 million people started in Ohio, crossed many state lines, and brought down a major portion of the eastern United States. What could the county in Ohio where the cascade failure began have done to prevent this incident? The answer: very little, because the electric power grid is distributed across many states; destruction of only a small portion of it is a national incident, not a local one.

The national strategy shifts the responsibility for responding to terrorist attacks from federal to state and local levels, because such attacks are considered "local." However, the interstate connectivity of most infrastructures pulls in the opposite

[5]The enclosed disk includes several simulation programs that demonstrate the concepts described here.

direction: Critical infrastructure protection is a national problem. It is doubtful that state and local governments and their emergency response units can have more than a minor influence on protecting the many vast interstate infrastructure sectors that the entire nation depends on. It will continue to be a national problem and, therefore, a national responsibility.

The alternative view of who should be responsible for critical infrastructure protection may never be implemented. But consider this: The global war on terrorism is a national problem, not a state and city problem. New York City cannot conduct an offensive war on terrorism any more than Honolulu could declare war on Japan. Similarly, critical infrastructure is vulnerable because it spans states and cities; the very nature of any interstate commerce is what makes it vulnerable. Countering the threat at a local level only addresses the vulnerability in piecemeal fashion. Whether the sector is power, energy, Internet, or telecommunication, the real damage will be felt across state and regional boundaries.

In summary, the current strategy consists of a Department of Defense "away game" denoted Homeland Defense, which has little to do with critical infrastructure within the United States, and a "home game" denoted Homeland Security, which is largely concerned with critical infrastructure, but because of federalism, stove-piped agencies, private sector ownership by competitive corporations, and general lack of information sharing, the national strategy will be extremely difficult to implement.

In even the most optimistic scenario, the resource allocation challenges of critical infrastructure will be suboptimal as long as allocation decisions are made solely at the local level. This economic reality is well-known. Consider a simple example that illustrates suboptimal allocation of a fixed resource, say, $105 million, which is equally allocated to three cities—$35 million each. The mayor of each city is invited to identify the most vulnerable asset within his or her city along with an estimate of the cost to protect the most vulnerable asset in the city. The first mayor identifies a transportation depot requiring $40 million to protect; the second mayor identifies a harbor requiring $25 million; and the third mayor identifies a power plant in need of $40 million to protect. Given $35 million each, the first and third mayors cannot fully protect their assets and the second mayor is allocated an excess of $10 million! Therefore, local allocation of resources wastes $10 million and does not sufficiently protect the other two.

A top-down allocation of $40, $25, and $40 million protects all three cities and does not waste any resource. In fact, if the damages associated with the loss of each asset vary widely, the suboptimal allocation can be exacerbated even further. But a top-down allocation reduces all three risks to zero. Performing vulnerability and risk analysis of national assets at the local level will generally lead to waste and ineffective use of resources.

Add suboptimal resource allocation to the fact that infrastructure is especially vulnerable to asymmetric threats, and you have the ingredients of a wicked problem. Wicked problems are usually solved by "thinking out of the box" or, more precisely, rethinking the problem from a different angle. This is exactly the next step in understanding the inherent problems identified here; strategists need to think differently about infrastructure protection.

FIVE STRATEGIC PRINCIPLES OF CRITICAL INFRASTRUCTURE PROTECTION

In late 2003, President Bush reiterated the policies of his administration when it came to critical infrastructure protection, stating once again:

> (7) It is the policy of the United States is to enhance the protection of our Nation's critical infrastructure and key resources against terrorist acts that could:
>
> (a) Cause catastrophic health effects or mass casualties comparable to those from the use of a weapon of mass destruction;
> (b) Impair Federal departments and agencies' abilities to perform essential missions, or to ensure the public's health and safety;
> (c) Undermine State and local government capacities to maintain order and to deliver minimum essential public services;
> (d) Damage the private sector's capability to ensure the orderly functioning of the economy and delivery of essential services;
> (e) Have a negative effect on the economy through the cascading disruption of other critical infrastructure and key resources; or
> (f) Undermine the public's morale and confidence in our national economic and political institutions.[6]

This declaration reiterates the goals of the 2003 National Strategy, but it does not suggest a specific strategy to pursue. Nor does this high-level strategy provide any guiding principles for constructing concrete polices that state and local jurisdictions can apply. Indeed, the best strategies for protecting the vast and complex infrastructure sectors of the United States are not known at this time, but we can distill several basic concepts into strategic principles that policy makers can use as a guide. These principles are given without justification; you will have to read the following chapters to understand the underlying rationale. Nonetheless, it is possible to see the outlines of a coherent approach to critical infrastructure protection in these principles. They will be turned into recommendations as you progress through this book and gain deeper knowledge of how each sector works.

It Takes a Network to Fight a Network

This principle has two meanings: (1) because the threat is organized as a network instead of a hierarchical command structure, the organizations designed to combat terrorism must also be organized as a network, and (2) because infrastructure sectors are so vast and complex, only network analysis will work; we cannot afford to protect every component and every location in the country that contains infrastructure. Network structures must be protected using strategies that derive from network analysis.

Hierarchical command structures as shown in Figure 1.1(a) have been steadily undergoing replacement by *disintermediated command hierarchies* over the past

[6]The Whitehouse: Homeland Security Presidential Directive/HSPD-7, December 17, 2003.

(a)

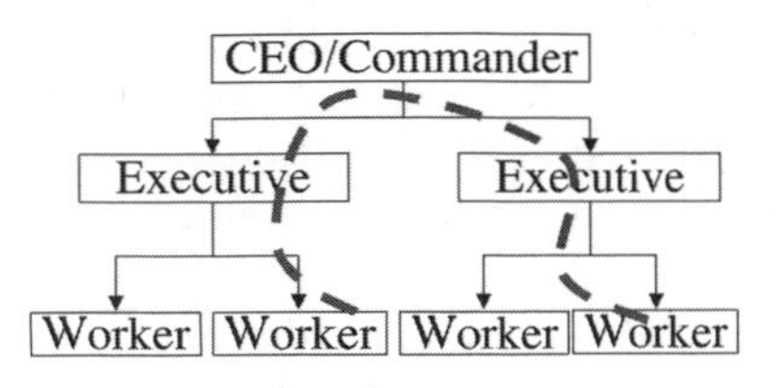

(b)

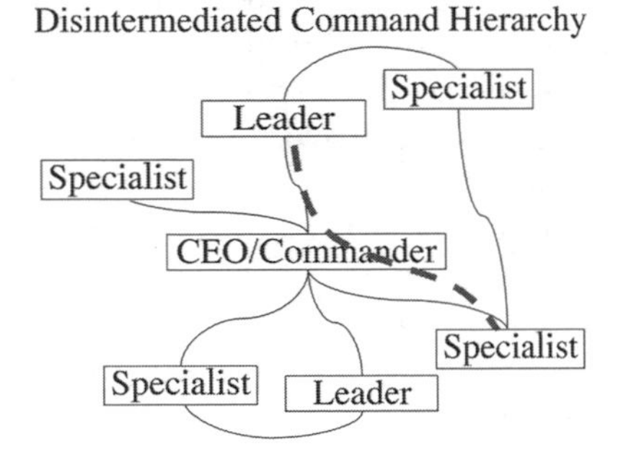

(c)

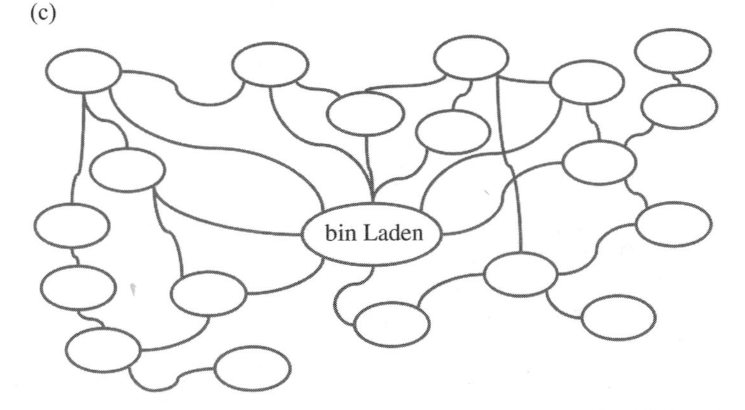

Figure 1.1. Hierarchical command versus disintermediated command hierarchy. (a) Hierarchical command structure shows a typical decision chain. (b) Disintermediated command hierarchy shows a typical decision chain. (c) Social network of Osama bin Laden and the hijackers of 9/11.

several decades. This is largely because of technology. Large organizations, for example, have increased productivity, decreased product development cycle times, and cut overhead largely by disintermediation of the middleman. A cogent example is the widespread use of e-mail that cuts across departments, hierarchies, and titles. Today's large-scale enterprises have removed much of the middle-tier management structure that reacts far too slowly, serves to prevent the free flow of information throughout the organization, and results in an increase in operational cost. The hierarchical command structure of yesterday has morphed into the disintermediated command hierarchy of today [Figure 1.1(b)].

During the 1990s, businesses learned to adapt the strict hierarchy shown in Figure 1.1(a) to the faster and less encumbered network structure shown in Figure 1.1(b). The problem with the hierarchical command structure was that it took too long for information to move from one worker to another. Decisions were based on data, discussions, and expertise that had to percolate up the hierarchy to the top, or nearly the top, and then trickle down along another path to another worker. The dotted lines in Figures 1.1(a) and 1.1(b) illustrate this flow.

In a disintermediated command hierarchy, decision makers are embedded within a network structure that encourages point-to-point movement of data, discussions, and decisions. Commanders can access the specialists possessing the right expertise as easily as the next in command executive or leader. Therefore, decisions are made more quickly because information is shared more widely. The middle tier is "flattened" or done away with, because modern communication technology removes barriers to the flow of information. Disintermediation means to flatten an organization by removing layers, as shown in Figure 1.1(b).

The overall effect of disintermediated command hierarchies on performance is startling. Old-style command hierarchies cannot compete with disintermediated command hierarchies because they are too slow, uninformed, and costly. The terrorists who attacked the United States in September 2001 are an unfortunate example. No hierarchical organization chart for Al-Qaida exists. Rather, the terrorists involved in 9/11 formed a disintermediated command hierarchy with Osama bin Laden as a hub or focal point; see Figure 1.1(c). The network of Figure 1.1(c) looks more like the network of Figure 1.1(b) than Figure 1.1(a) because both are flat. In other words, they are disintermediated—there are few middlemen.

When the flexibility, speed, and decision-making capability of the network shown in Figure 1.1(c) was positioned against the organization charts that looked like Figure 1.1(a) on 9/11, the result was disastrous. Seams between agencies, low reaction times, and poor coordination within the hierarchical command structure of the U.S. intelligence and law enforcement organizations created ample opportunity for the attackers to succeed even when they made several mistakes.

How does this apply to critical infrastructure protection? Quite simply, one prominent unresolved issue in infrastructure protection is the problem of information sharing. The lack of information ebb and flow among and between agencies in the hierarchical command and control systems currently responsible for prevention of attacks on the homeland lack the speed and adaptability of a disintermediated command hierarchy. Had the various intelligence agencies been a network, rather than hierarchically organized around "stovepipe" structures, there is clear evidence that the plans of the 9/11 terrorists would have been foiled, and the attack prevented. This is why the 2003 National Strategy emphasizes, "... protection-related information sharing among private entities within sectors, as well as between government and private entities." But this goal is impossible unless we learn to fight a network with a network. This means disintermediating existing hierarchies.

The second meaning of this principle applies even more directly to the various critical infrastructure sectors. Surprisingly, the sectors analyzed in this book form networks with a common feature: They are concentrated around critical nodes called "hubs." An example of a network-structured critical infrastructure is shown in Figure 1.2. Here the network is the national transmission and distribution pipeline system that provides nearly all of the gas and oil across the United States. Disruption of this network would have disastrous consequences, because oil products such as gasoline and heating oil could not reach consumers.

Figure 1.2 shows the network and identifies 14 hubs—concentrations of supply, storage, or transmission "intersections" in the vast oil supply chain. Note that this network is nonhierarchical and nonrandom. That is, it has structure. The hub structure is obvious in Figure 1.2; it is possible to identify hubs, visually. The point of this analysis is this: Any strategy aimed at protecting such a vast network as this cannot afford to protect every mile, every segment of pipe, and every flow control value in the sector. Rather, the strategy must aim to protect the most critical nodes—the hubs.

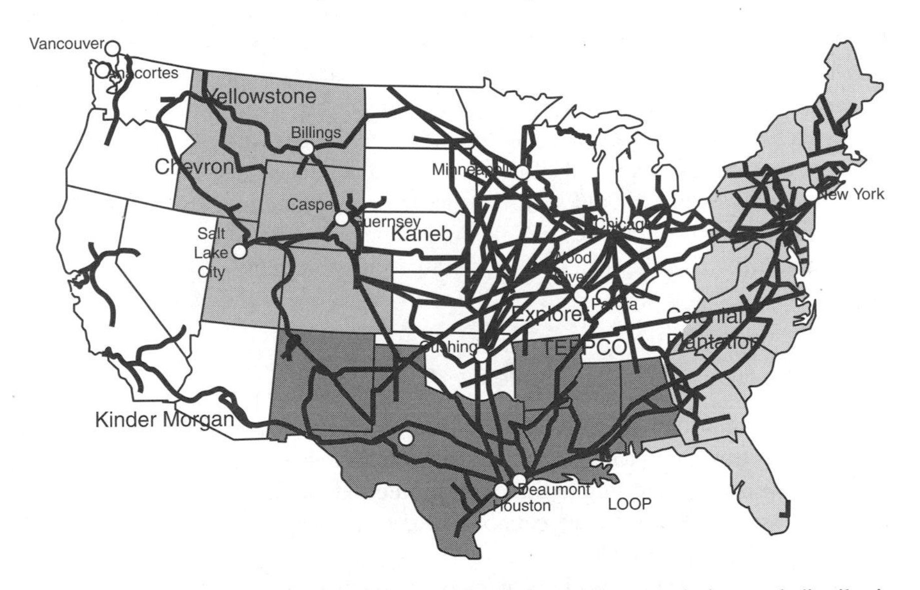

Figure 1.2. Network architecture of the major refined oil transmission and distribution pipelines in the energy sector. Courtesy of C. J. Trench (ctrench@concentric.net), "How Piplelines Make the Oil Market Work—Their Networks, Operation and Regulation," A memorandum prepared for the Association of Oil Pipe Lines and the American Petroleum Institute's Pipeline Committee, December 2001. http://www.aopl.org. Reprinted with permission.

This principle applies to many infrastructure sectors as shown in subsequent chapters. It is a general principle.

Hubs (a.k.a. critical nodes) are used to advantage in vulnerability analysis, because hubs are the most important points of failure. A hub is the one component within a sector where an attack can do the most damage. Conversely, it will be shown that adequate protection of hubs provides the greatest benefits. If we can protect the hubs, we can optimally protect the entire sector with the least cost! This fact will be used to great advantage, because it reduces the problem of size and complexity to a more manageable problem of selective investment.

Critical node analysis comes down to identifying hubs in networks. By carefully identifying and protecting these critical nodes, we can get the best return on our investment and prevent major disasters, which are two major objectives of the policies given by the national strategy.

This principle is easy to apply when critical infrastructure sectors are viewed as networks. In addition to the energy sector example in Figure 1.2, transportation systems such as the New York subway form a network with hubs. Protect the hubs, and you protect the critical nodes of the subway system. The water supply system of San Francisco sustains 2.4 million people, and its most vulnerable components are hubs in the network consisting of water collection, treatment, and distribution components or nodes. The "wheeling" architecture of the electrical power grid that failed in 2003 is a critical node of the network that spans the Eastern Power Grid. Had this node been hardened, the entire grid would have been hardened. The national telecommunications network and the Internet are all networks with hubs that can be identified and protected using this simple idea: Model each sector as a network, and then analyze the network and find its hubs.

Network analysis is an "out-of-the-box" approach to critical infrastructure vulnerability analysis that should be the centerpiece of any strategy for critical infrastructure protection. Indeed, this approach leads naturally and logically to subsequent principles that can be incorporated into an effective strategy for nearly all sectors.

Secure the Hubs, Not the Spokes

This principle is a direct consequence of the first principle. Critical infrastructure sectors are organized as networks with hubs. The hubs are the critical nodes, so the next step is to protect the hubs. Given limited resources and the fact that most sectors are extremely large, we cannot afford to protect everything, so we opt to protect only the critical nodes.

For example, the Internet is known to contain approximately 250 million servers; all are important, but only a few are critical. The current strategy of protecting each and every server is not effective and is very expensive. Information technology managers are spending far too much time and money on cyber-security, anti-viral software, and restrictive operating procedures.

An asymmetric alternative or counter-strategy to the current approach is to protect the hubs of the Internet. These are the servers with the largest connectivity to the Internet.[7] In fact, the Internet is highly clustered around fewer than 250 servers—the top 250 hubs. What happens to cyber-security when these 250 or so servers are hardened against computer worms and viruses? They stop spreading, and eventually the malicious worm or virus dies out.

By securing the hub servers of the Internet, we protect all servers. This surprising result is actually intuitive if you think asymmetrically about it. Consider this: Most traffic, and thereby most worms and viruses, are propagated by the most active servers, the hubs. If these "promiscuous" servers are protected, they cannot spread worms and viruses, and if they stop the spread of malicious software, nearly all propagation halts. This is intuitively obvious, but it will be demonstrated in a rigorous manner in Chapter 13.

The critical node strategy can also be turned into network warfare by using hubs to purposely spread "killer-virus" software. This software behaves just like a worm or virus, but instead of damaging other computer systems and destroying important information, a "killer-virus" destroys all other viruses. In other words, network hubs can be used as an offensive weapon. In the case of the Internet, we can release killer-viruses "into the wild" and let them hunt down and kill the malicious viruses. The most effective way to do this is to launch them from hubs. Therefore, critical nodes in the telecommunications and information sector can be used for good or evil. Why not use the network structure of most critical infrastructure sector to launch a counter-attack?

Spend 80% on 20% of the Country

Continuing the logical progression from network organizations to network-structured infrastructures leads to the next principle: Critical infrastructure is not spread evenly across the country. Rather it is concentrated, typically around hubs. Moreover, these hubs are often geographically clustered, typically around a small number of metropolitan areas. For example, Manhattan, New York, has a high concentration of components in the banking and finance sector. This is why it was attacked on 9/11. In addition to the New York Stock exchange, the largest Federal Reserve Bank is located in Manhattan. Manhattan is a geographical hub that is critical to the banking and financial sector.

The largest concentration of energy refineries and major source of refined gas and oil products for distribution throughout the United States is located in Galveston Bay, Texas. Fairfax County, Virginia, is the home to a large concentration of Internet servers, and Chicago, Illinois, is a national hub for transportation and logistics. Most of the 6 million cargo containers that form the backbone of U.S. trade flow through three ports; most of the energy mined to supply fuel for coal-powered power plants is concentrated in Wyoming, and most of the Industrial Defense base is concentrated in two or three areas of the United States.

[7]The "most popular" or "busiest" Internet servers are easily found by a variety of methods, not the least of which are "crawlers" or "spiders"—software programs that travel through the Internet on their own and report back on what they find.

These examples suggest the 80–20% rule: 80% of the investment in critical infrastructure protection should be spent on 20% of the country. This, of course, is a political impossibility, but if we are to think asymmetrically about the challenges facing critical infrastructure, we must face reality: Target hardening is too expensive to do everywhere. Instead, the national strategy must be focused on funding those areas of the country where the vulnerabilities lie. Equal funding to all parts of the country short-changes the vulnerable areas and creates attractive targets for terrorists.

In reality, critical infrastructure is concentrated in a very small number of communities. For example, the following communities rank at the top of the 80–20% list:

- Manhattan-NY Harbor (Banking, Energy, Telecom, National Monuments)
- DC-Fairfax Co (Energy, Telecom, Federal Government, Internet)
- Galveston Bay-Gulf Coast (Energy, Military Exit, Chemical Industry)
- LA-Long Beach (Defense Industry, Import/Export Port, Energy)
- Silicon Valley (Defense Industry, Internet, Telecom, Air Transportation)
- Chicago (Air, Rail, Truck Transportation, Telecom)

These communities should receive the lion's share of funding. Does the United States have the political will to do the right thing?

Think Asymmetrically

Several of the foregoing principles are nonintuitive or at least nonobvious without careful study of each sector. This is because they are asymmetric—meaning they are the result of thinking asymmetrically about critical infrastructure protection.[8] Using computer viruses to fight malicious computer viruses, identification and protection of network hubs, and reorganizing the bureaucracy to operate more as a disintermediated command hierarchy are examples. But even more nonintuitive examples of asymmetric thinking will be discovered as each sector is dissected in greater detail.

One of the most obvious examples of asymmetric thinking in the electrical power sector is the counter-notion that the electrical power sector is vulnerable because its power plants are vulnerable. Common wisdom says to prevent attacks on these large and expensive facilities. Indeed, nuclear power plants have been singled out as a "key asset" by the national strategy. But consider this: None of the major power outages over the past 40 years have been attributed to power plant failures. Even the partial meltdown of the Three Mile Island nuclear power plant did not disable the power grid nor disrupt the flow of electrical power to consumers.

[8]Asymmetric warfare is the art of force multiplication using nontraditional techniques such as converting jumbo jets into bombs, using the public broadcast media to spread propaganda and gain public support for a cause, and creating terror and mass destruction or economic disaster through focused attacks on infrastructure that millions of people depend on.

So what is the primary vulnerability in the power grid? The North American Electric Reliability Council (NERC) was established after the 1965 Northeast power blackout that plunged the northeast into darkness after a chain reaction in the distribution network. In August 1996, 11 western states and several Canadian provinces were unplugged when a tie line failed near the Oregon–California border. And in 2003, the "once-in-a-thousand year" blackout was caused by a software failure complicated by a tripped power line. None of these massive failures were caused by a power plant failure.

The largest single source of power generation in the United States is Grand Coulee Dam in Washington State. What would happen if it failed? Less than 1% of the national supply of electric power would go off line. A change of less than 1% would not even cause the lights to dim or an air conditioner to stall, because the grid can redirect power from one region to another without much advance notice. In fact, gas-fired peaker plants are constantly standing by to accommodate much greater variations in demand as consumers switch on their air conditioners and electric lights throughout the day.

In other words, the electric power sector is not vulnerable because of its power plants. Power plants are important, but not critical nodes in the power grid network. What then is the major vulnerability of the power sector? Suppose we think asymmetrically about strategies to protect this important sector.

The four power grids that supply all electrical power to North America (Canada and the United States) are *vulnerable in the middle*—the portion of the grid that transmits and distributes electricity to consumers. Historically, power grid failures have occurred because of failures in this middle. In other words, a modern electric power grid is vulnerable to an asymmetric attack on its transmission and distribution network—the so-called middle of the grid. The network analysis techniques described here can be used to find these vulnerabilities and suggest an optimal strategy for protecting the critical nodes of the middle.

This nonintuitive strategy for electrical power grid protection is an example of asymmetric thinking. A small amount of effort can lead to a large amount of security, if we take time to consider asymmetric methods of bolstering the critical nodes of electrical networks. But first, we must discard preconceived ideas like the false notion that power plants are critical nodes.

Think Dual-Purpose

Up to this point the importance of financial limitations on infrastructure protection have not been discussed. In fact the national strategy says nothing about how to pay for critical infrastructure protection. And as the private sector owns and operates most infrastructures, they are not motivated to invest in target hardening. So what strategy leads to prevention through costly security enhancements? If we can learn to think asymmetrically about the architecture of infrastructure sectors, why not think asymmetrically about how to finance these needed improvements?

The basic idea here is to "Think Dual-Purpose." It leads to the notion of enhancing productivity while improving security. That is, can an investment in security serve a dual purpose of also improving productivity? For example, can a private infrastructure sector company reduce operational costs by enhancing the security of its critical nodes? Is it economically feasible to reduce insurance premiums by decreasing theft and increasing security at ports? Can the telecommunications sector improve throughput and reliability, which in turn increases the number of telephone calls per hour they can bill to customers by building redundant networks that handle surge capacity during peak periods and provide a backup when not in use? Does redundancy in telecommunications also improve the security and reliability of the Internet?

Dual-purpose systems typically achieve greater security through redundancy. That is, redundancy provides a cushion against both heavy loading and failure. Extra "standby" telecommunication switches and alternate optical fiber lines may seem expensive if not used all the time, but they also provide a high degree of reliability because the system can switch to a backup when needed. Redundant components improve reliability and fill the gap during periods of surge in demand.

Dual-purpose thinking is nothing new to some critical infrastructure sectors such as the electrical power industry. The grid has two major weaknesses: (1) It is weak in the middle because of inadequacies in its transmission and distribution network, and (2) the nation is too heavily dependent on foreign sources of energy. Both problems can be solved simultaneously by converting the grid distributed generation",4>to distributed generation using alternative sources of energy such as wind and solar. Technology has made it economically feasible—using small wind and solar generators—to colocate power generation with factories and shopping malls. That is, the power is generated near to where it is consumed, removing the need to transmit it over long distances. If factories and shopping malls have their own power, they do not need to rely on the grid, except perhaps, during short periods of time when the wind is not strong enough or the sun is not shining.

As technology improves power generation through conversion of wind and solar energy, it may be possible to colocate a micro-scale generator at each home or office. The more of these generators there are, the less dependence on the grid. Just as the personal computer has reduced dependence on large centralized mainframe computers, distributed generation may be able to decrease and perhaps eliminate dependence on the unreliable and vulnerable middle of the grid.

Rethinking the power grid in terms of distributed generation is asymmetric thinking, because it reverses the century old concept of centralized power plants connected to the consumer through an extensive and complex transmission and distribution network. Over the past 40 years, we have learned that the larger the grid is, the harder it falls. Distributed generation can reduce this consequence to an insignificant concern.

Another example from the energy sector should help to solidify this principle. Nearly all of the energy—gas and oil—that supplies the Northeast United States comes from refineries located in the Galveston Bay area, and travels through a single pipeline to storage facilities located in the NY-Elizabeth harbor. These storage tanks are also supplied by oil tankers and miscellaneous other pipelines, but the point is that a large stockpile of this important asset ends up in the same place. Concentration of the gas and oil supplies in one location is an obvious single point of failure, because destruction of the storage facility puts an end to the winter supply of heating oil, aircraft fuel for the major airports on the eastern seaboard, and gasoline for millions of consumers. This is yet another example of a network-structured sector with a hub.

In this case, the energy sector can be protected through the application of redundancy. Instead of concentrating all storage in one place, why not distribute the transmission, storage, and distribution facilities across the northeast? Storage facilities could be located closer to the consumer, and the transmission pipelines connecting them could be duplicated so that a failure in one would not shut off the oil supply completely.

This strategy would simultaneously reduce distribution costs (but require a large capital investment) and improve security through the application of redundancy. It is an expensive strategy, but one that removes a major vulnerability in the energy sector. Any other strategy is likely to result in only a minor improvement and leave the storage facility of NY-Elizabeth harbor partially vulnerable to asymmetric attack.

ANALYSIS

The foregoing survey of various strategies leaves several issues unresolved, but the DOD and DHS strategy gives clear guidance at the highest level. The five principles provided by the author suggest secondary strategies to fill in some gaps, but there remain only partial answers for state and local homeland security professionals. This section addresses some of these gaps and uncertainties and sets the stage for the remainder of this book.

First is the issue of prevention versus response: Is the strategy biased more toward response as the 2003 National Strategy seems to suggest, or does it provide enough guidance on prevention? Critical infrastructure should be protected (prevention), because prevention is cheaper than suffering mass casualties, economic damage, psychological damage, or damage to national pride. Hence, the focus of this book is on methods of prevention.

A strategy of prevention leads to the following conclusions. First, there should be a set of principles that guide the formation of policies aimed at prevention. The five principles provide a start but not a final answer to how best to prevent successful attacks on critical infrastructure. These principles lean toward the asymmetric side of defensive warfare. Network organizations, network analysis, investing where it will do the most good, dual-purpose thinking, and nontraditional asymmetric thinking provide a foundation that will be refined as we get into the details of each sector.

Second, the "core competence" most needed to implement a strategy of prevention is to employ well-founded (scientific) vulnerability and risk analysis techniques, even though the results are often at odds with political will. But it is better to know what the right answer is, even if it is difficult to "sell" to politicians. The challenge here is that we still do not have a definitive nor standardized vulnerability and risk analysis technique that everyone uses. Thus, it is currently not possible to compare risk in one sector or part of the country with risk in another sector or region. Until the standards and tools are agreed on, political factors will trump scientific analysis.

Third, the problem of critical infrastructure protection is not only scientifically challenging, but it is also made daunting by size. Sectors are simply too vast. This has several consequences that work against the 2003 National Strategy. The most obvious consequence is that sector size obviates a comprehensive approach. We simply cannot afford to protect every mile, every building, every bridge, and so forth. The combination of size, complexity, and cost prevents us from protecting everything.

Network analysis is one way to cope with size and complexity within and across sectors. Network analysis is smarter than the comprehensive approach suggested by the 2003 National Strategy. It works as follows: Find the critical nodes using network analysis; analyze the critical nodes for vulnerabilities; and apply risk analysis to determine where to invest. The final step of this approach requires that we place a value on potential targets and then allocate the limited resources available to reduce or eliminate the risk. This multistep process is developed in greater detail in the following chapters.

The 2003 National Strategy calls for cooperation between the private corporations that own and operate most infrastructure and government agencies that regulate and oversee these corporations. But private ownership limits government involvement to such a great extent that state and local government agencies are likely to feel impotent. Even the creation of ISACs will not be enough to break down the barriers between competing corporations and the various federal, state, and local agencies. This is because profit-oriented organizations are motivated differently than public-good-oriented organizations. They simply have conflicting goals.

There is another way to gain cooperation with the private sector: Make the enhancement of infrastructure a profitable endeavor. That is, think dual-purpose. Dual-purpose programs such as reducing insurance premiums through increased safety and security; increasing surge capacity and thereby increasing security through redundancy; and industry-wide initiatives that cut across stove-piped companies are already used by some industries. But this strategy needs to be elevated to higher levels.

The 2003 National Strategy does not come to grips with the fundamental conundrum of the defender versus the attacker. The attacker has the luxury of optimizing when, where, and how it attacks. The defender does not. In fact, the defender must prevent any attack at any time and with any number of weapons. It would seem that the defender is heavily disadvantaged, regardless of strategy.

The attacker–defender paradigm is asymmetric. Therefore the defender must think asymmetrically too. Asymmetric warfare is an art, but it is an art that can be acquired and perfected. The example given in this chapter—launching killer-viruses from popular Internet hubs—is but one example of how to counter the cyber threat using asymmetric thinking. It is a counter-terrorist technique that can be applied quickly and inexpensively.

More ambitious strategies—such as converting the power grid from a network with vulnerabilities in its transmission and distribution "middle" to a system based on distributed generation—is much more expensive and will require much more time. But it solves several problems at once. It is an example of both asymmetric and dual-purpose thinking. Other examples will be described for each sector as you work your way through this book.

EXERCISES

1. What report was the first to use the term "critical infrastructure"?
   a. EO-13010
   b. The "Marsh Report"
   c. The Patriot Act
   d. The National Strategy for Homeland Security
   e. The National Strategy for the Physical Protection of Critical Infrastructures and Key Assets.
2. Which of the following industrial sectors is NOT in the list of critical infrastructures proposed by the 2003 National Strategy?
   a. Computer industry
   b. Telecommunications industry
   c. Defense industry
   d. Power and energy industries
   e. Transportation industry
3. Homeland Defense differs from Homeland Security, because it is aimed at fighting terrorism:
   a. In Iraq, Afghanistan, and other Muslim countries.
   b. Along the borders of the United States.
   c. There is no difference.
   d. Outside the country, and inside the country under civilian control.
   e. On the oceans and in the air.
   f. Using a layered strategy.
4. The 2003 National Strategy for homeland security is centered on:
   a. Preparedness, response, and research.
   b. Consequence management and funding of state and local governments.
   c. Hardening critical infrastructure targets.
   d. Defining the 11 sectors and 5 key assets.
   e. Education and training.
5. Which of the following is NOT a responsibility of the Department of Homeland Security?
   a. Border and transportation security.

b. Domestic counter-terrorism.
c. Defending against catastrophic terrorism.
d. Emergency preparedness and response.
e. Hardening state and local critical infrastructure.

6. Which of the following established the first legislation in the United States to cope with weapons of mass destruction?
   a. PDD-63
   b. HSPD-7
   c. Federal Emergency Management Act of 1988
   d. Nunn-Lugar-Domenici Act of 1999
   e. The National Strategy for the Physical Protection of Critical Infrastructures and Key Assets of 2003
7. Before 2003, most governmental organizations responsible for counter-terrorism suffered from organizational vulnerabilities called:
   a. Seams among agencies
   b. Intelligence errors
   c. Lack of sufficient funding
   d. Asymmetric thinking
   e. Inadequate preparation
8. According to the 2003 National Strategy, whose responsibility is it to establish standards and procedures for critical infrastructure?
   a. DOD
   b. DHS
   c. State and local governments
   d. Private sector companies that own most of the infrastructure
   e. National Institute of Standards and Technology
9. According to the 2003 National Strategy, whose responsibility is it to establish information sharing among competitors in each infrastructure sector?
   a. Voluntary ISACs
   b. DHS
   c. National Institute of Standards and Technology
   d. FBI
   e. Private-sector corporations that own most of the infrastructure
10. According to the 2003 National Strategy, whose responsibility is it to conduct demonstration projects and pilot programs?
    a. DHS
    b. State and local governments
    c. National Institute of Standards and Technology
    d. FBI
    e. Voluntary ISACs

11. According to the 2003 National Strategy, whose responsibility is it to identify and promote sector-specific risk management policies?
    a. State and local governments
    b. DHS
    c. FBI
    d. National Guard
    e. National Institute of Standards and Technology
12. According to the 2003 National Strategy, whose responsibility is it to respond to terrorist attacks on critical infrastructure?
    a. State and local governments
    b. DHS
    c. FBI
    d. National Guard
    e. National Institute of Standards and Technology
13. What was the organizational structure of the hijackers who crashed three airliners into buildings on 9/11?
    a. Hierarchical, from bin Laden down to each hijacker
    b. Disintermediated command hierarchy
    c. Top-down organization chart, from bin Laden down to each hijacker
    d. Carefully orchestrated cell structure emanating from bin Laden
    e. Small-world network with many (secret) links
14. The most prominent feature of most critical infrastructures is:
    a. They are unprotected.
    b. They have too many components.
    c. Protecting them is too expensive.
    d. They have critical nodes.
    e. They are nonrandom structures with hubs.
15. Geographically, critical infrastructure is concentrated around a few locations, which argues for:
    a. Investing to protect dense population centers.
    b. Hardening the top 12 metropolitan areas.
    c. Investing 80% of the money to protect 20% of the concentrations.
    d. Investing all of the money to protect the top 12 metropolitan areas.
    e. Distributing the generation of power to factories and shopping malls.
16. Dual-purpose strategies for coaxing investment in infrastructure protection from the companies that own and operate most infrastructure is defined as:
    a. Enhancing productivity while improving security.
    b. Forcing companies to lower insurance policies to pay for improvements.
    c. Taxing Internet companies to stop the spread of viruses.

d. Using redundancy to increase volume.
e. Spreading the components of an infrastructure across large geographical areas.

17. The major difficulty with vulnerability and risk assessment is:
    a. Lack of money
    b. Lack of training
    c. Lack of standard methods of assessment
    d. Lack of leadership
    e. Lack of sleep
18. The attacker–defender conundrum is asymmetric, because:
    a. Attackers use suicide.
    b. Attackers break the law.
    c. Attackers have the luxury of surprise.
    d. Defenders must prepare to defend against any weapon, applied at anytime, on anything, anywhere.
    e. Defenders do not have sufficient funding to protect everything.
19. The major reasons why critical infrastructure should be protected, rather than making preparations to respond to attacks on infrastructure, are:
    a. Successful attacks can lead to mass casualties and mass economic damage.
    b. Successful attacks mean a loss in major capital equipment.
    c. Successful attacks mean loss of power and energy.
    d. Consequence management would be too expensive.
    e. Emergency response capability does not exist for coping with such attacks.
20. The virtual city in cyber-space called San Lewis Rey has devoted $100 million to protect its critical infrastructure sectors, which consist mostly of a robust subway system, major telecommunications network, electrical power grid, and water utility. After an initial study performed by an engineering firm, the city council realized that these four infrastructures would be very costly to protect. Estimates to provide hardening of the physical components of each were:
    a. $1.5 billion: Subway System.
    b. $500 million: Telecommunications Network.
    c. $250 million: Electrical Power Grid.
    d. $3.3 billion: Water utility components.

    The total for all four sectors amounts to more than $5.5 billion or 55 times more than the city's $100 million budget! What should be the San Lewis Rey strategy for protecting its four major infrastructures?
21. For each of the five strategic principles described in this chapter, show how each principle may be applied to San Lewis Rey to provide the best

counter-terror protection possible, given limited funds. That is, apply each of the following strategies to this challenge:

a. It takes a network to fight a network
b. Secure the hubs, not the spokes
c. Spend 80% on 20% of the country
d. Think dual-purpose
e. Think asymmetric

CHAPTER 2

Origins

In this chapter we trace the development of the new discipline called *critical infrastructure protection* over several decades and note that it has evolved from initial awareness to the centerpiece of the national strategy for combating terrorism. There have been at least five distinct phases in this evolution:

1. Recognition: No such field of study existed before the mid-1900s. Although recognition began in 1962 with the Cuban Missile Crisis, it was nearly 30 years before the term *critical infrastructure protection* (*CIP*) was even defined. Throughout these 30 years, the roles and responsibilities of governmental agencies as well as the definition of CIP changed as the field evolved through many stages.
2. Natural Disaster Recovery: In the beginning, CIP was nearly identical to consequence management—recovery from disasters such as floods, hurricanes, and earthquakes. Terrorism was not a factor in CIP in the beginning. It would take a decade of attacks before CIP was linked with terrorism in the United States.
3. Definitional Phase: The term "critical infrastructure" did not exist before the 1990s. Then, from 1997 through 2003, the definition of what constituted a critical infrastructure expanded from 8 to 13 sectors plus five *key assets*. Today it is difficult to identify sectors of the national economy that are not critical.
4. Public–Private Cooperation: The role of the private sector in CIP was slow to take root until the late 1990s. But most national assets are in the hands of corporations—not local, state, or federal government. Even today, the federal government and private sector owners of infrastructure are not clear on their respective roles and responsibilities with respect to CIP.
5. Federalism: Because terrorists attack at the local level, the solution to the problem must also come from the local level—states and cities. The future of homeland security rests in the hands of local governments, and yet the local level lacks sufficient technology, skills, and funding to cope with global terrorism.

Critical Infrastructure Protection in Homeland Security: Defending a Networked Nation, edited by Ted G. Lewis

THE DAWN OF CRITICAL INFRASTRUCTURE PROTECTION

Before the dramatic and horrific attacks of September 11, 2001 (9/11), the U.S. public had little awareness of terrorism or how it could impact them personally. Attacks on the homeland were something that happened in other countries, not the United States. But a growing number of "national security emergencies" culminating in 9/11 exposed terrorism for what it is: a national challenge to the people of the United States. Even before 9/11, however, a few policy makers were busy formulating various strategies and policies that culminated in the formation of a national strategy for homeland security. A major part of this national strategy involves critical infrastructure protection—the protection of basic infrastructure sectors such as water, power, telecommunications, health and medical services, the Internet, and transportation systems. The early work of this small group peaked in the late-1990s, which marks the origins of what we now call "homeland security." During this same time, critical infrastructure and CIP emerged as a key element of homeland security.

Although CIP was defined and recognized as a major component of national security rather late in the game (1996), it really began with the creation of the National Communications System (NCS) in 1963 after communication problems between the United States and the Soviet Union threatened to interfere with negotiations during the Cuban Missile Crisis[1]:

> In October [1962], President John F. Kennedy, on national television, revealed that the Soviets had placed nuclear missiles in Cuba. As a result of this aggressive action, he ordered quarantine on all offensive military equipment under shipment to Cuba until the Soviets removed their weapons. . . . For nearly a week, the Nation was transfixed by the actions of Soviet Premier Nikita Khrushchev and President Kennedy. During this time, ineffective communications were hampering the efforts of the leaders to reach a compromise. Without the ability to share critical information with each other using fax, e-mail, or secure telephones such as we have today, Premier Khrushchev and President Kennedy negotiated through written letters. Generally, Washington and Moscow cabled these letters via their embassies. As the crisis continued, hours passed between the time one world leader wrote a letter and the other received it. Tensions heightened. On October 27 and 28, when urgency in communications became paramount, Premier Khrushchev bypassed the standard communication channels and broadcast his letters over Radio Moscow.
>
> Following the crisis, President Kennedy, acting on a National Security Council recommendation, signed a Presidential memorandum establishing the NCS. The new system's objective was 'to provide necessary communications for the Federal Government under all conditions ranging from a normal situation to national emergencies and international crises, including nuclear attack.'
>
> At its inception on August 21, 1963, the NCS was a planning forum composed of six Federal agencies. Thirty-five years later, it is a vital institution comprising 23 member

[1]http://www.ncs.gov/about.html.

> organizations that ensure NS/EP (National Security/Emergency Preparedness) telecommunications across a wide spectrum of crises and emergencies. ... During the 1980s and 1990s, the NCS expanded its focus to develop Government wide NS/EP procedures and enhancements to the Nation's public networks and information infrastructures.

The role of the telecommunications infrastructure grew more important as the United States entered the information age. In 1978, two telecommunication regulatory agencies (Department of Commerce Office of Telecommunications and the Whitehouse Office of Telecommunications) were combined into the National Telecommunications and Information Administration (NTIA) by Executive Order 12046. NTIA handles the process of selling spectrum to telephone, radio, and TV networks. It also has the distinction of being the federal agency that oversaw the commercialization of the Internet in 1998–1999. The NCS was formally assigned responsibility for the telecommunications infrastructure in 1984 by Executive Order 12472.

In 1982 President Reagan established the National Security Telecommunications Advisory Committee (NSTAC) by Executive Order 12382. This important Presidential advisory body is made up of the CEOs of the major telecommunications companies. This body is perhaps the first organization to advise the President on critical infrastructure protection.

The Department of Homeland Security (DHS) absorbed the NCS in February 2003. The NCS and the NSTAC were the first critical infrastructure agencies within the U.S. government. Twenty years would pass before the term *critical infrastructure* would be defined and the entire U.S. population would become aware of its importance in their daily lives.

DAWN OF TERRORISM IN THE UNITED STATES

Although the NCS and NSTAC were active throughout the 1970s and 1980s, terrorism was still on the back burner as far as CIP was concerned. The Federal Emergency Management Agency (FEMA) was created in 1978 to fight hurricanes and earthquakes. Soon after its creation, FEMA was assigned the responsibility of fighting terrorists by Executive Order 12148 in 1979[2]:

> All functions vested in the President that have been delegated or assigned to the Defense Civil Preparedness Agency, Department of Defense, are transferred or reassigned to the Director of the Federal Emergency Management Agency.
>
> All functions vested in the President that have been delegated or assigned to the Federal Disaster Assistance Administration, Department of Housing and Urban Development, are transferred or reassigned to the Director of the Federal Emergency Management Agency, including any of those functions redelegated or reassigned to the Department

[2]http://www.archives.gov/federal_register/codification/executive_order/12148.html.

of Commerce with respect to assistance to communities in the development of readiness plans for severe weather-related emergencies.
All functions vested in the President that have been delegated or assigned to the Federal Preparedness Agency, General Services Administration, are transferred or reassigned to the Director of the Federal Emergency Management Agency.

All functions vested in the President by the Earthquake Hazards Reduction Act of 1977 (42 U.S.C. 7701 *et seq.*), including those functions performed by the Office of Science and Technology Policy, are delegated, transferred, or reassigned to the Director of the Federal Emergency Management Agency .... For purposes of this Order, 'civil emergency' means any accidental, natural, man-caused, or wartime emergency or threat thereof, which causes or may cause substantial injury or harm to the population or substantial damage to or loss of property.

FEMA was confronted by perhaps the first major terrorist attack on U.S. soil in Oregon in 1984. Members of the politico-religious commune founded by Bhagwan Shree Rajneesh[3] attempted to influence a political election by poisoning voters with salmonella.[4]

In a bizarre plot to take over local government, followers of Bhagwan Shree Rajneesh poisoned salad bars in 10 restaurants in The Dalles in 1984, sickening 751 people with salmonella bacteria. Forty-five of whom were hospitalized. It is still the largest germ warfare attack in U.S. history. The cult reproduced the salmonella strain and slipped it into salad dressings, fruits, vegetables and coffee creamers at the restaurants. They also were suspected of trying to kill a Wasco County executive by spiking his water with a mysterious substance. Later, Jefferson County District Attorney Michael Sullivan also became ill after leaving a cup of coffee unattended while Rajneeshees lurked around the courthouse. Eventually, Ma Anand Sheela, personal secretary of the Bhagwan, was accused of attempted murder, conspiracy, arson, and other crimes and disowned by the Bhagwan. Convicted of the charges against her, she spent 29 months in federal prison, then moved to Switzerland.[5]

The salmonella incident in Oregon was an attack on one of many infrastructure sectors identified as critical over the past decade: agriculture. But in 1984 there was no generally accepted definition of *infrastructure*, nor any recognition of what sectors belonged to the list of national CIs. The importance of infrastructure was beginning to dawn on the federal government when in 1988 President Reagan issued Executive Order 12656. This order alludes to "essential resources"

[3] http://www.religioustolerance.org/rajneesh.htm.
[4] "The group settled on the 65,000 acre '*Big Muddy Ranch*' near Antelope, Oregon, which his *sannyasins* had bought for six million dollars. The ranch was renamed *Rajneeshpuram* ("Essence of Rajneesh"). This '*small, desolate valley twelve miles from Antelope, Oregon was transformed into a thriving town of 3,000 residents, with a 4,500 foot paved airstrip, a 44 acre reservoir, an 88,000 square foot meeting hall ...*'" http://www.clui.org/clui_4_1/lotl/lotlv10/rajneesh.html.
[5] http://home.att.net/~meditation/bioterrorist.html.

and places responsibility for their protection in the hands of the heads of federal departments:

> The head of each Federal department and agency, within assigned areas of responsibility shall:
> **Sec. 204.** *Protection of Essential Resources and Facilities.*
> (1) Identify facilities and resources, both government and private, essential to the national defense and national welfare, and assess their vulnerabilities and develop strategies, plans, and programs to provide for the security of such facilities and resources, and to avoid or minimize disruptions of essential services during any national security emergency;
> (2) Participate in interagency activities to assess the relative importance of various facilities and resources to essential military and civilian needs and to integrate preparedness and response strategies and procedures;
> (3) Maintain a capability to assess promptly the effect of attack and other disruptions during national security emergencies.

Even in the early 1990s the terrorism trend was subtle—it had not yet reached a point where it was of national interest. But by 1993–1995 the rate and severity of terrorism was increasing and becoming more alarming to the federal government. The 1993 attack on the World Trade Center by Ramzi Yousef, the capture of the Unabomber (1995), the devastating attack on the Federal Building in Oklahoma City, Oklahoma (1995), and the Sarin gas attack in a Tokyo subway in 1995 suggested a trend. Acts of violence by nongovernmental organizations were increasing and, as a consequence, raising the level of public awareness. Soon these acts would be attributed to terrorists. And by 1995, terrorism was reaching a threshold that deserved national attention. It was the dawning of the terrorism era in the United States. Within 5–6 years, this would become known as the *Global War on Terrorism* (GWOT).

WHAT IS A CRITICAL INFRASTRUCTURE?

The modern origin of homeland security, and its corollary, CIP, can be placed somewhere between 1993 and late 1995. In fact, 1995 is a reasonable start date because of the flurry of activity aimed at protecting national infrastructure and key national assets after 1995. Presidential Decision Directive 39 (PDD-39) issued by President Clinton in 1995 set the stage for what was to come—a new Federal Department of Homeland Security. PDD-39 essentially declared war on terrorists[6]:

> It is the policy of the United States to deter, defeat and respond vigorously to all terrorist attacks on our territory and against our citizens, or facilities, whether they occur domestically, in international waters or airspace or on foreign territory. The United States regards all such terrorism as a potential threat to national security as well as a criminal act and will apply all appropriate means to combat it. In doing so, the U.S. shall pursue vigorously efforts to deter and preempt, apprehend and prosecute, or assist other governments to prosecute, individuals who perpetrate or plan to perpetrate such attacks.

[6]http://www.fas.org/irp/offdocs/pdd39.htm.

> We shall work closely with friendly governments in carrying out our counterterrorism policy and will support Allied and friendly governments in combating terrorist threats against them. Furthermore, the United States shall seek to identify groups or states that sponsor or support such terrorists, isolate them and extract a heavy price for their actions. It is the policy of the United States not to make concessions to terrorists.

The criticality of national infrastructure and corresponding assets became an important issue when President Clinton issued Executive Order EO-13010 in 1996. It established a Presidential Commission on Critical Infrastructure Protection (PCCIP). This commission was chaired by Robert Marsh and hence became known as the "Marsh Report."[7] This report defines "critical infrastructure" in terms of "energy, banking and finance, transportation, vital human services, and telecommunications." It was the first publication to use the term "critical infrastructure" and has become one of the foundational documents of CIP history.

The "Marsh Report" and Executive Order EO-13010 provided the first definition of infrastructure and loosely described an infrastructure as "a network of independent, mostly privately-owned, man-made systems that function collaboratively and synergistically to produce and distribute a continuous flow of essential goods and services." And a CI as "an infrastructure so vital that its incapacity or destruction would have a debilitating impact on our defense and national security." According to Executive Order 13010[8]:

> Certain national infrastructures are so vital that their incapacity or destruction would have a debilitating impact on the defense or economic security of the United States. These critical infrastructures include telecommunications, electrical power systems, gas and oil storage and transportation, banking and finance, transportation, water supply systems, emergency services (including medical, police, fire, and rescue), and continuity of government. Threats to these critical infrastructures fall into two categories: physical threats to tangible property ("physical threats"), and threats of electronic, radio frequency, or computer-based attacks on the information or communications components that control critical infrastructures ("cyber threats"). Because many of these critical infrastructures are owned and operated by the private sector, it is essential that the government and private sector work together to develop a strategy for protecting them and assuring their continued operation.

The work of the PCCIP resulted in PDD-63 (Presidential Decision Directive of 1998), which defined critical infrastructure more specifically and identified eight basic sectors, shown in Table 2.1. According to PDD-63:

> Critical infrastructures are those physical and cyber-based systems essential to the minimum operations of the economy and government. They include, but are not limited to, telecommunications, energy, banking and finance, transportation, water systems and emergency services, both governmental and private.[9]

[7] "Critical Foundations: Protecting America's Infrastructures," The Report of the President's Commission on Critical Infrastructure Protection, October 1997.
[8] http://www.fas.org/irp/offdocs/eo13010.htm.
[9] http://www.fas.org/irp/offdocs/pdd/pdd-63.htm.

TABLE 2.1. The Basic Critical Infrastructure Sectors Defined by PDD-63 (1998).

Sector	Description	U.S. Department
1. Banking and Finance	Banking and Stock Markets	Treasury
2. Emergency Law Enforcement Services	Justice/FBI	Justice
3. Emergency Services	Emergency Fire and Continuity of Government	FEMA
4. Energy	Electric Power, Gas and Oil production and storage	Energy
5. Information and Communications	Telecommunications and the Internet	Commerce
6. Public Health Services	Public health, surveillance, laboratory services, and personal health services	HHS
7. Transportation	Aviation, Highways, Mass Transit, Rail, Pipelines, Shipping	Transportation
8. Water Supply	Water and its Distribution	Environmental Protection Agency

The definition of CI in PDD-63 went through rapid evolution and expansion after the attacks of 9/11. The office of the President of the United States released the "National Strategy for Homeland Security" in July 2002 and then rapidly followed up with an expansion of the definition of critical infrastructure sectors in February 2003 with the release of "The National Strategy for the Physical Protection of Critical Infrastructures and Key Assets."

In addition to the list of sectors shown in Table 2.2, the 2003 National Strategy lists five key assets:

- National Monuments and Icons
- Nuclear Power Plants
- Dams
- Government Facilities
- Commercial Key Assets

The year 1998 was a year of ramping up counter-terror programs. Major initiatives besides PDD-62 (Countering Terrorism), PDD-63 (Critical Infrastructure Protection), and PDD-67 (Continuity of Government) were as follows:

- National Infrastructure Protection Center established in the Department of Justice
- Chemical Safety Board formed
- National Domestic Preparedness Office created in Department of Justice
- Critical Infrastructure Analysis Office (CIAO) established
- Counter-Terror Coordination Unit in National Security Council formed

TABLE 2.2. The CI Sectors as of 2003.

Sector	Agency
Agriculture	Dept. of Agriculture
Food:	
Meat and Poultry	Dept. of Agriculture
All other food products	Dept. of Health & Human Services (HHS)
Water	EPA: Environmental Protection Agency
Public Health	Dept. of HHS
Emergency Services	Dept. of Homeland Security
Government:	
Continuity of government	Dept. of Homeland Security
Continuity of operations	All departments and agencies
Defense Industrial Base	DOD
Information and Telecommunications	Dept. of Homeland Security
Energy	Dept. of Energy
Transportation	Dept. of Homeland Security (TSA)
Banking and Finance	Dept. of the Treasury
Chemical Industry and Hazardous Materials	EPA
Postal and Shipping	Dept. of Homeland Security
Nat'l Monuments and Icons	Dept. of the Interior

- Congress earmarks $17M for Special Equipment and Training Grants
- Attorney General announces creation of National Domestic Prep. Office (NDPO)

By 1999 it became clear that most infrastructure in the United States is owned by the private sector, not by government. The Internet had just been commercialized in 1998, and the Telecommunications and Electrical Power sectors were in the process of being deregulated. Control of most public utility infrastructure was in the hands of corporations. Thus, in 1999, President Clinton established National Infrastructure Assurance Council (NIAC) to bring industry and government closer together. According to Executive Order 13130, NIAC was established to facilitate the partnership through Public Sector Information Sharing and Analysis Centers (PS-ISACs), which subsequently have become sector-specific ISACs[10]:

> By the authority vested in me as President by the Constitution and the laws of the United States of America, including the Federal Advisory Committee Act, as amended (5 U.S.C. App.), and in order to support a coordinated effort by both government and private sector entities to address threats to our Nation's critical infrastructure, it is hereby ordered as follows:
> **Section 1.** *Establishment.* (a) There is established the National Infrastructure Assurance Council (NIAC). The NIAC shall be composed of not more than 30 members appointed by the President. The members of the NIAC shall be selected from the private sector, including private sector entities representing the critical infrastructures

[10]http://www.archives.gov/federal_register/executive_orders/1999.html#13130.

identified in Executive Order 13010, and from State and local government. The members of the NIAC shall have expertise relevant to the functions of the NIAC and shall not be full-time officials or employees of the executive branch of the Federal Government.
(b) The President shall designate a Chairperson and Vice-Chairperson from among the members of the NIAC.
(c) The National Coordinator for Security, Infrastructure Protection and Counter-Terrorism at the National Security Council (National Coordinator) will serve as the Executive Director of the NIAC.
(d) The Senior Director for Critical Infrastructure Protection at the National Security Council will serve as the NIAC's liaison to other agencies.
(e) Individuals appointed by the President will serve for a period of 2 years. Service shall be limited to no more than 3 consecutive terms.
Section 2. *Functions.* (a) The NIAC will meet periodically to:
(1) enhance the partnership of the public and private sectors in protecting our critical infrastructure and provide reports on this issue to the President as appropriate;
(2) propose and develop ways to encourage private industry to perform periodic risk assessments of critical processes, including information and telecommunications systems; and
(3) monitor the development of PS-ISACs and provide recommendations to the National Coordinator and the National Economic Council on how these organizations can best foster improved cooperation among the PS-ISACs, the National Infrastructure Protection Center (NIPC), and other Federal Government entities.
(b) The NIAC will report to the President through the Assistant to the President for National Security Affairs, who shall assure appropriate coordination with the Assistant to the President for Economic Policy.
(c) The NIAC will advise the lead agencies with critical infrastructure responsibilities, sector coordinators, the NIPC, the PS-ISACs and the National Coordinator on the subjects of the NIAC's function in whatever manner the Chair of the NIAC, the National Coordinator, and the head of the affected entity deem appropriate.

CIP IS RECOGNIZED AS BEING A CORE COMPONENT

The final act before the creation of the DHS came a few weeks after the terrorist attack of 9/11. By Executive Order 13231 (October 2001) President Bush created the President's Critical Infrastructure Protection Board (PCIPB), with the primary responsibility to protect the information infrastructure of the federal government. EO 13231 recognized the growing importance of the telecommunications and Internet infrastructure as well as its interdependency with other sectors. Without information systems, the U.S. Federal Government could not continue to operate in the event of an attack:

> Consistent with the responsibilities noted in section 4 of this order, the Board shall recommend policies and coordinate programs for protecting information systems for critical infrastructure, including emergency preparedness communications, and the physical assets that support such systems.

In 2002 President Bush signed the Homeland Security Bill, establishing the new DHS. It began operation in February 2003 and incorporated 22 agencies that had been spread throughout the federal bureaucracy. This included the NCS, CIAO, and Department of Justice Office of Domestic Preparedness, along with several other agencies such as the INS, Border Patrol, and Coast Guard. Thus, protection of critical infrastructure is now the responsibility of the DHS.

In December 2003, President Bush replaced PDD-63 with HSPD-7 (Homeland Security Presidential Directive #7), which rewrote the list of sectors and who is responsible; see Table 2.3. HSPD-7 departs from the list of sectors given by the National Strategy, and clouds the issue of which department or agency is responsible for energy, power, and the Information and Telecommunications sector. Indeed, HSPD-7 does not specify who is responsible for several of the sectors previously identified as "critical." It appears that HSPD-7 was written to address in-fighting among departments and agencies that may have felt left out of the National Strategy. Alternatively, the purpose of HSPD-7 may have been to include departments and agencies that have expertise in fields such as cyber, chemical, and nuclear security. For whatever reason, HSPD-7 leaves some responsibilities unspecified, and spreads others across multiple departments.

TABLE 2.3. Critical Infrastructure Sectors and Responsibilities as Defined by HSPD-7.

Sector (Industry)	Responsible Department/Agency
Agriculture/Food (meat, poultry, eggs)	Department of Agriculture
Public Health/Food (other than meat, poultry, eggs)	Department of Health and Human Services
Drinking Water and Treatment Systems	Environmental Protection Agency
Energy (production, storage, distribution of gas, oil, and electric power, except for commercial nuclear power facilities)	Department of Energy
Nuclear Power Plants	Department of Homeland Security and Nuclear Regulatory Commission, and Department of Energy
Banking and Finance	Department of the Treasury
Defense Industrial Base	Department of Defense
Cyber-Security	Department of Commerce and Department of Homeland Security
Chemical	Not specified
Transportation Systems, including mass transit, aviation, maritime, ground/surface, and rail and pipeline systems	Department of Transportation and Department of Homeland Security
Emergency Services	Not specified
Postal and Shipping	Not specified
National Monuments	Department of the Interior
Key Assets: dams, government facilities, and commercial facilities	Not specified

ANALYSIS

CIP has grown to encompass most of the U.S. economy. It is difficult to identify sectors that are not critical. Indeed, CIP has come to embrace just about every aspect of society, from communications, power, and health care, to the food we eat, water we drink, and work we do. If CIP embraces nearly everything, perhaps it has lost its focus. What then is the main goal of CIP? According to the National Strategy document produced by the Department of Homeland Security[11]:

> The **first** objective of this *Strategy* is to identify and assure the protection of those assets, systems, and functions that we deem most 'critical' in terms of national-level public health and safety, governance, economic and national security, and public confidence. We must develop a comprehensive, prioritized assessment of facilities, systems, and functions of national-level criticality and monitor their preparedness across infrastructure sectors.

Although the stated objective of the national strategy is to protect the obvious, it is not so obvious how to do so. The quotation above defines "critical" in terms of the government's national-level responsibility. It does not claim to protect infrastructure at all levels, e.g., how "big" must an event be in order to warrant the designation of "critical?" This detail is not specified in the strategy, but HSPD-7 clearly identifies attacks involving weapons of mass destruction as the threat. Suffice it to say, CIP aims to head off terrorist attacks that could have national impact. That is, major casualties, major economic disruption, and major detrimental effects on the psychology of Americans. Under this definition, the Federal Building in Oklahoma City would not be considered critical.

Uncertainty remains at the national and local levels of government as to what is critical and what is not. In late 2003, President Bush signed HSPD-7: Critical Infrastructure Identification, Prioritization, and Protection. At the time of this writing, HSPD-7 was the latest declaration by the federal government on CIP. It attempted to narrow the definition of critical components and sharpen the roles and responsibilities of federal departments and agencies. For the first time, HSPD-7 declared that it is impractical to protect everything and focused effort on major incidents—ones that cause mass casualties comparable with the effects of using weapons of mass destruction:

> While it is not possible to protect or eliminate the vulnerability of all critical infrastructure and key resources throughout the country, strategic improvements in security can make it more difficult for attacks to succeed and can lessen the impact of attacks that may occur. In addition to strategic security enhancements, tactical security improvements can be rapidly implemented to deter, mitigate, or neutralize potential attacks . . . Consistent with this directive, the [DHS] Secretary will identify, prioritize, and coordinate the protection of critical infrastructure and key resources with an emphasis

[11] The National Strategy for the Physical Protection of Critical Infrastructures and Key Assets, February 2003, Department of Homeland Security. http://www.whitehouse.gov/pcipb/physical.html.

on critical infrastructure and key resources that could be exploited to *cause catastrophic health effects or mass casualties* comparable to those from the use of a *weapon of mass destruction.*[12]

CIP has come of age. It is now at the core of homeland security. It has evolved over several decades and gone through several phases:

1. Recognition: The 1962 Cuban Missile crisis revealed two important facts about infrastructure. First, modern telecommunications technology is an essential element of national policy and diplomacy. The formation of the NCS and its subsequent colleague organizations such as the NSTAC was recognition of the criticality of modern communication infrastructure. Second, the United States had entered the information age, and this meant it had become dependent on a technology—telecommunications—that did not exist when the Founding Fathers wrote the U.S. Constitution and organized the federal government. The implications of the transition to a technological society were not so clear in 1962, but by 1995, it became obvious that the United States would not only become dependent on the Internet, computers, and database software in its war on terrorism, but that the future success of CIP depends to a large extent on application of new technology.
2. Natural Disaster Recovery: Early in the evolution of CIP, the emphasis was on recovery from disaster—floods, hurricanes, and earthquakes. FEMA was formed to handle these events, and CIP was initially the province of FEMA. Under FEMA, CIP was a component of National Security/Emergency Preparedness (NS/EP). The threats to CI have escalated from natural disasters, disease, and pestilence to manmade hazards and terrorist attacks. This has complicated the solution so that, today, CIP is a "wicked problem" whose solution may be years away.
3. Definitional Phase: By mid-1997 the obvious became a driving factor in homeland security: The country is tremendously dependent on CIP. But what was CIP? From 1997 through 2003, the definition of CIP expanded from an initial 8 sectors to 13 sectors plus five *key assets*:
   - National Monuments and Icons
   - Nuclear Power Plants
   - Dams
   - Government Facilities
   - Commercial Key Assets

   By the time the new DHS began operation in February 2003, CIP had become one of the pillars of the new DHS. In 2003 the nation knew what CIP was, but few experts knew what to do about it. The vastness and level of sophistication of these diverse sectors posed such a challenge that it would be years before

[12]The Whitehouse, "Homeland Security Presidential Directive/Hspd-7," December 17, 2003.

progress toward prevention of attacks would be made. Yet the journey had begun by 1998.

4. Public–Private Cooperation: During the 1998–2003 period, a second obvious driving factor began to appear in the thinking of policy makers—the role of the private sector in CIP. According to the 2003 National Strategy[13]:

   > The second major objective is to assure the protection of infrastructures and assets that face a specific, imminent threat. Federal, state, and local governments and private-sector partners must collaborate closely to develop thorough assessment and alert processes and systems to ensure that threatened assets receive timely advance warnings. These entities must further cooperate to provide focused protection against the anticipated threat.

   These words of wisdom have, for the most part, fallen on deaf ears, as corporate America must remain profitable regardless of the potential threat. In their view, it is better to buy insurance than increase spending on expensive CIP, an investment that may never be needed. On the governmental side, the question is whether the new DHS will become a regulator of critical infrastructure businesses or simply recommend standards and practices. The question of who is going to pay for increased security is one of the imposing uncertainties that lie ahead.

5. Federalism: Finally, the 2003 National Strategy points out the importance of a federalist approach to CIP. But separation of powers, state's rights, and local control of local problems may become a limiting factor in the GWOT. Terrorists aim to do national-level damage, but because their attacks occur at the local level, the solution to the problem must also come from the local level—states and cities. At least this is the current strategy. In reality, major attacks on critical infrastructure would have a national impact. Hence CIP remains a national problem that will require a federal-level involvement.

   > *Homeland security*, particularly in the context of critical infrastructure and key asset protection, is a shared responsibility that cannot be accomplished by the federal government alone. It requires coordinated action on the part of federal, state, and local governments; the private sector; and concerned citizens across the country . . . The 50 states, 4 territories, and 87,000 local jurisdictions that comprise this Nation have an important and unique role to play in the protection of our critical infrastructures and key assets. State and local governments, like the federal government, should identify and secure the critical infrastructures and key assets they own and operate within their jurisdictions.[14]

Perhaps the longest phase of CIP development will be the gradual shift of roles and responsibilities from predominately federal to predominantly state and local governments. In the long run, CIP may become the responsibility of states, corporations, and federal regulatory agencies. The history of the United States suggests just the opposite: National problems percolate up to the national level. It would be out of

[13]Ibid.
[14]Ibid.

character for a nation that has centralized public education, public health, national defense, national emergency response, and intelligence gathering and dissemination at the national level to suddenly reverse its course and decentralize homeland security. It would require a major reversal of the flow of authority and money back to the states and cities.

Most critical infrastructure sectors span multiple states, which means they fall under the regulatory powers of interstate commerce. The huge energy pipeline system that supplies most of the gas and oil to the United States is regulated by interstate commerce rules. The electrical power grids, telecommunications networks, Internet and computer networks, water supply networks, and food, chemical, and industrial networks all cross state boundaries. Hence, the federal government will continue to exercise its powers as interstate regulator far into the future.

Perhaps there is a middle course that avoids these two extremes. The DHS could evolve its "business model" to emulate existing federal departments. In one model, the states and cities are the first line of defense, whereas the federal government focuses on national assets that cannot be protected by the states and cities. This two-tiered approach would require rethinking interstate commerce regulation, constitutional law, and domestic intelligence, information sharing, and distributed law enforcement and public safety.

Another business model is that of the standard-setter and regulator. In this case, the DHS provides scientific expertise and regulatory pressure to motivate the private sector to improve security, share information, and make the investment needed to prevent attacks on infrastructure. This model is reminiscent of the role played by the EPA and other regulatory agencies.

The DHS will evolve over time and be shaped by future events. Whatever its ultimate business model is, one thing seems certain: Although the new manmade threat is the most recent to join the list of natural, chemical, biological, and radiological threats facing all sectors, it is here to stay.

EXERCISES

1. What organizational structure was designed to handle the public–private sector partnership in homeland security (select one)?
   a. ISAC
   b. NIAC
   c. ENIAC
   d. NIPC
   e. PCIPB
2. What sector is not on the list of Table 2.2: Critical Infrastructure Sectors as of 2003 (select one)?
   a. Agriculture
   b. Internet and the Web
   c. Water

   d. Transportation
   e. U.S. Postal and Shipping
3. What organization was the first in the United States to advise a U.S. President on critical infrastructure issues (select one)?
   a. NCS
   b. NSTAC
   c. NIAC
   d. PCCIP
   e. FEMA
4. What federal government agency was the first to be assigned the responsibility of fighting terrorists in the United States?
   a. NCS
   b. NSTAC
   c. NIAC
   d. PCCIP
   e. FEMA
5. When and where was the first bio-terror attack on U.S. soil? Who perpetrated it?
   a. 2001: New York City; Al-Qaeda
   b. 1993: New York City; Ramzi Yousef
   c. 1984: Oregon; Ma Anand Sheela
   d. 1995: Oklahoma City; Unibomber
   e. 1995: Oklahoma City; Timothy McVeigh
6. When was critical infrastructure acknowledged as a major component of homeland security? By what document?
   a. 1995: PDD-39
   b. 1996: EO-13010
   c. 1998: PDD-63
   d. 2002: National Strategy for Homeland Security
   e. 2003: National Strategy for the Physical Protection of Critical Infrastructures and Key Assets
7. How many critical infrastructure sectors were defined in PDD-63 in 1998?
   a. 8
   b. 5
   c. 11
   d. 13
   e. 14

8. How many critical infrastructure sectors are defined in the 2003 National Strategy for the Physical Protection of Critical Infrastructures and Key Assets?
   a. 8
   b. 5
   c. 11
   d. 13
   e. 14
9. The NIAC was formed in 1999 by EO-13130. What does NIAC mean?
   a. National Industry Advisory Council
   b. National Infrastructure Assurance Council
   c. National Information Assurance Council
   d. National Information Advisory Committee
   e. National Infrastructure Advisory Committee
10. Who in the government is responsible for responding to Anthrax incidents?
    a. U.S. Public Health
    b. EPA
    c. Department of Homeland Security
    d. NSTAC
    e. FBI
    f. Centers for Disease Control
11. The city of San Lewis Rey[15] elects its Department of Regional Homeland Security (DRHS) director every 4 years. The director determines the policies that are enforced—or not—in the region. Unfortunately, the candidates for this office typically take opposing positions on the question of how the agency should be run, which means the agency changes its direction every 4 years. The main issue concerns the question of the Department's roles and responsibilities, funding, and powers of enforcement. When the San Lewis Rey DRHS was formed, it had no firm "business model" or strategy for implementing CIP. As in past years, candidates in the coming election are hotly debating the role of DRHS. Some voters claim the DRHS is a waste of money. Others want it to regulate the private sector into making infrastructure safer. Still others want the public sector to take on the responsibility even if it means more taxes. The following exercise explores these roles and responsibilities relative to critical infrastructure protection.

    The four candidates are described, below, followed by some questions.

    a. Harold Mindmeld has served on the board of education, and if elected, he would run DRHS like the U.S. Department of Education. In his mind, the DRHS should set and enforce standards of security competence

[15]San Lewis Rey is a virtual city in cyber-space where all bad things happen. It is a pedagogical device for exercising the reader's ability to apply what he or she has learned from the text.

among the public and private sector organizations that own and operate the critical infrastructure sectors. He wants to use strict governmental regulation to make sure each sector adequately funds and implements target hardening and other preventative measures. For example, he wants to pass laws forcing all Web companies in the region to eliminate spam, reduce cyber-threats to 10% of what they are today, and maintain security standards within the backbone of the Internet. To do so, he would require network equipment manufacturers to meet certain cyber-security standards. His vision of the DRHS includes being authorized to fine companies that do not comply with standards. They say that some solutions to these problems are not even available yet. If elected, Mindmeld would add a 1% "security sales" tax to all retail sales to pay for more emergency response equipment and hire more people to respond to emergencies. But the private sector would have to fund its own research and development. The private sector argues that Mindmeld's policies would cost them billions and put them out of business. His campaign slogan is "An Ounce of Legislation is Worth a Pound of Prevention."

b. Susan Marcel-Wave made her fortune in the broadcast business. Her family owns most radio and TV stations in the region. Now she wants to enter politics, and this position is a good place to start. She believes the DRHS should be run like the FCC. San Lewis Rey should auction "CIP rights" to all owners and operators of critical infrastructure sectors. For example, the water, power, energy, transportation, Internet, and telecommunications sectors would be required to bid on the right (and rights-of-way) to operate in the region. All critical infrastructure sectors would be owned and operated by the private sector, but they would be charged for that right. The private sector claims this would ruin them and has spent $50 million to defeat her. If elected, the DRHS would use its money to fund emergency response equipment and hire more people to respond to emergencies. Susan's campaign slogan is "Infrastructure Belongs to the People: Don't Let Big Corporations Steal It From Us."

c. Bill "Slammer" Wiseman was the San Lewis Rey District Attorney for 15 years until he stepped down after allegations of improperly contracting services to the private sector (outsourcing). He was found innocent of all charges 2 years later. Some say he is running for this position as a stepping-stone back into the DA's office. His campaign platform is similar to the one he used to get elected to the DA's office. He believes the DRHS is essentially a law enforcement agency that should coordinate the fire, police, and emergency management organizations throughout the region. If he was elected, he would dramatically expand the number of fire, police, medical, and emergency management personnel, increase funding for training and equipment, and raise salaries. He believes funding should come from existing governmental sources. There should be no added taxes or burdensome regulation. Slammer's policies include the outsourcing of CIP functions to corporations that deal with vulnerability analysis and commercial security. His campaign slogan is "Keep Critical Infrastructure Honest: A Vote For Me, is a Vote Against Terrorism."

d. Hale Frequently is a retired college professor and has dabbled in politics for years. He was a citizen volunteer on the water board, school board, and governor's ethics committee. His campaign platform advocates a research and development "business model" for the DRHS: The agency should mainly fund research at universities and rely mostly on the CIA, FBI, National Guard, and existing fire and police departments for all responses to terrorism. He is skeptical that a new agency is needed, but if it is, the DRHS should focus its attention on solving the problems of CIP. The DRHS should be a research agency like the National Science Foundation (NSF). It should not interfere with private industry, but help it solve the problem of CIP. Meanwhile, let the fire, police, and emergency management departments do their job. Frequently wants the federal and state departments of homeland security to supply his agency with research funding, which he would pass on to university researchers through a competitive proposal process. He has no position on what the roles and responsibilities of the DRHS should be, or how the department should interact with the private sector, other than to keep a "hands-off" approach to industry. His campaign is largely based on the slogan: "Prevent Terrorism, Vote Frequently."

Q1: Which candidate(s) would most likely be favored by the private sector (select all that apply)?

a. Harold Mindmeld
b. Susan Marcel-Wave
c. Slammer Wiseman
d. Hale Frequently

Q2: Which candidate(s) would most likely be favored by the public sector (select all that apply)?

a. Harold Mindmeld
b. Susan Marcel-Wave
c. Slammer Wiseman
d. Hale Frequently

Q3: Which candidate is most likely to cooperate with fire and police, federal and state, and San Lewis Rey bureaus to develop a counter-terrorism and terrorism prevention strategy (select one)?

a. Harold Mindmeld
b. Susan Marcel-Wave

c. Slammer Wiseman

d. Hale Frequently

Q4: Which candidate is most likely to win approval from the university district and researchers and policy makers at the local and state level (select one)?

a. Harold Mindmeld

b. Susan Marcel-Wave

c. Slammer Wiseman

d. Hale Frequently

Q5: Which candidate(s) are most likely to develop a strategy for the DRHS that focuses on long-term prevention of successful terrorist attacks on critical infrastructure (select all that apply)?

a. Harold Mindmeld

b. Susan Marcel-Wave

c. Slammer Wiseman

d. Hale Frequently

Q6: Which candidate(s) would make the citizens of San Lewis Rey feel safer in the short run (select all that apply)?

a. Harold Mindmeld

b. Susan Marcel-Wave

c. Slammer Wiseman

d. Hale Frequently

Q7: Which candidate would cost the citizens the most (select one)?

a. Harold Mindmeld

b. Susan Marcel-Wave

c. Slammer Wiseman

d. Hale Frequently

Q8: Which candidate would cost the government (state and local) the most (select one)?

a. Harold Mindmeld

b. Susan Marcel-Wave

c. Slammer Wiseman

d. Hale Frequently

Q9: Which candidate would cost the private sector the most (select one)?

a. Harold Mindmeld

b. Susan Marcel-Wave

c. Slammer Wiseman

d. Hale Frequently

Q10: Which candidate would regulate the private sector the most (select one)?

a. Harold Mindmeld
b. Susan Marcel-Wave
c. Slammer Wiseman
d. Hale Frequently

CHAPTER 3

Challenges

There is currently no field of study called "critical infrastructure protection" (CIP). Rather there is a set of "wicked problems" associated with the defense of the homeland, and one of these problems has to do with the vast and vital infrastructure systems that modern society has become dependent on. Therefore, the study of critical infrastructure protection is the study of challenges to be met and solutions to be found. This chapter divides the "wicked problems" of critical infrastructure protection into the following seven general categories and explores possible approaches toward overcoming these challenges:

1. Vastness: Each sector in the United States is a vast network that is so large and complex that it is impractical to protect every component of each sector.
2. Command: The interdependency of government agencies, public and private sectors, as well as regulatory and economic drivers makes the problem of "who is in charge" a major barrier to critical infrastructure protection.
3. Information Sharing: As a consequence of a lack of a clear command structure, each agency has evolved into stove-piped organizational structures that hoard their information. This lack of information sharing causes inefficiencies and vulnerabilities in the war against terrorism. Add to this the fact that critical infrastructures are in the hands of thousands of companies, and you have a major challenge in simply collecting and correlating information.
4. Knowledge: The technology behind various critical infrastructures is vast and complex, and yet it is necessary to understand these underlying technologies before effective strategies and policies can be enacted. At this early stage, we still do not know enough about CIP.
5. Interdependencies: Critical infrastructure sectors are infinitely complex because of subtle interdependencies that mirror the organizations that created and operate them in addition to inherent technical interdependencies that exist. Dependencies are due to human organizational structures as well as subtle technical linkages between components of a single sector and among components of multiple sectors.

Critical Infrastructure Protection in Homeland Security: Defending a Networked Nation,
edited by Ted G. Lewis

6. Inadequate Tools: The study of CIP—analysis of vulnerability, etc.—has only begun; we do not yet know enough about critical infrastructures to propose general approaches, general tools, nor general solutions.
7. Asymmetric Conflict: The threat employs asymmetric warfare techniques that look for high-payoff targets that can be damaged by a small force. Infrastructure is particularly vulnerable to asymmetric attacks. The 9/11 al-Qaeda attack on the banking and finance sector is an example.

THE CHALLENGE OF SIZE

One thing stands out in Table 3.1: The size and scope of critical infrastructures is enormous. These systems encompass just about everything, embracing the bulk of the U.S. economy. In fact, it is more difficult to ask, "What is *not* a critical infrastructure?" Just about everything that keeps a modern technological society going is part of the critical infrastructure.

TABLE 3.1. Critical Infrastructure Sectors and Key Assets are Vast and Complex Systems.[1]

Agriculture and Food	1,912,000 farms
	87,000 food-processing plants
Water	1,800 federal reservoirs
	1,600 municipal waste water facilities
Public Health	5,800 registered hospitals
Emergency Services	87,000 U.S. localities
Defense Industrial Base	250,000 firms in 215 distinct industries
Telecommunications	2 billion miles of cable
Energy	2,800 power plants
	300,000 oil- and gas-producing sites
Transportation	5,000 public airports
	120,000 miles of passenger railroad
	590,000 highway bridges
	2 million miles of pipelines
	300 inland/costal ports
	500 major urban public transit operators
Banking and Finance	26,600 FDIC insured institutions
Chemical Industry and Hazardous Materials	66,000 chemical plants
Postal and Shipping	137 million delivery sites
Key Assets	5,800 historic buildings and icons
	104 commercial nuclear power plants
	80,000 dams
	3,000 government-owned facilities
	460 skyscrapers

[1]The National Strategy for the Physical Protection of Critical Infrastructures and Key Assets, Feb. 2003, p. 9.

In addition, each sector is big. Some are big because they cover a large geographical area; some are big because they have many components; and others are big because they are complex. The water sector is a good example. One piece of the water sector is the Colorado River Basin, which provides hydroelectric power and irrigation for 25 million Americans. It spans 250,000 square miles and is shared by seven state and local governmental entities, including Mexico and the water rights owned by Native Americans. The postal and shipping sector is big because it spans the entire United States and consists of 137 million delivery sites (homes and offices). Protecting even a single infrastructure is a daunting task. Protecting all components of all sectors seems to be impossible.

A sector-by-sector analysis of Table 3.1 pounds home the first major challenge of CIP: its *vastness*. There are so many sectors, and each one is so huge that it is impractical to protect every mile of the 45,000 miles of interstate highway, 250,000 square miles of the Colorado River Basin, or untold miles of power and telephone lines across the United States. Providing protection against a 9/11-style attack on the 104 nuclear power plants—perhaps the smallest "key asset"—would require an enormous investment.

The first major challenge is one of size—critical infrastructures are vast and complex—too vast to protect inch-by-inch, mile-by-mile, building-by-building, hospital-by-hospital, port-by-port, nuclear power plant-by-nuclear power plant, or farm-by-farm.

THE CHALLENGE OF COMMAND: WHO IS IN CHARGE?

In Executive Order (EO) 12656 (1988), President Reagan assigned major roles and responsibilities to the Department of Defense (DOD) in the event of a "national security emergency" or attack on homeland soil. Specifically, EO 12656 says the DOD must:

> (1) Ensure military preparedness and readiness to respond to national security emergencies;
>
> . . .
>
> (4) Develop and maintain damage assessment capabilities and assist the Director of the Federal Emergency Management Agency and the heads of other departments and agencies in developing and maintaining capabilities to assess attack damage and to estimate the effects of potential attack on the Nation;
>
> . . .
>
> (6) Acting through the Secretary of the Army, develop, with the concurrence of the heads of all affected departments and agencies, overall plans for the management, control, and allocation of all usable waters from all sources within the jurisdiction of the United States.
>
> . . .
>
> (7) In consultation with the Secretaries of State and Energy, the Director of the Federal Emergency Management Agency, and others, as required, develop plans and

> capabilities for identifying, analyzing, mitigating, and responding to hazards related to nuclear weapons, materials, and devices; and maintain liaison, as appropriate, with the Secretary of Energy and the Members of the Nuclear Regulatory Commission to ensure the continuity of nuclear weapons production and the appropriate allocation of scarce resources, including the recapture of special nuclear materials from Nuclear Regulatory Commission licensees when appropriate;
>
> . . .
>
> Coordinate with the Director of the Federal Emergency Management Agency the development of plans for mutual civil-military support during national security emergencies.[2]

Subsequent to this executive order and release of a DOD plan, the assignment of roles and responsibilities by various PDDs and EOs began to blur as the threat increased. The department or agency placed in charge has constantly changed over the period from 1988 to 2003. It continued to undergo modification even after the formation of the Department of Homeland Security. In the following historical account, it is clear that a major challenge of CIP is the command and control problem associated with sorting out who is in charge.

Soon after assigning DOD the major lead role in 1988, President Bush's EO-12673 assigned the "national security emergency" response function to the Federal Emergency Management Agency (FEMA) in 1989:

> Section 4-203. The functions vested in the President by the Robert T. Stafford Disaster Relief and Emergency Assistance Act, as amended (42 U.S.C. 5121 et seq.), except those functions vested in the President by Section 401 (relating to the declaration of major disasters and emergencies), Section 501 (relating to the declaration of emergencies), Section 405 (relating to the repair, reconstruction, restoration, or replacement of Federal facilities), and Section 412 (relating to food coupons and distribution), are hereby delegated to the Director of the Federal Emergency Management Agency.

In 1991, on the heels of Hurricane Hugo and the Loma Prieta Earthquake in California, the Government Accounting Office (GAO) issued a warning that state and local emergency management capability needed to be improved.[3] In response, the National Response Team (NRT) was created in 1992. This organization has no less than 16 partners. The Environmental Protection Agency (EPA) chairs the NRT, and the Coast Guard serves as vice-chair. The purpose of the NRT was to provide technical advice and coordination during consequence management.

In 1995 the implications of terrorism were beginning to be understood at the federal level, and yet there was no national strategy—no national plan—and little public awareness of the vulnerability of the United States. The alarming increase in severity and frequency of terrorism prompted President Clinton to reassert

[2] http://www.archives.gov/federal_register/codification/executive_order/12656.html.

[3] http://www.gao.gov. Disaster Assistance: Federal, State, and Local Responses to Natural Disasters Need Improvement, T-RCED-91-57, May 15, 1991.

the roles and responsibilities of the various U.S. Departments in PDD-39. PDD-39 heightened federal awareness but provided no funding. According to PPD-39:

> K. *Costs:* Agencies directed to participate in the resolution of terrorist incidents or conduct of counterterrorist operations shall bear the costs of their participation, unless otherwise directed by me. (U)[4]

In 1996 the attack on the U.S. Embassy and the bombing of the Khobar Tower highlighted the vulnerability of buildings, both military and civilian. Given that the DOD was still the lead department for "national security" at this time, it is not surprising that the main concern was weapons of mass destruction (WMDs), rather than small-scale attacks on buildings and infrastructure. In fact, the "Defense Against WMD Act of 1996" (a.k.a. Nunn–Lugar–Domenici Bill) allocated $100M in FY97 and directed the Secretary of Defense to carry out a program for civilian training and advice with respect to WMD.

The DOD seemed to be in charge of counter-terrorism, and the challenges of CIP were still lurking beneath the surface before 2001. CIP was not mentioned in PDD-39. And the roles and responsibilities of government were even more blurred, rather than sharpened, by language that spread responsibility across departments. Most importantly, there was no incremental funding to support increased vigilance.

In addition to the expansion of sectors considered "critical" by 2003, lead responsibilities for CIP had spread across additional departments, including the newly created Department of Homeland Security. PDD-63 sharpened the focus for a time, and then HSPD-7 contributed to ambiguity by nullifying PDD-63 and assigning multiple departments to a subset of the sectors identified in PDD-63. When the Department of Homeland Security was formed in 2003, the roles and responsibilities of the various federal, state, and local government agencies were left to the imagination.

Adding to the confusion is the fact that most critical infrastructure sectors are in the hands of private owners. The source of food, its processing, and the distribution chain for feeding America is owned by private companies. The public health system may be monitored by a public agency (U.S. Public Health), but many hospitals and pharmaceutical companies are owned and operated by private, not public, entities. Power is owned, operated, and distributed through a complex web of public–private cooperation. Most transportation companies (airplanes, ships, railroads, and trucks) are privately held. Telephone companies have been deregulated to stimulate investment from the private sector, and the Internet—created by the government through the Defense Advanced Research Projects Agency (DARPA)—has been commercialized. Private sector companies such as WorldCom/MCI and AT&T run its backbone. Shipping, airlines, and rail transportation—a portion of infrastructure that moves 90% of goods into and out of the United States—are all run by private companies—many not even registered as U.S. companies.

[4]http://www.fas.org/irp/offdocs/pdd39.htm.

So we come to the second major challenge of CIP: the fog that surrounds roles and responsibilities among the various players, including the private sector, federal, state, and local agencies. There is no central point of control in most of the infrastructure we consider critical. For several reasons, both regulatory and economic, most of the critical infrastructure that the United States depends on is beyond the reach of direct governmental control. And as demonstrated by the power blackout of 2003, which deprived electrical power to 50 million U.S. and Canadian citizens, federal and state emergency management systems were of no value to restarting the power generators and transmission lines, because they belong to corporations and deregulated monopolies.

Nunn–Lugar–Domenici allowed the President to transfer disaster management to another federal agency within two years of its enactment. It was assumed that FEMA would become the owner of "national security," but FEMA rejected the transfer for budgetary reasons: The DOD had a huge $268B budget, whereas FEMA had a mere $3.4B budget. Interagency politics began to appear. The year 1998 was a year of jockeying for bureaucratic position: A five-year Interagency Plan on Counter-Terrorism was prepared, the National Infrastructure Protection Center was established by the Department of Justice (DOJ), the Chemical Safety Board was formed, the National Office of Domestic Preparedness (ODP) was created in the DOJ, the Critical Infrastructure Analysis Office (CIAO) was established, the Counter-Terror Coordination Unit within the National Security Council was formed, and Congress earmarked $17M for "Special Equipment and Training Grants."

Pork continued to drive the agenda for homeland security and the hunt for a strategy to protect the United States. In 1999 the first of three Hart–Rudman reports described in detail some of the macro-scale problems facing "national security." The Gilmore Commission met and produced the first of three reports on terrorism. These reports suggested realignments in government but did not mention critical infrastructure. The Hart–Rudman III report stopped short of calling for a new Department of Homeland Security.

Despite the organizational chaos during this period, a strategy was beginning to emerge and money was beginning to flow. The Department of Health and Human Services (HHS) got $161M to begin stockpiling pharmaceuticals and do research in FY99. Their budget increased to $260M in FY2000 for preparing the Department to respond to a bio-terrorism attack.

The Nunn–Lugar–Domenici program was transferred to the DOJ, effective October 1, 2000. Thus, the DOJ became the lead agency for crisis management of terrorist incidents with a FY99 budget of $3.9B. At the same time, the DOD established a dozen WMD Civil Support Teams of 22 soldiers each to support domestic incidents.

The USS Cole bombing occurred in 2000. The attacks were getting bolder and causing more serious damage. President Clinton promised $2B in funding for protection of cyber-space. Part of the money was spent on the Scholarship-for-Service program, which sent students back to graduate school to study information security and learn how to protect the Internet from cyber-attacks.

An Interagency Coordination Center was formed, but the problem of coordination across intelligence agency lines was hampering defense. This deficiency became painfully clear after the attacks on the Pentagon and Twin Towers at the World Trade Center on September 11, 2001. The Whitehouse Office of Homeland Security and the Homeland Security Council was formed in rapid succession. The Transportation Security Administration was spun out of the Department of Transportation and then spun back into the new Department of Homeland Security in February 2003.

At the time of the formation of the new Department of Homeland Security, the national strategy for homeland security stood on three legs—high consequence events, WMD attacks, and preparedness at the state and local level[5]:

1. Domestic preparedness is focused on highest consequence, least-likely attack",4>highest consequence, least-likely attack, i.e., low-probability, high consequence WMD (Weapons of Mass Destruction) terrorism,
2. It is geared toward consequences of chemical/biological WMD attack, because WMD are becoming more accessible to terrorists,
3. It is geared toward Federal investments at the state and local level due to Federalism and the belief that attacks will be local, not national; the US is too large to maintain a national operational capability at the local level; Federalism gives states extensive rights and responsibilities; and the division of labor across local, state, federal jurisdictions was compatible with the Stafford Act.

THE CHALLENGE OF INFORMATION SHARING

The divisions among federal bureaucracies, the lack of a coherent strategy to deal with the vastness of critical infrastructures, and the separation of powers—across federal, state, and local jurisdictions—lead to the third major challenge in critical infrastructure, the lack of information sharing; e.g., if knowledge is power, then the lack of knowledge is a weakness. The shuffling of roles and responsibilities in government, at all levels, combined with the lack of a means for communicating data, plans, strategies, and policies became "the problem." The problem of CIP was confounded by the stove-piped nature of information gathering and dissemination.

Why is information sharing difficult? There is a technical and an organizational reason. First, the information technology infrastructure within government, at all levels, is flawed. Information is locked up in stove-piped, outdated, or just plain inaccessible systems. For the most part, these systems are incompatible or hold their information in databases that have different indexes, formats, and encodings. This creates interoperability problems. Interoperability of information systems is a major barrier preventing information sharing, but it is not the only barrier.

Much of the information needed to effectively prevent attacks on infrastructure is highly sensitive. Legal, cultural, and bureaucratic sensitivity runs high among

[5]Richard A. Falkenrath "Problems of Preparedness" Paper 2000-28, Belfer Center, Harvard University, December 2000.

agencies that need to share information. In particular, the FBI holds its own classified data tightly and, perhaps, jealously. The same story applies to the CIA, DIA, and NSA. Moving that data across agency boundaries is problematic. For example, consumers of the data may be cleared in one agency but not in another. This is known as the *multilevel security problem*, and it will take years to resolve.

THE CHALLENGE OF INADEQUATE KNOWLEDGE

We can add another dimension to the third major challenge: We do not as yet possess sufficient technical knowledge to fully understand how to protect the various sectors of critical infrastructure. The technology of CIP begins with an understanding of the technology of individual sectors. This requires an understanding of electrical power generation and transmission, the technology of telecommunications, the protocols of the Internet, the science of pharmaceuticals, genetic engineering, and models of epidemics. The inner workings of the intermodal transportation system, banking and finance, water and utilities, gas and oil pipelines, and so on must be fathomed before sensible strategies can be crafted and policies dictated. Therefore, sound comprehension of an array of technologies is a prerequisite to forging an effective strategy for protection of the nation's critical infrastructure.

THE CHALLENGE OF INTERDEPENDENCIES

Interdependencies within a sector, and across multiple sectors, complicate the problem of inadequate sector-specific knowledge. The power grid, for example, is so complex that it may be unfathomable. It seems that the simplest failure in a single power line can propagate like a contagion through a crowded population, toppling the entire Eastern United States! This occurred in 2003 even though power engineers have done their best to understand the Eastern Grid in great detail.

Add to the inherent complexity of such sophisticated infrastructures the fact that power is connected to almost all other sectors—telecommunications, water, agriculture, transportation, and so on—and all need power. Clearly, critical infrastructure sectors do not exist alone; they interact with other sectors. In fact, there is a hierarchy of dependencies among the fundamental sectors identified by the 2003 National Strategy. Sectors at the top of this hierarchy depend on sectors below, leading to a three-tier structure as shown in Figure 3.1.

Figure 3.1 illustrates some of the dependencies that exist among the 11 sectors described in the 2003 National Strategy. At the root is Level 1 consisting of fundamental infrastructure; all others depend on this level. Information and telecommunications, power/energy, and water support everything else. Hence, protecting these should be our highest priority.

Level 2 infrastructures—Banking & Finance, Transportation, and Chemical/Hazardous Materials Industry is supported by Level 1 and in turn supports Level 3. Transportation, in its broadest definition, is central to the sectors at Level 3.

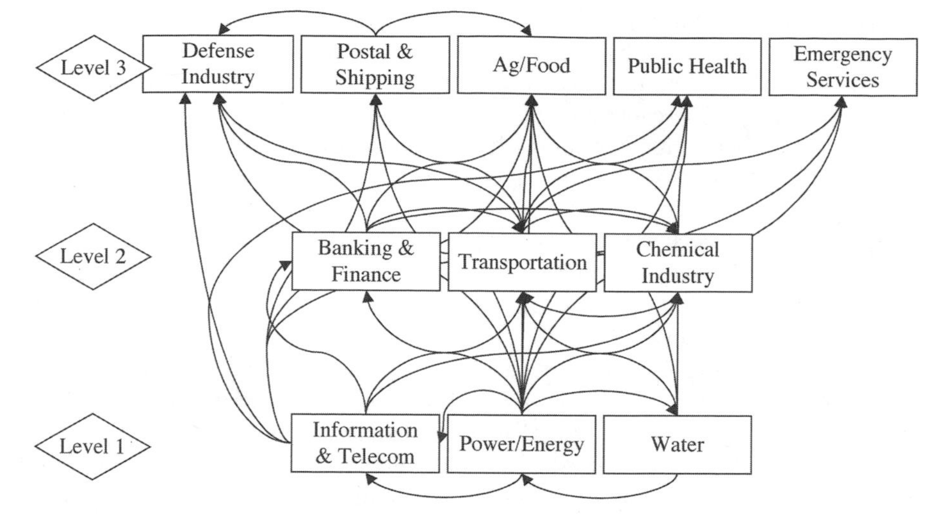

Figure 3.1. Levels and dependencies among the 11 critical infrastructure sectors.

Level 3 consists of the Defense Industries, Postal & Shipping Logistics, Agriculture & Food, on the one side, and Public Health and Emergency Services on the other side (see Figure 3.1). These are supported by Level 1 and Level 2 sectors. A failure in one linked sector may propagate to another. This is called an *interdependency failure*.

New York City Mayor Rudy Giuliani illustrates the complexities of interdependencies as he described the events of 9/11 in chapter one of his book, *Leadership*. The first tower was rammed during his breakfast, beginning a day of chaos as the tragedy unfolded. According to the Mayor, his first task was to regroup and establish a command post from which to direct the police, fire, and medical first responders as well as communicate with the governor and the White House. He needed a safe and connected communication center. But the World Trade Center (banking and finance sector) was intimately intertwined with the telecommunications and transportation infrastructures. In addition, the physical proximity of the City Hall to the World Trade Center made City Hall a poor choice as an emergency command center. Giuliani explains his dilemma:

> Our cell phones were all but dead. The landlines throughout Lower Manhattan were dead. Every entrance to the city was closed. No subways or buses were running, and there wasn't a taxi in sight. There was no way to find out what was going on—the World Trade Center towers held many of the antennae that broadcast cellular phone and television signals, both of which were reduced to minimal capacity . . . Our government no longer had a place to work. Not only was City Hall a likely target, but also it was only about a quarter mile from the World Trade Center. Covered in ash, it was nearly invisible. Bernie had sent a couple of guys ahead to the Tribeca Grand to set up phone lines and clear the place out, like a scouting party in an old Western.[6]

[6]R. W. Giuliani, Leadership, Hyperion Books, New York, 2002, pp. 14–15.

The leader needed to quickly regroup and rebuild, but he was cut off from the world by a series of failures caused by interdependencies among the various sectors. The Twin Towers were a logical place to put cellular network antennas, unless of course, they came tumbling down. The World Trade Center was also a logical place to position lower Manhattan's transportation hub, unless the hub became the target of terrorism. When the towers came down, so did the telecommunications and transportation sector.

Giuliani's transportation and communication infrastructure sectors were not the only sectors damaged by the chain reaction. After the towers collapsed, backup systems began to fail, too! Across the street at 140 West Street a major telecommunications switching network owned and operated by Verizon Communications was damaged by flooding. The fiber optical cable and switching gear for Verizon was in the basement of this *carrier hotel*. To make things worse, Verizon served not only Verizon customers, but also shared its telecommunications capability with other telecommunications companies. In fact, the switching gear in the basement of 140 West Street supported a large volume of electronic transactions emanating from the New York Stock Exchange and other financial institutions. It was a hub of telecommunications traffic that also included Internet messages. Because of the Verizon hub, failures due to interdependencies extended all the way to South Africa!

> Communications infrastructure located in the World Trade Center itself and nearby at the Verizon central office at 140 West Street, along with fiber-optic cables that ran under the Trade Center complex, was destroyed.[7]

The subtleties of failures due to interdependencies did not end with the destruction of optical cabling and antennas. In fact, the telecommunications sector is intimately connected to the power and energy sector. On 9/11, the backup power systems also failed:

> Poor operating procedure resulted in a facility's backup generator being shut off to conserve fuel . . . Fuel delivery problems, including delivery of the wrong type of fuel to one location, made it difficult to keep generators running. Communications equipment was allowed to continue operating even when electrical power necessary for cooling systems had been lost . . . as power outage extended over multiple days (past the planned life of the backup power systems), maintaining power became a serious issue as batteries expired and backup generators ran out of fuel.[8]

The loss of power in lower Manhattan also impacted one of the most important critical infrastructures—public health. Modern hospitals use wireless technology to speed the movement of laboratory and patient data. On 9/11 it turned out that one hospital had outsourced the operation of its wireless network capability to a company that depended on the Internet. Wireless devices used in the hospital

[7] *The Internet Under Crisis Conditions: Learning from September 11*, National Research Council, National Academies Press, Washington, D.C., 2001, p. 1.
[8] Ibid, p. 8; p. 25.

were connected to wire-line Internet services, which in turn were connected (indirectly) to the power and telecommunications failures surrounding the World Trade Center. This subtle dependency was not discovered until that fateful day.

Transmission facilities of 9 of the 14 local television stations and 5 local radio stations were lost when the North Tower collapsed. As a consequence, 6 networks relocated transmitters to the Empire State Building. Two remained at their backup site in Alpine, New Jersey. Loss of power resulted in 160 cell sites becoming useless (5% of the total). Restoration of these dependent sectors took weeks after 9/11.

Dependencies do not always crop up in the backyard of a disaster. Failure of a segment of the Internet in Manhattan disrupted service in South Africa! This time the dependency was traced to the design of the Internet itself. Internet addresses such as http://www.whitehouse.gov are translated into Internet Protocol numbers such as 131.200.40.5 by a special "telephone book" called a *Domain Name Server* (DNS). The Internet is structured in such as way as to minimize the effort needed to do this translation. After all, it has to be done each time an e-mail message is sent or a Web page fetched. Efficiencies are gained by minimizing the number and location of DNS computers. South Africa depended on a DNS that was accessed by routing through Manhattan; hence, when Manhattan's telecommunication infrastructure failed, the failure was felt in South Africa.

Examples of telecommunication infrastructure dependencies abound, but the challenge goes even deeper when other sectors are studied in great detail. Many infrastructure sector industries are regulated and closely monitored by oversight agencies. Nuclear power plants are regulated by the Nuclear Regulatory Commission; telephone companies by the Federal Communications Commission; electrical power utilities by the Federal Energy Regulatory Commission (FERC); and health care by the U.S. Public Health department. These agencies contribute to yet another form of dependency—the so-called *interagency dependency*.

Interagency Dependencies

It is not surprising then that the nature of regulated and near-regulated sectors mirrors the structure of their regulators. And the regulators are hampered by their own haphazard and overlapping structures. For example, the Anthrax biological terror attack launched shortly after 9/11 required the attention of the U.S. Public Health and Center for Disease Control and Prevention. But the responsibility for clean up rested with the Environmental Protection Agency (EPA), because Anthrax is considered a hazardous material. The EPA is responsible for responding to hazardous materials, and the Center for Disease Control is responsible for hunting down and eradicating communicable diseases. So, who should respond? Add to this confusion the fact that the anthrax attacks in 2002 were crimes, and you have at least three major agencies involved.

Similarly, the regulatory structure of organizations such as the FERC and dispersal of state and local regulatory agencies across all 50 states compounds the problem of unraveling interdependencies. One needs only compare the definition of who is responsible for what kind of incident described in HSPD-7 to realize that interagency dependencies are rampant in the national strategy. If it takes a network organization to fight a network organization, then these interagency dependencies must be dissolved. In short, *the interdependencies connecting overlapping and contradictory organizational structures have led to complex and poorly understood interdependencies in the various sectors themselves*. The structure of most infrastructure sectors mirrors the regulatory structure of the government. This adds to the complexity of CIP.

THE CHALLENGE OF INADEQUATE TOOLS

The challenge of complexity resulting from interdependencies derives from both organizational and technical complexities. To compound the problem, there are currently few tools to help. Tools and techniques are needed for modeling complex infrastructures, comprehending their interdependencies, analyzing their vulnerabilities, and finding optimal means of protection. For example, there is no computer software or manual procedure in existence today (2003) that could have predicted the impact of two airliners flying into the Twin Towers. Furthermore, there is no known methodology for estimating the likelihood of a subsequent attack of the same magnitude.[9]

Almost every aspect of infrastructure protection is sorely lacking in foundational theory, scientific proof of effectiveness, and applied tools and techniques. Indeed, the very basic task of identifying what is critical, and measuring its vulnerability, is currently inadequate and poorly understood by the people whose responsibility is to implement the national strategy.

The subject of inadequate tools and techniques is very immature, so the following list is preliminary, only.

1. Due to the vastness of each sector, an electronically accessible database that describes each component of each infrastructure needs to be created. This database needs to be compatible with several analysis tools that automate the process of vulnerability analysis, risk assessment, and optimization of investment.
2. Due to the complexity of interactions among single-sector components and components of related sectors, there needs to be computer models of each sector as well as the interdependencies among sectors. These models need to be able to detect sector-specific cascade failures like the August 2003 power blackout, and the cross-sector interdependency failures as demonstrated by the cross-sector failures that occurred on 9/11.
3. Due to the high cost of "target hardening" and the vastness of each sector, there needs to be tools for computing the best way to invest limited funds to remove or minimize vulnerabilities. We need quantitative methods of allocating limited resources to the most important assets of each sector. This is

[9]One can argue that the only reliable methodology is foolproof intelligence, which is an oxymoron.

especially important, because without a quantitative and objective method of evaluation, funding will become political "pork" rather than a solution to the problem.

4. Vulnerability analysis is an unsolved problem: Automated tools are needed, standard formats and standard selection criterion are needed, and practitioners need to be trained in how to use such tools. Currently, several incompatible, stove-piped, sector-specific vulnerability analysis and risk assessment tools exist. The problem is that they are ad hoc—there is no underlying scientific foundation—and they do not (generally) assist analysts in identifying what assets are important, and what are not. Such tools must operate at the local, state, regional, and national level, because as we know, attacks on infrastructure can have local, state, regional, or national impact.
5. Information systems are needed for various tracking and data-mining operations such as tracking immigrants, borders, intermodal shipping, and monitoring of various transportation systems, performing public health surveillance, and fusing intelligence data. These systems need to interoperate and seamlessly share information in real time. Information becomes more valuable as it is shared and aggregated. Currently, sharing and aggregation across infrastructure sectors is impossible. In the future, it will be an imperative.

Research is actively underway in many of these topics, but it will take many years to solve the complex and demanding problems posed by size, complexity, interdependencies, and cost. In the meantime, Chapters 4–6 describe elementary techniques for performing vulnerability analysis and risk assessment based on network theory. Although this technique is imperfect, it illustrates one approach to the challenges described here.

THE THREAT IS ASYMMETRIC

The theory of warfare has been under development for thousands of years. Only in the last 100 years has it become a quantitative discipline. Frederick Lanchester (1914) devised the earliest equations for estimating the likely outcome of conflict between two armies—attacker (red) versus defender (blue). His equations have been used since World War I to predict the outcome of battles. In fact, they were used during the Cold War to estimate troop strength along the Iron Curtain.

Lanchester equations have been shown to accurately model *symmetric conflict* between red and blue armies.[10] Combat is considered *symmetric* when the number of troops on each side largely determines the outcome. Superior troop strength wins over inferior troop strength. It is assumed both armies have roughly equal capabilities, denoted as *firepower*. For example, if the attacker has 10,000 soldiers and the defender has 5,000, and they both fight with bows and arrows, the most likely

[10] D. S. Hartley, "Topics in Operations Research: Predicting Combat Effects," Military Applications Society, INFORMS, informs@informs.org, 2001.

outcome is that the attacker will overwhelm the defender because it has more fighters. The strategy of overwhelming superiority in numbers (troop strength) has been the dominant doctrine for U.S. forces for most of the 20th century.

Conflict between red and blue armies is a numbers game. The winner is the army left standing; the loser's troop strength is driven to zero. This is called *attrition*. The goal of symmetric warfare is to reduce the other side to zero by attrition. Figure 3.2(a) illustrates Lanchester's theory of symmetric warfare and shows that the side with greater troop strength wins because it drives the opposition troop strength to zero.

But what happens when a small army with a huge advantage in firepower confronts a much larger army? When the firepower of one army far outstrips the firepower of the other army, asymmetries begin to show up in the Lanchester equations. That is, asymmetric conflict is characterized by major differences in firepower. In fact, firepower is used to compensate for troop strength. Figure 3.2(b) illustrates asymmetric warfare; a smaller army can win against a larger army if its firepower is much greater.

Figure 3.2 illustrates the application of Lanchester equations to symmetric and asymmetric battles. Lanchester equations attempt to predict the outcome of conflict when troop strength and/or firepower are varied. Figure 3.2(a) shows that a larger army can defeat a smaller army, simply because it is larger. This assumes the two armies have equal firepower.

Figure 3.2(b) shows that even a smaller army can be victorious over a larger army, if it has more firepower. A small band of soldiers with machine guns can defeat a much larger army of soldiers equipped with spears. Nineteen hijackers willing to sacrifice their lives in a collision with skyscrapers can overcome the defenses of a mighty U.S. military using the asymmetric "firepower" of fuel-laden jumbo jets.

Firepower may come in different forms. Superior weapons are a common form, but a suicide bomber possesses greater firepower than someone who fears for his or her life. This is why terrorism is a form of *asymmetric conflict*. Terrorists seek ways of making up for their inferior troop strength by increasing their firepower. They are often extremely clever in the methods used to offset the inferiority of their numbers.

Critical infrastructure vulnerabilities provide ample opportunity for attackers to magnify their firepower through asymmetric techniques. A few examples are listed as follows:

1. Critical infrastructure is a "good" target for an asymmetric-thinking foe because most sectors are relatively exposed, vast, and unprotected; they are "easy targets." An Internet search engine such as Google can be used by anyone to discover the locations (even pictures) of critical components.
2. Small forces can make a major impact, because the most valuable assets of most sectors are concentrated into a small number of critical components or locations. Significant attacks can be mounted with little force, because they require knowledge more than troop strength. For example, most telecommunications assets are housed in a relatively small number of buildings and

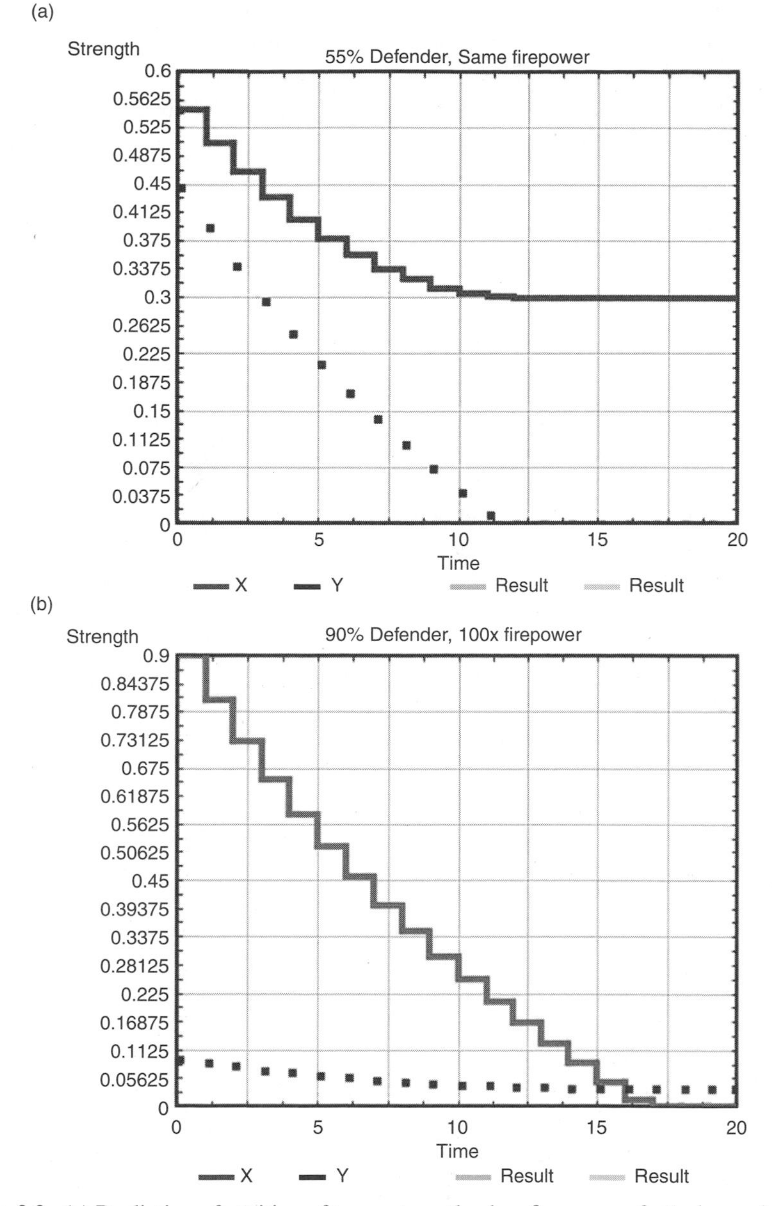

Figure 3.2. (a) Prediction of attrition of troop strength when firepower of attacker and defender are equal, and (b) prediction of attrition of troop strength when firepower of the attacker is 100 times greater than the defender, but the initial strength of the attacker is much less than that of the defender. (a) Attacker (dotted line) initial troop strength is 45% of total, and Defender (solid line) initial troop strength is 55% of total. The firepower of each army is the same. (b) Attacker (dotted line) initial troop strength is 10% of total, but its firepower is 100 times that of the Defender (solid line).

most Internet assets are controlled by fewer than 100 computers. These components can be attacked by a single person working from home using a computer and expert knowledge of how computer worms and viruses work.

3. Technological societies are heavily dependent on their infrastructures, because "everything" is powered by the Level 1 sectors: power/energy sector and the information/telecommunications sector. All other sectors depend on these; hence, this dependency magnifies the effects of an attack. In the Information Age, even geographical isolation by oceans is no longer a deterrent.

ANALYSIS

The challenges described in this chapter provide a clue as to how to approach the larger problem of CIP. For every challenge there is a solution or approach leading to a solution. Whether the answer is cost-effective or feasible in the practical world of budgets and politics is another question. But from the point of view of a discipline, identification of the challenges also suggest solutions.

1. Vastness: The geographical expanse of sectors such as telecommunications (wires and fiber optics), power grids (transmission lines, substations, and power plants), and water networks (large water basins and large number of consumers) at Level 1 in the CI hierarchy seems almost impossible to comprehend let alone protect. But these are fundamentally network systems, and we know a lot about networks, thanks to the Internet and over 100 years of engineering. Thus, we can use network theory to fathom what to do about such vast systems. In short, most networks have structures that lend themselves to efficient means of protection. For example, in the following chapters, we learn that the telecommunications, Internet, and power grid networks are structured around concentrations called *hubs*. Hubs are more important than most other components; hence, our attention should be focused on protecting hubs. And there are far fewer hubs than other components. This leads to the concept critical node",4>of critical node—the nodes of a sector that are critically important to the overall operation of the sector. This suggests an approach that is more manageable—identification and protection of critical nodes. Thus, we can recommend optimal strategies that aim to harden hubs and ignore most other components.
2. Command: The formation of the Department of Homeland Security began the process of consolidation of government agencies at the federal level, but much remains to be done. As this consolidation and clarification of roles and responsibilities evolves, its effects will trickle down to the state and local levels. Even so, the interface between the public and private sectors will remain problematic. Government will lean toward regulatory control as a means of forcing privately owned telecommunications, power/energy, banking/finance, and so on companies to improve their security. But these companies will be

reluctant to do so because of the cost. Who will pay for the improvements needed to comply with the new regulations?

One approach that may be tried is the so-called *dual-purpose approach*. Dual purpose means that a costly infrastructure improvement may pay for itself, if it also improves efficiency, lowers long-term costs (has a positive return on investment), or lowers insurance premiums. For example, an alarm system that reduces losses due to theft may also lower insurance premiums. Over time, the investment in the alarm system pays for itself through reduced insurance premiums.

Generally, government regulations have led to decentralized control, whereas economic forces have led to centralization of sector assets. For example, government regulation of the national telecommunications infrastructure has led to distributed command and control across many competitors. Hence, control is distributed rather than clustered. But economic forces have driven these competitors to colocate their switches and exchanges. That is, the assets have been concentrated into a small number of buildings throughout the United States. Concentration of assets means the infrastructure is clustered; the telecommunications network has evolved into a system of hubs. Concentration of assets provides the attacker with easy targets. Decentralization of command leads to difficulty in responding to attacks.

How is this problem solved? One approach is to bring the private sector entities responsible for the assets together in the form of consortia, councils, and industry associations. For example, the North American Electric Reliability Council (NERC) and the Internet Society span their industries and provide standards that tend to promote interoperability across different owners of the infrastructure components. This approach promotes increased security and reliability through existing councils and societies.

3. Information Sharing: The fact that critical infrastructures are in the hands of thousands of companies aggravates the challenge of collecting and correlating information that may be used to prevent attacks. This is why various Information Sharing and Analysis Centers (ISACs) were proposed in the national strategy. The Homeland Security Act of 2002 requires ISACs:

> Sector-focused ISACs provide a model for public/private sector information sharing, particularly in the area of indications and warnings. Numerous critical infrastructure sectors use this structure to communicate potential risks, threats, vulnerabilities, and incident data among their constituent memberships—includes exploring ways to expedite the conduct of necessary background checks and issuance of security clearances to those with a need to know.[11]

Table 3.2 lists the ISACs as of 2003. Note that not all critical infrastructure sectors had ISACs a year after the Homeland Security Act of 2002 was passed and the Department of Homeland Security began operation.

[11]National Strategy for the Physical Protection of Critical Infrastructures and Key Assets, February 2003.

TABLE 3.2. Official ISACs and Information Sharing Organizations as of 2003.[12]

Agriculture/Food	Food Industry ISAC
Water	Water ISAC
Public Health	An ISAC is in development
Emergency Services	Fire Services ISAC Law Enforcement ISAC
Government	State Government
Defense Industrial Base	None
Information and Telecommunications	Information Technology ISAC Telecommunications ISAC
Power/Energy	Electric Power ISAC (NERC) Energy ISAC (Oil and Gas)
Transportation	Surface Transportation ISAC
Banking and Finance	Financial Services ISAC
Chemical Industry and Hazardous Materials	Chemical Industry ISAC
Postal and Shipping	None
Information Sharing Organizations	Internet Security Alliance The Millennium Solution ISAC World Wide ISAC

4. Knowledge, Dependencies, and Tools: The general lack of knowledge of various critical infrastructure sectors, how dependencies complicate them, and the development of tools will all require an investment in research and education. This challenge was recognized by the creators of the Homeland Security Act of 2002, which dictated the formation of a research arm within the Department of Homeland Security—the Science and Technology division, also known as the Homeland Security Advanced Research Projects Agency (HSARPA).

> (Sec. 307) Establishes the Homeland Security Advanced Research Projects Agency to be headed by a Director who shall be appointed by the Secretary and who shall report to the Under Secretary. Requires the Director to administer the Acceleration Fund for Research and Development of Homeland Security Technologies (established by this Act) to award competitive, merit-reviewed grants, cooperative agreements, or contracts to public or private entities to: (1) support basic and applied homeland security research to promote revolutionary changes in technologies that would promote homeland security; (2) advance the development, testing and evaluation, and deployment of critical homeland security technologies; and (3) accelerate the prototyping and deployment of technologies that would address homeland security vulnerabilities. Allows the Director to solicit proposals to address specific vulnerabilities. Requires the Director to periodically hold homeland security technology demonstrations to improve contact among technology developers, vendors, and acquisition personnel.[13]

[12] http://www.ciao.gov/related/index.html#Critical Infrastructure Information Sharing and Analysis Centers.
[13] http://thomas.loc.gov/cgi-bin/bdquery/z?d107:H.R.5005.

The Act also provides for education:

> **Subtitle D: Academic Training**—(Sec. 1331) Revises agency academic degree training criteria to allow agencies to select and assign employees to academic degree training and to pay and reimburse such training costs if such training: (1) contributes significantly to meeting an agency training need, resolving an agency staffing problem, or accomplishing goals in the agency's strategic plan; (2) is part of a planned, systemic, and coordinated agency employee development program linked to accomplishing such goals; and (3) is accredited and is provided by a college or university that is accredited by a nationally recognized body.[14]

5. Asymmetric Conflict: When an attacker lacks the technology, force strength, and financial backing of a nation-state, the alternative is to use asymmetric conflict as a battleground strategy. The infrastructure of the United States is particularly vulnerable to asymmetric attacks for all of the foregoing reasons. Hence, the U.S. military establishment must rethink its Cold War strategy of matching force-on-force. It must develop counter-terrorism strategies for combating asymmetric warfare. This will be a difficult transformation because of the 50-year investment in symmetric warfare techniques, platforms, and thinking.

Some elements of this rethinking process are as follows:

- It takes a network to fight a network. Terrorism uses modern network warfare techniques that incorporate speed, flexibility, and global reach even though terrorist organizations are typically small.
- Good intelligence can substitute for force strength. If it was possible to know in advance where, when, and what the next attack is, then it would take far fewer resources to protect the vast infrastructure of the United States. Thus, human as well as technological intelligence collection is essential.
- Technology is a double-edged sword. Highly technological societies are vulnerable because they depend heavily on technology. On the other hand, technology can be used in innovative ways to protect itself. For example, so-called self-healing networks, hi-tech surveillance systems, face-recognition, biochemical detection equipment, and anti-missile protection on airlines are all technological advantages possessed by highly technological countries.

EXERCISES

1. How many miles of cable are there in the telecommunications sector?
   a. 2 million
   b. 100 million

[14]Ibid.

c. 1 billion
d. 2 billion
e. 10 billion

2. Before 9/11, which federal agency/department was responsible for combating terrorism within the United States?
   a. It depends on which year
   b. FEMA
   c. DOD
   d. DOJ
   e. DHS
3. Who is in charge of the ISACs?
   a. It depends on which one
   b. DHS
   c. EPA
   d. IEEE
   e. CIA
4. Who owns and operates most critical infrastructure sectors?
   a. DHS
   b. Various governmental departments and agencies
   c. Various private sector companies
   d. Congress
   e. Nonprofit corporations and public utilities
5. Interoperability is the ability to:
   a. Cross international borders
   b. Accommodate interdependencies among sectors
   c. Accommodate interdependencies within a sector
   d. Cooperate among agencies
   e. Share information across different systems
6. In general, critical infrastructure sectors are made more complex because they are interdependent with (select one):
   a. Power and energy
   b. The Internet
   c. Internal assets
   d. Internal assets and other sector assets
   e. Regulatory agencies
7. Interdependency failures are sector failures that (select one):
   a. Spread like falling dominoes
   b. Apply only to power grids

   c. Are caused by poor sector design
   d. Apply only to Level 1 sectors
   e. Cause the largest amount of damage
8. What is the main characteristic of asymmetric warfare? (select only one):
   a. The attackers are terrorists
   b. Small forces with large firepower
   c. Small firepower with large forces
   d. Large forces with large firepower
   e. Small forces with small firepower
9. There does not seem to be any uniform or standard way to structure the ISACs listed in Table 3.2. Recommend a general organizational model for structuring ISACs, and explain its advantages. The following list may provide you with some ideas:
   a. Professional associations like the IEEE Computer Society provide an extensive list of standards for the computer industry and a forum for interaction (conferences) with its governmental, industry, and academic members.
   b. The Water ISAC has taken on the role of vulnerability assessment educator and information disseminator for the water works utilities, cities, and industry. ISACs should be nonprofit organizations that educate and disseminate sector information.
   c. The Internet has no official ISAC but instead relies heavily on university "institutes" like the CERT to provide guidance and interaction among government, university, and industry players. ISACs should be all-volunteer organizations that have no central organization that oversees the sector.
10. Military operations are carried out through a hierarchical central command structure that is designed to provide "unity of command" during an operation. Joint strike forces involving Army, Navy, Marine, and Air Force components are all coordinated through the central command. But San Lewis Rey is a major metropolitan area that must cooperate with a plethora of military, nonmilitary, corporate, and noncorporate entities. For example, San Lewis Rey's water system is owned and operated by a pseudo-corporate public utility company; its telephone and Internet system is owned and operated by five different private-sector companies; and its electrical power infrastructure is regulated by a state-level Independent System Operator (ISO) that is overseen by the NERC.

    Design and document a "unity of command" CONOPS (Concept of Operations) structure for CIP operations in San Lewis Rey, and argue why it will work. Does it solve the problem of Unity of Command and break down barriers to Information Sharing? Include the following representation in your CONOPS:

Private- and public-sector organizations (telecommunications, power)
ISACs (IT-ISAC, Water-ISAC, EP-ISAC, etc.)
City department heads (fire, police, mayor)
National guard leadership
Intelligence community leadership/linkages
Public health, emergency managers
Others as you deem necessary

CHAPTER 4

Networks

This chapter develops a theory of critical infrastructure development called the *scale-free network theory* proposed by Albert-Laszlo Barabasi.[1] It then goes on to apply scale-free network theory to explain how critical infrastructure sectors have evolved over the past and continue to evolve, today. The theory not only explains how most sectors have come about, but it also establishes a basis for vulnerability analysis, because it reveals concentrations of assets that may be vulnerable to purposeful attack. Although somewhat theoretical in nature, we illustrate the theory with many examples from real critical infrastructure sectors such as transportation, telecommunications, and electrical power.

The following concepts are described in detail:

1. *Networks*: A fundamental understanding of the formal theory of networks provides the basis for comprehending the big picture because most critical infrastructure sectors are networks or can be modeled as a network of some kind. Thus, the structure of a critical infrastructure sector is modeled as a network: Nodes and links abstractly represent cities and roads, power generators and power lines, telephone switches and telephones, or sector assets and relationships among those assets.
2. *Concentration of Assets*: The most surprising property of most infrastructure sectors is that assets are highly concentrated. Vulnerabilities are easily identified in terms of these concentrations, thus revealing the critical nodes of each sector. Oil refineries and storage tanks in the energy sector are excellent examples of concentrated assets, but "concentrations as critical nodes" are observed in other sectors as well. Using an analogy with termites that stockpile wood chips, we show that most sectors have evolved into nonrandom networks that are characterized by a small number of *hubs* or clustering of nodes within a neighborhood. The hub structure of a sector's network architecture reveals its

[1] A.-L. Barabasi, *Linked: How Everything Is Connected To Everything Else and What It Means for Business, Science, and Everyday Life*, PLUME Cambridge Center, Cambridge, MA, 02142, 2003.

Critical Infrastructure Protection in Homeland Security: Defending a Networked Nation,
edited by Ted G. Lewis

most important assets as well as its vulnerabilities. This observation points the way to a practical strategy for critical infrastructure protection (CIP).

3. *Critical Node/Link Approach to CIP*: The hub structure of most sectors suggests a general theory of CIP: Their network representations are non-random. In fact, they fall into a class of networks that are either *scale-free* or *small world*. Indeed, many sectors of high importance such as energy, telecommunications, and water are *scale-free*—that is, their most critical components are hubs. Other sectors are not so perfectly clustered, but they are nonrandom, and in some cases *small world*. This observation provides a scientific basis for a homeland security strategy: Given that we cannot afford to protect everything equally, our prevention strategy should focus on the hubs of scale-free networks, or the clusters in small-world networks. These hubs or clusters are called *critical nodes*.
4. *Simple Critical Node Testing*: Fortunately, it is extremely easy to identify the critical nodes or links in a network. Simply count the number of links at each node and arrange these counts in a histogram showing the percentage of nodes with k links. If this histogram obeys a *power law* (the histogram drops off quickly as k increases), then the network is *scale-free*. Critical nodes are identified as the rare nodes with the most links. These appear at the extreme right end of the histogram.
5. *Cascade Networks*: Sector failure is often caused by a *cascade failure* in a network. A relatively small fault in one node can propagate to other nodes through a series of faults, until the entire network is rendered useless. We show that our strategy of protecting the hubs, and clusters, is the most effective means of preventing cascade failures. In the case of a scale-free network, protecting the high-degreed hub is the most effective prevention strategy. In general, it is best to protect the node or nodes with the most links in any network—a strategy that pays off in terms of the best hedge against cascade failures.
6. *Simulations*: The value of modeling and simulation in the study of networks is illustrated by several computer simulations: *Termites* illustrates the principle of emergent behavior; *PowerGraph* illustrates the power of organizing principles such as increasing returns; *ShortestLink* demonstrates how a random network can evolve into a small world; and *CascadeNet* proves that the best way to prevent cascade failures is to protect the highest-degreed nodes of any network.

CONCENTRATION OF ASSETS

The major challenge of CIP, which was described in the previous chapter, is one of size. Most sectors are simply too large and complex to be fully protected. It is not only impractical but also infeasible to protect every mile of the 2 billion miles of cable in the telecommunications sector. Similarly, other sectors are either too big because they span large geographical areas or because they are complex and diverse. Supervisory control and data acquisition (SCADA) systems are an example. They are found in almost all industrial processes. But because they are complex electronic data processing and communication systems, they are difficult to protect. Instead, protection of only the most critical parts of each sector is the only practical and feasible strategy.

A closer study of most infrastructure sectors reveals that they have a unique nonrandom structure that can be exploited for good or evil. Specifically, critical infrastructures are almost always organized around *critical nodes* containing a high concentration of assets. These asset concentrations are considered critical, because their destruction can cause great harm to the overall sector. For example, the switches and routing equipment of the Internet are largely clustered in special buildings called *carrier hotels* or *telecom hotels*. Energy refineries and storage stations are concentrated in a relatively small number of critical locations around the country. Utilities, banking, public health, and other critical infrastructure sectors all exhibit this peculiar tendency to cluster. This is both a blessing and a curse because concentration of assets makes it easy for terrorists to identify prime targets. It is a blessing, because concentration of assets makes it easier for the defender to identify and protect the most "critical nodes" of each sector's structure.

Concentration of assets "evolved" around some organizing principle that is specific to the sector. In the case of telecom hotels, the organizing principle is profit; telecommunications companies can reduce their operating costs, increase performance of their networks, and ease maintenance by concentrating their switches and routers in one place. To understand how these sectors are structured, it is helpful to understand how they evolved. We can do this by fathoming the *organizing principle* that shaped the sector. For the purposes of this book, an organizing principle will be a simple rule, operating on a local level that causes a sector-wide pattern to emerge over time. For example, a business decision by a telecommunications company to share expenses with its rivals, in order to reduce costs, is a simple organizing principle operating within a single company that causes the telecommunications industry to evolve telecommunications hotels. Over time, the telecommunications hotels have become critical nodes, because so many of the assets of the entire telecommunications industry are colocated in a small number of buildings.

Organizing principles of sectors such as water utilities, transportation systems, telecommunications, energy, and banking are often extremely simple. They are also often revealing of how the sector evolved, and how it might change when regulations and economics change. This is why it is important to understand how a sector evolved or emerged from humble beginnings. Once the organizing principle is understood, the policy maker can avoid policies that result in unintended negative consequences.

TERMITES

The following experiment may seem unrelated, but the simple behavior of termites explains a lot about the complex behavior of critical infrastructures. In addition, this experiment illustrates an important concept: emergence. Emergence is the result of applying an organizing principle over and over again, until a pattern appears.

In 1994, Mitchel Resnick illustrated how a "non-intelligent" system organizes itself into a complex "intelligent" system using very simple local rules of behavior.[2] He used an amusing but sobering example: termites roaming about a field containing wood chips. In Resnick's *Termite* simulation, "non-intelligent" termites acting independently and autonomously unknowingly construct an "intelligent" infrastructure of wood chips with unexpected structure as follows:

Rule 1. Walk around randomly—not purposefully—until encountering a chip.

Rule 2. If not already carrying a chip, pick up a chip whenever one is encountered.

Rule 3. If carrying a chip and a second chip is encountered, drop the chip and continue to roam at random.

Initially, the termites and chips are scattered around the field at random [see Figure 4.1(a)]. They roam around at random, repeatedly applying the three simple rules, forever. Termites have no goal, do not know what the "big picture" is, nor do they collaborate with one another. And yet they unknowingly apply an organizing principle that causes a pattern to emerge.

What emerges from this seemingly "dumb" behavior? Figure 4.1(b) suggests that the chips will all end up in one big pile after an undetermined length of time. In other words, structure will eventually evolve out of randomness. The randomly scattered chips end up in a few nonrandomly "organized" piles.

In fact, this is a profound experiment because the termites are acting on their own, and their behavior is random, or seemingly without purpose. And yet, the termite population produces a nonrandom pattern of chips. It is almost as if each termite "knows" what it is doing, but of course this termite world is only a computer simulation.

The simplicity of the *Termite* experiment and its relatively complex result is surprising. Indeed, these termites are not as sentient as real-life termites, and yet the computer model produces a result that is much like the stockpiling of chips by insects in the real world. This example illustrates one of the fundamental ideas of self-organizing systems: *emergent behavior.*[3] And, as we shall see, the termite system simulates what seems to happen in critical infrastructures—from randomness comes self-organization. And from self-organization comes concentration of assets. These concentrations are where the sector's vulnerabilities can be found and fixed.

[2]M. Resnick, *Turtles, Termites, and Traffic Jams: Explorations in Massively Parallel Microworlds*. MIT Press, Cambridge, MA, 1994.

[3]"Emergence is the process of deriving some new and coherent structures, patterns and properties in a complex system. Emergent phenomena occur due to the pattern of interactions between the elements of a system over time. Emergent phenomena are often unexpected, nontrivial results of relatively simple interactions of relatively simple components." From http://en.wikipedia.org/wiki/Emergence.

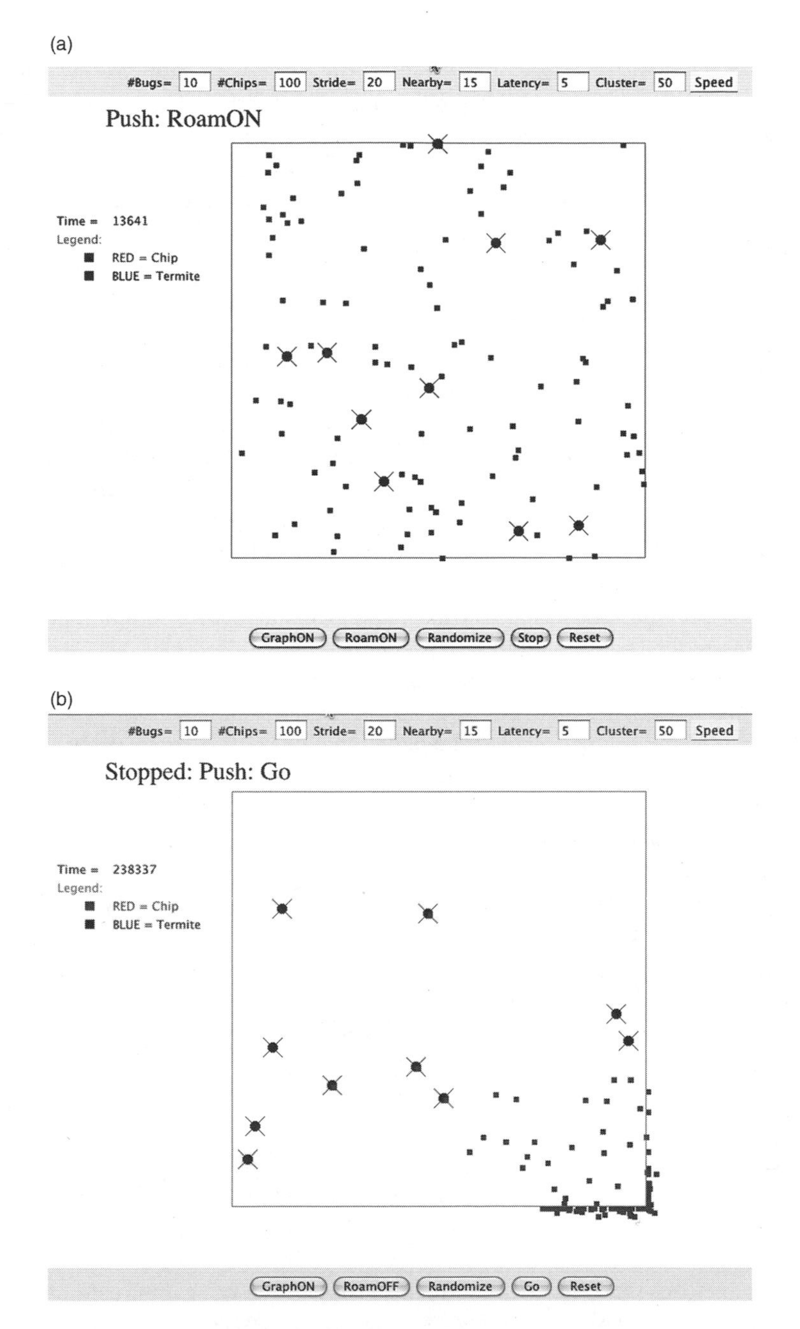

Figure 4.1. *Termite* simulation: (a) Initially 10 termites and 100 chips are placed at random on a rectangular field, and (b) piles of chips emerge in a nonrandom pattern after a long period of time. (a) Initial configuration of 10 termites and 100 chips, randomly placed in a rectangle. (b) After 238,265 iterations, most chips are in one pile.

Termite is a computer program that simulates a termite population that follows the three simple rules given above.[4] Interestingly, the randomness of chips and termites ends up producing a nonrandom pattern of chips. The chips—like assets in an infrastructure sector—all end up in a small number of piles! The assets are unknowingly, seemingly randomly, placed in orderly clusters.

This phenomenon is known as *emergent behavior*, and it explains many biological and physical phenomena in the real world. An "unintelligent" system can organize itself into a system with the level of complexity observed in the real world. Such self-organization is *emergent*, because the resulting order is not apparent at the level of the individual termite nor is it made apparent by the repeated application of simple rules. The pattern is revealed only after the passage of time.

The program *Termite* has six inputs and five controls for starting, restarting, and activating the simulated termites. The user may use the preset values at the top of the screen or enter his or her own values for #Bugs (number of bugs), #Chips (number of chips), Stride (size of each random step), Nearby (distance needed to cause a termite to pickup or putdown a chip), Latency (rest time after a termite drops a chip), Cluster (distance around a chip that defines a cluster), and Speed (Slow, Medium, or Fast simulation time). Pressing GO randomizes the termites and chips. Pressing RESET starts the simulated clock over again. The ROAM-ON/ROAM-OFF button toggles between activating the rules and deactivating them. The termites follow the pickup and putdown rules when ROAM-ON is pressed.

To see how the simulation program defines a cluster and to display a histogram of clusters, press the GRAPH-ON toggle button. This will cause the program to draw a histogram where the vertical axis displays percentage of chips in a cluster, versus the size of each cluster along the horizontal axis. It will also display concentric rings around each chip; the inner ring is the boundary that defines proximity of termite and chip (used for pickup and putdown), and a second ring that defines the boundary of a cluster. All chips that are inside of the second ring are considered part of the same pile of chips.

Termite exhibits an emergence property that is frequently found in nature when a population of individuals, each following local rules, produces an unexpected group behavior—one that pops out of the system without explanation. Some other examples are as follows:

- Crickets synchronize their mating calls—calling all at once at the same rate.
- Birds create large flocks that seem to behave as one.
- Investors buy and sell in synch, resulting in an economy that rises and falls.

What does this have to do with CIP? If the assets of a critical infrastructure sector were scattered randomly throughout a region, and all assets were equally critical, the problem of protecting assets would be untenable, indeed. But instead, it will be shown throughout this book that almost all sectors are organized much like chips

[4]*Termite* is available from the website or the disk accompanying this book.

in *Termite* world. Like these chips, assets are concentrated in piles, and these piles constitute the most important assets to be protected.

The general hypothesis of this chapter is that concentration of critical infrastructure assets leads to vulnerabilities. But this structure also suggests a strategy for prevention. Concentrations are critical nodes, and critical nodes can be hardened and protected at much lower cost than all assets taken together. In fact, it will be shown that the more we protect these concentrations—the critical nodes—the more "safe" is the entire sector. In other words, there is an economy of scale associated with protecting critical nodes.

We will use this critical node strategy over and over again: Protect the "piles" because they are the most important and therefore the most critical. Without this sector "feature," it would be impossible to protect every asset of a large sector.

A NETWORK NOTION

In this section we show that critical infrastructure sectors are naturally modeled as networks containing nodes (assets) and links (relationships between pairs of assets). This approach will reveal important properties of specific sectors, such as the concentration of assets, and cascade failure modes. It also leads to strategies for allocating protection funds to harden these nodes against successful attacks on the infrastructure.

The *Termite* simulation illustrates how critical infrastructure sectors, acting on their own, aggregate assets into a small number of vulnerable piles—the critical node or nodes. But *Termite* ignores one important aspect of a critical infrastructure—its connectedness. In practice, assets are connected to one another through a series of links. These links are important to the successful operation of the infrastructure, because no sector component operates completely on its own. That is, the analogy with termites has limitations. We need to understand more than autonomous behavior to understand how most critical infrastructures work.

A major claim of this book is that critical infrastructures can be understood, analyzed, and then protected using an approach called *network theory*. Networks are mathematical *graphs*. Graphs have been studied for hundreds of years, and their properties are well understood (see Definition 4.1). Network theory is a way to model critical infrastructure systems as abstract graphs containing nodes, links, and a *map* that tells which nodes are connected to other nodes in the network.

This is more than theory; it is actually a practical way to model, analyze, and harden potential targets in nearly every critical infrastructure sector. Specifically, infrastructures like electrical power, telecommunications, transportation, and banking can be formally modeled as networks. Then these networks can be analyzed rigorously and formally to identify assets that may be at risk.

Network theory seems to be an appropriate framework for analyzing sectors in the large because a network clearly identifies the sector's structure: its *architecture*. For example, electrical power grids can be modeled as a network of transmission lines weaving together power generators, switching substations, and consumers.

Telecommunications systems, the Internet, epidemics in public health, and obviously, highways, railways, shipping routes, and airline routes are easily modeled as a network. Most importantly, network theory provides a formal foundation for a *scientific* study of CIP.

DEFINITION 4.1. WHAT IS A GRAPH?

A *graph* is a collection of nodes (vertices) and links (edges). A graph has a mapping function that defines how the nodes and links are connected.

Simple graphs do not permit:

- Multiple links connecting a pair of nodes.
- Loops: Nodes that connect to themselves through a single link.

A *Path* from node x to node y is a sequence of nodes and links that lead from x to y.

- Path length: Number of links in a path.
- Connected Graph: there is at least one path from every node to every other node.
- Cycle: A path in which the first and last nodes are the same.
- Acyclic: Graph with no cycles.

A *Tree* is an acyclic connected graph.

- Forest: A group of trees.
- Spanning tree: Subgraph that contains all nodes of the graph but only enough links to form a tree.

A Complete graph is one in which every possible edge is present.

- Sparse: Graph with relatively few links.
- Dense: Graph with most edges present.

A Directed graph: Links have arrows indicating an ordering of the links.

Networks Defined

What is a network? In simple terms, a *network* is a collection of nodes and links that connect pairs of nodes. Nodes and links are abstract concepts: A node might represent a city, Internet switch, or single person. A link might represent a road that connects two cities, a fiber-optic cable that connects two Internet switches, or a relationship that connects two people in society. Network theory is general; it can be used to model a variety of real-world things. Thus, we can apply network theory to critical infrastructures and then apply known techniques to the resulting (abstract) network to gain insights and answers to questions like, "are these two cities connected by a critical highway," "does the Internet have single points of failure," and "are relationships between people in society responsible for the spread of certain epidemics?"

Network abstractions can be applied at different levels of detail. Nodes can be anything, and the links that connect them can be anything. Identification of nodes and links is important in terms of building an accurate model of the sector. For example, in modeling a transportation system as a network, are bridges links or nodes? In a network model of a public water utility system with rivers, treatment plants, and storage tanks, treatment plants can be either nodes or links; the decision is up to the modeler. In the analysis procedure about to be proposed here, sector components that we are interested in studying will always be nodes, and their connections and relationships will always be links. Therefore, if interest is in securing bridges, they are modeled as nodes and roads connecting them are modeled as links. However, if interest is in securing roads instead, roads are modeled as nodes, and the bridges, intersections, and ferry boats connecting them are modeled as links.

Once we have a network abstraction of an infrastructure, several analysis techniques can be used to yield important results and insights. For example, a network analysis of the system of railways, combined with information about the source of energy for 50% of the United States, reveals a single point of failure in the network. Over 50% of the coal used to fuel power generation facilities in the United States is moved from Wyoming over a single railway that crosses a certain bridge near Powder Ridge, WY. This critical node in the network cannot be easily nor quickly rebuilt/replaced. Remove the bridge, and fuel ceases to flow to the power plants supplying power to most of the Eastern United States. Network analysis of the flow of coal through the transportation network provides this insight.

This same analysis technique can be use to analyze vulnerabilities in shipping (most of the goods imported and exported to the United States pass through three ports), vulnerabilities in human populations (epidemics spread faster when socially promiscuous people are infected because they are connected by more social network links), and power grids (the transmission and distribution components of power and utility grids are more vulnerable than the power generation facilities). Network theory is powerful because of its generality and our ability to apply known analysis techniques to the network model.

PROPERTIES OF NETWORKS

More formally, networks are mathematical graphs. Hence the study of networks is rooted in the study of graphs. To illustrate some important concepts borrowed from graph theory, suppose we apply graph and network theory to a practical

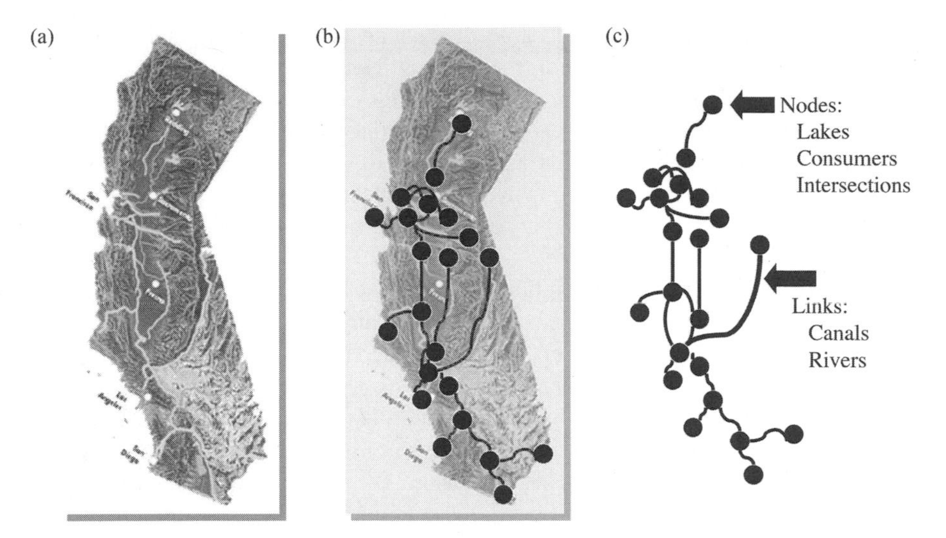

Figure 4.2. A mathematical graph is used to model the California Aqueduct as a network. (a) Map of California Aqueducts. (b) Network nodes and links layered on the California Map. (c) Network model of aqueducts as a graph.

example. Figure 4.2 illustrates the application of mathematical graphs to network models of infrastructure—in this case, the network of the State Water Project (aqueducts) of California.[5] How is graph theory useful in the analysis of California's aqueducts?

Figure 4.2(a) shows a highlighted map of the aqueducts of California. Figure 4.2(b) is the same map with nodes and links of a graph layered on top of the aqueducts. Finally, we have taken the map away in Figure 4.2(c), leaving the graph only. This graph is an abstraction of the aqueduct network, showing the important assets as nodes and links. Nodes represent lakes, reservoirs, consumers, and intersections. Links represent flows in the canals and rivers making up the aqueduct system.

The *degree* of a node is the number of links connected to the node. A node with degree of one means the node is *terminal*. A node with degree equal to three means there are three links connecting the node to the rest of the network. A node with degree equal to zero is an *isolated* node, because it is not connected at all.

The *size* of a graph is the number of nodes, N, it contains. A *complete graph*, or complete network, has the maximum number of links possible. Therefore, each node

[5]Stretching more than 600 miles from Lake Oroville in the north to Lake Perris in the south, **California's State Water Project** is one of the modern world's most ambitious public works projects. The State Water Project includes 20 pumping plants, 14 principal dams and reservoirs, 9 power-generating plants, 4 ancillary branches (the North Bay, South Bay, Coastal aqueducts, and the East Branch Pipeline), and the 444-mile-long Edmund G. "Pat" Brown California Aqueduct. In addition to water supply, the project provides flood protection, power generation, recreation, and fish and wildlife benefits. http://www.yvwd.dst.ca.us/water_stateproject.htm.

in a complete graph has (N − 1) links, one link for each of the other (N − 1) nodes. Since there are N such nodes, the total number of links is N × (N − 1)/2, because each link connects a pair of nodes. In a complete graph, the degree of all nodes is the same: (N − 1).

A *dense graph* or dense network is a graph with many links. A complete graph is dense. Typically, a dense network contains nodes with degree of N/2 or more. A *sparse* graph is a graph with relatively few links; each node has degree of (N − 1)/2 or fewer.

A network *path* is a sequence of links and nodes leading from one node to another node in the network. The *length of a path* is the number of hops (links) along the path. So, if node A is connected to node C through a link from A to B, then B to C, the path is of length 2. It takes two hops to go from node A to C.

If no path exists between one part of the network and another part, we say the network is divided or segmented into *components*. A component is a standalone network or island in the graph representing the network. A power grid separates into a graph with components or islands when a massive power outage occurs such as happened in August 2003. Portions of the power grid became disconnected from other portions.

A *directed graph* is a graph containing links with a direction. Directedness means that a graph can contain a link from node A to node B but *not* a link from B to A. For example, a *tree* is a directed graph containing directed links. A special node called the *root* is linked to the rest of the graph through outward directed links. Each link from the root is directed away from the root. A *binary tree* is a tree in which the root has two directed links, and every other node is connected via three directed links (one incoming and two outgoing), except for the terminal or *leaf* nodes. The degree of each terminal node of a binary tree is one, because terminal nodes are connected to the tree through a single directed link.

A binary tree with N nodes is considered sparse because no matter how large N becomes, each node is of degree one, two, or three. In fact, a binary tree with N nodes has at most (N − 1) links. And yet, it is possible to trace a path from the root to all other nodes of the tree. This is why hierarchical graphs such as an organization chart are considered efficient; they connect all nodes to one another with a minimum number of links.

Tree networks represent important organizations and infrastructures in the real world. For example, most hierarchical organizations, such as the military command structure, are trees. Because N people can be connected with only (N − 1) links, tree-structured command organizations are very efficient. However, information must flow up the tree to a common shared node and then back down again to the recipient. This takes almost (N − 1) steps, which can be slow.

The Internet naming system is tree-structured, as is much of the physical connectedness of the Internet. Each time a URL such as http://www.mycompany.com is used in an e-mail or Web access, a message must be sent up the tree-structured system of servers to an intermediary containing a list of URLs and their IP addresses such as 121.32.44.04. This can be time-consuming, but electrons travel at 186,000 miles per second, so most people do not notice the delay.

We will have a good reason to revisit sparse networks, trees, and hierarchies later in the analysis of various infrastructures. As it turns out, most critical infrastructure networks are sparse. Many are trees.

Figure 4.2 illustrates how network theory will be used throughout the remainder of this book. First, we model the infrastructure as a collection of nodes and links. Then we abstract away from the physical infrastructure and concentrate on the graph that represents the components of the infrastructure. This approach begins to pay off once we analyze the network as a graph and determine whether the corresponding graph is dense, sparse, tree-structured, and so forth. If the structure of the graph is nonrandom, we exploit that structure to advantage and focus on the concentration of links or cluster of highly connected nodes.

SCALE-FREE MEANS HUBS

Network theory has recently undergone a renaissance in the technical community because of the work of mathematicians such as Albert-Laszlo Barabasi who observed that the Internet—perhaps the largest network in terms of spanning the globe—is not what it seems. Before Barabasi's discovery, mathematicians thought networks such as the Internet were shaped more or less like random graphs. It was widely assumed that the Internet links (wires) between pairs of nodes (computers) were randomly distributed; e.g., each node randomly connected to one or more other nodes through a randomly distributed number of links. If this was true, the number of connections between computers on the Internet should be uniform (about the same for all) and not deviate much from the average. Everyone on the Internet would be connected to about the same number of websites as everyone else.

But Barabasi and his students discovered that most computers on the Internet are connected to a relatively small number of others through a relatively small number of links. Furthermore, a few rare nodes (Amazon.com, AOL.com, Ebay.com, etc.) are connected to a huge number of other computers. In terms of graph theory, a small number of nodes possessed a large degree and a large number of nodes possessed a small degree. The distribution of links to nodes was not random at all, but the number of nodes with degree k fell off sharply as k increased. The rate of decline in the percentage of nodes of degree k follows a power law, which has no variance (scale); hence, the name, *scale-free*, was given to such networks.

Barabasi and his students launched Web crawlers that mapped the nodes (servers) and links (communication lines) of the Internet, returning the map as a network that could then be analyzed. He discovered that most of the network is sparse (empty) but a few neighborhoods are dense (full of lots of links). Even more important was the observation that the most highly connected nodes, called hubs, got that way for an important reason. The "popular" nodes such as Ebay.com and AOL.com had more links than the average nodes, because of the economic, social, and technical forces shaping the Internet. These forces behaved much like the Termites described earlier.

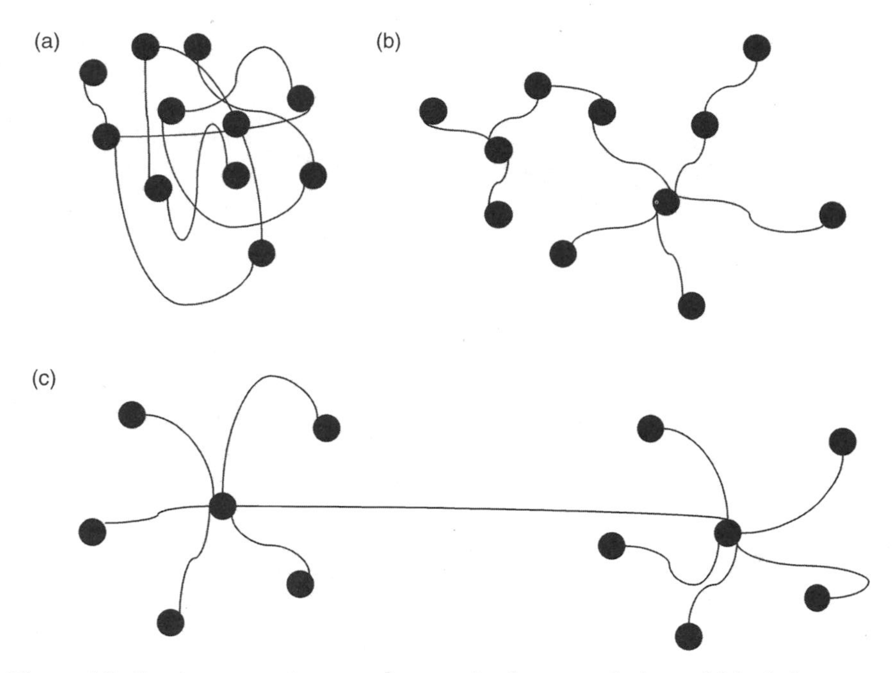

Figure 4.3. Random networks are unstructured, whereas scale free exhibits hub structure. Scale-free networks can be identified by their power law distribution. (a) Random network with no hubs. (b) Scale-free network with two hubs. (c) Small-world network with two neighborhoods. Any node can be reached from any other node by three or fewer hops.

That is, a simple organizing principle was operating on each node, which produced an Internet-wide pattern. The result is a *nonrandom network*—one with an unusually high concentration of links around a small number of nodes. This unexpected result suggested that the evolution of the Internet was not haphazard. The Internet's hidden structure is nonrandom. Indeed, it is scale free.

Barabasi makes a strong case against randomness in his groundbreaking book[6]:

> Would I be able to write this book if the molecules in my body decided to react to each other randomly? Would there be nations, states, schools, and churches or any other manifestations of social order if people interacted with each other completely randomly? Would we have an economy if companies selected their consumers randomly, replacing their salespeople with millions of dice?

Barabasi's work stimulated mathematicians to study the new field of *scale-free networks*—networks with a nonrandom structure that follows a *power law* distribution of links. These mathematicians seek to understand how structure evolves out of randomness—a seemingly contrary idea to the Second Law of

[6]Ibid, p. 23.

Thermodynamics.[7] Instead of running down like a mechanical clock, networks that represent organizations, infrastructure sectors, and biological organisms tend to become ordered. They evolve from disorder to order—from randomness to structure. Like the *Termite* simulation, critical infrastructure networks have evolved from disorder to order—from randomness to structure. This is the key to understanding their vulnerabilities as well as how they might be protected.

COUNTING LINKS

Figure 4.3(a) shows what a random network looks like, whereas Figure 4.3(b) shows what a scale-free network looks like. Note that the random network has no apparent structure, whereas the scale-free network has *hubs*—high-degree nodes with more than an average number of links. The rare nodes with a high degree are of most interest to us, because these are the critical nodes in any sector. We will return to this idea later when we analyze specific sectors.

How do we determine whether a network is scale free, random, or something in-between? Testing a network to see if it qualifies as scale free is extremely easy. Simply count the number of nodes with degree of one, degree of two, degree of three, degree k, and so forth, until the degree of all nodes has been counted. Then summarize these counts in a histogram of the percentage of nodes with *degree k* versus k to visualize whether a network is scale free.

Barabasi defines a network as scale free when the number of links at each node is distributed according to a power law that says the probability of a node having k links is proportional to (1/k) raised to a power that is greater than one.[8] Typically, the exponent of the power law is somewhere between 2 and 3. For example, the power law $(1/k)^2$ produces a histogram that declines rapidly: 1/4, 1/9, 1/16... as k increases from 2, and the power law $(1/k)^3$declines even more rapidly: 1/8, 1/27, 1/64 . . . .

Power law distributions are rapidly declining distributions, which means the networks they fit are lopsided, because only a select few nodes have a large number of links. Indeed, exponents of 2, 3, and higher may be too extreme for applications to critical infrastructure, so we will be less rigorous and exploit the "scale-free-ness" of any network that obeys the power law with an exponent greater than one. This lack of rigor is not important, as long as we can identify the hubs. After all, the highest-degreed nodes are the most important components to the overall network. Removal of a hub does the most damage to the network because it removes the largest number of links.

In plain language, a scale-free network is one that has a small number of high-degree nodes: Most nodes have a few links, and a few nodes have lots of links.

[7]The Second Law of Thermodynamics says entropy always increases whenever a physical process takes place. Hence, order tends to disorder—not the opposite.

[8]The term "scale free" does not refer to the size nor the breadth of the network. Instead, scale-free networks obey a power law that has no variance, and the links are distributed according to a probability density function that is proportional to $(1/k)^p$, where p is typically between 2 and 3.

The few nodes with most of the links are what we look for, because their removal does the most damage to the infrastructure. They are the critical nodes.

How do we know whether a certain network is scale free? The test is simple. We count the number of links attached to each node as follows:

Scale-Free Network Test

1. For every node of the network, count the number of links connecting the node to the network. This is called the *degree* of the node.
2. Count the number of nodes with degree of 1, 2, 3, and so on. Divide these counts by the number of nodes in the entire network. This is the *frequency* of nodes with a certain number of links.
3. Plot the node frequencies as a histogram starting with the frequency of nodes with 1 link, then 2 links, 3, and so on. Figure 4.4 illustrates this.
4. The resulting histogram has a shape; that is, the frequency counts decline as the number of links increases. If the rate of decline approximates the curve, $(1/k)^p$, where p is greater than one, the network is scale free.[9]

The histogram obtained by applying this test to the 11 nodes of the network of Figure 4.3(b) is shown in Figure 4.4. The power law $1/(k+1)^p$ with $p = 1.3$ fits best, so we conclude that this is a scale-free network. The most common nodes have 1 link; there are six such nodes; hence, 6/11 is 54.5%. One rare node has five links. That is, only $1/11 = 9.1\%$ of the nodes are of degree equal to 5. Notice that the most common nodes have a small number of links (1) and the least common of

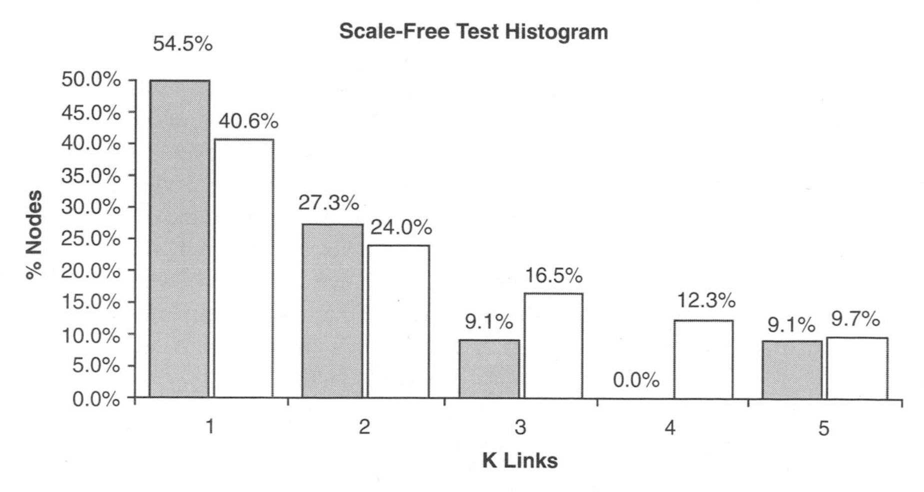

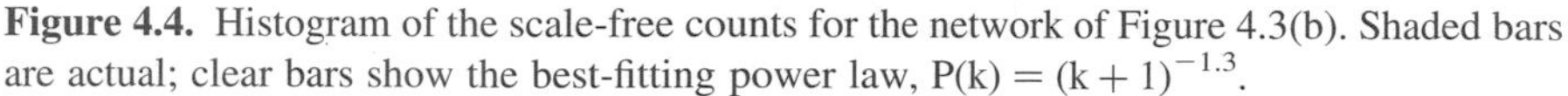

Figure 4.4. Histogram of the scale-free counts for the network of Figure 4.3(b). Shaded bars are actual; clear bars show the best-fitting power law, $P(k) = (k+1)^{-1.3}$.

[9]Barabasi and others define a scale-free network as one with a power law p between 2 and 3. However, for our purposes, any value of p greater than 1.0 will do.

all nodes has the most number of links (5). We are most interested in the rare, but high-degree node or nodes, because removal of such nodes is likely to segment the network into disconnected components. These rare nodes are *critical nodes*.

INCREASING RETURNS

What forces act on a network to make it scale free? If left to its own devices, a system should decay into disorder—not order! But instead, scale-free networks emerge out of randomness to become highly ordered. Energy goes in, and structure comes out! In this section we show that the economic *law of increasing returns* leads to scale-free structure when applied to a random network. In this case, increasing returns means that the more links a node has, the more it gets, as the network evolves over time.

Increasing returns explains why many critical infrastructure sectors have evolved into scale-free networks. It also connects economics to emergent behavior as it applies to critical infrastructure. For example, in the telecommunications sector, the more carriers there are in a telecom hotel, the less it costs each carrier to provide services to its customers. This "forcing function" has caused telecom hotels to evolve out of the 1996 Telecommunications Act.

The following simulation program called *PowerGraph* illustrates how a scale-free network forms.[10] Suppose a network consisting of N nodes and L links begins to form by repeatedly inserting a link between randomly chosen nodes. That is, we randomly select node A, then randomly select node B, and if A−B are distinct and not already linked, connect them with a link. Repeat this until all L links have been inserted into the growing network.

Figure 4.5(a) shows a network formed by linking randomly selected pairs of nodes. This random network has $N = 50$ nodes and $L = 100$ links, so it stops forming when all 100 links have been inserted. In Figure 4.5(a), nodes are labeled with their degree and displayed in a circular pattern. Larger degreed nodes move toward the center of the circle, so it is easy to identify the nodes of highest degree. In Figure 4.5(a), the nodes all have approximately the same degree, which is to be expected from a random graph. It is also worth mentioning that the graph is sparse because the average number of links per node is 4, which is much smaller than $N = 50$. The lowest degree is 1, and the highest degree is 8. The distribution of nodes versus their degree is shown in the bar chart located in the lower left-hand corner of *PowerGraph's* display window.

Formally, the nodes of a random graph are distributed according to the Poisson distribution.[11] Poisson distributions model the behavior of telephone call traffic,

[10]PowerGraph.class and PowerGraph.html can be found on the disk and the website under *Scale-Free Demo*.

[11]The **Poisson distribution** is a discrete probability distribution for X, where X takes on the values $X = 0, 1, 2, 3, \ldots$. Its shape is determined by one parameter λ, which is the expected value of X. If the average number of random occurrences per interval = $\boldsymbol{\lambda}$, the probability **P** of **X** occurrences in the interval is $P(X) = e^{\wedge}[\lambda^{X}/X!]$.

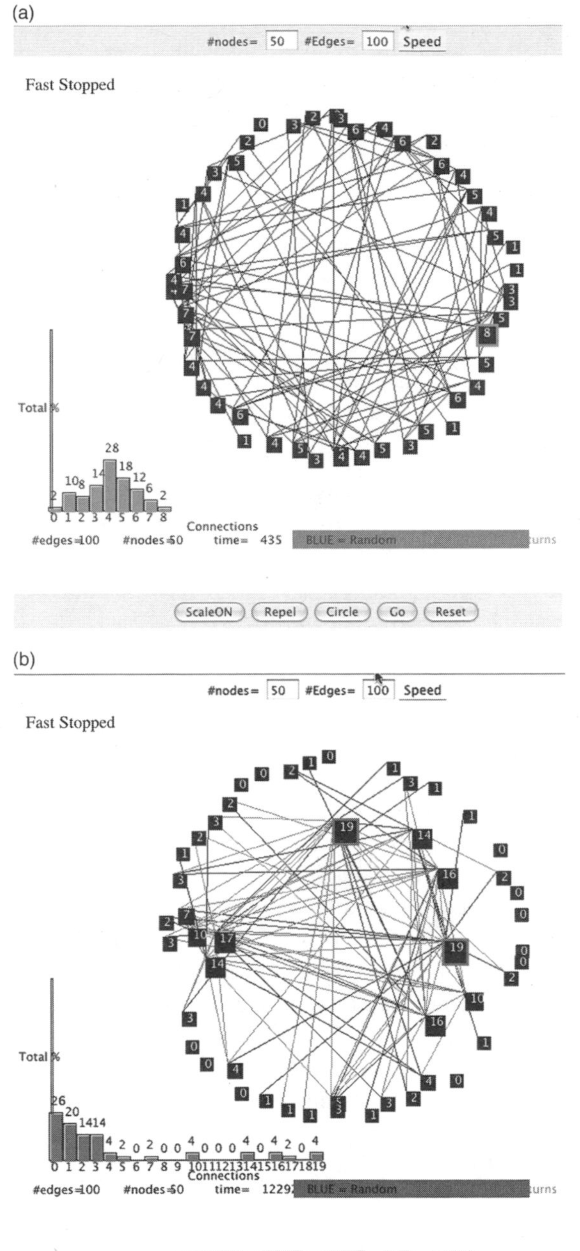

Figure 4.5. Output from *PowerGraph* shows (a) a random graph and the Poisson distribution of links and (b) a scale-free network formed by applying the increasing returns rule. (a) Random network with N = 50 nodes and L = 100 links. (b) Scale-free network that emerged from (a).

biological processes, and so forth. For example, the interval between telephone calls has been observed to be a random variable whose distribution is shaped like the one in Figure 4.5(a). The main point, however, is that the distribution of a random network is not a power law.

Now suppose the SCALE-ON button is pressed in *PowerGraph*. The simulator shifts from random node selection to the following *organizing principle*:

1. Randomly select link R, connecting nodes A and B together.
2. Randomly select a third node C.
3. Identify which node, A, B, or C, has the highest degree.
4. Disconnect L from the network and reconnect it so that it links together the two nodes with the highest degree.
5. Repeat steps 1–4 forever.

For example, in Figure 4.5(b), suppose a link is selected at random and its nodes have degree 3 and 5, respectively. In step 2, suppose another node is selected at random with degree of 8. The link is disconnected from 3 and 5 and reconnected between 5 and 8. The overall effect of this is to increase the degree of node C from 8 to 9 and to decrease the degree of nodes with degrees of 3 and 5 to degrees of 2 and 4.

Figure 4.5(b) shows one scale-free network that emerged from a random network of Figure 4.5(a) after applying the organizing principle thousands of times. How do we know it is scale free? By counting the nodes with degree 1, 2, and so on, until reaching the maximum node of degree 20, and plotting these counts as a histogram as shown in the lower left-hand corner of Figure 4.5(b), we obtain a power law histogram.

The scale-free network of Figure 4.5(b) is still sparse, but now there are many nodes with few links, and a few nodes with many links. This can be observed by looking at Figure 4.5(b) and noticing that several nodes have many links, and many nodes have only a few links. It is not necessary to compute the power law that best fits this histogram, unless we want to be rigorous. Indeed, nodes range in degree from 1 to 20. Although 20 is still much smaller than $N = 50$, it is significantly larger than the highest degree in the random network. Something has changed. The network is no longer random. It now has hubs. Scale-free networks have hubs—highly connected nodes—that random networks lack.

PowerGraph can be run with any number of nodes and any number of links, but regardless of how many nodes and links we use, the result is always the same. Scale-free networks emerge from random networks whenever the law of increasing returns is applied.

The overall effect of applying the law of increasing returns is to increase the connectedness of highly connected nodes and to decrease the connectedness of lower degreed nodes. The "popular" nodes get more popular, and the less popular nodes become less popular. The "rich get richer" and the "poor get poorer." This is the *law of increasing returns* at work in networks. In economic terms, the law of increasing returns states that a commodity becomes more valuable as it becomes more plentiful. In contrast to the *law of diminishing returns*, which states that a commodity loses value as it becomes more common, the law of increasing returns increases both the value and the supply of a commodity. It is contrary to the famous law of supply and demand.[12]

The law of increasing returns operates in economic systems such as the software industry where Microsoft products increase in value as they proliferate. The more Microsoft Office software people own, the more other people want to own them too. This is due to a number of reasons, such as training, compatibility, interoperability, and marketing prowess on the part of Microsoft.

The law of increasing returns has been around for a long time. Infrastructures such as water, power, and telecommunications obey this law. Known as Metcalf's Law in network systems such as the Internet, the invisible force of increasing returns can lead to monopolies such as Microsoft, and in earlier times, AT&T Long Distance Lines.[13] During the rise of the water utility, telecommunications, and power industries, the federal government's policy was to yield to the power of increasing returns and declare these industries regulated or *natural monopolies*. In a sense, many of today's Level 1 critical infrastructure sectors have emerged from a policy designed to ride increasing returns to its natural monopolistic conclusion. Only recently, during the 1990s, has this trend been reversed as the federal government has attempted to deregulate these so-called *natural monopolies*. After decades of being ruled by the law of increasing returns, the telecommunications, energy, and power sectors have evolved into scale-free networks, just as *PowerGraph* turns random networks into scale-free networks.

PowerGraph demonstrates several key concepts in the formation of critical infrastructures:

1. Scale-free structure can emerge from unstructured, random networks.
2. Emergence is the result of a simple, local, organizing principle.
3. Localized organizing principles such as increasing returns do not need central control to cause macro-scale changes in the overall network.
4. Simple organizing principles can produce complex patterns in networks.
5. The law of increasing returns leads to "winner take all" consequences.

AN EXAMPLE: HIGHWAYS AND RAILWAYS

The previous theory can be applied to a real-world infrastructure to illustrate the technique of scale-free network analysis. A close examination of the Interstate Highway map reveals a hub-like structure that suggests it is scale free. As it turns out, most interstate highway nodes have a small number of links, e.g., two

[12]Increasing returns eventually plays out, leading to an S-shaped adoption curve. Most technological products follow this curve where market share (as a percentage of total market demand) is plotted on the vertical axis, and time is plotted on the horizontal axis. Consumer product adoptions, from the horseless carriage (automobiles), TV, VCR, telephone, to home computer, have followed this curve for the past 100 years.

[13]Metcalf's Law states that the value of a network increases proportional to the square of the number of users.

links (coming into a city and leaving the city), or a cross-road intersection with four links (North, South, East, West). Only a few rare cities have more than six or seven links. That is, only a small number of U.S. cities are interstate hubs.

Below is a list of U.S. cities with six or more links to the interstate highway system.[14]

City	Number of Links
Atlanta	6
Ann Arbor	6
Cleveland	6
Nashville	6
Oklahoma City	6
St. Louis	6
Indianapolis	7
Dallas/Ft Worth	7
Chicago	10

Figure 4.6 shows a histogram that compares the actual link count versus a scale-free histogram produced by the *Scale-Free Network Test* described above. The similarity with a scale-free network is obvious—the interstate network is *not* random—it has structure. Chicago has the highest number of links of all cities in the network; hence, it is an important hub—a critical node.

A similar analysis of the Amtrak network reveals the same result; Chicago is a rare node with high network connectivity. It has more Amtrak railway links than any other U.S. city. After a few more transportation sector analyses, it becomes clear that Chicago is a critical node in the transportation infrastructure. When it comes to this sector, Chicago is more important than Atlanta or Nashville.

Clearly, the first beneficial result of scale-free analysis is critical node identification. Could it be that nodes with lots of links are special? Take out the hub node and most of the network becomes fragmented, perhaps even disconnected. This phenomenon also occurs in banking and finance, health care and epidemiology, transportation, telecommunications, and perhaps all critical infrastructure sectors. (This assertion may be too strong—the application of scale-free network theory has not been fully explored for all sectors). Hubs are obvious targets. How do we analyze these obvious targets to gain a deeper understanding of their vulnerabilities?

IT'S A SMALL WORLD

Increasing returns is a general organizing principle that explains why several infrastructure sectors have evolved into scale-free networks. Scale-free infrastructure networks evolved through repeated application of this simple organizing principle.

[14] *The Road Atlas Deluxe, RAND McNally: United States, Canada, and Mexico*, 2003. Only one and two digit interstate routes were counted—three-digit routes are beltway or spur roads in the system.

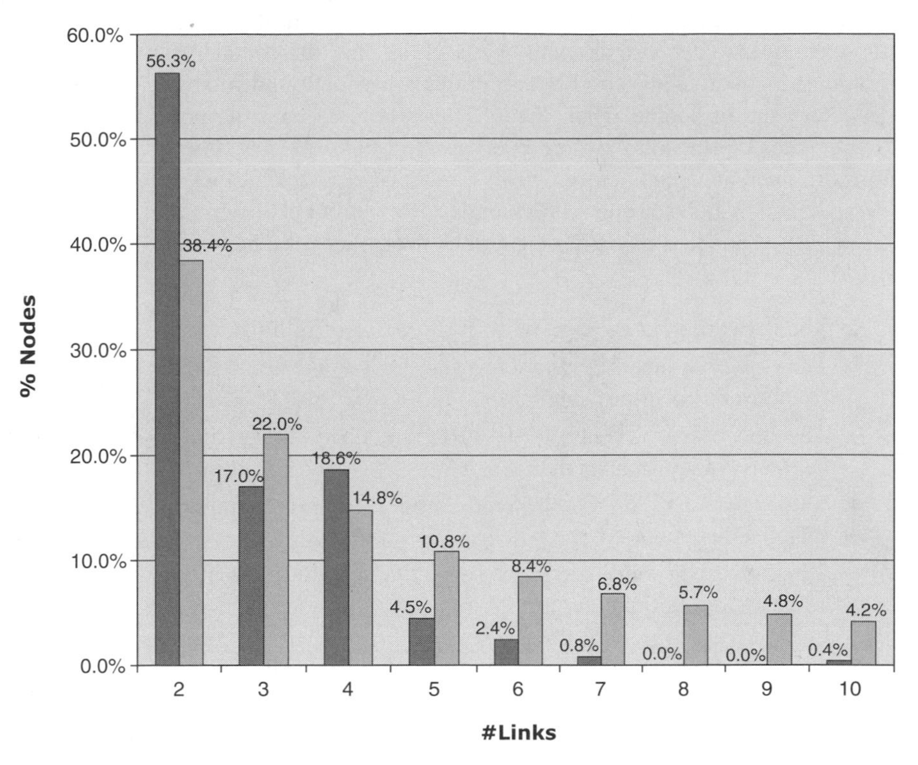

Figure 4.6. Histogram shows the percentage of U.S. cities with interstate highway links versus the number of links. Dark bars are actual data. Light bars are best-fit power law distribution values with an exponent of 1.38. [Probability (k) = $k^{-1.38}$.]

This is an example of *emergence* because a meaningful pattern emerges from a random, unstructured system. Complex structures emerge from simple ones—order emerges out of disorder. This begs the question, "Are there other organizing principles that lead to emergent behavior besides the law of increasing returns?" The answer is "yes."

Small worlds are networks that form around clusters or neighborhoods of nodes, not merely single hubs. Neighborhoods are groups of nodes that can be reached by one another through a relatively small number of hops. If node A is connected to node B, which is connected to node C, then A can reach C through two hops. Generally, nodes in a small world can reach each other in several hops that is much less than the size of the network. Figure 4.3(c) illustrates a simple small-world network. Every node is reachable from any other node through three hops. Three is much smaller than 12—the size of the network. In addition, the network of Figure 4.3(c) is sparse; the short-hop separation of nodes from one another is not a function of the number of links in the network. Rather, it is a function of the structure of the network.

Small-world networks were studied by Strogatz and Watts, who noted that everyone knows someone who in turn knows someone else in a rather small number of hops among people. Strogatz and Watts claim that the social network of human relationships is a small world where nodes are people and links are relationships between pairs of people. Their theory attempts to explain the popular concept of "six degrees of freedom," where every person knows every other person through no more than six acquaintances. [This concept goes back further than the work of Strogatz and Watts, starting with chain letter experiments.]

Duncan Watts lists five strict conditions necessary for a network to form a small world[15]:

1. The network is large; e.g., N is on the order of billions.
2. The network is sparse in the sense that the average nodal degree k is hundreds of thousands of times smaller than N.
3. The network is decentralized—there are no dominant hubs—the maximal degree is much smaller than N.
4. The network is highly clustered—nodes form neighborhoods of strongly linked subnetworks.
5. The network is connected in the sense that any node can be reached from any other node in a finite number of hops.

For our purposes, it is sufficient that N be large, but it does not need to be in the billions. Rather, most critical infrastructure networks have perhaps thousands of nodes. But in reality, most critical infrastructure networks are sparse and the nodes are connected to the remaining network through a relatively small number of links. So the first three conditions above easily match the real world. The idea of a small-world network is important, but we do not need to be as rigorous as Watts in defining the difference among small-world, scale-free, and random networks. For our purposes, we only need to know when a network contains clusters of nodes that form neighborhoods that act like a hub. Removal of such a cluster would cause great damage to the network.

In a small world, network nodes tend to form clusters (local neighborhoods) where there are enough links between neighboring nodes to bind them together as a community. This may or may not represent reality. In addition, condition 5 says that clusters are connected to every other cluster through a relatively small number of hops. For example, in six-degrees-of-freedom lore, six hops connect everyone to everyone else in the world. These two conditions may or may not exist in a real-world infrastructure network.

Figure 4.3(c) shows a small-world network with two neighborhoods connected by a single link. No node is more than three hops from any other node. And yet, it is clear that the two neighborhoods form a cluster and all nodes in one cluster can be reached from any node in the network.

[15]D. Watts, "Networks, Dynamics, and the Small-World Phenomenon," *American Journal of Sociology*, vol. 105, no. 2, pp. 493–527, Sept. 1999.

SHORTESTLINK SIMULATION

How does a small-world network emerge from a random network? What organizing principle dictates this emergent behavior? The emergence of a small world from a random network is illustrated by *ShortestLink*, which works much like *Termite* and *PowerGraph* combined. Network nodes travel like termites in a random walk, while long links are repeatedly replaced by shorter links. *ShortestLink* repeats the following simple organizing principle, forever, or until the user stops it.

Shortest Link Organizing Principle

1. Randomly select a link L (L connects nodes N and M together) and a node P.
2. Compute the distances between pairs: N and M, N and P, and M and P. Let the shortest link be S.
3. Replace L with a new link that connects the two nodes with the shortest distance between them, S.
4. Repeat this forever.

For example, if L is 6 hops long and the distance between P and M is 5, then L is dropped and a new link between P and M is created. A link of length 6 is replaced by a shorter link of length 5, so over time, the links become shorter and shorter. The network maintains the same number of nodes and links, but as links are replaced by shorter and shorter links, a small world emerges from the random network.

ShortestLink simulates the emergence of a small world from a random network as shown in Figure 4.7. Nodes randomly move around like termites in the *Termite* simulation. At each time step a link is selected at random and replaced; thus, the nodes in the network become slightly more clustered. But because the number of links and nodes remains constant, the network remains sparse.

Notice in Figure 4.7 how the network forms three or four clusters. These clusters form and dissipate as the nodes perform their random walk around the rectangular area. Like rain clouds on a balmy summer day, clusters build up, scatter, and form again. The patterns formed by *ShortestLink* are nonintuitive—almost surprising. For example, the network becomes disconnected some of the time, and at other times, nodes "flock together" like a flock of birds.

Also notice the histogram in the lower left-hand corner. It is not a power law, and it is not strictly a Poisson distribution. In fact, the link distribution histogram of a network formed by *ShortestLink* is more like a random network than a scale-free network. But this is deceptive, because the histogram smoothes over the detail of clustering that is clearly a consequence of the length of links rather than the degree of the nodes.

Small-world networks form neighborhoods of critical nodes rather than single hubs. They model the real world of spatially oriented assets. For example, in an electric power grid, the power generation nodes, substations, and power lines are spatially fixed. Power stations do not move. But the economics of power dissemination favors shortest-distance connections (shortest links) over long-distance links.

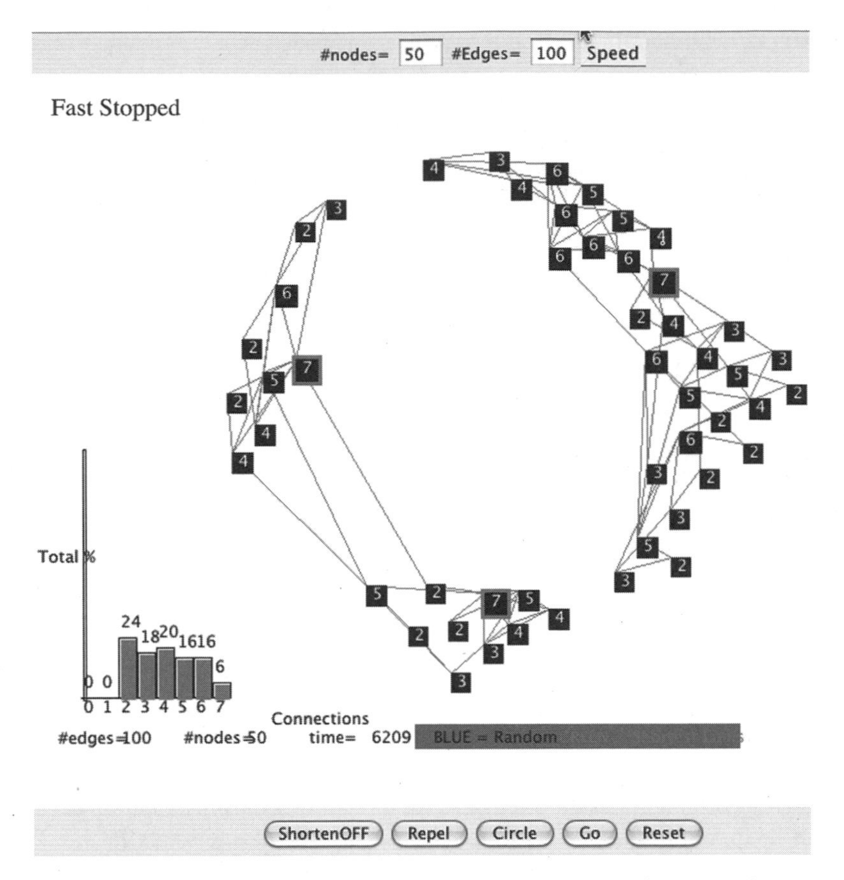

Figure 4.7. Program *ShortestLink*: A small-world network emerges from a random network by repeating the shortest length rule.

Hence, the power grid has evolved according to the small-world organizing principle. Although extremely simple, this principle explains why most spatially fixed networks are often small-world networks. It also accounts for the architecture of the power grid and will lead us to an understanding of its vulnerabilities. In fact, major portions of the Eastern Power Grid have been studied and found to be small-world networks. This is not a coincidence but a consequence of the spatial organizing principle explored by *ShortestLink*.

CRITICAL INFRASTRUCTURES ARE EMERGENT NETWORKS

What do emergent networks have to do with critical infrastructures? As it turns out, many (but not all) critical infrastructure sectors are the result of simple organizing principles like the law of increasing returns. These simple organizing principles produce networks with hubs or clustered critical nodes. These hubs and neighborhoods are particularly vulnerable because they provide the major connectivity of the sector. Here are some examples:

- The Internet (hardware) is scale free.
- The World Wide Web (software) is scale free.
- The Amtrak railway system is scale free with a major hub at Chicago.
- The interstate highway system is scale free with a major hub at Chicago.
- The intermodal shipping network is scale free with major hubs at the ports of Long-Beach/LA, Elizabeth-New Jersey, and Seattle-Tacoma.
- The concentration of telecommunication assets in "Telecom Hotels" suggests that telecommunications networks are scale free or perhaps small worlds.
- The Federal Reserve banking network is scale free.
- The Electrical Power Grid is clustered into small worlds because they grew up around the organizing principle of shortest distance among power generation stations, substations, and the consumer.
- The social networks responsible for the spread of sexually transmitted diseases such as AIDS are scale free.

Is this an amazing coincidence, or is some simple organizing principle at work in each case? Economic factors such as amortization of costs, economies of scale, and convenience have been at work for decades. What once began as a random network has in many cases evolved over time into a scale-free or small-world network. For example, the Telecom Hotel phenomenon exploded during the rise of the Internet much like the emergence of a scale-free network. The Telecommunications Act of 1996 accelerated this emergence because it allowed competitors to share resources. This also created vulnerable nodes such as 140 West Street, New York, NY.

Here is how emergence shapes the telecommunications sector, for example. First, an electronic switch is used to connect users to the backbone of the Internet—to achieve fast access to the Internet. Next, it attracts others because of its performance. Then the cost of colocating more and more switching equipment is amortized over more tenants, thus lowering the overhead for all. The more popular the shared facility is, the more popular it becomes. This vicious circle feeds on itself. Increasing returns powers growth until each shared facility becomes critical to the operation of the entire Internet. Failure of a portion of the switches in the hub at 140 West Street shortly after the collapse of the North Tower on 9/11 led to failures in network access in South Africa!

A scale-free Internet emerged from a random Internet. It used perhaps the simplest possible rule: Drop unpopular links in favor of links that connect to more popular nodes. After repeated application of this simple rule, hubs emerged. And hubs are points of vulnerability. In Chapter 1 we even suggest that Internet hubs be used against malicious programs such as worms and viruses. Therefore, hubs can be used for good or evil. We will employ asymmetric strategies to turn these "hub liabilities" into an advantage.

LINKS CAN BE CRITICAL TOO

The links of a network can be critical too. Suppose network theory is used to model a system of highways. If there happens to be only one highway that connects the Western United States to all roads in the Eastern United States, this link is critical because it is essential to the connectivity of the entire road network. Eliminating the critical link (road) separates the network into components—pieces of the overall network that no longer connect to one another. When it is impossible to travel from one network neighborhood to another neighborhood, we say the network is separated into *disjoint or disconnected components.*

Consider, for example, the Western Interconnection power tie lines that connect the Northwest component of the Western Power Grid to the California component. These are critical to the overall operation of the entire Western Grid, because disconnecting one from the other results in a cascade of failures. The two components are dependent on one another because demand shifts along time zones as well as seasonal timelines—power flow peaks in Portland, Oregon, during the winter months and in Los Angeles during the summer months. The critical link between the two allows Bonneville Power to sell electrical power to Los Angeles in the summer and buy it back from the Colorado Basin during the winter. But if the tie lines break, the rapid drop in power trips a series of power shut-downs that ripples through the entire 11 Western states and part of Canada, bringing the whole interchange to a halt. One critical link can start a chain reaction that leads to a cascade failure that sweeps the entire network. This is exactly what happened in 1996, leading to a major disruption of power to the West.

CASCADE NETWORKS

How do failures spread through an infrastructure, causing the entire sector to fail as in the Western Power Grid Blackout of 1996? *Cascade failures* are system-wide failures that begin with a relatively insignificant fault, which propagates throughout a major portion of the infrastructure, ending in calamity. Typically, a small event, such as a power line shorting out, escalates to a bigger fault, such as shutting down a power generator, which leads to an even bigger fault, such as shutting down an entire power subgrid. As the components of the grid are shut down, the network becomes disconnected leading to more shut-downs and more disconnections.

Cascade failures can be modeled as the propagation of *faults* in networks. Faults move through a *cascade network* much like an epidemic moves through a human/animal population. Indeed, communicable diseases spread through human-to-human contact. Nodes are modeled as humans and relationships (contacts) are modeled as the links in the network. The pathogen spreads at a rate determined by its incubation period. After a certain delay, an adjacent (connected) node is infected, and so on, until all nodes are affected. Thus, it takes time for the contagion to spread through a relationship network. In the meantime, it is possible that some nodes recover, or that some nodes are immune to the disease. Incubation time, natural immunity, and recovery work against rampant propagation of the pathogen, so it is possible that the disease can die out before it infects all nodes.

Cascade failures in critical infrastructure networks spread like diseases. Hence we can model their propagation as an epidemic. This has been done for human populations, and Internet viruses, which have been shown to behave like a human contagion. This gives us confidence that we can apply the lessons learned from human and Internet viruses to the problem of cascade networks.

Let a *cascade network* be a network of nodes and links containing a single fault (virus, pathogen, failure) that spreads with probability λ along links. That is, the fault jumps from one node to another through links only. In the case where faults are repaired (cured) at one rate, say χ, and incurred at another rate, say ϖ, the spread-rate $\lambda = \varpi/\chi$ is the ratio of infection rate to cure rate. Intuitively, if this ratio is greater than one, the fault continues to spread, and if it is less than one, it eventually fades. But this depends on another parameter called the epidemic threshold λ_c that determines whether faults persist or die out after some length of time.

A fault begins with one node becoming infected with a fault, followed by a series of infections that spread via the connecting links. At each node of the network, the fault either takes hold or not (with probability ϖ), and after a sufficient latency time δ, the fault spreads further along links connecting the infected node with noninfected nodes.

The question is, "what can be done to halt the spread of a fault, in a cascade network?" Here is what we know about epidemics in networks and, therefore, cascade networks:

1. Some networks, such as the Internet, exhibit oscillating or reverberating faults. They are called Susceptible-Infected-Susceptible (SIS) networks, because the infection, virus, or fault starts, spreads, and then declines; starts again, spreads, and so on.
2. In a random cascade network, an epidemic will die out if the spreading rate (probability that an infected node A will infect node B, if A and B are connected by a link) is less than some threshold λ_c. Otherwise, the epidemic will persist.
3. In a scale-free cascade network, $\lambda_c = 0$, so regardless of the spreading rate, the epidemic will persist.[16] Reducing the spread-rate will not prevent the collapse of the scale-free cascade network!
4. Epidemics will not spread throughout a scale-free cascade network if the hubs are protected such that λ_c is greater than or equal to λ. That is, hub-protected scale-free networks may or may not be at risk of cascading. It depends on

[16] R. Pastor-Satorras and A. Vespignani, "Epidemic spreading in scale-free networks," *Physical Review Letters*, vol. 86, p. 3200, 2001.

whether the hubs are protected and the spread-rate is below the epidemic threshold.[17]

The significance of a scale-free structure is obvious; it is possible to halt the propagation of a fault in a scale-free network, if we protect the hubs. That is, critical infrastructure sectors that are scale free can be hardened against cascade failures by protecting their hubs. This is a very important tool for protecting vast networks such as the four power grids that supply electrical power to the United States and Canada. In fact, Dezso and Barabasi have explored this property of scale-free networks and pathogens:

> The finding that the epidemic threshold vanishes in scale-free networks has a strong impact on our ability to control various virus outbreaks. Indeed, most methods designed to eradicate viruses—biological or computer based—aim at reducing the spreading rate of the virus, hoping that if λ falls under the critical threshold λ_c, the virus will die out naturally. With a zero threshold, while a reduced spreading rate will decrease the virus' prevalence, there is little guarantee that it will eradicate it. Therefore, from a theoretical perspective viruses spreading on a scale-free network appear unstoppable. The question is, can we take advantage of the increased knowledge accumulated in the past few years about network topology to understand the conditions in which one can successfully eradicate viruses?

The answer to the Dezso–Barabasi question above is "yes," and it is accomplished by hardening the most popular nodes—the ones with the greatest number of links—which in the case of scale-free networks means the hubs. This claim is powerful. The following simulation supports this conjecture.

FAULT-TOLERANT NETWORKS

A *fault-tolerate network* is one that continues to function even when some of its nodes are "killed." It may function at a lower capacity, but it continues to operate at some level of proficiency because some of its nodes remain intact. How can we construct a fault-tolerant network?

Here is the strategy: The more links a node has, the more we should protect it. This strategy is almost intuitive, but is it supported by reality? We can answer this question using a simulation of random versus scale-free cascade networks. Program *CascadeNet*[18] implements random and scale-free networks using the same technique as program *PowerGraph*. But *CascadeNet* goes further by allowing the user to experiment with several fault-tolerant network strategies.

Figure 4.8(a) shows the results of running *CascadeNet* for a scale-free network that has been seeded with a fault in a single, randomly selected node. The spread

[17] Z. Dezso and A.-L. Barabasi, "Halting viruses in scale-free networks," *Physical Review E*, vol. 65, p. 055103-(R), May 2002.

[18] CascadeNet.jar and CascadeNet.html can be found on the website or accompanying disk.

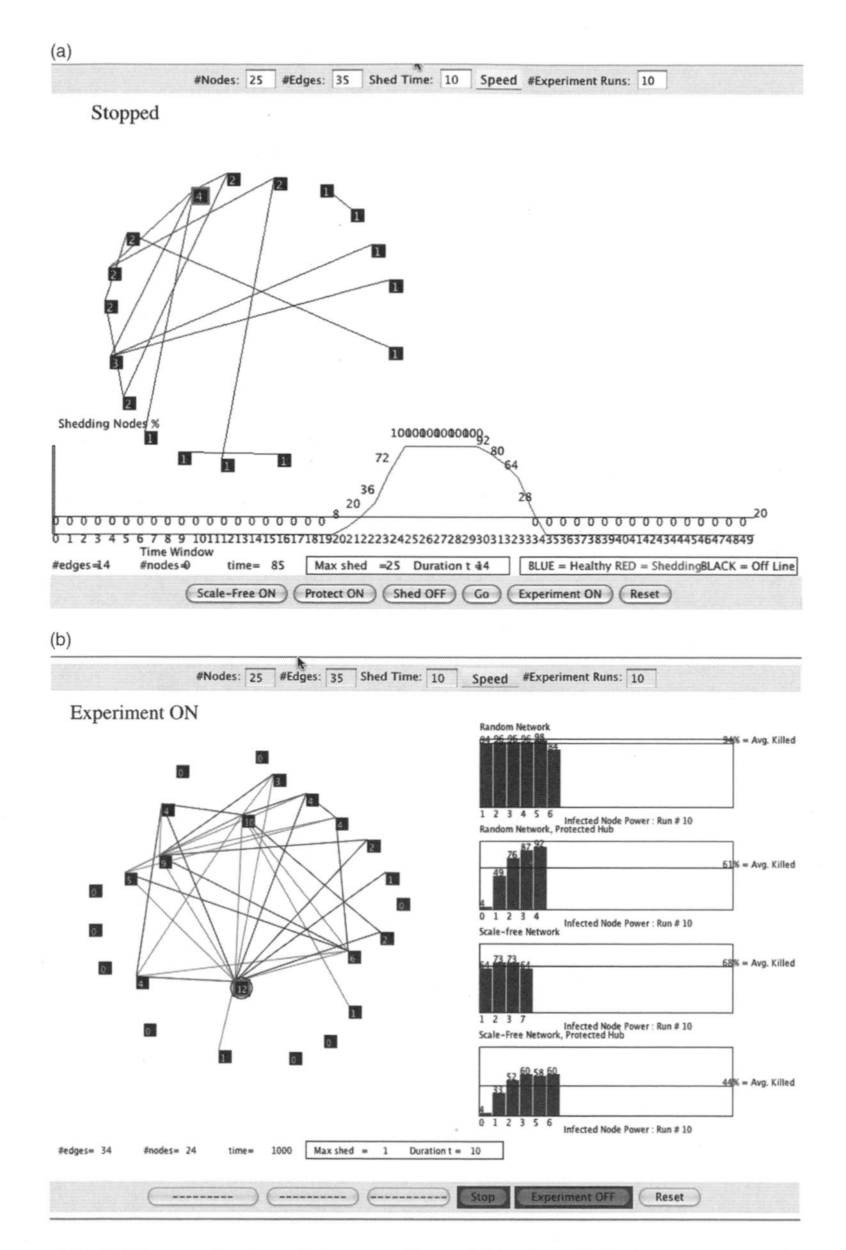

Figure 4.8. (a) Screen display of program *CascadeNet* for a fault that propagates through a network and (b) Screen display showing four experiments: random network, random network with highest degree node protected, scale-free network, and scale-free network with its highest degree node protected. (a) Graph showing the propagation of a fault through a network. (b) Fault-tolerant experiments showing the percentage of nodes that are "killed" by the spread of a random fault injected into a single node of the network.

rate is 1.0, so all but the protected nodes are eventually infected. This is a worst-case scenario: a single fault propagates with certainty to adjacent nodes. Adjacent nodes pass the fault on, in turn, until the fault runs its course. The question is, "How fault-tolerate is a random network versus a scale-free network?" Also, "Is a network with hardened hubs fault-tolerant?"

Figure 4.8(a) animates the spread of a fault and plots a graph of the percentage of infected nodes versus time, for the most recent 50 time intervals. This is shown as a moving curve along the bottom of the display. The simulation can be slowed down by selecting a simulation speed from the SPEED menu at the top of the display window. Each simulation run is activated by pressing GO, followed by pressing SHED ON, thus initiating the contamination of a randomly selected node.

CascadeNet constructs a random network by default. Pressing SCALE-FREE ON causes *CascadeNet* to transition from a random to a scale-free network using the *increasing returns organizing principle*. The hub of this network will be colored green. The user can also pick one or more nodes (using the mouse) and then press PROTECT ON to protect the selected node. A protected node is immune to the fault and remains intact throughout the simulation.

A fault or contagion spreads through links—hopping from node to node. As it spreads, each infected node is colored red to indicate that it is infected. After a delay of SHED TIME simulation time units, each red node turns black, indicating that it is "dead." Dead nodes are removed, along with the links connected to it.

Figure 4.8(b) shows the results of performing four comprehensive experiments. The first experiment estimates the percentage of nodes that are "killed" by an initial network fault in a random network. A node is selected at random, and because the spread rate is 1.0, the fault spreads to adjacent nodes—all of which are "killed." This manner of propagation is repeated at each time step. Only the nodes that are not connected at all (degree of zero) are spared the contagion.

The percentage of nodes killed versus the degree of the initially infected node is plotted as a bar chart in the display window. The horizontal axis is the degree of the node that is initially selected and infected. The vertical axis is the number of nodes killed, expressed as a percentage of the number of nodes initially in the network. As expected, most nodes are killed, because only the nodes with a degree of zero (not connected) survive, unless they are selected initially. Therefore, the bars in the bar chart for this experiment are all nearly 100%, so the average value across all bars is also close to 100%.

The four experiments shown in Figure 4.8(b) are as follows:

1. Random network, randomly selected initial node.
2. Random network, highest-degree node is selected and initially infected.
3. Scale-free network, randomly selected initial node.
4. Scale-free network, highest-degree hub is selected and initially infected.

The results shown in Figure 4.8(b) assume 25 nodes, 35 links, SHED TIME of 5 units, and 20 runs per experiment. As we might expect, the percentage of nodes "killed" by a fault decreases as we protect highest degree nodes. This is shown by the results of the experiments shown in Figure 4.8(b), which produced the following results (using the inputs given above)[19]:

1. Random network: 99% of the original nodes are killed on the average.
2. Random network, protected highest-degree node: 94% killed on the average.
3. Scale-free network: 78% killed on the average.
4. Scale-free network, protected hub: 66% killed on the average.

Another way to interpret these results is that the random network has only a 1% chance of surviving a random fault; a scale-free network has a 22% survivability rate; and both kinds of networks increase their survivability by protecting the nodes with the most links. Intuitively, protecting the highest degree nodes improves the survivability of any network. Less intuitive is the observation that a scale-free network with a hardened hub is the most fault-tolerant of the four types of networks. Why? There are two major explanations:

1. Most nodes have fewer than an average number of links; hence, they spread the fault less than a more "popular" node.
2. It is unlikely that a random attack on the hub will occur, relative to the likelihood that an attack will occur on another node, but when an attack does occur (by infection), the hub is protected. Therefore, it does not fail and it does not contribute to the further propagation of the fault. Protected nodes stop the fault from spreading further.

These results are only enhanced when more than a single high-degree node is protected. Program *CascadeNet* allows the user to manually select nodes and "harden them" by pressing the PROTECT ON button.

These results are applicable to critical infrastructure protection because there are insufficient resources to protect all nodes of critical infrastructure networks. We must invest in the highest payoff strategy. Therefore, a strategy focused on critical node identification and protection is the only practical solution.

ANALYSIS

This chapter describes a scientific process for analyzing vast infrastructure architectures using modern network techniques. The centerpiece of this approach is the relatively

[19]The bar chart values obtained from each run are averaged by simply taking the sum of the two most-recent bar chart values, and dividing by two. So the estimates are running averages—not mathematically exact averages. In addition, the bar charts of each experiment are averaged to estimate the survivability of nodes in each experimental network. For example, if run #3 produces a kill count of 66% and run #4 produces a kill count of 50% for an initial node of degree 8, the bar chart value for degree 8 is $(66\% + 50\%)/2$ or 58%.

new concept of a scale-free and small-world network. The advantages and disadvantages of using network theory to drive the policy of prevention are analyzed below.

Infrastructure as a Network: This idea originated with Barabasi, Watts, and others while studying the architecture of the Internet. Extrapolation to other infrastructure sectors such as the power grid and telecommunications seems to be appropriate and reveals insights into their structure. However, this is still a new idea, and the question remains, "is the theory of networks—especially scale-free networks—appropriate for all sectors?" We believe it is, but at this early stage, the application of network theory must be understood and its limitations appreciated. Moreover, its broad applicability to all sectors remains a research topic.

Non-scale-free network theory has been used by military strategists to decide how best to cut off enemy supply lines by taking out *cut sets* (least number of pieces to remove from the network to render it useless) of roads, railways, power grids, and so forth. When used in this way, the network of nodes and links represents a flow of some kind. The objective of the military commander is to disrupt the flow of supplies, weapons, or information by taking out the minimum number of links or nodes. For example, destroying certain critical bridges or telephone lines using a minimum number of bombs, troops, and equipment. Called *interdiction*, this strategy is used to disrupt the enemy's network of railroads and telephone lines that provide weapons to the front line.

Network interdiction theory uses exotic algorithms to find the best cut set. It yields accurate results, but as the size of the network grows, the time it takes to compute the optimal strategy grows even faster. Hence, large problems require elaborate software and extensive computing power.

On the other hand, scale-free network analysis is extremely simple: Count the number of links at each node and rank them from highest to lowest. If the result fits a power law, or nearly so, the network is either scale free or small world. In either case, the most likely candidates for an attack are easily identified, because they are critical to the operation of the entire network (most of the time). Although scale-free network theory sounds daunting, it is actually simple and effective.

However, scale-free network analysis does not model flow networks as does interdiction. If vulnerability analysis of network flow is needed, then network interdiction should be used instead of scale-free network analysis.

Critical Node/Link Approach to CIP: Given that we cannot protect everything equally, our prevention strategy is to focus on the hubs identified by scale-free analysis. Is this sufficient? One can argue that links may be more important than hubs. One can also argue that it is insufficient to examine one node out of perhaps hundreds when the cost of failure is high. Might it be better to analyze everything? The efficacy of critical node analysis has yet to be proven through experience with its application to real-world situations.

How do we know that narrowing down the search for faults to a handful of nodes or links is sufficient? We can only guess, but complex systems such as large software programs, large-scale mechanical systems, and so forth are prone to similar complex behavior. For example, large software systems are known to contain highly clustered faults—sections of the program that are much more likely to contain errors than the average. We also know that power laws as observed in scale-free networks behave similar to Pareto's principle—the 80%/20% rule[20]:

> The 80:20 rule originated from Vilfredo Pareto, an Italian economist who studied the distribution of wealth in a variety of countries around 1900. He discovered a common phenomenon: about 80% of the wealth in most countries was controlled by a consistent minority—about 20% of the people. Pareto called this a "predictable imbalance." His observation eventually became known as either the "80:20 rule" or "Pareto's Principle." The 80:20 rule has been expanded far since it's first economic use. Whilst one might quibble about the 80% or 20% (it is sometimes 60:40 or 90:10) the insight is broadly applied to leadership and management. The '80:20 rule' has become one of the best known "leadership shorthand terms" reflecting the notion that most of the results (of a life, of a program, of a financial campaign) come from a minority of effort (or people, or input).

Why should critical infrastructure sectors be any different? The method described here assumes that complex systems such as power grids, oil pipelines, roads, airline routes, and other Level 1 infrastructures are no different than many other complex systems in this respect. Again, we will show this to (generally) be the case as we study each sector in greater detail in subsequent chapters.

Cascade Networks: Are real infrastructure systems anything like cascade networks? Do cascade failures behave like epidemics? We believe this characterization, although novel, is indeed correct. In the past, major failures in electrical power grids have behaved like epidemics spreading through a population. One needs only to observe the August 1996 failure in the Western Power grid and the August 2003 failure in the Eastern Power grid to be convinced of the similarity.

The results of *CascadeNet* simulations suggests an even stronger strategy: hardening of high-degreed nodes. This strategy was completely untried and unproven at the time this chapter was written. So it remains a theoretical result. However, proposed alternatives cannot guarantee the prevention of future cascade failures in power grids, for example. Currently, there is no affordable strategy for protection of power grids short of replacing much of the existing infrastructure. The cost of replacement is estimated to exceed a trillion dollars! A more modest approach would be to improve the critical nodes.

In summary, critical node analysis and the theory proposed here leads to the following conclusions:

1. For networks that do not fail due to cascade failures:
    a. In structured networks (scale free, small world, Pareto, etc.), protection of a relatively small number of critical nodes (high degree, high damage value, or both) is the best policy.
    b. In random networks, single component failures have a small impact on the overall network, but protection of the entire network is expensive, because a large number of nodes and links must be hardened.

[20]http://www.paretolaw.co.uk/principle.html.

2. For networks that are prone to cascade failures (power grids, Internet contamination, etc.):
    a. In structured networks, protection of a relatively small number of critical nodes (high degree, high damage value, or both) is the best policy, just as it is for non-cascade networks.
    b. In random networks, single component failures may have a large impact on the overall network, because they can spread throughout the entire network. Protection of the entire network is expensive, because a large number of nodes and links must be hardened. This is the worst-case situation, because cascade networks are responsible for the most dramatic failures, and prevention is the most costly.

EXERCISES

1. What is emergent behavior?
    a. A psychological trait of terrorists.
    b. Patterns of order derived from simple rules and disorder.
    c. A theory of how termites stockpile chips.
    d. Result of applying the law of increasing returns.
2. What is the degree of a network node?
    a. The number of links connected to the node.
    b. The number of hubs in the network.
    c. The size of the network.
    d. The number of links in a network.
3. What is a scale-free network?
    a. A network with hubs.
    b. A hierarchical network.
    c. A dense network with hubs distributed as a Poisson law.
    d. A network whose node degrees are distributed as a power law.
4. When a power grid fails, its network is:
    a. Crippled by an electrical short in a power line.
    b. Separated into components.
    c. Crippled by having to turn off generators.
    d. Separated into hubs with critical links.
5. A binary tree is a network with:
    a. Scale-free hubs.
    b. A critical node at the top.
    c. One root node with two sub-trees; each sub-tree is a binary tree.
    d. One root node with two links to other nodes with degree of 1 or 2.

6. The law of increasing returns says that:
   a. The more links a node has, the more it gets.
   b. A random network evolves into a scale-free network.
   c. A random network evolves into a small-world network.
   d. Termites tend to move chips into piles.
7. A property of small world networks is:
   a. They are dense.
   b. They have components with short links.
   c. They contain highly clustered subnetworks.
   d. They contain dominant hubs with shortest links.
8. In a cascade network:
   a. A single fault propagates to all linked nodes.
   b. Components form "islands" in the network.
   c. Components are connected by a single link.
   d. Random networks fail less often than scale-free networks.
   e. A scale-free network is less fault-tolerant than a random network.
9. In a fault-tolerant cascade network:
   a. The more links a node has, the more vulnerable is the network.
   b. Random networks are more fault-tolerant than scale-free networks.
   c. Network hubs fail more often than random nodes.
   d. Once a fault starts, it cannot be stopped.
10. The best strategy for protecting a scale-free cascade network is to:
   a. Harden the links.
   b. Harden the nodes.
   c. Harden the hub.
   d. The more hubs you harden, the more secure is the network.
11. The telecommunications sector in the city of San Lewis Rey[21] has evolved and reorganized over the decades because of changing state regulations and economic forces beyond the control of any single telephone company or mayor. In the early years, the industry was dominated by "mom and pop" telephone companies serving a few thousand households. During this formative stage—called the "blue period"—all telephone calls were local. But then technology made it economically feasible for the independently owned "blue companies" to support long-distance calls. However, they had to agree on intercompany call standards, and pay one another a fee to use the long-distance lines. This created a combination of cooperation and competition that led to a period of consolidation—the so-called "red period," because the leaders were designated as "reds." The red "fast

[21]San Lewis Rey is a virtual city in cyber-space where all bad things happen.

movers" became larger and dominated the smaller less aggressive "greens." In fact, the red companies got so big and powerful, they would only cooperate or connect with each other (red links to red) or buy out other less dominant companies (red links to blue). Over time, the assets of the telecommunications sector began to form a sparse network, but it was not a scale-free network. What kind of network emerged?

To understand the San Lewis Rey telecommunications structure, consider the following network abstraction:

In the beginning, the San Lewis Rey telecommunications industry network was populated by BLUE nodes only. These represent the "mom and pop" companies that provided local service. But then the new technology of long-distance dialing arrived and so the nodes had to cooperate (link together). A node linked to another node when the two telephone companies cooperated to take advantage of the long-distance service. Thus, red linked to red, and blue linked to blue. But the pair of nodes changed color as a result of linking. The power of each node is defined as the number of links (degree) of each node. When a pair of nodes linked, the more powerful node changed its color to red. The less powerful node changed to green.

More formally, the sector evolved according to the simple organizing principle, below:

Repeat the following forever:

1. Nodes N and M are selected at random.
2. If nodes N and M are both BLUE, or both RED, link them and color the most POWERful node RED and the less POWERful node GREEN.
3. If one node is RED, and the other is BLUE, do the same: Link them and color the most POWERful node RED and the other one GREEN.
4. If N or M is GREEN, do nothing (do not link them).
5. In case of a POWER tie, randomly select the most POWERful node.

After an infinite length of time, what kind of telecommunications network emerges? Is it a small world?[22]

[22]Program *TreeLink*, a.k.a RGB.html",4>RGB.html, is available on the website or the accompanying disk.

CHAPTER 5

Vulnerability Analysis

This chapter introduces the model-based vulnerability analysis (MBVA) process that zeros in on vulnerabilities of critical nodes for the purpose of quantitatively evaluating the vulnerability of a sector. MBVA combines network analysis with fault analysis tools used by reliability engineers to analyze faults in complex systems and, in Chapter 6, risk assessment and resource allocation. Taken together, these tools provide a complete process for identification of sector weaknesses, estimating vulnerabilities, computing risk, and then allocating funds to improve the security of the sector's critical nodes.

Risk assessment and resource allocation answers the question, "Given there is only a limited budget for protecting the sector, how best should the money be spent?" The answer to this question will be deferred until Chapter 6, but it is part of the MBVA process.

The following analytical tools are described in detail:

1. *Model-Based Vulnerability Analysis*: MBVA is a comprehensive method of analysis that combines network, fault, event, and risk analysis into a single methodology for quantitatively analyzing a sector component such as a hub. In MBVA, hubs are identified, hub vulnerabilities are organized and quantified using a *fault tree*, all possible events are organized as an *event tree*, and then an optimal investment strategy is computed that minimizes *risk*. MBVA gives the policy analyst a top-to-bottom tool for achieving critical infrastructure protection (CIP) under budgetary constraints.
2. *Network Analysis*: The first step in the MBVA process is to identify all assets in the sector—roads, bridges, intersections, ports, ships, airplanes, airports, telecommunications centers, power lines, power stations, and so on. This step also categorizes the assets: Are they nodes or links in the sense of a network? How are they connected? What are the dominant components? Network analysis was described in Chapter 4.
3. *Fault Tree Analysis*: A fault tree is a tree containing vulnerabilities and a model of how vulnerabilities work together to create a fault or failure in a

Critical Infrastructure Protection in Homeland Security: Defending a Networked Nation, edited by Ted G. Lewis

critical infrastructure component or hub. The root of the tree is at the top and represents the entire sector or major component, and the leaves of the tree are at the bottom and represent the threats arrayed against the components of the sector. Fault trees use logic and probabilities to estimate the likelihood that one or more failures will occur in a component. The output from a fault tree is a list of vulnerabilities along with the probability that each will occur.

4. *Event Tree Analysis*: The outputs from a fault tree are entered as inputs into an event tree. An event tree is an enumeration of all possible events obtained by single and multiple combinations of faults. Event trees are binary trees that consider all YES/NO (TRUE/FALSE) possible events, for each potential fault, that may occur in the fault tree. Once again, the root of the event tree is at the top, and the leaves are at the bottom (or side). The leaves represent all possible events that can occur, including failures and nonfailures. The output from an event tree is a list of faults (vulnerabilities) and their probability of occurrence, expressed as a fault probability histogram. This method of analysis leads to high assurance that we have fully explored all vulnerabilities—both singly and in combination. In addition, event tree analysis yields a quantitative result; we obtain the vulnerabilities and their probability of occurring.
5. *Event Matrix Analysis*: The number of events enumerated by an event tree grows exponentially with the number of potential faults. A fault tree with k faults produces an event tree with 2^k events. An event matrix reduces this number to far fewer—$k(k+1)/2$—events. Therefore, an event matrix is more compact, but it enumerates only single and double faults. It should only be used where the probability of triple and higher order faults is so low that they can be ignored.
6. *Risk Assessment and Resource Allocation*: Once events are identified, we can allocate funding to harden the components of a sector to protect it against the vulnerabilities identified by the foregoing steps. This step of MBVA will be studied in detail in Chapter 6.

MODEL-BASED VULNERABILITY ANALYSIS

Vulnerability is defined as the probability of a successful attack on a component. That is, vulnerability is a measure of the strength of a component in the face of a threat. It ranges from 0% to 100%. Sector vulnerability is the probability that an entire sector will fail if attacked. Sector and component vulnerability is sometimes interchangeable, but for the purposes of the following analysis, component vulnerability will differ from sector vulnerability. This difference will become apparent when we construct a fault tree model of the sector.

Vulnerability analysis (VA) is the process of calculating sector vulnerability from estimates of component vulnerabilities. Vulnerability analysis is perhaps the most important skill needed to practice CIP. Without VA, policy makers are merely making wild guesses about what to protect and how best to invest limited funds to harden potential targets against attacks. This important skill involves several difficult steps: identification of essential components (critical nodes), understanding the linkages and relationships among critical nodes (network analysis), and focusing on what is critical and what is desirable to protect.

The task begins with asset identification, but in addition, the analyst must identify vulnerabilities, estimate the likelihood of each, and perform a financial analysis of the impact of investing in target hardening versus the anticipated improvement in sector security. Thus, VA involves at least the combination of sector modeling, vulnerability modeling, financial modeling, and planning.

Vulnerability is not the same as risk. Risk analysis is described in Chapter 6. Vulnerability is a probability, whereas risk is measured in terms of financial risk, casualty risk, equipment risk, and so forth. Risk is typically computed as vulnerability multiplied by cost $-V \times D$, where D is an estimate of damages and V is a probability ranging from 0 to 1.0.

It is important to separate the calculation of vulnerability from risk, because vulnerability reduction achieves a different goal than risk reduction. In simple terms, vulnerability reduction attempts to limit the likelihood of undesirable incidents, whereas risk reduction attempts to limit the cost associated with an undesirable incident. For example, an automobile accident may occur with probability of say 50%, but one accident may cause $100 of damage, whereas another may cause $1000 of damage. The vulnerability is the same in both cases, 50%, but the risk is $50\% \times \$100 = \$50$ in one case and $50\% \times \$1000 = \$500$ in the other case. Risk is ten times greater for one accident than the other.

In addition, vulnerability is never absolute. Instead, vulnerabilities differ depending on the threat. For example, two automobiles are both vulnerable, but one is 25% vulnerable to a head-on collision, whereas the other is 75% vulnerable to rear-end collision. We cannot say they are both equally vulnerable, because it depends on the threat. The first automobile is less vulnerable, but only in relation to the threat—head-on collision. Indeed, both automobiles may be vulnerable to both threats, but in different proportions, depending on the size and safety of each.

The foregoing examples point out an important fact: VA is complicated by several factors, such as the nature of the threat, the likelihood of successful attacks, and the interplay among components that make up the critical infrastructure sector. This sophistication requires a sophisticated approach involving probability, logic, and modeling. The predominant tool for constructing such models is the fault tree—a logic and probability model of the infrastructure's critical nodes. We will populate fault trees with vulnerability estimates to derive the overall sector vulnerability.

In this section we describe one of many possible methods of VA: MBVA. MBVA borrows from network theory, reliability engineering, and financial risk assessment. It combines known methods commonly used in other disciplines. It is called

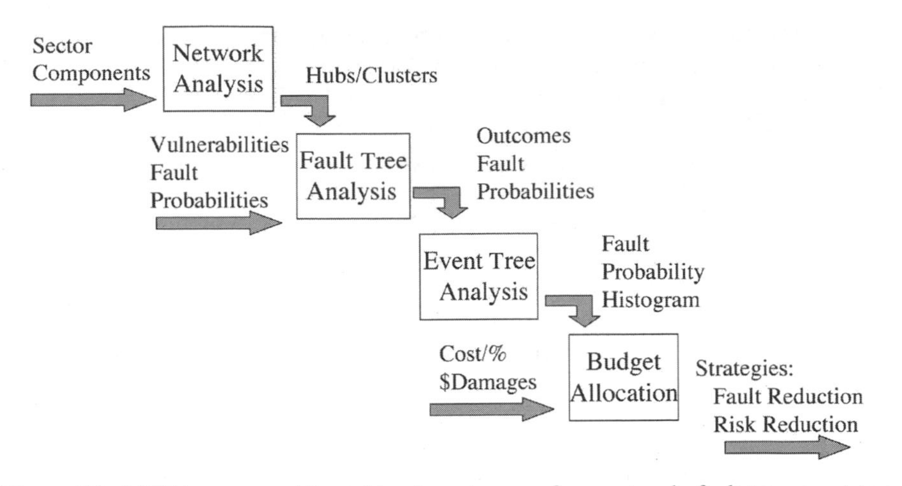

Figure 5.1. MBVA process: After taking inventory, perform network, fault tree, event tree analysis, and budget allocation.

"model-based" because we build a simple model of the sector components of interest. Specifically, MBVA combines network analysis with fault tree modeling to derive vulnerability, risk, and resource allocation strategies that tell the decision maker how best to allocate resources. This method will be used in the remainder of this book.

The steps of MBVA are as follows (also see Figure 5.1):

MBVA

1. List assets—Take inventory
2. Perform network analysis—Identify hubs
3. Model the hubs as a fault tree
4. Analyze the fault tree model using an event tree
5. Budget analysis—Compute optimal resource allocation

Step 1: List Assets—Take Inventory

The assets of an infrastructure are the components of its network structure—telecommunications switching equipment, power substations, bridges, important banks, and key shipping port's cargo handling, and airport baggage handling processes, and so on. There is nothing unusual about assets, but it is important to take inventory and describe each asset in terms of its capabilities, links to each other, and attributes such as replacement cost.

Assets—sector components in Figure 5.1—are represented as nodes and links in a network. But what is a node and what is a link? Deciding what is a node and a link is not always apparent in MBVA; it requires a thorough understanding of the sector. In most cases, links will be relationships of some sort—the relationship between people in a social network, the pipes in an energy supply chain, the cabling in a telecommunications network, and so forth. Assets may be major junctures in a sector such as the confluence of rivers, roads, pipelines, and telecommunication switches. They may be physical or cyber, e.g., the information in a database.

In general, assets that we want to protect should be nodes, and the relationships between assets should be links. For example, in a transportation system containing roads, bridges, and trucks, we might model the roads and bridges as nodes and the truck routes as links. This way, both roads and bridges can be studied for their vulnerability to damage, and the routes used to establish which roads and bridges are critical and which ones are not. If only bridges are designated nodes, then the vulnerability of the roads might be overlooked.

Step 2: Perform Network Analysis—Identify Hubs

The first box shown in Figure 5.1 says to perform network analysis so that the critical nodes of the network model can be identified. Determine if the abstract network is scale free, small world, or random. If it has hubs, zero in on them, as they are critical nodes in the following VA. If the network is a small world, then focus on nodes linked to neighborhoods or the neighborhoods themselves. Neighborhoods with highly connected nodes are also called "clusters."

Figure 5.2 illustrates this step using the Amtrak Rail System as the infrastructure. The network model straightforwardly defines rail stations as nodes and the railways

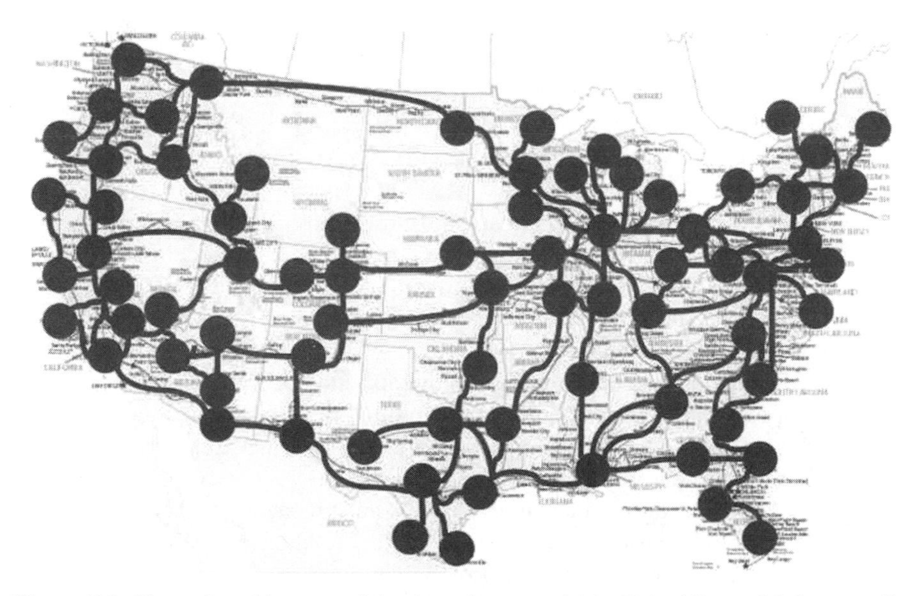

Figure 5.2. Network architecture of the Amtrak system in the United States: Nodes are railroad junctions (stations), and links are railroad tracks.

as links. Thus, the sector components of interest are stations and connecting railroads, but because we designated rail stations as nodes, they are the components that are studied for vulnerabilities. If the railways connecting stations are the subjects of the VA, then they would have to be designated as nodes in the network model. Similarly, railway bridges might be of interest, in which case we would also model them as nodes.

In Chapter 4, we analyzed a similar network (interstate highways) and identified a hub at Chicago. Chicago is also a hub of the Amtrak system as shown in Figure 5.2. The Chicago node is a rare node with ten links. In addition, there are other high-degree nodes in the Amtrak system that can be visually identified in Figure 5.2. Complete analysis of this network is left as an exercise for the reader.

Step 3: Build a Model Using a Fault Tree

Reliability engineers have developed methods of failure analysis based on *fault trees*. In this section we show how to adapt fault tree analysis to critical infrastructure analysis. The terminology of reliability engineering, "fault," and the terminology of critical infrastructure analysis, "vulnerability," can be used interchangeably. A *fault* is a failure or failure mode caused by a malfunction, natural, or manmade event, and *vulnerability* is a measure of the likelihood that a fault will occur. A *threat* is an entity that wants to attack a sector component such as a terrorist group that wants to bomb a building or infect the Internet with a virus. The definition of a threat as an entity versus a weapon is not important here: We use them interchangeably. Therefore, a threat can be either the terrorist group or the bomb it uses.

Vulnerability is defined as the probability of a threat-induced fault. In other words, vulnerability is identical to fault probability and is expressed as a number between 0 and 1.0 or as a percentage between 0% and 100%:

$$V(i) = \text{Probability}(i) = \text{Probability that an attack by threat } i \text{ will succeed.}$$

As an illustration, suppose a bridge can be destroyed by a bomb or rendered unusable by a traffic jam. Furthermore, suppose experts estimate the likelihood of a bomb destroying the bridge as 5%, and that a traffic jam could block the bridge with probability of 90%. The threat is either the criminals who might perpetrate the bombing or the bombing itself and either the commuters who jam the traffic or their cars. In either case, the bridge is vulnerable to two threats, either a bomb or a traffic jam. Its vulnerability to bombing is 5%, and its vulnerability to closure due to traffic congestion is 90%. In other words, vulnerability is a probability that measures the susceptibility of a component to a threat.

A model of the bridge and its vulnerabilities must be able to capture the threats, the vulnerabilities as probabilities, and the logical relationship between the bomb and the traffic jam threats. Note that nothing can happen, OR a bomb, OR a jam, OR both events are possible. In other words, a complete model that enumerates all possibilities is needed to fully capture the threats to the bridge. The logical relationships among threats and the vulnerabilities of components in a sector can all be captured using a fault tree.

A fault tree is simply a model of the components of a critical node or sector organized as a hierarchy or tree-structured graph. The nodes in the tree are called *components*, *logic gates* (AND/OR), and *threats*. These three kinds of nodes are typically layered as shown in Figure 5.3. A component is any major asset of the sector such as a water treatment plant, electrical power transformer, and telecommunications hotel. The root of the fault tree is a special component that represents the entire sector, e.g., power, telecommunications, and energy.

A threat is any physical or cyber-threat to a sector component. Threats are represented as terminal nodes in the tree-structured fault tree (see Figure 5.3). A fault occurs when a threat is activated—an attack—and successfully damages one or more components of the sector. The purpose of the fault tree is to model what happens to the sector when a threat turns into a fault.

A *logic gate* is a node in the fault tree that determines how faults propagate up the tree. Logic gates are diamond-shaped nodes in the fault tree. Three logic gates are shown in Figure 5.3. One is an OR gate, and the other two are AND gates. In the case of an OR gate, the occurrence of one or more faults causes a fault to propagate up the tree. In the case of an AND gate, all threats connected to the AND gate must occur for the fault to propagate up the tree.

The branches of the fault tree connect threats to logic gates and logic gates to components as shown in Figure 5.3. Faults initiated by one or more threat work their way up the fault tree according to the branches and logic gates. The sector fails only when a fault reaches the root of the fault tree. So the

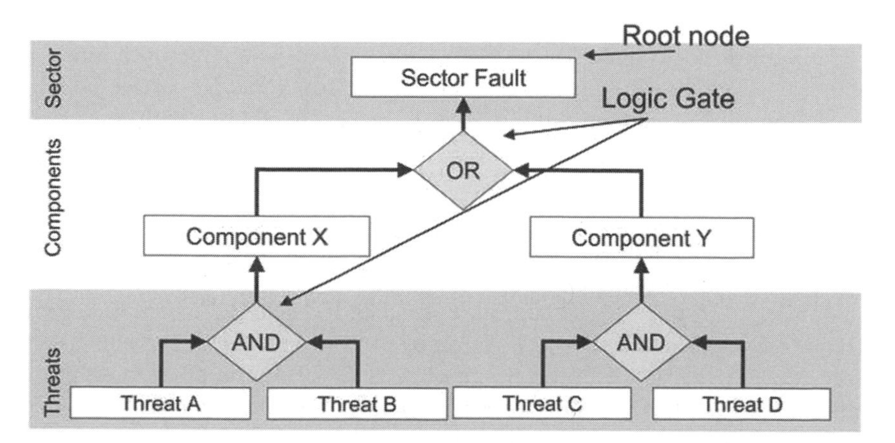

Figure 5.3. A "standard" fault tree has three layers: the root representing the sector, the intermediate component layer, and the threat layer.

root of the fault tree represents sector-wide failure—should a fault propagate to the root of the tree. In the forthcoming analysis, we will be interested only in faults that reach the root node. All others will be ignored. Therefore, a sector failure is defined as one or more faults that propagate all the way to the root of the fault tree.

For example, the fault tree of Figure 5.3 can only fail if either Threat A AND Threat B together cause a fault, OR Threat C AND Threat D cause a fault, or all four cause a fault. The sector does not fail due to a fault if only one of the four threats succeed.

Logic Gates

Logic gates model the relationships among faults in a fault tree. They determine how faults emanating from threat nodes propagate up the layered fault tree and whether the entire sector fails. For example, a bomb attack and a cyber-attack may or may not lead to a sector failure, depending on the logical structure of the components in the sector. Because logic is sometimes counter-intuitive, suppose we examine some simple fault trees and see how logic gates determine the impact of faults on the overall sector.

Figure 5.4 illustrates two simple fault trees—one with an AND logic gate, and the other with an OR logic gate. In Figures 5.4(a), a bomb attack AND cyber-attack must both occur before the sector fault occurs. Thus AND equates with *both*. If only one or the other threat succeeds, the sector remains secure. There is no sector fault.

Alternatively, in Figures 5.4(b), a bomb attack OR a cyber-attack, or both, leads to a sector failure. Only one of the conditions must be TRUE for sector failure to occur. The logic gate OR means **one or the other or both** conditions must be true. If both are FALSE, then OR means there is no propagation of failure to the next level in the fault tree.

The Boolean logic of AND and OR is summarized by simple truth tables (see Table 5.1).[1] In the case of an AND relationship, both A and B must be TRUE. In the case of an OR relationship, only one must be TRUE. But both may also be TRUE, which leads to the same event.

We assume events A and B occur independently. That is, one does not depend on the other, nor does the probability of one occurring affect the probability of another occurring. In fact, the sum of all vulnerabilities in a fault tree may exceed 100%, because they are independent events.

In Table 5.1, A is evaluated and B is evaluated. If A and B are both true, then (A AND B) is evaluated as true. This is shown by the truth table entry at the intersection of the row, "B is true" with the column, "A is TRUE." The other three possible evaluations of (A AND B) are shown as entries in the AND truth table. Similarly, the OR truth table enumerates all possible evaluations of (A OR B). Note that

[1]Named after George Boole, 1815–1864, Ireland. Boolean variables A and B can take on only one of two values: true or false.

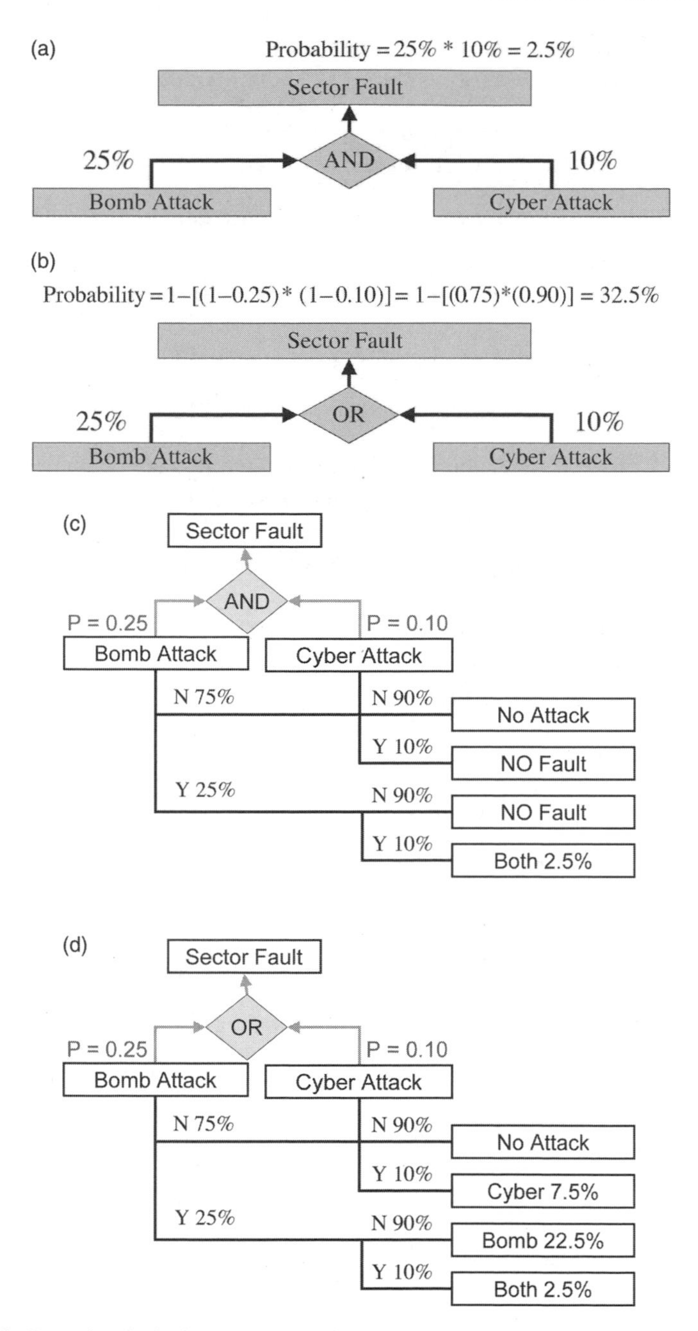

Figure 5.4. Two simple fault trees and their event trees: (a) Two threats connected by an AND logic gate, (b) two threats connected by an OR logic gate, (c) event tree obtained from (a), and (d) event tree obtained from (b). (a) AND means both faults must occur to propagate up the tree. (b) OR means one, two, or all faults may occur to propagate up the tree. (c) Event tree for the fault tree of (a). (d) Event tree for the fault tree of (b).

TABLE 5.1. Truth Tables for AND and OR Logic: Two Events A and B are Independently Considered to be Either TRUE of FALSE. Thus, Four Possibilities Exist.

AND	A is TRUE	A is FALSE
B is TRUE	TRUE	FALSE
B is FALSE	FALSE	FALSE

OR	A is TRUE	A is FALSE
B is TRUE	TRUE	TRUE
B is FALSE	TRUE	FALSE

both A and B must be false for (A OR B) to be false. Both must be true for (A AND B) to be true.

Boolean variables like A and B in Table 5.1 can only take on one of two values, TRUE and FALSE. Truth values, TRUE and FALSE, may be substituted by YES and NO, meaning the threat either occurs or does not. Similarly, binary arithmetic is a modified form of Boolean logic where variables take on values of only 0 and 1.0.[2] All assignment of values in Boolean logic, whether they are TRUE/FALSE, YES/NO, or 0/1.0, produce logical results according to Table 5.1. We will use TRUE/FALSE and YES/NO interchangeably here.

What is the motivation for this formalism? As we shall soon demonstrate, the interactions among threats can produce complex combinations that a simpler analysis will not uncover. Only formal logic is guaranteed to reveal all possible combinations of faults that may lead to sector-wide failure. Fault tree analysis may seem too complicated, but it will reward the analyst with a complete answer. No combination will be overlooked.

Probability

Faults do not occur with certainty. In most cases, they are considered unexpected events that occur with a degree of uncertainty as measured by the vulnerability of a component to a certain threat. Thus, the fault tree needs to also capture the probability—vulnerability—of each threat succeeding.[3] In most cases, fault probabilities will be small fractions—expressed as a percentage—quantifying each vulnerability as a fault probability. The fault trees of Figure 5.4 illustrate the use of fault probability as a measure of each vulnerability. In this section we will use these probabilities of individual threats causing single faults to compute the overall sector vulnerability. That is, we can calculate the sector vulnerability from the individual component vulnerabilities and the logic gates embedded in the fault tree.

[2] $0+0=0$, $0+1=1$, $1+1=0$ with a carry of 1, $1+0=1$ in binary arithmetic.

[3] We interpret *probability* as a "partial truth," that is, a number between zero (FALSE) and 1.0 (TRUE), which measures the level of certainty that an event will occur. A probability of 50% means we are equally uncertain as to whether an event will occur. A probability of 25% means we are relatively sure that an event will not occur.

TABLE 5.2. Probability Formulas for AND and OR Joint Probabilities: Two Events A and B Independently Occur with Probability P(A) and P(B).

AND	P(B)
P(A)	P(A) × P(B)

OR	P(B)
P(A)	1.0 − [(1.0 − P(A)) × (1.0 − P(B))]

Vulnerabilities propagate—level-by-level—starting from the threat nodes of the fault tree through logic gates and intermediate levels, to the top of the tree. At each gate, we apply one of the rules summarized in Table 5.2. In the case of an AND gate, the probabilities of independent vulnerabilities are multiplied to get the vulnerability at the next level. In the case of the OR gate, the arithmetic is more complicated, because the vulnerability of (A OR B) propagating to the next level is the sum of the probabilities of all combinations of A, B, and (A AND B). But a shortcut to calculating this sum—using *De Morgan's Law*—leads to the vulnerability at the next level, much quicker. **Augustus De Morgan** was a 19th-century English mathematician who taught at University College London.

Here is an explanation of De Morgan's Law.[4] Whenever an OR logic gate is used, we must consider all possible events generated by independent faults. All enumerated possibilities must include what can happen, as well as what cannot happen! Suppose there are only two independent faults, A and B:

I. A occurs with probability, P(A); B occurs with probability, P(B)
II. A occurs with probability, P(A); B does not occur with probability, 1.0 − P(B)
III. A does not occur with probability, 1.0 − P(A); B occurs with probability, P(B)
IV. A does not occur with probability, 1.0 − P(A); B does not occur with probability, 1.0 − P(B)

TOTAL = 1.0

The first three of these events apply when A or B is TRUE, because if A, B, or both are TRUE, then (A OR B) is TRUE. Only the last event, IV, fails this test. In other words, the probability of (A OR B) is the sum of the probabilities in I, II, and III above. Another way to compute this is to subtract the probability of IV above from 1.0, because the sum of probabilities of I, II, III, and IV is 1.0. That is, the

[4]De Morgan's Law says that (A OR B) is logically equivalent to NOT[NOT(A) AND NOT(B)]. In probabilistic terms, this is 1.0 − [Probability(NOT(A)) × Probability(NOT(B))]. Substitution of 1 − Probability(A) in place of Probability(NOT(A)) and similarly for Probability(NOT(B)), then Probability(A OR B) = 1 − [(1 − Probability(A)) × (1 − Probability(B))].

probability of I, II, and III sums to 1.0 minus the probability of IV:

$$\begin{aligned}\text{Probability (A OR B)} &= \text{Probability (I, II, III)} = 1.0 - \text{Probability (IV)} \\ &= 1.0 - [(1.0 - \text{Probability (A)) AND } (1.0 - \text{Probability (B))}] \\ &= 1.0 - [(1.0 - \text{Probability (A)}) \times (1.0 - \text{Probability (B)})].\end{aligned}$$

Table 5.2 summarizes the calculations used to label the entire fault tree with fault probabilities—starting from the bottom (threats) and working up to the root or apex of the tree. As an example of using Table 5.2 in a rudimentary application, suppose two coins, a nickel and a dime, are tossed, and we ask, "What is the probability that the nickel is a HEAD AND the dime is a HEAD?" Assuming each coin is fair, then a HEAD occurs with probability of 50% and a TAIL occurs with probability 50%, for both coins. Tossed at the same time, the two coins (Nickel, Dime) will produce one of four events: (HEAD, HEAD), (HEAD, TAIL), (TAIL, HEAD), or (TAIL, TAIL). Each of these occurs with an equal likelihood of 25%. But only one (HEAD, HEAD) occurs with a probability of 25%. Using the formula for (A AND B) in Table 5.2, we calculate the Probability (Nickel = Head AND Dime = Head) as 50% × 50% or 25%.

Now suppose we ask, "What is the probability that the nickel is a HEAD OR the dime is a HEAD?" In this case, Table 5.2 says to calculate the Probability (Nickel = HEAD OR Dime = HEAD) = 100% − [(100% − 50%) × (100% − 50%)] = 100% − 25% = 75%. Here is why: Three of the possible events produce the desired result of (Nickel = HEAD) OR (Dime = HEAD). They are (HEAD, HEAD), (HEAD, TAIL), and (TAIL, HEAD). Only the (TAIL, TAIL) event is false, because neither (Nickel = HEAD) nor (Dime = HEAD). Three out of four equals 75%.

Figure 5.4(b) shows the result of applying De Morgan's Law to the example in Figure 5.4(a). Given the vulnerability to a bomb attack of 25% and the vulnerability to a successful cyber-attack of 10%, what is the vulnerability of the sector? De Morgan's Law is used because of the OR logic gate that connects the two threats to the root of the fault tree. Therefore, the sector vulnerability is $1.0 - (1.0 - 0.25)\ (1.0 - 0.10) = 1.0 - (0.75)(0.90) = 1.0 - 0.675 = 0.325$, using decimals instead of percentages. Converting back to percentages, the sector vulnerability is 32.5%.

How is it possible for the sector vulnerability to be greater than either one of the threat vulnerabilities? The sector vulnerability of 32.5% is greater than 25% or 10% because OR logic means there are three events possible: either the bomb attack, the cyber-attack, or the combination of bomb and cyber-attacks. It is intuitively obvious that the probability of three events occurring should be greater than the probability of one or the other single event happening. This subtlety must not be overlooked in analyzing vulnerabilities, because in reality, it is possible that one, two, three, or more threats can simultaneously occur. After all, the attacks of 9/11 were three threats that were carried out at the same time!

Step 4: Analyze the Fault Tree Model Using an Event Tree

The foregoing analysis illustrates the complexity of logic even when dealing with a very simple fault tree. It is easy to overlook a combination of events and even easier to make a mistake when calculating combination probabilities. For this reason, all events of a fault tree need to be formally enumerated using an *event tree*. An event tree is simply a binary tree that enumerates all possible combinations of events generated by the fault tree. It is a binary tree, because each fault is considered either TRUE (YES it occurred) or FALSE (NO it did not occur).

Figures 5.4(c) and 5.4(d) show the event trees for the fault trees in Figures 5.4(a) and 5.4(b). Notice how *all possible events are enumerated by the event tree*. That is, to be complete, we must consider all single events (BOMB, CYBER), as well as double events (BOMB and CYBER), as well as the possibility that no event will occur. All possible events are enumerated in Figures 5.4(c) and 5.4(d) as the leaves of a binary tree extending from the upper left (BOMB) down and across to the right-hand side of the event tree. The event tree is binary, because each branch represents TRUE (YES, Y), or FALSE (NO, N), to the question, "did the threat occur?" Thus, the BOMB threat occurs with a probability of 25%, but it does not occur with a probability of 75%. Similarly, the CYBER event occurs with a probability of 10%, but it does not occur with a probability of 90%.

The event tree also reflects the logic of the fault tree. For example, in Figure 5.4(c), the AND gate in the fault tree means that both BOMB and CYBER attacks must succeed for the sector fault tree to fail. In other words, a fault only occurs when BOMB AND CYBER attacks succeed. Therefore, only one event (BOTH 2.5%) is meaningful, because it is the only event that leads to a fault.

The AND gate in Figure 5.4(c) means that only one event leads to sector failure. In addition, the failure is caused by a combination event: both BOMB and CYBER, with a probability of 2.5%. This number matches the sector vulnerability computed in Figure 5.4(a).

Figure 5.4(d) is quite different because the event tree of an OR-Fault tree allows many more possibilities. The event tree in this case contains two single events and one double-event outcome: CYBER, BOMB, and CYBER + BOMB. The most likely event is a BOMB attack, whereas the least likely event is both.

In addition, notice in Figure 5.4(d) that the single events BOMB and CYBER occur with lower probability than the vulnerabilities listed in the fault tree (22.5% versus 25% for BOMB, and 7.5% versus 10% for CYBER). Why is this? The event tree outcomes include all possible single, double, triple, and so on combinations as well as the possibility that no event occurs. In other words, when a BOMB-only event occurs, it means that a CYBER attack does not occur. The probability of a nonevent must be factored into the event tree outcomes, which tends to lower the single-event probabilities.

The sector vulnerability computed in Figure 5.4(b) is 32.5%. The sum of all event probabilities in the event tree must also add up to 32.5%. This is a way to check the

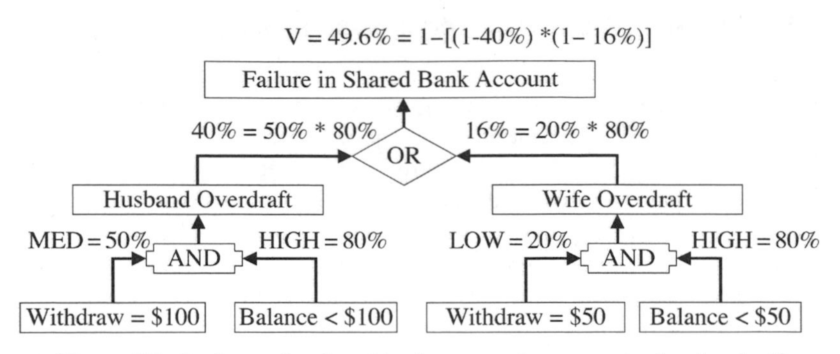

Figure 5.5. Fault tree for shared bank account between a husband and wife.

correctness of the event tree calculation:

$$32.5\% = 7.5\% + 22.5\% + 2.5\%.$$

The event tree enumerates all possibilities including nothing, single, double, and multiple events. It can give surprising results. For example, what is the most likely event to occur in Figure 5.4(d), if the vulnerability to a BOMB is 70% and the vulnerability of a CYBER attack is 80%? In this case, the event probabilities are changed:

$$\begin{aligned} \text{CYBER} &= 80\% \times 0.3 = 24\% \\ \text{BOMB} &= 70\% \times 0.2 = 14\% \\ \text{BOTH} &= 0.7 \times 0.8 = 56\% \leftarrow \text{Most Likely Event.} \end{aligned}$$

In this case, the most likely outcome is a combination event: both BOMB and CYBER. This result may be nonintuitive, but it is reality. The probability of a single event is reduced by the probability of the other event NOT occurring, which is a small number when vulnerabilities are high (over 50%). The terrorist attacks of 9/11 were combination events. If it had been known that each of the multiple attacks was more than 50% likely to succeed, the corresponding event tree would have indicated that multiple attacks were most likely to occur.

Now we use a more challenging fault tree to show how an event tree is an extremely effective means of analyzing *all possible combinations of events* that may occur in a realistic situation. Although the fault tree of Figure 5.5 seems relatively simple, it will soon become clear that it models a much more complex system than is immediately obvious. It will also reveal shortcomings in the fault tree method.

The Shared Bank Account Fault Tree

The simple fault tree of Figure 5.5 models a common infrastructure used by married couples that share the same bank account. As long as the shared bank account balance stays positive, the couple's "banking and finance sector" is secure. But if the bank balance drops below zero, the "sector" fails. So a fault in this system is defined as a negative balance.

In this somewhat amusing example, the infrastructure is modeled as a system containing withdrawal components and threats in the form of transactions that result in a negative bank balance.[5] A negative balance "fault" can occur if one or the other, or both, partners withdraw money from the account, and the balance becomes negative—an overdraft. But even more subtle "failure modes" are difficult to identify unless we use an event tree to analyze all possible events. Although somewhat tedious, event tree analysis is important because it is exhaustive.[6]

The fault tree of Figure 5.5 uses AND/OR logic to model what can happen to the shared account. For example, if the husband withdraws $100 AND the balance is less than $100, the account succumbs to a fault called "Husband Overdraft" (H-Over). Similarly, the "Wife Overdraft" (W-Over) fault occurs when the wife withdraws $50 AND the balance is less than $50. These faults are called *events* or *outcomes*. Events occur with a certain probability as described in Tables 5.1 and 5.2. These probabilities correspond with vulnerabilities when modeling an infrastructure, and the actions of both husband and wife (withdrawals) correspond with threats.

The goal of fault tree analysis is to expose all possible events to determine if they lead to sector failure. Given a set of conditions such as the starting balance of the joint account and the actions of husband and wife, we work through the fault/event tree to explore all possible events. Once we have enumerated them we can decide if they are faults (overdrafts) and, furthermore, if they bring the entire sector down (lead to damages).

For each set of initial conditions, we work all the way through the fault/event tree before considering another set of conditions. That is, we do not allow "simultaneous" actions that happen at precisely the same time. In addition, we assume independence among the threats—each fault probability is independent of all others. This will lead to some limitations, as shown. However, it is a simplification that is necessary for fault/event tree analysis to work.

The vulnerability (fault probability) of each component in the fault tree is computed and attached to the fault tree as shown in Figure 5.5, using the rules of AND and OR logic described previously. For example, if the probability of the husband withdrawing $100 is 50% and the probability of the initial bank balance being less than $100 is 80%, then the probability of a "Husband Overdraft" (H-Over) is $(50\%)(80\%) = 40\%$. Both must occur to reach the "Husband Overdraft" (H-Over) component because of the AND logic. Similarly, "Wife Overdraft" (W-Over) is true with a probability of 16%, because of the AND logic connecting the "Withdraw" and "balance < $50" threats.

[5]OK, so it is not so amusing when your bank account is overdrawn, but this real-world example serves to illustrate subtleties in even simple systems.

[6]It is not necessary to perform manual calculations as done here. Program *FTplus.html* (FT.jar) computes event tree events automatically.

Now move up a layer in the fault tree of Figure 5.5. A fault propagates up the tree when **either** husband **or** wife **or** both overdraft the account. This is modeled as an OR gate in the fault tree. Applying the OR formula from Table 5.2:

$$\text{Probability (A OR B)} = 1.0 - [(1.0 - \text{Probability (A)}) \times (1.0 - \text{Probability (B)})]$$

where A = Husband Overdraws; B = Wife Overdraws.

Probability (Husband Overdraws) OR Probability (Wife Overdraws) = 100% − [(100% − 40%) × (100% − 16%)] = 1.0 − [0.6 × 0.84] = 49.6, or approximately 50%. In other words, the probability that the shared bank account will fail is approximately 50%. There are three failure modes: Husband Overdraft, Wife Overdraft, or both overdraft. That is, the bank account fails—one way or another—with a probability of 49.6%, which may seem counter-intuitive, because 40% + 16% add up to 56%. But the probability of *H-Over* OR *W-Over* is *not* the sum of each independent event. It is the sum of three events: H-Over, W-Over, and both H-Over and W-Over.

But the fault caused by one or the other—or both husband and wife withdrawing too much from their shared bank account—is only three of several failure modes in this banking sector. The question is, "what other faults are there, and how do we find them all?" To enumerate them all, we need to transform the fault tree into an event tree.[7]

An *event tree* is a binary tree (two branches at each node) that enumerates all possible events of a fault tree. Because each threat can take on the logic values of TRUE or FALSE, we must exhaustively consider all combinations of TRUE and FALSE logic value combinations in the fault tree. This is done by brute force—each threat is alternately considered to be TRUE and then FALSE. And for each TRUE event we must consider all TRUE and FALSE combinations of all other threats. Similarly, for each FALSE, we must also consider both TRUE and FALSE events for all other threats. Taken all together, there are 2^k combinations of k threats, assuming each can be either TRUE or FALSE. The exhaustive enumeration of all combinations can be visualized as a binary event tree (see Figure 5.6).

Figure 5.6 contains the same data as the fault tree of Figure 5.5, but with every possible combination of TRUE/FALSE (YES/NO) enumerated and shown. This is a "binary tree within a binary tree," because each level of the tree divides further into another binary tree and so on until all possible combinations have been enumerated. The left-most threat (H-Withdraw = $100) is placed at the root of the event tree, and then one branch represents (H-Withdraw = $100 is TRUE or YES) and the other branch represents (H-Withdraw = $100 is FALSE or NO). This pattern is repeated for the second threat (Balance < $100), and so on, until all threats have been enumerated. Thus, each additional threat adds two more branches: one for YES and one for NO.

[7] A truth table could be used instead. An event tree was chosen because it is visual—the reader can see all possible events produced by assuming each fault is alternately true and then false.

MED = 50% | HIGH = 80% | LOW = 20% | HIGH = 80%
H-Withdraw = $100 | Balance < $100 | W-Withdraw = $50 | Balance < $50

H-Withdraw = $100	Balance < $100	W-Withdraw = $50	Balance < $50	Vulnerability
N 50%	N 20%	N 80%	N 20%	OK
			Y 80%	OK
		Y 20%	N 20%	OK
			Y 80%	W-Over 1.6%
	Y 80%	N 80%	N 20%	OK
			Y 80%	OK
		Y 20%	N 20%	OK
			Y 80%	W-Over 6.4%
Y 50%	N 20%	N 80%	N 20%	OK
			Y 80%	Impossible
		Y 20%	N 20%	Unknown 0.4%
			Y 80%	W-Over 1.6%
	Y 80%	N 80%	N 20%	H-Over 6.4%
			Y 80%	H-Over 25.6%
		Y 20%	N 20%	H-Over 1.6%
			Y 80%	Both-Over 6.4%

Figure 5.6. Event tree for the fault tree of Figure 5.5 enumerates all possible events.

Each branch in the event tree is labeled with the probability that the fault will occur (YES) or not (NO). These probabilities come from the vulnerabilities listed in the fault tree. The probability of YES plus the probability of NO must add up to 100%, so the YES branch is taken with probability equal to the vulnerability associated with the threat, and the NO branch is taken with probability of one minus the threat vulnerability. For example, the probability of (Balance < $100) is 80%, so the probability that this is FALSE or NO is (100% − 80%) = 20%. The entire event tree and the probability of YES and NO branches are computed in this way, until all branches are labeled with a probability.

A binary event tree with k threats generates 2^k events, because there are 2^k possible combinations. These are the events we want to investigate to see whether they cause a sector fault. In other words, all possible events must be studied to determine whether the logic of the fault tree allows any one of the event tree events to propagate to the root of the fault tree. Although there are 2^k events in an event tree, not all lead to a failure of the entire sector. That is, not all events represent a fault that propagates to the root of the fault tree.

The probability of each of the 2^k events occurring is computed by multiplying the probabilities along a path from the event tree root to the event. For example, in

Figure 5.6, the probability of the event labeled "W-Over 6.4%" is computed by multiplying (50%)(80%)(20%)(80%) = (0.5)(0.8)(0.2)(0.8) = 0.064 = 6.4%. Therefore, this combination event occurs with a probability of 6.4%.[8]

The calculation of event probabilities is important but subtle, so let us examine the other events shown in Figure 5.6. In each case, we compute the probability of each event by following a path from the top left-hand corner of the event tree through the layers of branching, to the event of interest. Consider these detailed examples, starting with the "W-Over 1.6%" event in Figure 5.6 and working down the vulnerability column:

Event	Path from Root	[Comment]
W-Over 1.6%	(N 50%) (N 20%) (W-Withdraw $50 20%) (Balance < $50 80%)	
W-Over 6.4%	(N50%) (Balance < $100 80%) (W-Withdraw$50 20%) (Balance < $50 80%)	
Impossible	(H-Withdraw 100%) (Balance > = $100 20%) (N 80%) (Balance < $50 80%)	 [Balance >= $100 AND Balance < $50!]
Unknown	(H-Withdraw 100%) (Balance >= $100 20%) (W-Withdraw $50 20%) (Balance >= $50 20%)	[Deduct $100] [Deduct another $50] [Balance >= $100 AND Balance >= $50!]
W-Over 1.6%	(H-Withdraw 100%) (N 20%) (W-Withdraw $50 20%) (Balance < $50 80%)	
H-Over 6.4%	(H-Withdraw 100%) (Balance < $100 80%) (N 80%) (N 20%)	
H-Over 25.6%	(H-Withdraw 100%) (Balance < $100 80%) (N 80%) (Balance < $50 80%)	

[8]A combination event is an event made up of a single, double, or triple event all at the same time.

Event	Path from Root	[Comment]
H-Over 1.6%	(H-Withdraw 100%)	[Independent Events]
	(Balance < $100 80%)	
	(W-Withdraw $50 20%)	[Independent Events]
	(Balance >= $50 20%)	
Both-Over 6.4%	(H-Withdraw 100%)	
	(Balance < $100 80%)	
	(W-Withdraw $50 20%)	
	(Balance < $50 80%)	

The probability and logic values come from the fault tree, but the event probabilities come from the event tree. Every "yes = TRUE" and "no = FALSE" decision must add up to 100%, so events that occur with, say, a probability of 80% do not occur with probability of 100 − 80 = 20%. Furthermore, the event tree probabilities are always computed as the product of all probabilities along the path from upper left corner to event. De Morgan's Law only applies to the fault tree, never to the event tree.

The event tree for this example was constructed by drawing the first two branches from the left-most threat—"H-Withdraw = $100." One branch is labeled NO with a probability of 100% − 50%, and the other is labeled YES with a probability of 50% taken from the fault tree. Moving left to right in Figure 5.6, the threat, "Balance < $100" occurs with a probability of 80% and does not occur with a probability of 100% − 80% = 20%. So we draw two trees below this threat—one for the branch labeled NO and another identical subtree for the branch labeled YES. At this point, the event tree has four events. But we have two more threats to consider.

Continuing left to right, a binary tree is drawn for each threat and duplicated for each branch preceding it. This is repeated until all threats have been considered. This event tree contains 16 events, because there are 4 threats. Had it contained 5 threats, there would have been 32 events. In general, an event tree derived from a fault tree with k threats has 2^k events. In other words, there are 2^k branches at the lowest level of the event tree.

Now look at the events listed in the far right column of Figure 5.6. This column enumerates all possible events of the fault tree. There are no more or less than these events, because the event tree enumerates all events. Note that the first *OK* in this column means that no fault propagates along the path from the threat: (H-Withdraw = $100 is FALSE, Balance < $100 is FALSE, W-Withdraw = $50 is FALSE, Balance < $50 is FALSE), to the top of the fault tree. The shared bank account sector does not fail because of this event.[9]

Continuing down the column in Figure 5.6, note the event labeled *Impossible*. Why is this event impossible? It must be reached from the path: (Husband

[9]Perhaps now the reader realizes that a combination event is simply a single, double, triple, and so on combination of one, two, three, and so on threats.

Withdraw = \$100) is FALSE, (Balance < \$100) is FALSE, (Wife Withdraw = \$50) is FALSE, (Balance < \$50) is TRUE. But, it is not possible for (Balance < \$100) to be FALSE AND (Balance < \$50) to be TRUE, because the balance cannot be greater than \$100 and less than \$50 at the same time. So this event is *Impossible*. This points out why event trees are so important. It is unlikely that a casual examination of the fault tree would expose logical inconsistencies such as this.

A Fault Tree Weakness

Before we get over-confident in the power of event tree analysis let us consider the event labeled *Unknown*. *Unknown* illustrates one of the deficiencies of fault tree and event tree analysis. Suppose both husband and wife withdraw money at the same time from the account when it has a Balance of more than \$100. The path (Husband Withdraw = \$100) is TRUE, (Balance < \$100) is FALSE, (Wife Withdraw = \$50) is TRUE, (Balance < \$50) is FALSE leads to an indeterminate event. That is, we do not know what happens in this case. Here is why.

Both husband and wife simultaneously withdraw \$100 and \$50, respectively, so the total withdrawal amount is \$150. Assume the account balance is greater than \$100, which means it is also greater than \$50. Furthermore, assume that we do not know if the balance is at least \$150, which is enough to cover the combined withdrawals of \$150. It could be \$125, \$149, \$151, or more and still make (Balance < \$50) and (Balance < \$100) FALSE. If the starting balance is greater than \$150, then the event is OK. If it is greater than \$100 but less than \$150, the event is either H-Over or W-Over, but we do not know which! This event is *Unknown*.

So the existence of a fault propagation along the path (Husband Withdraw = \$100) is TRUE, (Balance < \$100) is FALSE, (Wife Withdraw = \$50) is TRUE, (Balance < \$50) is FALSE leading to the top of the fault tree is unknown, because we do not know whether the original account balance is more or less than \$150. To prove this, the reader should trace the path from the threat to the event using an initial balance of \$125. Then trace it again using a starting balance of \$200. Why do these two cases give different results?

Fault Distribution Histogram

Most of the events listed in an event tree will be "OK"; that is, they will not lead to a fault or failure in the overall fault tree. Some lead to the same failure; e.g., "W-Over" and "H-Over" appear several times in Figure 5.6. Even though there are 2^k events from an event tree, many of them will be OK or identical to one another. The next step, then, is to collapse the events into a more compact form—the *fault distribution histogram* as shown in Figure 5.7. A fault distribution histogram summarizes the results obtained from the event tree analysis. It is displayed in the form of a histogram, because we want to use it in the final step of MBVA, which aims to reduce the probability of each event to zero, or at least to minimize these probabilities.

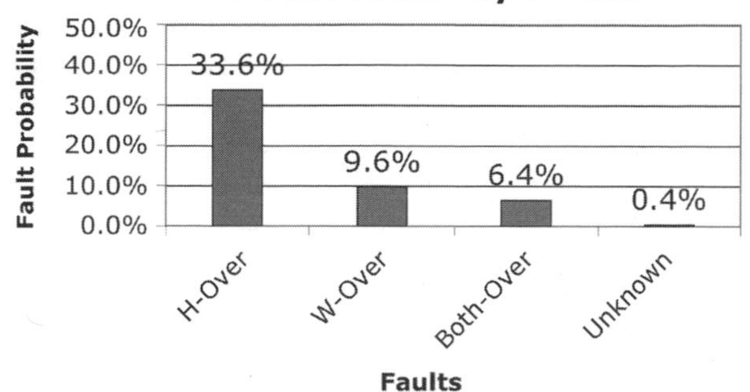

Figure 5.7. A fault distribution histogram derived from the event tree of Figure 5.5 lists all faults and/or threats and their probability of occurring.

The fault histogram plots the fault probability (combination event probability) on the vertical axis versus the event that leads to a sector fault along the horizontal axis. In fact, the sector fault may be due to a single, double, triple, and so on event, so the fault distribution histogram will contain all single and combination events that make the sector fail. Figure 5.7 lists all faults and their probabilities of occurring, i.e., the threats confronting the shared bank account.

A fault distribution histogram is constructed by combining similar or identical events from the event tree into distinct faults, and eliminating the events that are of no interest, such as *OK* and extremely low-probability events. In the shared bank account example, the event tree events are *OK*, *Impossible*, *Unknown*, *H-Over*, *W-Over*, and *Both-Over*. We can eliminate the *OK* event because it represents no fault. This leaves four faults to consider further: *Unknown*, *H-Over*, *W-Over*, and *Both-Over*. Even though there were 16 possible events enumerated in Figure 5.6, only 4 came out of the analysis. The fault distribution histogram incorporates only these 4 faults.

The next step is to consolidate the events of interest by adding together the probabilities shown in the event tree. For example, the *H-Over* events are consolidated into one *H-Over* fault with a probability equal to the sum of the *H-Over* events shown in Figure 5.6: *H-Over* 33.6% = 6.4% + 25.6% + 1.6%. Similarly, *W-Over* 9.6% = 1.6% + 6.4%+ 1.6%. Ranking the events in order from highest to lowest yields:

W-Over: 36.0%

H-Over: 9.6%

Both-Over: 6.4%

Unknown: 0.4%

The results shown in Figure 5.7 are what we have been looking for—a list of shared banking account vulnerabilities. In this example, the shared bank account sector is 33.6% likely to fail due to the "Wife Withdraws $50" threat; 9.6% due to "Husband Withdraws $100"; 6.4% vulnerable to "Both Over" threat; and 0.4% vulnerable to an "Unknown" threat. This information sets the stage for creating a strategy of prevention based on investing a fixed amount of money to fix these vulnerabilities.

VA leads to the fault histogram shown in Figure 5.7. It ties together threats and vulnerabilities, plus their interactions with one another by modeling the sector as a fault tree. But this is only one half of the complete MBVA process. We are not done yet!

The next step in MBVA is to form an investment strategy based on the results obtained and summarized in the fault distribution histogram. But before leaving this discussion, we will expand on this introduction to fault trees and examine several interesting fault and event trees for the purpose of gaining greater insight into their usefulness.

MATHEMATICAL PROPERTIES OF FAULT TREES

This section contains a rigorous analysis of some common types of fault/event trees, but because of its mathematical nature, the reader may want to skip to Chapter 6. Four results can be summarized from this analysis:

1. The vulnerability of a fault tree (sector vulnerability) is computationally identical to the sum of the vulnerabilities of its corresponding event tree. This fact is useful for checking the correctness of calculations.
2. The vulnerability of an OR-tree[10] is computed as $V = 1.0 - \Pi_1^k(1.0 - p_i)$, where Π is the product of all terms multiplied together, and p_i is the probability associated with each threat. As the number of threats, k, increases without bound, V tends toward 100%, or certainty. If at least one threat occurs with probability of 100%, the fault tree vulnerability is also 100%.
3. The vulnerability of an AND-tree[11] is computed as $V = \Pi_1^k(1.0 - prob_i)$. As the number of threats, k, increases without bound, V tends toward 0%, or complete security. As any one of the threat probabilities tends to zero, the fault tree vulnerability tends to zero also. Hence, an AND-tree converges toward complete security "faster" than does an OR-tree.
4. In an arbitrary fault tree, the lower AND gates are in the tree, the more secure (less vulnerable) is the sector. Conversely, the higher OR gates are in a fault tree, the more secure is the sector. This fact is used to make any system more secure by adding redundancy, because redundancy is the same as adding an AND gate between identical components in a fault tree.

[10]An OR-tree is a tree with one OR gate connecting k threats to the root of the fault tree.
[11]An AND-tree is a tree with one AND gate connecting k threats to the root of the fault tree.

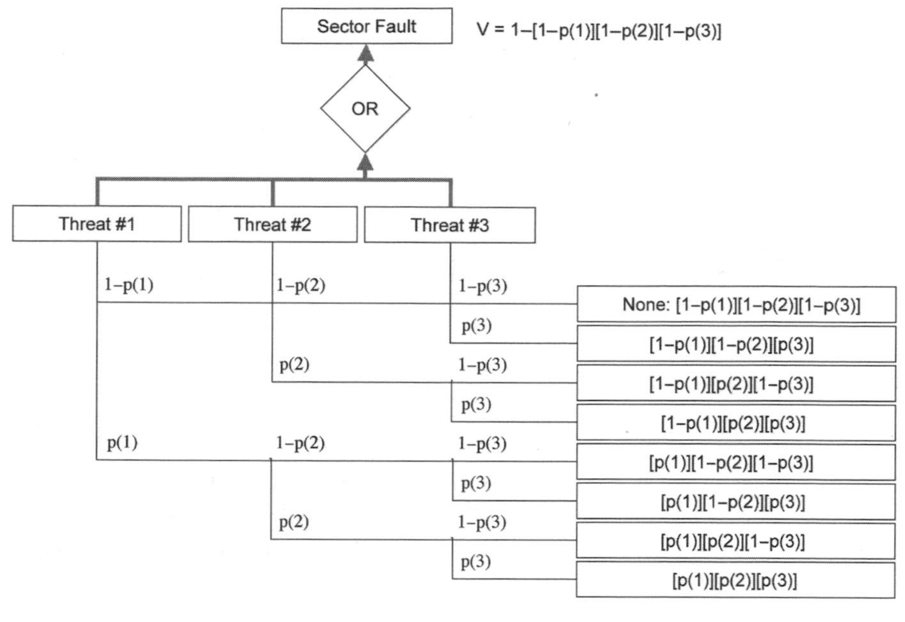

Figure 5.8. OR-tree and its associated event tree containing three threats and the probabilities of all events.

These useful "rules-of-thumb" can be applied to actual sectors analyzed later in this book. The following falls short of a mathematical proof of each of these claims, but it should provide reassurance that these are valid claims. Figures 5.8 and 5.9 will be used to illustrate each claim.

Fault Tree Vulnerability Equals Event Tree

The overall vulnerability of a sector is equal to the vulnerability of the fault tree used to model it. This calculation can be performed in two ways: by adding the fault probabilities of all events emanating from the event tree, and by working up the fault tree. These two results should be identical, which means you can use one to check on the correctness of the other.

Figures 5.8 and 5.9 show this result. In Figure 5.8, the fault probabilities of the event tree, should sum to 100%. In an OR-tree, all but one event leads to a sector failure. Therefore, we subtract the non-failure event probability from 1.0 and obtain the same result as computed from working up the fault tree. Therefore,

$$V = 1.0 - [1.0 - p(1)][1.0 - p(2)][1.0 - p(3)].$$

In general, the vulnerability of an OR-tree is 100% minus the product of all non-failure probabilities. For k threats, each with fault probability of p_i, for i = 1,2,3, . . . k:

$$V = 1.0 - \Pi_1^k(1.0 - p_i).$$

The vulnerability of an OR-tree decreases as the fault probability associated with each threat decreases. The minimum vulnerability, V = 0, occurs when all fault probabilities are zero. Minimization of vulnerability is achieved through minimization of each threat probability.

We define *availability* of component i as follows:

$$a_i = (1.0 - p_i).$$

Availability is the complement of fault probability or vulnerability and is a measure of the dependability or security of a component. An availability of 100% means the component is 100% dependable, whereas an availability of 0% means a component is completely undependable. In terms of availability, the vulnerability of an OR-tree is:

$$V = 1.0 - \Pi_1^k a_i.$$

In the case of an AND-tree, only one event leads to a sector-wide failure; therefore,

$$V = p(1)\,p(3)\,p(3).$$

This is generalized for k threats:

$$V = \Pi_1^k p_i.$$

Each probability is a number between zero and one, so multiplication results in a smaller number than either of the product terms. This means that sector vulnerability gets increasingly smaller and smaller as the number of terms increases. So, in the case of an AND-tree, vulnerability shrinks to zero as the number of vulnerabilities increase! AND-tree vulnerability can be minimized in three ways: *increasing* the number of threats with fault probabilities less than 100%, *decreasing* one or more fault probabilities, and forcing *at least one fault probability to zero*, AND-tree vulnerability is zero if one or more fault probabilities is zero. It is also zero, if the fault tree is extended to include an infinite number of threats of less than 100% fault probability!

OR-Tree Vulnerability Increases with Number of Threats

The OR-tree of Figure 5.8 can be generalized by allowing the number of threats to increase without bound. As k increases, notice what happens to the sector's vulnerability. The term $\Pi_1^k(1.0 - p_i)$ becomes smaller and smaller, because each $(1.0 - p_i)$ is less than one. This means the product term $\Pi_1^k(1.0 - p_i)$ goes to zero, and the sector vulnerability goes to 1.0. Therefore, the vulnerability of an OR-tree sector tends toward 100% (certainty) as the number of threats increases. Even if all vulnerabilities are extremely small, the accumulation of OR-tree vulnerability tends toward certainty as the number grows without bound.

Furthermore, all fault probabilities must be zero to assure the sector of complete safety. Any nonzero fault probability results in an overall OR-tree sector vulnerability of nonzero. The only way to minimize vulnerability in a pure OR-tree is to reduce fault probabilities "across the board."

This effect results in a nonintuitive result when the fault probabilities of individual threats are high, typically greater than 50%. Consider the case of two threats, A and B, with a probability of successful attacks of 70% and 80%, respectively. What is the most likely event in the fault histogram? Is it event B, because 80% is greater than 70%? No! It is the combination of both A and B, because event "A-only" occurs with a probability of $(0.7)(1.0 - 0.8) = (0.7)(0.2) = 14\%$, and event "B-only" occurs with a probability of $(0.8)(1.0 - 0.7) = (0.8)(0.3) = 24\%$. But the probability of event "both A and B" is $(0.7)(0.8) = 56\%$, which is larger than all others. This counter-intuitive result suggests the following practical result: Whenever fault probabilities are large and the fault tree is an OR-tree, expect combination events to be the most likely to occur.

AND-Tree Vulnerability Decreases with Number of Threats

When the AND-tree of Figure 5.9 is generalized by allowing the number of threats, and therefore the number of terms in the equation for its vulnerability, to increase without bound, a nonintuitive result is observed—the AND-tree sector becomes *more* secure—not less! Here is why.

The vulnerability of an AND-tree is computed by multiplying numbers less than 1.0 together:

$$V = \Pi_1^k p_i.$$

The more numbers we multiply together, the smaller V becomes, until it approximates zero. In other words, a sector that is properly modeled as an AND-tree, becomes more secure as we add threats, because each threat has a probability less than 1.0 of occurring, and it takes ALL threats COMBINED to cause a sector failure. This only holds if all fault probabilities are less than one, but greater than zero.

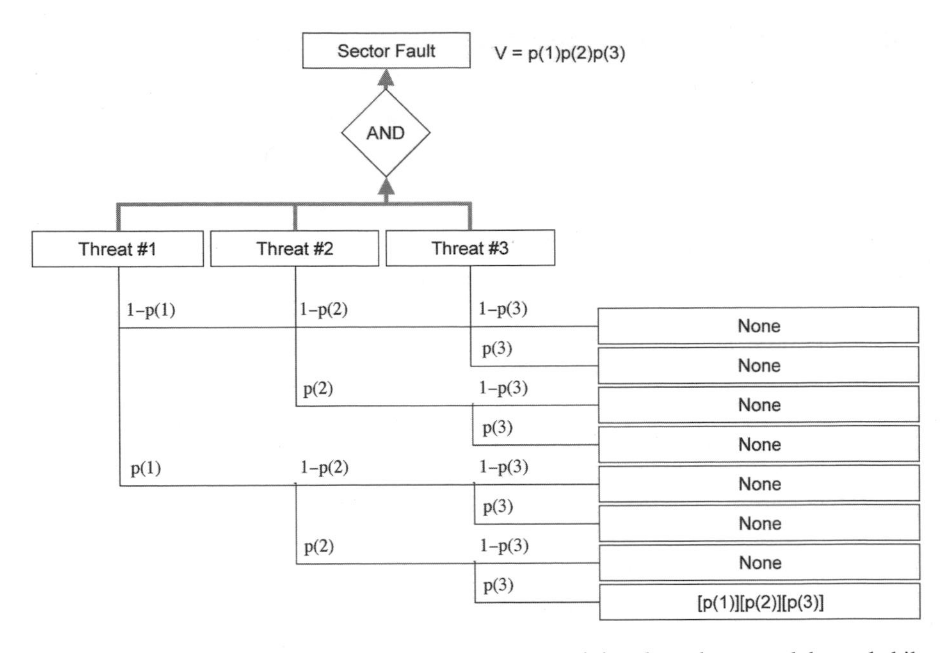

Figure 5.9. AND-tree and its associated event tree containing three threats and the probabilities of only the event that leads to sector failure.

However, if all threats are certain ($p_i = 1.0$), then all terms in the product are 1.0, and multiplication yields 100%; the sector is certain to fail! But if any term is zero, the vulnerability of the sector is once again zero.

Consider a redundant sector whereby a certain component—say a bridge connecting two parts of a city separated by a river—is replicated k times. When $k = 1$ there is only one bridge connecting the two parts of the city, and when $k = 2$ there are two bridges, and so on. Suppose further that the probability of a failure in any of the bridges is identical, say, 50%. The vulnerability of this transportation system is obviously 50% when one bridge is used. When two identical bridges span the river ($k = 2$), sector vulnerability is 50% of 50%, or 25%. When $k = 3, 4, 5$, and so forth, the vulnerability decreases by 50% each time a redundant bridge is added, so when the 10th bridge is added, the vulnerability is one part in a thousand, or approximately one tenth of a percent. As the number of redundant bridges increases, the vulnerability of the bridge system decreases until it approximates 0%. Redundancy is used in this way to dramatically reduce a sector's vulnerability. In many cases, redundancy may be the best way to protect a critical infrastructure, because it may be less expensive to duplicate a component than to harden it.

AND-Gates Are Most Effective at the Lowest Level

What is the effect of combining AND gates with OR gates in the construction of a fault tree? Figure 5.10 will help to answer this question. Which of the two trees

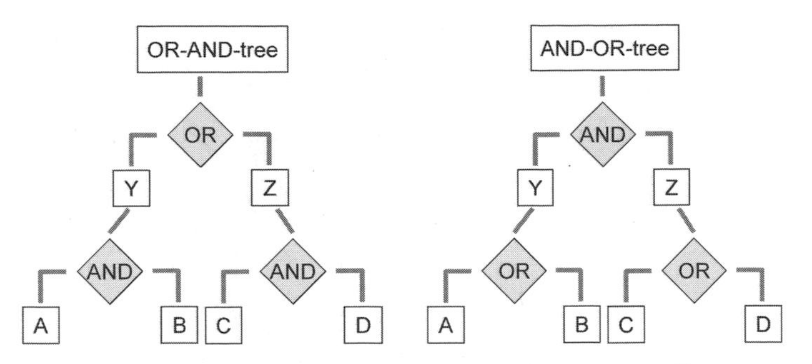

Figure 5.10. Example of an AND-OR-tree and an OR-AND-tree.

shown in Figure 5.10 has the lowest vulnerability, assuming fault probabilities are identical? As it turns out, the lower the AND gates are placed within a fault tree, the lower is the sector's vulnerability. Redundancy does the most good when it is used closer to the threats than at the top of the fault tree.

With some loss of generality, we show that the vulnerability of an OR-AND-tree is equal to or lower than the vulnerability of an AND-OR-tree (see Figure 5.10). In other words, redundancy at the lowest levels of a fault tree does more to reduce vulnerability than redundancy at the top of the tree.

Assume an equal probability of fault for all threats: p_0. The vulnerability of the OR-AND-tree shown on the left of Figure 5.10 is $1.0 - (1.0 - p_0^2)(1.0 - p_0^2)$, and the vulnerability of the AND-OR-tree on the right is $[1.0 - (1.0 - p_0)^2] \times [1.0 - (1.0 - p_0)^2]$. We claim that the vulnerability of the OR-AND-tree is less than or equal to the vulnerability of the AND-OR-tree, so:

$$[1.0 - (1.0 - p_0^2)(1.0 - p_0^2)] <= [1.0 - (1.0 - p_0)^2][1.0 - (1.0 - p_0)^2].$$

Expanding the right-hand side and simplifying the term on the left yields:

$$1.0 - (1.0 - p_0^2)^2 <= 1.0 - 2(1.0 - p_0)^2 + (1.0 - p_0)^4.$$

The "1.0" terms on the left and the right cancel, multiplying by "-1" on left and right reverses the inequality, and $(1.0 - p_0^2)$ factors into $(1.0 - p_0)(1.0 + p_0)$, leading to an expression containing $(1.0 - p_0)$ terms:

$$(1.0 - p_0)^2(1.0 + p_0)^2 > 2(1.0 - p_0)^2 - (1.0 - p_0)^4.$$

Divide by $(1.0 - p_0)^2$ to simplify:

$$(1.0 + p_0)^2 > 2 - (1.0 - p_0)^2.$$

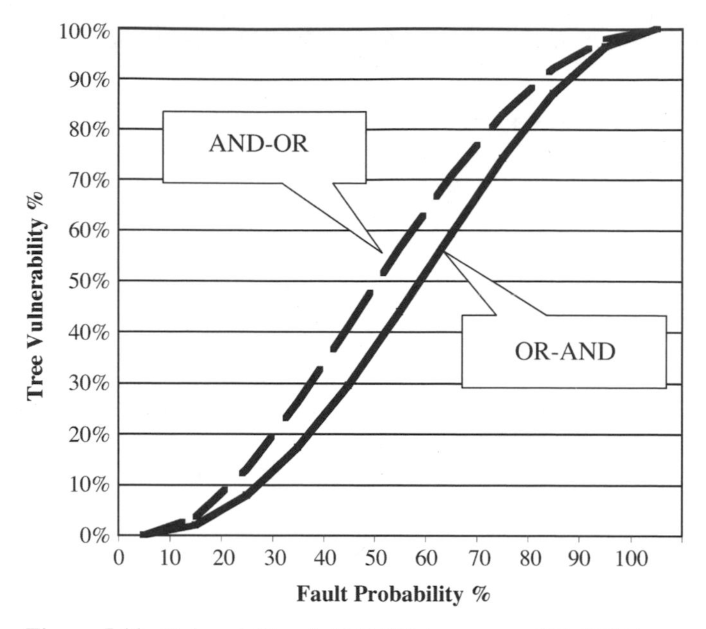

Figure 5.11. Vulnerability of AND-OR-tree versus OR-AND-tree.

Expanding and gathering terms yields:

$$2\,p_0^2 > 0$$

or

$$p_0 > 0.$$

Therefore, the conjecture is true when $p_0 > 0$, but what about $p_0 = 0$? In this case, both fault tree vulnerabilities are zero. How do the two fault tree vulnerabilities compare when $p_0 = 1.0$? Once again the two are equal. But for all values in between 0 and 1.0, the OR-AND-tree is more secure than the AND-OR-tree. Therefore, the lower the AND gate in the fault tree, the more secure is the sector.

Figure 5.11 shows how the vulnerability of the OR-AND-tree and the AND-OR-tree vary as fault probability p_0 varies from zero to one. The OR-AND-tree represents a more secure sector, assuming the fault probabilities are the same for both.

EVENT MATRIX ANALYSIS

Event trees grow geometrically as the number of threats grows linearly. A fault tree with two threats generates an event tree with 4 events; a fault tree with four threats generates 16 events—four times as many. Doubling the number of threats in a fault tree quadruples the number of events that must be evaluated in the event tree. Although a computer can handle the explosion in the number of events that must be considered, the results become unwieldy. In some cases, the probability of an event is so small as to contribute almost nothing to the analysis. Suppose only the most likely events are evaluated and the obscure events are ignored?

In this section we use an abbreviated event tree technique that considers only single and double combination events. All higher order combination events are ignored. This technique works when fault probabilities are small, so that multiplication of three or more results in an even smaller probability. For example, if single faults occur with a probability of 1%, then two occur with a probability of (1%)(1%) or 0.01%. The odds that three occur simultaneously are even more rare: one in a million! Such a rare occurrence may justify the following technique.

Consider the OR-tree of Figure 5.8, again. The fault tree has three threats so there are $2^3 = 8$ events to be analyzed. Of course, one is nonthreatening, so we are left with seven events as enumerated below. Furthermore, suppose the vulnerabilities of each threat are small, and identical, say 5%. The events for the event tree are computed by plugging 0.05 into the equations of Figure 5.8:

1. Threat #3 ~4.5%
2. Threat #2 ~4.5%
3. Threat #2 & Threat #3 ~0.24%
4. Threat #1 ~4.5%
5. Threat #1 & Threat #3 ~0.24%
6. Threat #1 & Threat #2 ~0.24%
7. Threat #1 & Threat #2 & Threat #3 ~0.0125%

The full event tree contains single, double, and triple combination events. But notice that the triple combination event is extremely unlikely; it is only one part in 8000. This unlikely event can be ignored (at our peril), and the analysis can be simplified by considering only single and double combination events. This is known as an *event matrix* analysis, because instead of a tree, we need only a two-dimensional matrix. The event matrix corresponding to Figure 5.8 is shown in Table 5.3.

All threats are listed along the top row and down the left-most column of an event matrix. The vulnerabilities are computed as before, but only for single and double events. Therefore, it is only necessary to evaluate along the diagonal (single events) and across the upper half of the matrix (double events). Triple and higher order combination events are ignored, so we do not need more than a two-dimensional matrix to capture them.

In general, given N threats, the total number of events for single faults is N; for double faults is $N(N+1)/2$; and for triple faults, $N(N^2+5)/6$ (see **Derivation 5.1**). For example, an event tree with $2^6 = 64$ events simplifies to $N(N+1)/2 = 6 \times 7/2 = 21$ events. Although the event tree grows geometrically, the event matrix grows as the square of the number of threats in the fault tree.

TABLE 5.3. General Form of an Event Matrix and the Event Matrix Corresponding to Figure 5.8 Showing Single and Double Combination Events Only.

Threats		Threat #1	Threat #2	Threat #3
	Fault%	p1	p2	p3
Threat #1	p1	p1 × (1 − p2)(1 − p3)	p1 × p2(1 − p3)	p1 × (1 − p2)(p3)
Threat #2	p2		(1 − p1) × (p2)(1 − p3)	(1 − p1) × (p2)(p3)
Threat #3	p3			(1 − p1) × (1 − p2)(p3)
Example		Threat #1	Threat #2	Threat #3
	Fault%	5%	5%	5%
Threat #1	5%	4.50%	0.24%	0.24%
Threat #2	5%		4.50%	0.24%
Threat #3	5%			4.50%

An event matrix is easy to construct: Simply list all threats across the top and down the left-hand column of the matrix. The elements in the matrix representing all $N(N + 1)/2$ single and double combination events are computed as shown in Table 5.3. Single fault probabilities are computed and stored along the diagonal of the matrix, and because the matrix is symmetric, only the upper half elements containing double faults need be computed and stored.

The vulnerability of the fault tree will be higher than the event matrix, because the matrix ignores higher order event combinations. For example, fault probabilities in Table 5.3 sum to 14.22%, whereas the fault tree vulnerability is 14.26%. A significant difference between the two suggests that the event matrix has missed an important event.

DERIVATION 5.1: NUMBER OF EVENTS IN AN EVENT TREE WITH N THREATS.

The number of events of an event tree with N threats is 2^N, which can be derived by noting that the number of combinations of N things, taken in groups of i is:

$$C_i^N = N!/(i!(N - i)!).$$

For example, suppose a certain event tree has $N = 3$ threats: A, B, and C. How many groups of two can be arranged from A, B, and C? Here $i = 2$, so the answer is:

$$C_2^3 = 3!/(2!(3 - 2)!) = 3.$$

They are AB, AC, and BC.

An event tree enumerates all possible single, double, and so on combinations; hence:

$$\#\text{Events(N,N)} = \sum_{i=1}^{N} C_i^N = \sum_{i=1}^{N} C_i^N N!/(i!(N-i)!) = 2^N.$$

But suppose we want to count the number of events for 1, 2, 3-tuples, and so on up to only K-tuples, instead of counting all 2^N events. Then the formula is:

$$\#\text{Events(N,K)} = \sum_{i=1}^{K} C_i^N = \sum_{i=1}^{K} C_i^N N!/(i!(N-i)!).$$

Here are a few:

$K = 1$: $\#\text{Events(N,1)} = \sum_{i=1}^{1} C_i^N \, N!/(i!(N\text{-}i)!) = N!/(1!(N\text{-}1)!) = N.$
$K = 2$: $\#\text{Events(N,2)} = \sum_{i=1} C_i^N N!/(i!(N\text{-}i)!) = N + N!/(2!(N\text{-}2)!)$
$= N(N + 1)/2.$
$K = 3$: $\#\text{Events(N,3)} + C_3^N = N(N^2 + 5)/6.$

ANALYSIS

MBVA gives the strategist a rational, scientific approach to policy making, because VA can be justified on sound principles. The policy maker can justify his or her budget based on these principles, knowing full well what the assumptions are and how reliable they are. In other words, fault tree analysis is a scientifically valid approach to VA.

The practitioner should be aware of alternative methods of performing VA as well as be aware of the limitations of the MBVA technique described here. Each technique has its strengths and limitations. The following analysis suggests caution; MBVA is not perfect.

Model-Based VA: The fundamental assumption underlying MBVA is that all sectors yield to the same method of analysis. Although this is a strong assumption, it is backed up by experience. MBVA is used throughout this book and has been applied to all Level 1 infrastructures. It has not been applied to all sectors, and so it remains unproven in the broadest sense.

MBVA is an adaptation of existing techniques—fault tree and event tree analysis—used by reliability engineers. This seems like a good fit, because sector reliability—under stress—is what vulnerability analysis is all about. Fault tree and event tree analysis seem altogether appropriate as a tool for analyzing the reliability of infrastructures—just as it is useful as a tool in other disciplines.

The event tree visually enumerates all possible events of a fault tree. Event tree analysis is exhaustive. The tree will have 2^n possible events for n events. The events occur with probabilities computed by taking the product of all probabilities listed along the branches of the tree—from the root to the event. This could be a lengthy calculation. An abbreviated technique, event matrix analysis, is a shortcut that ignores higher order combination events. This may or may not be a good idea. Event matrix analysis should only be used when the number of events is large and the fault probabilities associated with threats are small.

Fault tree and event tree analysis is static. That is, neither one models the dynamic behavior of a system. The *Unknown* event in the husband–wife shared bank account example illustrates this limitation. Bank accounts do not allow simultaneous withdrawals, so either the husband or wife withdrawal would occur first, followed by a second withdrawal. The order is important because withdrawals could cause either a husband or wife overdraft if the total amount in the account was, say, $125. Yet the fault tree and event tree do not properly represent this timing dependency. Which person caused the fault, the husband or the wife? In terms of critical infrastructure protection, we want to know which threat causes a fault, and in what order. This limitation may not be important in the analysis of infrastructure sectors such as water and energy, but it is extremely important when analyzing sectors that fail due to failures in synchronization of events. For example, the cascade failures observed in the electrical power grid are often caused by timing faults. We do not elaborate on this limitation but merely point it out as a weakness in the fault tree/event tree analysis technique.

One can argue that MBVA lacks the expressive power of other techniques because fault trees are static—they do not account for the passage of time—and the event tree easily becomes so large that it is useless as a tool.[12] These are valid criticisms of MBVA. In fact, they are limitations of MBVA. Which means that MBVA should not be used in all cases. Rather, MBVA is an easy-to-apply method to use when seeking a first-order estimate of vulnerabilities. MBVA trades completeness for ease of use.

More powerful techniques such as system modeling and simulation, Petri network theory, and statistical reliability theory should be used when dynamic components are involved. In fact, elaborate simulation models may be necessary to account for all vulnerabilities in complex systems such as the electrical power sector. MBVA should not be used in such cases.

EXERCISES

1. Which of the following are major steps in MBVA?
   a. Threat analysis—what are the threats?
   b. Gap analysis—what is the funding gap in the infrastructure strategy?
   c. Network analysis—identify hubs.
   d. Rank vulnerabilities by cost and likelihood.
   e. Give vulnerabilities a score.

[12]Program *FTplus.html*, a.k.a *FT.jar*, described in Chapter 6, does much of the arithmetic needed to handle fault and event trees, which makes it possible to analyze fault/event trees containing thousands of events.

2. Network hubs are (select one):
   a. The centers of network wheels.
   b. Places where Amtrak stations are located.
   c. Centers of small worlds.
   d. Highly active people in an epidemic.
   e. Node with the highest degree.
3. A fault, as defined in a fault tree, is:
   a. A failure or failure mode.
   b. A node that can fail.
   c. An event that causes damage.
   d. A highly possible event or incident.
   e. A bomb or cyber-attack.
4. Logical (A AND B) means:
   a. Faults A and B must both occur for (A AND B) to be TRUE.
   b. Faults A and B must both be TRUE, or both be FALSE for (A AND B) to be TRUE.
   c. P(A) × P(B) must be TRUE.
   d. Faults A and B always happen.
   e. Faults A and B must occur with a probability of 100%.
5. Logical (A OR B) means:
   a. (1.0 − P(A) × P(B)) is TRUE.
   b. Faults A and B must both be FALSE for (A OR B) to be FALSE.
   c. Faults A and B must both be TRUE for (A OR B) to be TRUE.
   d. Faults A and B sometimes happen, but not always.
   e. Faults A and B must each occur with a probability of 50% for (A OR B) to be TRUE.
6. If A occurs with P(A) = 33%, and B occurs with P(B) = 67%, then what is the probability of occurrence of (A OR B)?
   a. 100%
   b. 33%
   c. 67%
   d. 50%
   e. 78%
7. If three coins are tossed at the same time, how many events contain at least one HEAD and at least one TAIL?
   a. 4
   b. 3
   c. 16

d. 6

e. 8

8. What are the OR-tree and AND-tree sector vulnerabilities of a transportation system consisting of two bridges A and B with fault probabilities of 1/3, each?

   a. 5/9 and 1/9, respectively.

   b. 8/9 and 1/9, respectively.

   c. 2/3 and 1/9, respectively.

   d. 1/9 and 2/3, respectively.

   e. 4/9 and 2/3, respectively.

9. How many events are there to evaluate in an event matrix containing 10 threats?

   a. $2^{10} = 1024$.

   b. $2^{10} - 1 = 1023$.

   c. $10(11)/2 = 55$.

   d. $9(10)/2 = 45$.

   e. $10^2 = 100$.

10. The city of San Lewis Rey wants to protect a system of bridges in their transportation infrastructure against BOMB, SCADA, and POWER faults (because they are drawbridges that need power to operate). The bridges of San Lewis Rey were designed and built by a descendant of the engineer that built the famous bridges of Koenigsberg [see Figure 5.12(a)].[13] The seven bridges connect an island in the middle of the city, so their destruction could bring traffic to a halt and even render the island inaccessible.

    A VA was conducted by the city to determine the best way to protect the bridges. Figure 5.12(b) shows the fault tree produced by the analysis. The faults and their fault probabilities are summarized as follows:

Fault and Fault Probability Table.

Faults	Probability
POWER outage	5%
SCADA attack	10%
BOMB attack	25%

Q1: What components should be designated as nodes and links in the network model of Figure 5.12(a)?

   a. Landmasses are nodes, and bridges are links.

   b. Bridges are nodes, and landmasses are links.

[13]Sisma, Pavel, "Leonhard Euler and the Koenigsberg Bridges," *Scientific American*, pp. 66–70, July 1953. http://mathforum.org/epigone/math-history-list/skodwendjerm.

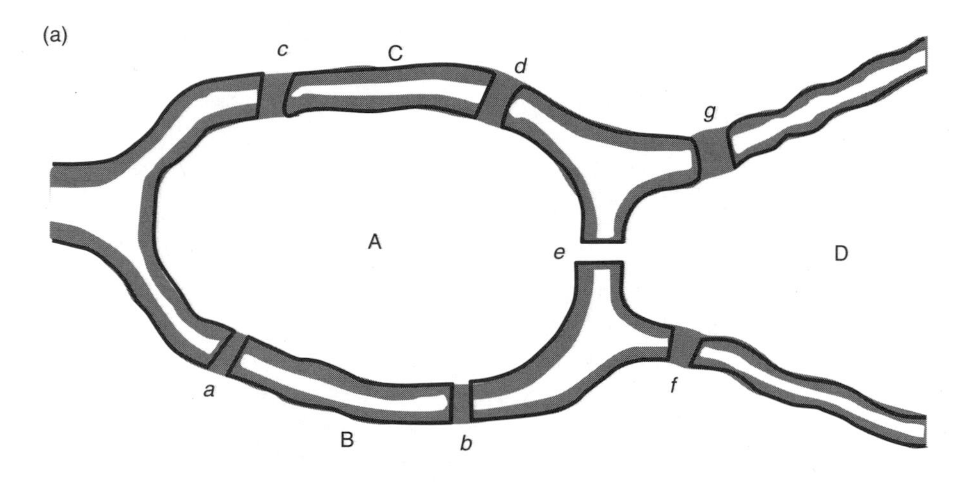

Seven bridges of Koenigsberg crossed the River Pregel

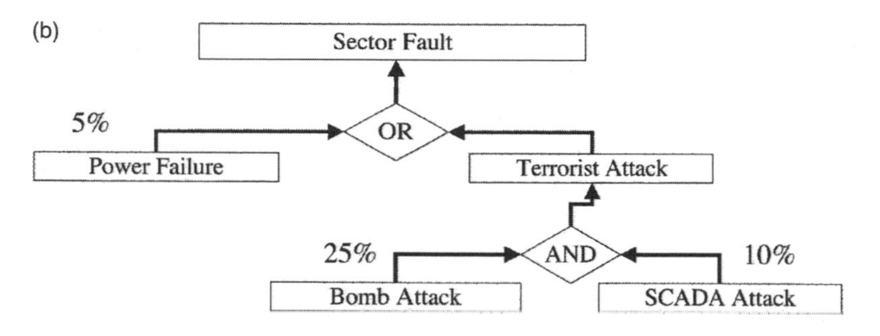

Figure 5.12. The seven bridges of San Lewis Rey are identical to the seven bridges of Koenigsberg, studied by Leonhard Euler. (a) Bridges are labeled a, b, c, d, e, f, and g, and land masses are labeled A, B, C, and D in the map of Koenigsberg. (b) Bridge fault tree from San Lewis Rey Transportation VA.

c. It does not matter.
d. This infrastructure cannot be modeled as a network.
e. Both bridges and landmasses are nodes: Adjacencies are links.

Q2: What is the architecture of the network model of Figure 5.12?
a. It is random with no discernable hub.
b. It is scale free with landmass A as the hub.
c. It is a small-world network.
d. It is an exponentially distributed network.
e. None of the above.

Q3: What is the probability that the sector will fail, given the fault tree of Figure 5.12(b)?
a. 7.38%
b. 40%
c. 7.5%
d. 1.625%
e. 0%

Q4: How many events that lead to a sector fault are there for the event tree corresponding to Figure 5.12(b)?
a. 8
b. 5
c. 4
d. 2
e. 1

Q5: If the AND gate in Figure 5.12(b) is exchanged with the OR gate, what would be the vulnerability of the resulting AND-OR-tree?
a. 7.38%
b. 40%
c. 7.5%
d. 1.625%
e. 0%

Q6: How many bridges must fail to isolate the hub of the network obtained by modeling the landmasses and bridges as nodes?
a. None
b. 5
c. 4
d. 7
e. 3

Q7: Does a BOMB-only or SCADA-only failure cause the sector to fail, according to the fault tree of Figure 5.12(b)?
a. It does not cause sector failure, because of the AND logic gate.
b. It does because of the OR logic gate.
c. It does because BOMBs cause physical damage.
d. It does because SCADA damage can ruin the drawbridge electronics.
e. This fault tree is not formed correctly.

Q8: What is the probability that *all bridges* will fail, assuming the fault probabilities of Figure 5.12(b)?
a. $(0.074)^7$
b. 7(1.625%)

c. 1.625%/7

d. Zero

e. 1 − (1.0 − 0.01625) (1 − 0.01625) (1 − 0.01625) (1 − 0.01625) (1 − 0.01625) (1 − 0.01625) (1 − 0.01625)

Q9: [Advanced]. If a backup power supply is furnished to each bridge, and the fault probability of the backup supply is the same as the primary supply (5%), what is the probability of both power supplies failing?

a. 10%

b. 2.5%

c. 25%

d. 0.25%

e. 1%

Q10: [Advanced]. If an infinite number of redundant power supplies was added to the fault tree of Figure 5.12(b), what would be the vulnerability of the sector?

a. 10%

b. 2.5%

c. 25%

d. 0.25%

e. 1%

CHAPTER 6

Risk Analysis

This chapter completes the description of the model-based vulnerability analysis (MBVA) process. Step 5 takes cost and damage estimates in, and produces an investment strategy for reducing vulnerability or risk. In Chapter 5, we showed how MBVA zeroed in on vulnerabilities of critical nodes; in this chapter, we find several solutions to the problem of resource allocation. We answer the question, "Given there is only a limited budget for protecting the sector, how best should the money be spent?" There is no single answer to this important question. Instead this chapter provides five strategies for infrastructure protection investment: network-wide, ranked vulnerability/risk, apportioned vulnerability/risk reduction, optimal vulnerability/risk reduction, and manual vulnerability/risk reduction. After reading this chapter you should be able to perform a vulnerability analysis, risk assessment, and investment strategy—all based on sound principles.

The following analytical tools are described in detail:

1. *MBVA Step 5*: MBVA's final step defines how best to allocate a given budget, M, to protect an infrastructure against damage. But there is no clear-cut solution because of conflicting objectives; is it better to reduce the worst damage that can occur, reduce vulnerability, or reduce expected damages? Five strategies are examined and compared.
2. *Network-Wide Investment*: The components of a critical infrastructure network are nodes and links. Given a damage estimate for every node and link, what is the best way to allocate funding to nodes and links such that the risk to the entire sector is minimized? We show that the best strategy is to spend the limited budget on the highest degreed and highest damaged nodes, in descending order of "risk." The program *NetworkAnalysis.html* (a.k.a. *NA.jar*) automates the allocation procedure using emergence and a model of network risk developed in this chapter.
3. *Failure Modes*: Failure in a network is often modal; portions of a network survive an attack, whereas other portions fail, which is one surprising result of network analysis. Failure mode analysis shows how partial target hardening

Critical Infrastructure Protection in Homeland Security: Defending a Networked Nation, edited by Ted G. Lewis

may lead to modal failures, and how too much investment in a network may go to waste because of these modes. The program *NetworkAnalysis.html* (*NA.jar*) identifies and simulates failure modes in infrastructure sectors that can be modeled as a cascade network.

4. *Ranked Allocation*: This strategy is perhaps the most common strategy used by practitioners as it funds the highest-ranking components first, the second-highest next, and so on until funds are fully expended. Ranking is defined in terms of vulnerability or risk. Vulnerability reduction attempts to reduce the probability of a fault, and risk reduction attempts to reduce financial risk.
5. *Apportioned Allocation*: Apportioned resource allocation is a method of allocating limited funds to protect the infrastructure by reducing the likelihood that faults occur—across the entire fault tree. The apportioned allocation formula provides a justification for investing something in each infrastructure component. Two strategies are analyzed: *Fault reduction* attempts to reduce the likelihood of a fault, and financial risk reduction attempts to reduce overall financial risk.
6. *Optimal Allocation*: This strategy minimizes vulnerability or risk, where vulnerability is defined as the overall fault tree vulnerability (probability of a fault that leads to sector failure), and risk is defined as the overall financial loss due to fault tree failure. Optimal allocation may or may not produce the same allocation as the other strategies.
7. *Manual Allocation*: Finally, this strategy simply computes the vulnerability or risk that results when allocation is done by hand—the amount of money invested to reduce vulnerability or risk is set by the policy maker. This strategy may override the others because of intangible considerations such as unknown psychological damages implied by the destruction of national icons such as the Statue of Liberty.
8. *FTplus*: The program *FTplus* (a.k.a. *FT.jar*) automates the fault tree analysis described in Chapter 5 and extended here. It supports on-screen editing of fault trees and calculates and displays the event tree outcomes. *FTplus* calculates how much of the budget to allocate to each threat of the fault tree using one of four allocation strategies developed in this chapter: *ranked* vulnerability/risk reduction, *apportioned* vulnerability/risk reduction, *optimal* vulnerability/risk reduction, and *manual* vulnerability/risk reduction.

STEP 5: BUDGET ANALYSIS—COMPUTE OPTIMAL RESOURCE ALLOCATION

Chapter 5 covered four of the five steps of MBVA, as repeated below. All but the final steps were developed and illustrated by example. In this chapter, we examine the final step in great detail, because there is no clear-cut solution to the question, "what is the best way to protect a critical infrastructure sector, given that funds are limited, and the cost of protecting everything is too great?"

MBVA

1. List assets—Take inventory
2. Perform network analysis—Identify hubs
3. Build a model using a fault tree
4. Analyze the fault tree model using an event tree
5. Budget analysis—Compute optimal resource allocation

The final step requires the derivation of an investment strategy that removes or diminishes the likelihood of faults occurring. That is, we want to know how much money to spend on each threat to minimize the probability of faults occurring—across all faults. In fact, the policy maker may opt for one of many strategies. He or she might try to minimize the probability of any event, minimize the probability of the worst event, minimize the probability of the expected event, or determine the vulnerability remaining after manually allocating funds. Indeed, the policy maker may want to reduce risk as opposed to vulnerability; these two different objectives often lead to different investment strategies. All of these policy decisions come with a price tag.

Budgeting is complicated by the fact that the cost of reducing vulnerabilities varies with different kinds of failure modes, and the total bill for reducing all vulnerabilities to zero might be too high. For example, a cyber-security fault may be completely remedied for $5 million, whereas physical security may cost $10 million to improve, but not eliminate. In addition, although the total cost to reduce both cyber- and physical vulnerabilities may add up to $15 million, the organization may have only $1 million to spend. Generally, policy makers and planners have limited funds, so they have to allocate these funds in the best way possible. This raises the question of what is "best."

Here is how the technology for solving this problem will be further developed in this chapter. First, terms will be defined and clarified, because this is a new discipline that lacks a common terminology. Next, we analyze an arbitrary infrastructure network and show how to generally improve its security by allocating funds to nodes and links such that network risk is minimized. Maximizing *network availability* will do this, because maximum availability equals minimum risk. *Availability* is a reliability engineering term meaning the fraction of time the network components are available. An availability of 100% means it is always available—faults never occur. An availability of 0% means the network never works. Availability of a single component is 100% minus the probability that the component will fail. The general goal of all critical infrastructure protection is to maximize availability, which has the effect of minimizing the probability that it will fail.

Third, we show how the vital components of a network—its hubs—are modeled as a fault tree and then turned into an investment strategy for improving the security of the critical node. Actually, fault tree modeling reveals some interesting properties of infrastructures. In particular, we discover that there are multiple answers to the question of which investment strategy is best. This result is a consequence of the fact that different policy goals are met by different investment strategies. Only the policy maker will be able to determine which answer to use.

The programs *NetworkAnalysis* and *FTplus* are provided to do the laborious calculations. *NetworkAnalysis* implements an optimal resource allocation algorithm that computes the best allocation of limited funds to reduce network risk. *FTplus* computes event tree outcomes from a fault tree, applies one of the four allocation strategies to the fault and event tree outcomes, and produces an allocation for each strategy. *FTplus* implements ranked vulnerability/risk reduction, apportioned vulnerability/risk reduction, minimal vulnerability/risk reduction, and manual vulnerability/risk reduction.

Definitions

We begin with definitions of terms that will be turned into equations, then algorithms, software, and finally, strategies. Once again the critical infrastructure is modeled as a network N, with n nodes and e links. The basic building block of the infrastructure network is shown as a "barbell" in Figure 6.1. Every network link connects two nodes, except for isolated nodes of degree zero or one.[1] This property is used to construct a vulnerability model of an arbitrary network, but first we need to define terms.

Vulnerability v_i is the probability of a fault, as defined in Chapter 5. It is expressed as a probability p_i that component i will fail: $v_i = p_i$. For each fault there is an associated cost expressed in terms of casualties, loss of productivity, loss of capital equipment, loss of time, and so forth. This is called the *component's value* or, simply, *damage* d_i, which when multiplied by the fault probability

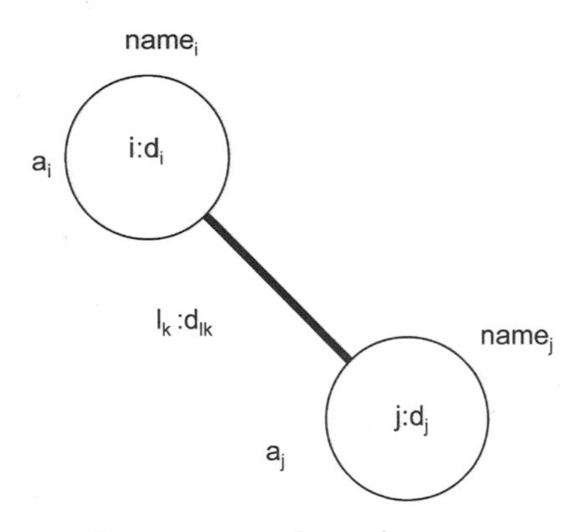

Figure 6.1. Network "barbell" component. Legend: name = name of node, $i{:}d_i$ = node number and damage for node i; a_i = availability of node i; $l_k{:}d_{lk}$ = link number and damage estimate.

[1]We will come back to this later: Isolated nodes and nodes of degree one are shown to have little impact on the resulting investment strategy.

produces risk. Therefore, *risk* r_i is the product of vulnerability times damage for component i.

Risk = fault probability times damage: $r_i = p_i \times d_i$ for component i.

Availability is $(1 - p_i)$, which simply means that availability is the complement of fault probability. Think of availability as the probability that a component is operational at any point in time. Therefore:

$v_i = p_i$ = component vulnerability (ranges from 0% to 100%).
d_i = fault damage (typically expressed in terms of dollars).
$r_i = p_i \times d_i$ = risk of component i (typically expressed in terms of dollars).
$a_i = (1 - p_i)$ = availability of component i (probability component does not fail).
g_i = degree of node i (number of links connecting node i to the network).

These quantities define properties of individual components of a sector but not the entire sector. Sector-wide vulnerability and risk is defined as the sum of these component vulnerabilities and risks.[2] In particular, we are interested in defining sector-wide vulnerability and risk in terms of network vulnerability and risk. Intuitively, a sector with n components might have an overall vulnerability and overall risk is defined as:

$$V = (1/n)\Sigma_i^n v_i = (1/n)\Sigma_i^n p_i.$$
$$R = \Sigma_i^n r_i = \Sigma_i^n p_i \times d_i.$$

Given these definitions of V and R, suppose all probabilities and damages are the same. For example, assuming p's equal 1.0 and d's equal 100, the network will fail with certainty. Specifically, $V = (1/n)n = 100\%$ and $R = 100n$. This worst-case scenario is intuitive: If the likelihood of each fault is 100%, then the overall network vulnerability should also be 100%. Similarly, the overall network risk should be the sum of individual component risk, which in this case is 100n, because each component damage estimate is the same: 100. Generally, risk R will be defined in terms of the expected loss to the entire network, should one or more of its nodes or links fail.

What about the best case? Suppose all p's equal zero—a completely secure sector—then $V = 0\%$ and $R = 0$. Once again this is intuitive: Network vulnerability should range from 0% (secure) to 100% (certain to fail), and risk should range from 0% (no risk) to the sum total of all possible damages as well as their combinations.

Financial risk is measured in terms of dollars, human risk in terms of casualties, and psychological risk in terms of fear. The quantification of these measures may be difficult to estimate exactly, but at least we can rigorously define each, using this approach.

[2]These definitions will change for fault trees, because of AND/OR logic in the fault tree model.

NETWORK AVAILABILITY STRATEGY

Now we are in a position to apply these definitions to an important problem: What is the best allocation strategy for securing the components—nodes and links—of a network infrastructure such as the power grid, water supply network, or telecommunications infrastructure network? In the following it will be shown that nodes are far more important than links, and that the nodes with the highest degree (most links) or highest damages associated with them are the highest in priority to be protected. Once again, Figure 6.1 will be used to derive this result.

More formally: Given M dollars to prevent damage to a critical infrastructure sector, what is the best way to invest M? Suppose M is converted into A availability points; each point represents 1% of availability, and because availability is the complement of fault probability, each point represents a reduction in fault probability of 1%. The idea is to "buy security" by increasing the availability of each component of the network. It is assumed that money can buy an improvement, and that improvement is measured in terms of increased availability. We also assume a linear relationship between investment and improvement; investing twice as much in a node reduces its vulnerability by one half. This assumption is only an approximation at best, but it will suffice for the following derivation.

This problem now becomes an optimization problem: What is the best way to distribute A points across a network of nodes and links, such that the sector-wide risk is minimized? **Derivation 6.1** contains the mathematical derivation of network risk, R, given the barbell model of Figure 6.1, and its fault tree model is shown in Figure 6.2. The mathematically inclined reader may want to study **Derivation 6.1** in great detail. Other readers will simply want to use the programs *NetworkAnalysis* or *NA.jar* and let them perform these calculations, automatically.

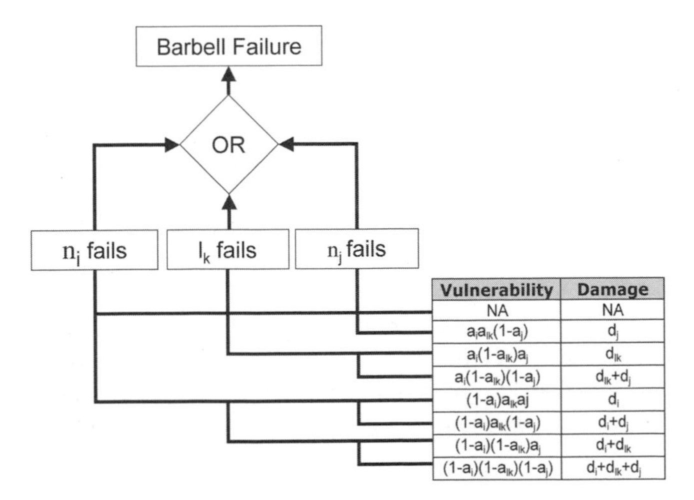

Vulnerability	Damage
NA	NA
$a_i a_{lk}(1-a_j)$	d_j
$a_i(1-a_{lk})a_j$	d_{lk}
$a_i(1-a_{lk})(1-a_j)$	$d_{lk}+d_j$
$(1-a_i)a_{lk}aj$	d_i
$(1-a_i)a_{lk}(1-a_j)$	d_i+d_j
$(1-a_i)(1-a_{lk})a_j$	d_i+d_{lk}
$(1-a_i)(1-a_{lk})(1-a_j)$	$d_i+d_{lk}+d_j$

Figure 6.2. Fault tree for barbell of Figure 6.1.

DERIVATION 6.1: NETWORK RISK MODEL

Consider the barbell network fragment of Figure 6.1 and its fault tree model of Figure 6.2. What is the risk formula for a network N with n nodes and e links, each with the following associated values:

A: total number of availability points, expressed in terms of percentage points.

a_i: availability of node/link i.

d_i: damage associated with failure of node/link i.

$\Sigma a_i = A$: the total number of availability points, A, cannot be exceeded.

Table 6.1 shows how we obtain the risk of a single barbell, r_i. To obtain the overall network risk, we must sum r_i over all links, because there is one barbell for every link in the network (e = number of links):

$$
\begin{aligned}
R = \Sigma_1^e r_{lk} = \Sigma_{lk=1}^e [a_i a_{lk}(1 - a_j)d_j + a_i(1 - a_{lk})a_j d_{lk} \\
+ a_i(1 - a_{lk})(1 - a_j)(d_{lk} + d_j) + (1 - a_i)a_{lk}a_j d_i \\
+ (1 - a_i)a_{lk}(1 - a_j)(d_i + d_j) + (1 - a_i)(1 - a_{lk})(1 - a_j)(d_i + d_{lk}) \\
+ ((1 - a_i)(1 - a_{lk})(1 - a_j)(d_i + d_{lk} + d_j)].
\end{aligned}
$$

As a sanity check, suppose we compute R when all a's = 1. R = 0, because every term in this ungainly equation contains at least one (1 − a), which is zero. Now, when all a's = 0:

$$
\begin{aligned}
R = \Sigma_1^e[d_i + d_{lk} + d_j] = \Sigma_i^n g_i d_i + \Sigma_1^e d_{lk} \\
= \text{sum of (node degree)(node damage)} \\
+ \text{sum of link damages.}
\end{aligned}
$$

As an example, consider the case of all damages equal 100: R = 100e + 100 sum of degrees, which simplifies because the sum of degrees is twice the number of links: R = 100e + 200e = 300e. Similarly, if all a's = 0.5, then R = 50% the foregoing, or 150e.

TABLE 6.1. Risk Calculation for Figure 6.2.

Event	Vulnerability	Damage	Risk
No fault	NA	NA	NA
node j fails	$a_i a_{lk}(1 - a_j)$	d_j	$a_i a_{lk}(1 - a_j)d_j$
link fails	$a_i(1 - a_{lk})a_j$	d_{lk}	$a_i(1 - a_{lk})a_j d_{lk}$
link, j fail	$a_i(1 - a_{lk})(1 - a_j)$	$d_{lk} + d_j$	$a_i(1 - a_{lk})(1 - a_j)(d_{lk} + d_j)$
node i fails	$(1 - a_i)a_{lk}a_j$	d_i	$(1 - a_i)a_{lk}a_j d_i$
i, j fail	$(1 - a_i)a_{lk}(1 - a_j)$	$d_i + d_j$	$(1 - a_i)a_{lk}(1 - a_j)(d_i + d_j)$
i, link fail	$(1 - a_i)(1 - a_{lk})a_j$	$d_i + d_{lk}$	$(1 - a_i)(1 - a_{lk})a_j(d_i + d_{lk})$
all faill	$(1 - a_i)(1 - a_{lk})(1 - a_j)$	$d_i + d_{lk} + d_j$	$(1 - a_i)(1 - a_{lk})(1 - a_j)(d_i + d_{lk} + d_j)$
			r_{lk} = sum

Note: This formulation of risk, R, assumes each node is counted g = degree times. In other words, it is larger than the sum of all damages. Given a node n, with degree g, and damage d, n contributes $g \times d$ points to R.

A risk model that answers this "optimal allocation" question is derived by modeling the network as a collection of barbells, computing the formula for risk r_{lk} in the barbell (see Figure 6.2), and then summing barbell risk over all barbells that make up the network. Each barbell in a network corresponds to a link, and hence, the summation of risk is taken over all links.

Figure 6.2 shows a fault tree and corresponding event tree model of the barbell component of Figure 6.1. There are three components: the two nodes connected by a link and the link. These are designated as n_i, n_j, and l_k. Barbell failure results when any single, double, or triple combination event occurs, so the fault tree is an OR-tree.

We know from Chapter 5 that the vulnerability of the OR-tree of Figure 6.2 is:

$$V_{lk} = (1 - (1 - p_i)(1 - p_{lk})(1 - p_j)) = (1 - a_i a_{lk} a_j).$$

For example, if all a's are zero (certain to fail), $V_{lk} = 100\%$, and if all a's are 1.0 (completely secure), $V_{lk} = 0\%$. Summing V_{lk} over all barbells yields the vulnerability of the entire network. This is left as an exercise for the reader.

Risk is defined as the product of vulnerability times damage, so the risk of this barbell is the sum of its component risks as shown in the "RISK" column of Table 6.1. The network risk is simply the sum of all barbell risks over all links of the network, e.g., the sum of r_{lk}; Table 6.1, Figure 6.2, and **Derivation 6.1** produce a generalized risk formula for any arbitrary network.

The risk formula expresses the overall network risk in terms of availability and damages. The formula does not care whether the network is random, scale free, or small world. Indeed, in the end, it does not matter whether a network is scale free, because optimal allocation of availability to nodes and links automatically seeks to distribute points to where they are needed most. In some cases, this will be a hub, and in other cases, this will be a node with high damage values. In this way, we accommodate "high value targets" in the network analysis.

Allocation by Emergence

Now that the formula for network risk, R, is known, we can return to the question of allocation of the a_i such that R is minimized. The answer to the question of how best to allocate availability points to the nodes and links of a network is to simply let the allocation pattern *emerge* from the network.

Recall that emergence is a method of solving problems in complex systems that are too difficult to solve with analytical techniques because we lack the necessary explicit knowledge of complex behavior. Instead, we find a simple organizing principle that can be employed locally to each node or neighborhood of a network and watch as a pattern emerges.

In this case, allocation by emergence is achieved by applying the following organizing principle to any network that is annotated with link and node damage estimates, as shown by Figure 6.3. The following organizing principle evolves a network-wide solution to the problem of availability point allocation.

Availability Point Organizing Principle

Initially, distribute the A availability points to nodes and links. This is done like dealing cards to card players, until all A points have been dealt. In most cases, this means nodes and links get an equal number of points, or approximately the same number. This initial allocation satisfies the constraint, $\Sigma a_i = A$. To stay within this constraint, the organizing principle must allocate the same number of points it takes from one network node or link to another node or link. These randomly selected nodes and links are called DONORS and RECIPIENTS:

Repeat forever:

1. Select a node or link at random, called the DONOR; select another node or link at random, called the RECIPIENT.
2. If DONOR has at least one availability point and RECIPIENT has less than 100 availability points, deduct 1 point from DONOR and add 1 point to RECIPIENT.
3. If risk R is not decreased, take back the donation; otherwise keep it.

Like all organizing principles in this book, this is so simple it is difficult to imagine what pattern emerges. The formula for R is complicated enough that the

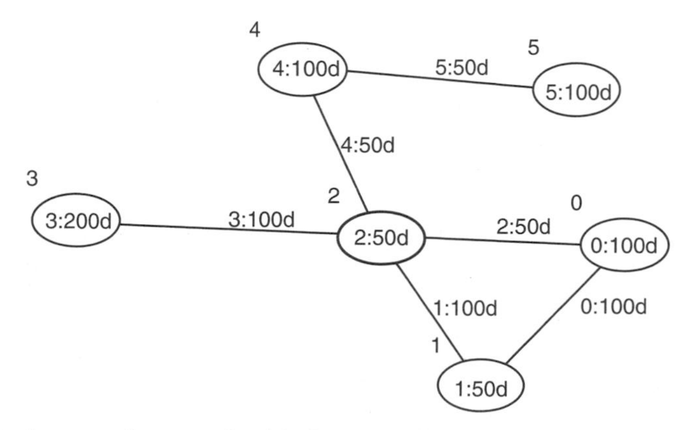

Figure 6.3. An example network with damage and availability point annotations. Given damages, what is the best allocation of availability points?

solution cannot be easily found through traditional analytical techniques such as constrained optimization, but the solution is intuitive. In general, the organizing principle will produce the following result:

- Links lose availability points until they have zero points (100% vulnerable), in the case where there are not enough points to go around.
- Critical nodes accumulate availability points: The nodes with the highest degree (hubs) and the highest value (highest damages) accumulate the most points; all others lose points to hubs and high-value nodes.
- It the network has nodes of degree one, links will be favored over these nodes.

Availability points are distributed to where they do the most good: hubs and high-value nodes. This pattern is intuitive, but why do links decrease—sometimes to zero availability—in most cases? Consider the network in Figure 6.3 more carefully. Suppose all damage d_i and availability values a_j are identical for all nodes and links. Now suppose one node or link is destroyed: Which one has the greatest impact on the network, if removed?

First consider links. Removal of any link causes damage d, say. No other node or link is affected. Now consider the nodes. Removal of node n_1, for example, causes damage 3d, because it brings down two links in addition to itself. The same result is obtained for all other nodes with degree equal to two. But node n_3 has degree of four, and so its removal causes damage equal to 5d. In other words, ranking the order of damage done by removal of a node or link, from highest-to-lowest, it is easy to see that the hubs are most important, followed by non-hub nodes, followed by single links.[3]

In a zero-sum game where there is a fixed number of availability points—and not enough to go around—allocation of points to one node or link takes away from another node or link. Because links have the lowest impact on availability, they are the first to lose points. Only nodes with degree of one are lower in rank. This, of course, assumes the damage values are all the same. When they are not, the ranking must include high-value (high damage) nodes or links.

Network Analysis

In most critical infrastructures, the best utilization of scarce availability points follows the ranking described above: from hubs to non-hubs and finally to links. But hubs are not the only critical nodes; high damage value nodes are critical too. They compete with hubs for limited resources. Program *NetworkAnalysis* (a.k.a. *NA.jar*) implements the organizing principle described above, and this principle emerges as the best possible allocation is made to both hubs and high-damage nodes and links.[4]

[3]In the case of a node with degree of one (a terminal node, for example), the amount of damage caused by its removal is less than the removal of its link, so the single-degreed node will lose points to its link.
[4]Network Analysis also performs flow analysis, but this topic will not be covered here. See the Network Analysis User Guide for descriptions of other Network Analysis features.

NetworkAnalysis experiments show that the best allocation is one that favors hubs and high-value components. But more significantly, *NetworkAnalysis* yields important insights that tell the policy maker where to focus resources to maximize protection.

Figure 6.4 shows the display window of *NA.jar* along with a simple network.[5] This window contains a row of buttons at the top that are used to open/save network files, select from built-in example networks, and edit nodes and links. A second row of buttons at the bottom of the window contains controls for entering the budget (number of points to allocate), repair time for simulating faults and their repair time, default damage value associated with each link or node ($Damage), and several control buttons. The display shown in Figure 6.4 is one of several displays. A frequency histogram and "best-fit" power law curve are shown in Figure 6.4.

Network Analysis also performs flow maximization or link analysis. Pressing the LINK radio instead of the NODE radio (in the bottom row of controls) causes the program to perform a flow analysis that allocates points to links and nodes such that flow from source to sink nodes is maximized. Flow analysis may be more appropriate when water, traffic, oil, or products flow in one direction, only. But this topic will not be described in detail, here.

The FILE menu contains standard File-Open, File-Save, and File-Save As items. The EXAMPLES menu contains examples, including the example shown in

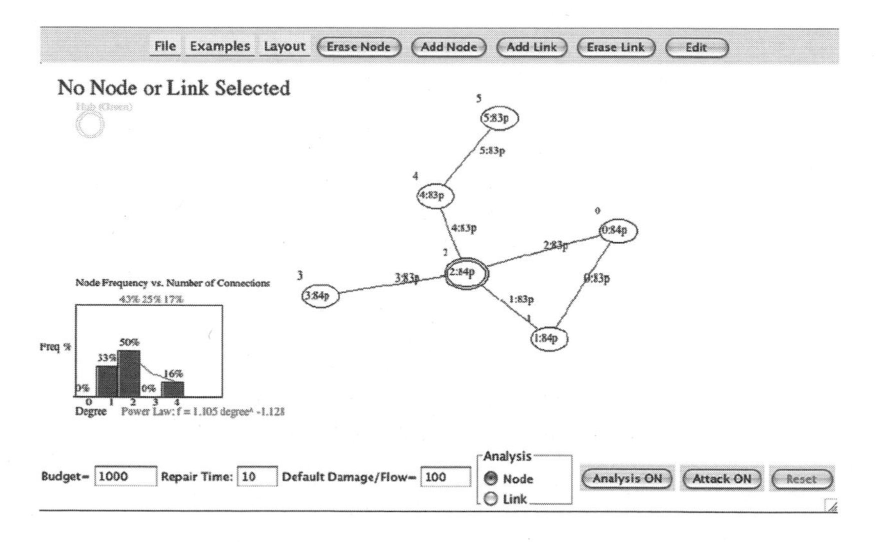

Figure 6.4. Display window of program *NA.jar* (*a.k.a. NetworkAnalysis*). This simulator finds the optimal allocation of A points to links and nodes such that risk is minimized.

[5]*NA.jar* is the desktop version, and *NetworkAnalysis.html* is the Web version. *NA.jar* can save and open local disk files, but *NetworkAnalysis.html* cannot.

Figure 6.4, networks used in end-of-chapter exercises, and network models of actual infrastructures. The LAYOUT menu contains several network layout options, plus a speed control that allows you to run the program in one of three speed levels: SLOW, MEDIUM, and FAST.

LAYOUT includes[6]:

REMEMBER THIS LAYOUT merely remembers the current network layout. This is useful if you have rearranged the network by dragging nodes around the screen, and you want *NetworkAnalysis* to recall this arrangement, later. This works as long as the program is running, but does not work if you return to the program later after quitting.

RECALL SAVED LAYOUT simply recalls the layout that you saved by selecting REMEMBER THIS LAYOUT.

AROUND HUB(S) arranges the current network around the hubs. Layout tries to place the hub in the center of the screen and radiate the other nodes outward, like spokes in a wheel. The higher each node's degree, the closer it is to the center. The lower the degree, the further away from the center.

AROUND CENTER(S) arranges the current network around the network center. A center is a node that is the closest to all other nodes in the network. More formally, the center nodes are the ones possessing the minimum of the maximum distance from it to all other nodes. Of course, an arbitrary network may have more than one center. Center node(s) are red rectangles, and the outer-most nodes are red diamonds. The nodes are marked with their number (10:) and the distance (R = 3) it is from all other nodes. The letter "R" means "radius."

AROUND DAMAGES arranges the current network according to the damage value of each node. Nodes with highest damage value are moved toward the center of the display, and nodes with lowest damage values are moved away from the center. This layout makes it easy to pick out the highest value targets.

AROUND ALLOCATIONS arranges the current network around the current allocation of points to nodes and links. The nodes with the highest allocation move toward the center of the display, and the nodes with the lowest allocation move away. When this layout is selected and the ALLOCATE button is pressed, the network will continually rearrange itself so that the highest allocation nodes tend toward the center of the screen.

CHECKERBOARD GRID arranges the current network in rows and columns like the squares of a checkers board. This layout is useful because it often makes it easier to view the structure of the network.

The ATTACK-ON button starts a simulation of a cascade failure caused by a single fault in the network. If a node or link is selected before pressing ATTACK-ON, that node or link is disabled and the fault is propagated to connecting links. If no link or node is selected, then the software randomly selects a link or node for you. Disabled links and nodes are shown in orange. The repair time field set by the user is the number of time steps that must elapse before a disabled link or node

[6]The users manual for *NetworkAnalysis* has many more details than provided here.

becomes operational, again. Cascade failure simulations reveal overall network availability as well as failure modes.

Network nodes and links are displayed in the middle of the window along with their index number and points or damage values. A colon separates the two numbers assigned to each node or link: (#: nx), where # is the index of the link or node, np means the node or link has been allocated n points, and nd means the node or link has been assigned a damage value of n.

New nodes and links are added to the network by selecting a node first and then pushing ADD NODE or ADD LINK from the buttons at the top of the window. When ADD NODE is selected, a dialog will appear to request the name and damage value assigned to the new node. When ADD LINK is selected, a link dialog will appear so that you can enter its damage value.

An EDIT button is also located in the top row of buttons. A node can be selected by clicking on it, and then its name and damage value changed by clicking on EDIT.

The graphical display in the lower left-hand corner of Figure 6.4 plots the network histogram or node frequency. The percentage of nodes with k links will be shaped like a power law if the network is scale free. In fact, *NetworkAnalysis* will do its best to fit a power law curve to this histogram and display the exponent, p, of the power law. Recall that the network has a structure similar to a scale-free network when p is greater than one, and it is formally scale free when p is between 2 and 3. The highest-degree node is indicated in the display by circling it with a green border.

When ALLOCATE-ON is pressed, a thermometer appears in the left side of the display window. This thermometer starts with an initial value of risk and declines to the minimum value as the organizing principle is repeatedly applied. At some point in time, the simulator cannot decrease risk any further, and the thermometer will settle in to a minimum value. Other displays will appear corresponding with LAYOUT, ALLOCATE-ON, and ATTACK-ON processing. These graphs and charts, described in the operations manual, will not be repeated here.

For example, consider the network shown in Figure 6.4 obtained by pulling down the EXAMPLE menu and selecting EXAMPLE CH6. Given 600 points to allocate, and damage values for each node and link, the software produces an optimal allocation as follows:

Node#	Damage	Allocation	Link#	Damage	Allocation
0	100	100	0	100	37
1	50	80	1	100	20
2	50	100	2	50	0
3	200	100	3	100	60
4	100	100	4	50	0
5	100	3	5	50	0

Two generalizations can be made about this allocation: (1) Nodes with high degree and nodes and links with high damage values are favored over other nodes, and (2) nodes are generally favored over links. For example, nodes 0, 2, 3, and 4 have either high-damage values or high-degree values (node 2 has the highest degree of all nodes). Links receive a total of only 117 points (20%), whereas nodes receive 80% of the 600 points. In other words, nodes are favored over links.

What can be learned from *NetworkAnalysis*? First, we learn that regardless of whether the network is scale free, point allocation favors hubs, high-value nodes, and lastly, links. Most points go to highest degreed nodes and highest damaged nodes first. Only then are links given points. In general, this result once again validates the network analysis approach adopted by MBVA.

NetworkAnalysis provides an answer to the question of how to allocate funding to critical infrastructure components. Obtaining this answer is a major achievement on its own, because we cannot afford to fund everything. But what is even more remarkable is that such a simple organizing principle easily finds the optimal allocation by emergence. This allocation is carried forward to the fault tree corresponding to the critical node or nodes identified by this analysis.

Properties of Network Availability

Networks can model almost any kind of critical infrastructure. In fact, network modeling has been successfully used to model the San Francisco water supply (Hetch Hetchy), top Internet service providers, major energy pipelines, transportation systems, and electrical utility grids.[7] Examination of many of these sectors reveals some interesting properties of networks. These properties are summarized, here:

1. *Node Degree Tipping Point*: Most critical infrastructure networks exhibit a tipping point whereby the number of availability points allocated to a node abruptly changes as the total number of points increases. Allocation percentages transition abruptly from one node to another. There is no smooth or gradual shift in allocation.
2. *Damage Value Tipping Point*: A similar tipping point is observed as the number of availability points shifts from a node with low-damage value to one with high-damage value.
3. *Failure Modes*: Most networks exhibit distinct and multiple failure modes. That is, the nodes and links affected by a random failure in one node or link propagate to subsets of the entire network, depending on where the first failure occurs. Thus, one failure mode may affect one half of the nodes and links, and another failure mode may affect the other half. This is especially true of networks that are partially hardened.

As an example, suppose we explore the behavior of the availability organizing principle when applied to the simple barbell network of Figure 6.1 to gain a better understanding of these concepts. Assume the damage value of all nodes

[7]These examples can be found in the *NetworkAnalysis* software provided along with this book.

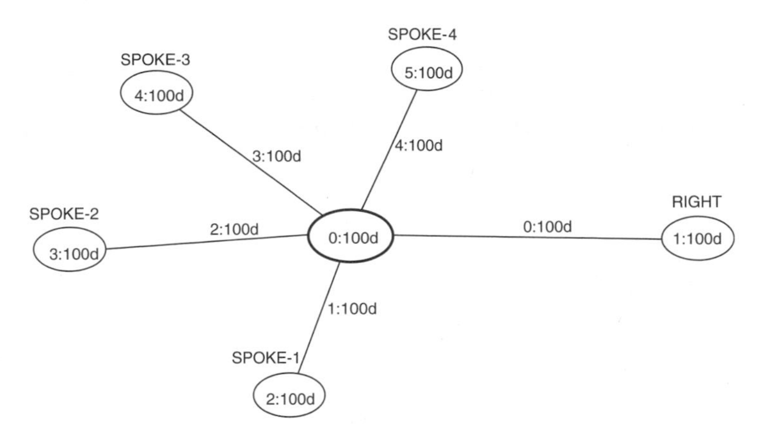

Figure 6.5. A simple network consisting of a hub with five spokes. The damage values are all set to 100, except for the RIGHT node.

and links is equal to 100, and the total budget of points is also 100, which is not enough to harden all three components of the barbell. What is the allocation of 100 points to this network? As you might expect, the optimal risk reduction allocation strategy is to evenly distribute one third of the points to each of the two nodes and single link.

Now suppose the barbell is enhanced as follows: A hub with five links, each link connecting one node to the hub, is formed as shown in Figure 6.5. The degree of the hub is 5, and the damages of all links and nodes is 100, except for the node labeled RIGHT, which is assigned a range of damage values. Initially, RIGHT is assigned a damage value of 100, then 200, 300, 400, and so on. What is the allocation of points to HUB and RIGHT nodes as the damage value of RIGHT increases?

The total number of points in the budget is 100, so there are not enough points to harden all nodes and links. The allocation process must choose which of the nodes and links to harden. Running the availability allocation organizing principle built into *NetworkAnalysis*, we get the following allocations to the HUB and RIGHT nodes (everything else receives zero points).

Allocation of 100 Points.

Damage Value of RIGHT	HUB Allocation	RIGHT Allocation
100	100	0
200	100	0
300	100	0
400	100	0
500	50	50
600	0	100
700	0	100
800	0	100
900	0	100

As the damage value of RIGHT increases to 400, the allocation favors the HUB, because it has five links. At exactly DAMAGE = 500, the points are divided between HUB and RIGHT. This is the tipping point. When the damage value of RIGHT is assigned 501, and above, all 100 points are shifted to RIGHT! In other words, the allocation organizing principle changes its allocation from favoring HUB to favoring RIGHT.

This simple example illustrates the first two concepts: High-degreed nodes are favored due to their high connectivity to the network until nodes and links with high-damage values are taken into consideration. But when damage values reach a level that equals the degree of the HUB, points tip toward the high-damage nodes and links. The transition is abrupt, not gradual.

Now, suppose we consider a more realistic and complex example. Figure 6.6 shows the network model of the Washington, D.C. water system—WASA. This

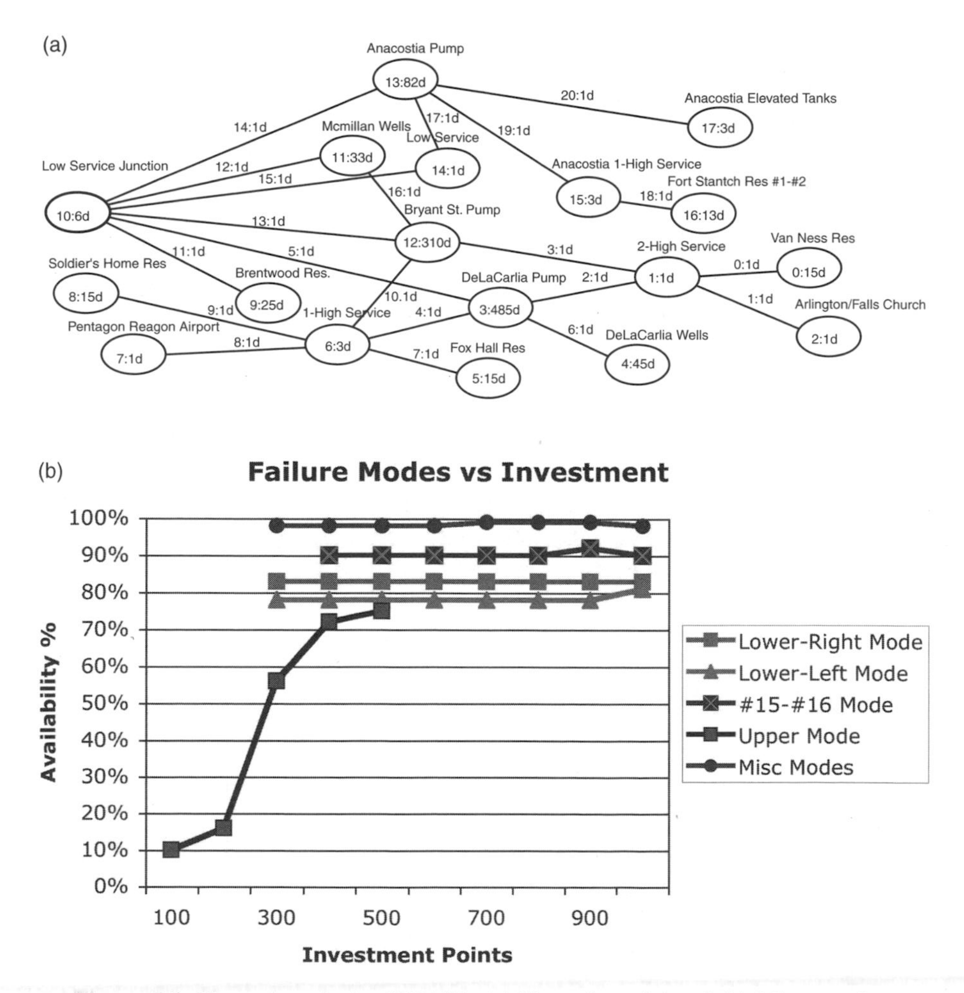

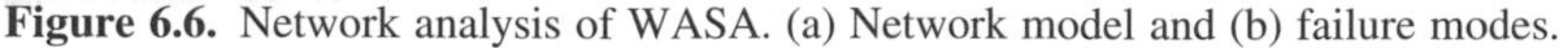

Figure 6.6. Network analysis of WASA. (a) Network model and (b) failure modes.

water distribution and pumping system is a comprehensive network of reservoirs, pipes, and pumping stations. Each pump is given a damage value equal to its pumping capacity, and each reservoir is given a damage value equal to its storage capacity—measured in millions of gallons. Links model water flows (pipes) and are assigned a low-damage value ($d = 1$), because they are easily repaired if damaged.

The critical nodes of the WASA network are now determined by a combination of node degree and damage value. The number of nodes and links and their differing damage values and node degree values complicates discovering the tipping points for this network. What is the best allocation of points to nodes and links? As it turns out, the Bryant Street and Delacarlia pumping stations are critical nodes. They receive points before any other nodes and links.

Figure 6.6(b) shows the results of computing the WASA network availability versus budget, as the budget ranges from 100 to 1000 points. Notice two things: (1) The availability of the WASA network increases as the number of points increases—up to a point, and (2) increases in allocation expose five failure modes—subnetworks that fail even when other parts of the network do not.

The first observation shows that it is possible to spend too much money (resources) on making further improvements in availability. At approximately 500 points, the availability of the WASA network no longer improves. It does not pay to enhance WASA beyond this point of diminishing returns.

The WASA network illustrates the third concept: Critical infrastructure networks can fail in a variety of ways—these are called failure modes. A *failure mode* is a cascade failure that affects part of a network while leaving other parts whole. Each failure mode corresponds with a subset of the network that fails as a result of a cascade failure. A mode is essentially a portion of the overall network. The network is partitioned into modes whenever allocation of points hardens one or more links, thus setting up barriers that prevent fault propagation beyond the partition.

As investment in availability points increases, the number of failure modes may also increase, because more barriers are established. But once every node is hardened (100 points allocated), the availability of the entire network tends toward 100%. But such an ideal may never be reached, due to limits on the budget. Indeed, it may not be necessary to harden all nodes and links to achieve near-perfect availability.

The second summary observation is also obvious from Figure 6.6(b): The manner in which WASA fails depends on where the failure point is, and how this point affects connected parts of the network. Five failure modes are shown in Figure 6.6(b) as separate plots of availability versus investment. Each plot represents the availability of a portion of the WASA network—a subset that continues to operate when one component within the network fails.

Failure modes in WASA correspond to backups of water and sewage through portions of Washington, D.C.[8] In this example, a major sewage backup would perhaps

[8]WASA is an example in the *FTplus* examples menu.

lead to the pollution of the Potomac River as WASA flushes raw sewage to purge the network.

In summary, network allocation is an extremely effective way to obtain useful information for the policy maker and planner:

1. Critical nodes are the ones that receive the most points.
2. Allocation tells the policy maker where to invest.
3. Simulated attacks on the allocated network often reveal failure modes that tell the policy maker what can happen to a partially hardened network.

CRITICAL NODE VULNERABILITY REDUCTION

The combination of network analysis with network allocation of availability zeros in on the critical nodes and where to invest to get the best return on security. This approach is how we can cope with the vastness and complexity of most infrastructure sectors.

Once the critical node or clusters of nodes have been identified and investment is focused on the best place to get maximal protection, the next step in MBVA is to find the best way to allocate funds to protect the node or cluster. This is called *critical node vulnerability reduction*, because it reduces the probability of a fault in the critical node or critical cluster.[9] Similarly, *critical node risk reduction* reduces the risk associated with each critical node or critical cluster.

Given a critical node, and a limited budget, what is the best strategy for investing these funds to protect the node? This is the problem we will solve in this section. Unfortunately, there is no simple answer. Rather, there are four strategies that merit consideration:

1. Manual risk reduction—Establish your own allocation.
2. Ranked order risk reduction—Reduce the worst case.
3. Optimal risk reduction—Mathematically minimal.
4. Apportioned risk reduction—Spread allocation across threats.

In manual risk reduction, the policy maker decides the best way to allocate funds, perhaps based on social, political, or economic factors. The Ranked Order strategy funds the highest ranked components of a node first, working down the ranked order list, until money runs out. Various ranking schemes are used in practice, such as a weighted sum of damages, vulnerability, and casualties. In the following, only two objectives are considered: vulnerability (probability of a fault) and risk (financial damages). The optimal risk reduction strategy reduces the mathematical vulnerability or risk by allocating funds such that vulnerability or risk is minimized. Hence, this strategy is also called Minimal Risk Reduction. And finally, the

[9]Small-world networks will contain critical clusters—neighborhoods of nodes that are critical.

apportioned risk reduction strategy attempts to spread the money across all threats, reducing each one by an amount that maximizes the sum-of-squares of the difference between risks before and after allocation.

Why so many strategies? Each strategy described here has its advantages and disadvantages. For example, the optimal risk reduction strategy guarantees the absolute minimum risk, but it may leave some vulnerabilities unfunded. The ranked order risk reduction strategy gives more weight to the worst-case event, at the expense of the average or expected event. Apportioning funds across all threats is a middle-of-the-road approach that protects everything to some extent, but nothing fully. Which is best? Policy makers must decide what their objectives are, and select the strategy that addresses their objectives. This decision requires a full understanding of these strategies.

All strategies have the same general goal: to fund improvements in the components of a critical node such that the vulnerability of the node is reduced. The basic underlying assumption is that we can "buy protection." That is, we assume there is a linear relationship between investment to harden a component and reduction in the vulnerability of the component.

For example, if a certain component has a 50% chance of being successfully attacked by a cyber-terrorist, and this vulnerability can be completely removed for $2 million, then we assume the 50% vulnerability can be reduced to 25% for $1 million, and 0% for $2 million. Partial reduction is assumed to be linear; doubling the investment doubles the protection.[10]

This assumption underlies all strategies described here, but it may not hold in practice. For example, the vulnerability of a certain component may be reduced by 80% for, say $100,000, but the final 20% may cost another $100,000. Indeed, it may not even be possible to reduce vulnerability to zero, no matter how much we invest. This reality will be ignored in the following derivations. Instead, a linear relationship between investment and fault probability reduction is assumed. Partial investment to secure a component will result in partial security.

Figure 6.7 contains a fault tree that will be used to provide a concrete illustration of each strategy. This is the *shared bank account* example (husband and wife sharing the same bank account) explored in Chapter 5. We will extend it to the problem of resource allocation called for by step 5 of MBVA.

Figure 6.7 has been modified to include two columns: a vulnerability calculation column and a risk calculation column. The vulnerability column is identical to the event tree outcomes calculated in Chapter 5. The risk column shows how to calculate risk for this fault tree.

For the purposes of illustration, suppose the cost of writing a check is $1, and the penalty for over-drawing the account is $20. In fault analysis terms, *cost* and *damage* associated with (H-Withdraw = $100) is $1 and $0, respectively; (Balance < $100) is $100 and $20; (W-Withdraw = $50) is $1 and $0, and (Balance < $50) is $50, and $20, respectively. Note that the damage associated with both partners

[10]This simplification is an approximation of reality, because we know it is more costly to eliminate the last 1% of vulnerability than the first 1%.

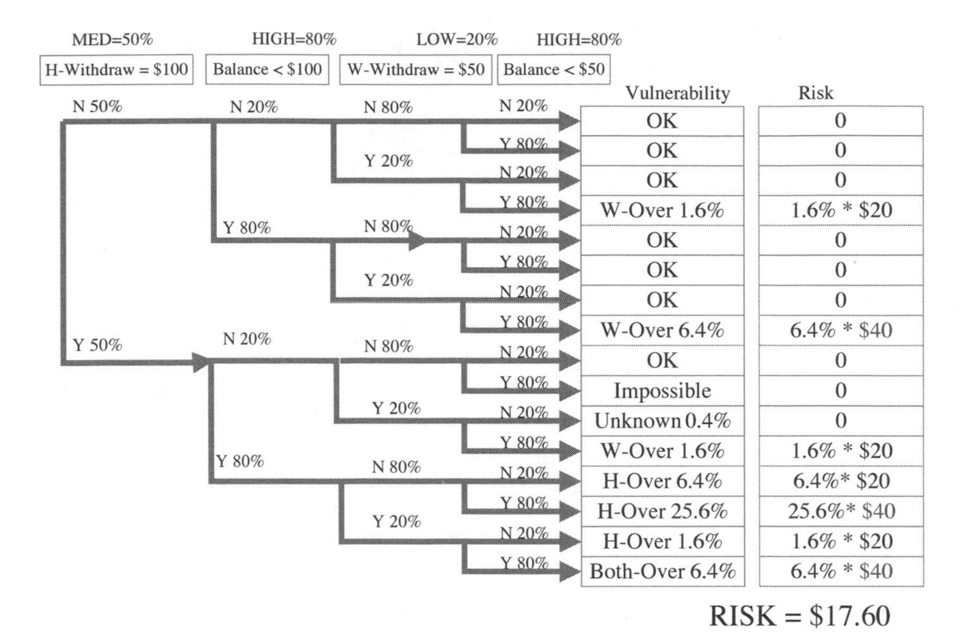

Figure 6.7. Example fault tree (see Figure 5.6) showing vulnerability and risk calculations.

over-drawing at the same time is $40. This example assumes there is no penalty unless the bank account balance goes negative, which only occurs when a withdrawal AND a Balance less than $50 or $100 occurs. The model in Figure 6.5 associates this damage with (Balance < $100) and (Balance < $50), and not with (H-Withdraw = $100) or (W-Withdraw = $50)—a simplification that is not entirely accurate. In the case in which we do not know (Unknown), it is assumed that the penalty is $0, because it is not possible to estimate the damages due to an unknown event.

Recall that risk equals vulnerability multiplied by damage, so the products of all fault probabilities times damages, as shown in Figure 6.7, are summed to obtain the risk. The financial risk to husband and wife is $17.60. What does this mean? Risk is a kind of *expected loss*, or average over all probable outcomes. It takes the likelihood of each outcome into consideration. Although the worst-case loss is $40 (both overdraft), and the best-case loss is $0, the expected loss is somewhere in between: $17.60.

Suppose the two marriage partners want to reduce their risk to avoid recrimination by their bankers and save an average of $17.60? What strategy should they employ? Does it make sense to purchase overdraft insurance for $200 per year, pad their bank account with an extra $150 to cushion against overdrafts that may never happen (and suffer the loss of use of the cash), or open separate accounts? Each of these has a cost. The "correct answer" depends on the partner's strategy.

An *investment strategy* is defined as an allocation of a fixed amount, M, for the purpose of reducing vulnerability or risk in the components of a critical node. For example, suppose the husband–wife team want to spend $100 to reduce their risk of overdraft penalties. What is their investment strategy? How can they reduce their risk from $17.60 to some lower amount by allocating $100 to the components of Figure 6.7?

Manual Risk Reduction

The manual risk reduction strategy requires that the policy maker determine, through manual means, the best investment strategy to use to prevent faults. Given the fault tree model shown in Figure 6.7, what is the best way to allocate $100 to reduce the couple's risk?

In addition to financial damage associated with each threat, we need to know how much it costs to remove the vulnerability. The investment needed to reduce a vulnerability to zero is simply called *cost*, in contrast to the damage that can be caused by a threat. Let the costs and damages associated with each of the threats in Figure 6.7 be:

#	Threat	Cost	Damage
1.	H-Withdraw = $100	$1	$0
2.	Balance < $100	$100	$20
3.	W-Withdraw = $50	$1	$0
4.	Balance < $50	$50	$20

Now that we know how much it costs to nullify each threat, an investment can be made against these costs. If we have enough money, the obvious approach is to invest an amount equal to cost, for each threat. For $152, all vulnerabilities can be eliminated from the shared bank account, because the sum of all costs is $152. This reduces the vulnerability to zero, which eliminates the threat. But when we do not have enough money to completely eliminate all threats, it is assumed that a partial investment reduces vulnerability in a linear fashion. That is, vulnerability is cut in half if we invest twice as much to reduce it.[11]

For example, the enterprising couple might allocate only $100 to reduce their risk of over-drafting the account—$75 to reduce the vulnerability in Figure 6.7 designated as (Balance < $100), and the remaining $25 to reduce the vulnerability designated as (Balance < $50). This is not enough to eliminate all vulnerabilities, but if we assume a linear relationship between investment and vulnerability reduction, an allocation of $100 will reduce the overall vulnerability. The other two threats, (H-Withdraw = $100) and (W-Withdraw = $50), receive nothing, because it is assumed check writing in general causes no damages. However, as

[11]As mentioned, this may not be a valid assumption, but it is a simplification.

stated earlier, the bank charges $1 per check, so there is a cost associated with checks. Cost and damage are different; vulnerability reduction is associated with cost, whereas risk is associated with damage.

This allocation results in an overall fault tree vulnerability of 17% and risk of $4.40 (Table 6.2). These calculations are left as an exercise for the reader. [The program *FTplus* may be used to perform these calculations. *FTplus* contains this example under the heading of Examples > Shared Bank Account.]

Manual allocation may be a good strategy, especially if political or intangible considerations take precedence, but as we shall see, it rarely gives the best result if risk minimization is the objective. In fact, we will see that it is possible to reduce vulnerability to zero and, hence, reduce risk to zero through other allocation strategies.

Ranked Order Risk Reduction

Perhaps the most common strategy used by vulnerability analysts is the ranked order method of allocation. In this method, the highest vulnerability or risk is reduced first, followed by the next, and so on, until the budget is depleted. The sector's vulnerabilities or financial risks are computed using some kind of weighted sum of factors. The results are sorted into order by value, and the highest-valued vulnerability or risk is given the highest funding priority. If there are sufficient funds, the next highest in order is funded, and so on, until funding runs out.

Ranked order strategies typically employ some kind of weighting algorithm to compute the vulnerability or risk. For example, the following formula might be used:

$$\text{Risk} = 25\% \times \text{Financial Damage} + 40\% \times \text{Casualties} + 35\% \times \text{Economic Damages}.$$

Risk is calculated for each component of the sector and then ranked from highest to lowest. For example, assume there are four components as listed below, plus the cost to prevent each from being successfully attacked:

Component #1: Bridge, Protection = $80 million.
Component #2: Railway Station, Protection = $50 million.
Component #3: Tunnel, Protection = $25 million.
Component #4: Roadway, Protection = $10 million.

TABLE 6.2. Summary of Risk Reduction Allocation Strategies Applied to the Shared Bank Account for a Budget of M = $100.

Strategy	H-Over	Balance < $100	W-Over	Balance < $50	Risk
Manual	$0	$75	$0	$25	$4.40
Ranked Order	$0	$50	$0	$50	$4.00
Optimal	$1	$49	$1	$49	$0.00
Apportioned	$1	$58	$1	$40	$0.00

Now assume there is $100 million available to protect this miniature transportation sector from successful attacks. The ranking given above and the budget limitation of $100 million leads to the following allocation strategy:

Component #1: Bridge, Investment = $80 million.

Component #2: Railway Station, Investment = $20 million.

Component #3: Tunnel, Investment = none.

Component #4: Roadway, Investment = none.

This strategy funds the must vulnerable components at the expense of the least vulnerable components. But it is not optimal in general. That is, it does not guarantee that the money will be invested in such a manner as to minimize risk. Instead, it removes the highest risks.

The ranked order strategy applied to Figure 6.7 (and Figure 5.8) produces the following allocation and reduces the vulnerability of the shared bank account system from 49.6% to 20%, and the risk from $17.60 to $4.0. See Table 6.2. This is an improvement over the manual strategy, but it does not reduce risk to zero.

Using the ranked order strategy, the thrifty husband–wife team allocates $0 to threat (H-Withdraw = $100), $50 to threat (Balance < $100), nothing to the threat posed by (Wife Withdraw = $50), and the remaining $50 to reduce the vulnerability due to threat (Balance < $50). This allocation reduces vulnerabilities from those given in the fault tree of Figure 6.7 as follows: 50% to 50% (no change), 80% to 40%, 20% to 20% (no change), and 80% to 0%.

Because the cost of removing the vulnerability due to threat (H-Withdraw = $100) is only $1, and similarly removing the vulnerability due to (W-Withdraw = $50) costs only $1, it may seem somewhat contrived to allocate $0 only to these vulnerabilities. For a total investment of $2, the shared bank account would be exempt of all risk, because of the AND gate in the fault tree. But the ranked order strategy overlooked this, because it seeks to reduce the highest risk threats, first. The $2 vulnerabilities were not the highest risks. In fact, they are the lowest risk threats!

This example illustrates one limitation of the ranked order strategy. Ranked order allocation ignores fault tree logic, as illustrated in this case, and does not guarantee the minimum risk allocation. Ranked order is easy to apply, but not always the best strategy. What is?

Optimal Risk Reduction

The optimal strategy allocates funding such that the mathematical risk is reduced. That is, it finds the minimum risk by solving the following problem:

Given M dollars and a fault tree with its associated vulnerabilities $prob_i$, $cost_i$, and damages d_i, what is the allocation $alloc_i$ that minimizes R?

M = total budget, not to be exceeded.
n = number of threats in fault tree.
k = number of events in the event tree; $k = 2^n - 1$.
$v_i = prob_i$ = component vulnerability (ranges from 0% to 100%).
d_i = fault damage (typically expressed in terms of dollars).
$r_i = prob_i \times d_i$ = risk of component i (typically expressed in terms of dollars).
$cost_i$ = cost to remove vulnerability threatening component i.

Find:

$alloc_i$ = amount to invest in each vulnerability, to reduce it.
$= p_i \times (cost_i / prob_i)$.
p_i = reduced fault probability: $(0 <= p_i <= prob_i)$.

Subject to the constraint: $\Sigma_1^n alloc_i <= M$. **Derivation 6.2** contains the details on how the following objective function, R, is obtained:

$$R = \Sigma_{j=0}^{k} FT(j)[\Sigma_{i=1}^{n}(d_i \times b_i(j))\Pi_{i=1}^{n}(1 - p_i)^{\bar{b}_i(j)} \times p_i^{b_i(j)}]$$

where $\bar{b}_i(j)$ is the complement of $b_i(j)$, which is simply the ith bit of the binary expansion of j.[12]

This formidable equation compactly models the risk of any fault tree. But it also poses a challenge to minimize, because it is a highly nonlinear polynomial of order n. In addition, the presence of exponent $\bar{b}_i(j)$ means the volume defined by R in n-dimensional space has many "sharp corners," which makes it particularly difficult to solve using traditional mathematical optimization techniques. In more formal mathematical terms, the convex set formed by R is not smooth enough to lend itself to traditional continuous mathematics steepest-descent techniques.

Fortunately, we live in an age of fast and inexpensive computers, so this problem can be solved using the familiar emergence technique used earlier. All that is needed is a fast and inexpensive computer and an organizing principle as before.[13] Here is an organizing principle that finds the best allocation of M, for an arbitrary fault tree, such that R is minimized:

Optimal Risk Reduction Organizing Principle

Initially, distribute the M dollars to the n threats. This is done like dealing cards to card players, until M has been allocated. In most cases, this means threats get an equal number of dollars, or approximately the same number. This initial

[12]Binary expansion of $j = 6$ is 110. Counting from left to right, the bits are $b_1(6) = 1$, $b_2(6) = 1$, and $b_3(6) = 0$.

[13]The program *FTplus* incorporates the optimal risk reduction technique described here.

allocation satisfies the constraint, $\Sigma alloc_i = M$, which must be maintained throughout all steps of the emergence process. To stay within this constraint, the organizing principle must move the same number of dollars from one threat to another. The randomly selected fault tree threats are called DONORS and RECIPIENTS:

To find:

$$alloc_i = p_i \times (cost_i / prob_i).$$
$$p_i = \text{reduced fault probability: } (0 <= p_i <= prob_i).$$

Repeat forever:

1. Select a threat at random, called the DONOR; select another threat at random, called the RECIPIENT.
2. If $(0 < p_i <= prob_i)$ of DONOR and $(0 <= p_i < prob_i)$ of RECIPIENT, deduct \$1 from DONOR and add \$1 to RECIPIENT.
3. If risk R is not decreased, take back the donation; otherwise keep it.

Repeated application of this organizing principle will result in an allocation of M to each threat such that R is minimized. But the answer may not be satisfactory, because optimality, like the ranked order strategy, often leaves some vulnerability unfunded, whereas others are completely funded.

Returning to the shared bank account, the inputs to the optimal risk reduction organizing principle are:

$$M = \$100.$$
$$n = 4.$$
$$k = 15.$$
$$prob_i = 50\%, 80\%, 20\%, 80\%.$$
$$d_i = \$0, \$20, \$0, \$20.$$
$$cost_i = \$1, \$100, \$1, \$50.$$

Initially, the organizing principle allocates \$100 as follows:

$$alloc_i = \$25, \$25, \$25, \$25.$$

After several iterations of the organizing principle, the sector risk is driven to zero, and the optimal risk reduction strategy emerges[14]:

Husband Withdraw = \$100:	alloc = \$1	v = 0%.
Balance < \$100:	alloc = \$49	v = 41%.

[14]Emergence is a random process, so the answer may vary from application to application. In this case, the author obtained an allocation of \$1, \$49, \$2, and \$48.

Wife Withdraw = \$50: alloc = \$1 v = 0%.
Balance < \$50: alloc = \$49 v = 2%.

Why is the sector risk reduced to zero if two vulnerabilities remain greater than zero? Once again, the AND logic in the fault tree prevents these two nonzero vulnerabilities from contributing to the overall fault tree vulnerability. Even though the probability of (Balance < \$100) is greater than zero, the AND logic prevents this vulnerability from propagating up the tree to its root. The organizing principle accommodates the fault tree logic through the FT(j) term of R (**Derivation 6.2**).

DERIVATION 6.2: OPTIMAL RISK REDUCTION

Optimal risk reduction finds the best allocation of M dollars to reduce the risk in a fault tree with n threats by reducing the fault probabilities $prob_i$ associated with each threat i. It essentially "buys down" vulnerability, according to a linear rule that equates dollars with fault probabilities, $prob_i$.

More formally, find two numbers for each threat:

$$alloci = p_i \times (cost_i / prob_i).$$
$$p_i = \text{reduced fault probability: } (0 <= p_i <= probi).$$

This minimization is subject to the constraint:

$$\Sigma_1^n alloc_i <= M.$$

The objective function, R, is the mathematically exact expression for risk:

$$R = \Sigma_{j=0}^{k} FT(j)\ [\Sigma_{i=1}^{n}(d_i \times b_i(j))\ \Pi_{i=1}^{n}(1 - p_i)^{bi\wedge(j)} \times p_i^{b_i(j)}].$$

The $b_i^\wedge(j)$s are the complement of $b_i(j)$, which are simply the ith bit of the binary expansion of j. Each $b_i(j)$ is either 0 or 1. If $b_i(j) = 0$, then $b_i^\wedge(j) = 1$, and the reverse. FT(j) is zero or one, depending on whether event tree outcome j propagates up the fault tree to the top, and d_i are the damages associated with each threat.

To understand the objective function, it is necessary to understand how FT(j) and the $b_i(j)$s are constructed. Consider the fault trees developed in Chapter 5 (Figures 5.3, 5.8–5.10, and 5.12) as illustrative examples.

The term $b_i(j)$ is obtained by numbering the event tree outcomes from zero to $k = 2^n - 1$, as follows. Number the event tree outcomes as j = 0, 1, 2, 3 . . . k, and convert each j to its binary number equivalent $b_1(j)b_2(j)b_3(j) \ldots b_k(j)$ = 000, 001, 010, 011 . . . for k = 7. Each b corresponds to TRUE/FALSE in the event tree. For example, in Figure 5.3, Threat A = TRUE corresponds with $b_1(j) = 1$.

Threat A = FALSE corresponds with $b_1(j) = 0$. As there are $n = 4$ threats in Figures 5.3 and 5.8, the values of j and $b_1(j)$ are as follows:

j	b_1	b_2	b_3	b_4
0	0	0	0	0
1	0	0	0	1
2	0	0	1	0
3	0	0	1	1
4	0	1	0	0
5	0	1	0	1
6	0	1	1	0
7	0	1	1	1
8	1	0	0	0
9	1	0	0	1
10	1	0	1	0
11	1	0	1	1
12	1	1	0	0
13	1	1	0	1
14	1	1	1	0
15	1	1	1	1

The binary expansion of j is exactly equivalent to the binary event tree. Each bit represents YES/NO or TRUE/FALSE just as in the event tree. Every path through the event tree corresponds with a binary expansion $b_i(j)$. This explains why $b_i(j)$ and $b_i^\wedge(j)$ are used as exponents in the product term of R. Raising $(1 - p_i)$ and p_i to a zero power produces 1, regardless of the value of p_i. Similarly, raising them to the power of one does nothing to change their value but simply includes them in the product term. Therefore, the term in R for vulnerability: $(1 - p_i)^{b_i^\wedge(j)} \times p_i^{bi(j)}$ is either $(1 - p_i)$ or p_i. Multiplying over all threats results in product terms like those shown in Figure 6.2.

For each binary expansion of j, the truth value of the fault tree FT(j) is either TRUE or FALSE. Let TRUE be modeled as "1" and FALSE as "0." Therefore, if FT(j) = 1, event j propagates to a sector fault, and if FT(j) = 0, the fault does not propagate to the top of the fault tree. What is the nature of FT(j)?

Every fault tree can be expressed as a *logical expression* consisting of $b_i(j)$s and OR and AND operations. For example, the logical expressions for the following Chapter 5 fault trees are:

Figure 3: FT = (Threat A AND Threat B) OR (Threat C AND Threat D) = $b_1(j)b_2(j) + b_3(j)b_4(j)$.
Figure 8: FT = (Threat #1 OR Threat #2 OR Threat #3) = $b_1(j) + b_2(j) + b_3(j)$.
Figure 9: FT = (Threat #1 AND Threat #2 AND Threat #3) = $b_1(j)b_2(j)b_3(j)$.

For the jth outcome of an event tree, convert j into binary number $b_1(j)b_2(j)b_3(j)\ldots$ and substitute the value of each bit into the expression FT(j). FT(j) will evaluate to zero or one, which corresponds to a sector fault or not, depending on the logic of the fault tree. Binary arithmetic is addition and multiplication in modulo 2. OR logic converts to addition:

$$0 + 0 = 0.$$
$$0 + 1 = 1.$$
$$1 + 0 = 1.$$
$$1 + 1 = 0.$$

AND logic converts into multiplication:

$$0 \times 0 = 0.$$
$$0 \times 1 = 0.$$
$$1 \times 0 = 0.$$
$$1 \times 1 = 1.$$

The only unexplained term in R is $\Sigma_{i=1}^{n}(d_i \times b_i(j))$. Once again, each $b_i(j)$ is either zero or one, and each d_i corresponds to the damage caused by vulnerability i. If the event tree outcome is a single fault, d_i damage is done; if it is a double combination fault, $d_i + d_j$ damages result; and a triple combination fault causes $\Sigma_{i=1}^{n}(d_i \times b_i(j))$. Therefore, the jth event of the event tree contributes $\Sigma_{i=1}^{n}(d_i \times b_i(j))$ times event probability risk to the overall fault tree risk.

These equations are implemented in program *FTplus*. It is not necessary to calculate them by hand. In fact, manual calculation is perhaps too error-prone to leave to a human!

Apportioned Risk Reduction

Manual allocation may not yield a low risk, ranked order allocation may not be optimal, and the optimal allocation strategy may not satisfy politicians. A fourth strategy is to spread the money across as many threats as possible, while reducing risk as much as possible. This is the concept underlying the apportioned allocation strategy described here.

As shown in Table 6.2, the apportioned risk reduction strategy tends to spread investment more evenly across threats while reducing risk. In this case, it reduces the sector risk to zero just like the optimal strategy. Apportioned reduction is a "middle of the road" strategy that meets two objectives: (1) reduce risk and (2) fund as many counter-threat target hardening projects as possible.

Equations for apportioned reduction will be developed for two substrategies: vulnerability reduction and risk reduction. Vulnerability reduction attempts to reduce fault probabilities—the likelihood of any successful attack—whereas the risk reduction equations attempt to reduce risk—the financial liability implied by a successful attack.

Apportioned risk reduction equations maximize the difference between initial fault probability (vulnerability) and reduced fault probability, thereby reducing the risk contribution from each threat. Funding is allocated to reduce fault probabilities and thereby to reduce vulnerability and risk. In each case—vulnerability reduction or risk reduction—the formula is based on the least-squares sum of these differences.

VULNERABILITY REDUCTION

Vulnerability reduction attempts to reduce the probability of a successful attack across all threats given in the fault tree. Its purpose is to judiciously spread funding across these faults such that the *total* improvement is maximized. It does this by minimizing the sum-of-squares objective function FR, given by:

$$FR = \Sigma_{i=1}^{n}[\text{prob}_i - p_i]^2 \text{ subject to } = \Sigma_{i=1}^{n}\text{alloc}_i = M.$$

This is a classic optimization problem. The inputs to this problem are the fault probability distribution prob_i, the estimated costs to eliminate each fault cost_i, and the total budget available M. The outputs are p_i and alloc_i for all n threats. The allocation cannot exceed M, so the computation of each p_i in this function must be constrained by M.

More formally, for a fault tree with n threats and a budget of M:

Given: prob_i = probability of fault due to threat i.

cost_i = the cost to reduce fault probability i to zero

Note: $c_i = \text{cost}_i/\text{prob}_i$.

M = the budget (total funding available), where i = 1 to n

Find: p_i = the reduced probability of fault i occurring after the investment

alloc_i = dollar allocation to reduce fault probability i. This is computed by multiplying $p_i \times c_i$ for each threat i = 1 to n.

The optimal budget allocation is given by the following formulas (these formulas follow directly from application of classic optimization techniques; see **Derivation 6.3**):

$$\text{Slack} = \Sigma_{i=1}^{n}[\text{prob}_i \times c_i] - M.$$

$$c2 = \Sigma_{i=1}^{n}c_i^2.$$

DERIVATION 6.3: FAULT REDUCTION OPTIMIZATION

Let fault i occur with probability prob(i). Find the optimal decrease p(i) such that the total fault reduction is maximized. Use the following:

Given:

M = budget total.
n = number of faults.
i = threat number; i = 1 to n.
$prob_i$ = likelihood that threat i occurs.
c_i = cost to reduce vulnerability posed by threat i by 1% point = $cost_i / prob_i$.

Solve for:

p_i = amount of reduction in fault i obtained by investment $alloc_i$
$alloc_i$ = portion of budget M Allocated to remedy threat i; i = 1 to n. This is computed by multiplying the amount of fault probability reduction by cost per point: $c_i \times p^i$.

Step 1: Form an objective function FR:
Let the difference between the needed reduction $prob_i$ and the affordable reduction p_i be computed as the sum of squares: $\Sigma_{i=1}^{n}[prob_i - p_i]^2$, which can be thought of as the vulnerability remaining in the infrastructure after M has been invested.

Let the budget be fully allocated, which means $\Sigma_{i=1}^{n} alloc_i = \Sigma_{i=1}^{n} c_i \times p_i = M$. This places a constraint on the sum of squares above, so that the objective function can be written as a combination of sum of squares and budget constraint as follows:

$$FR = \Sigma_{i=1}^{n}[prob_i - p_i]^2 + 2\lambda[M - \Sigma_{i=1}^{n} c_i \times p_i],$$

where λ is the *Lagrangian multiplier*—a *slack* quantity that must be solved for in the next step.

Step 2: Minimize FR to obtain λ and p_i:
Do this by setting the partial derivative of FR to zero with respect to λ and all n unknowns p_i and by solving a system of n + 1 simultaneous equations.
Note: δ is the partial derivative operator.

I. $\delta FR/\delta\lambda = [M - \Sigma_{i=1}^{n} c_i - p_i] = 0.$
II. $\delta FR \delta p_i = -2[prob_i - p_i] - 2\lambda \times c_i = 0.$
III. Solving II. for p_i: $p(i) = prob_i + \lambda \times c_i$.

Now, substitute III into I and solve for λ:

$$M = \Sigma_{i=1}^{n} c_i \times [\text{prob}_i + \lambda \times c_i] = \Sigma_{i=1}^{n} c_i \times \text{prob}_i + \lambda \Sigma_{i=1}^{n} c_i^2.$$

Rearrange terms and solving for λ:

$$\lambda = [M - \Sigma_{i=1}^{n} c_i \times \text{prob}_i]/\Sigma_{i=1}^{n} c_i^2.$$

Continuing the quest for p(i), substitute the expression above for λ into III. To get:

$$\text{IV. } p_i = \text{prob}_i - c_i \times [\{\Sigma_{i=1}^{n} c_i \times \text{prob}_i - M\}/\Sigma_{i=1}^{n} c_i^2].$$

Step 3: Consider cases:
Let Slack $= \Sigma_{i=1}^{n} c_i \times \text{prob}_i - M$ be the amount of money needed to reduce all fault probabilities to zero, minus the budget M. Slack represents the shortfall in funding. If this shortfall is greater than zero, it must be spread across all faults.

Case I: Slack less than or equal to zero: $p_i = \text{prob}_i$.
Case II: Slack is greater than zero: $p_i = \text{prob}_i - c_i \times [\text{Slack}/\Sigma_{i=1}^{n} c_i^2]$.

Step 4: Calculate the allocations, alloc_i:

$$\text{alloc}_i = c_i \times p_i, \text{ for all } i = 1 \text{ to } n \text{ threats.}$$

Note: Objective equation FR does not constrain the reduced probabilities p_i to positive numbers. This limitation can be remedied by noting that negative values of p_i are not realistic, and hence, the negative p_i's must be set to zero and the "excess" reallocated to other faults. This is done as follows:

Let $T = \Sigma_{i=1}^{n} \text{alloc}_i$ for only the alloc_i's that are positive. That is, ignore the negative allocations in summing T. Then adjust each remaining alloc_i as follows:

If alloc_i is greater than zero, set alloc_i to **min**$\{M, \text{alloc}_i \times M/T\}$; otherwise, set $\text{alloc}_i = 0$.
The fault reduction probabilities are adjusted:

$$p_i = \text{alloc}_i/c_i, \text{ for all } i = 1 \text{ to } n.$$

For i = 1 to n:

If slack > 0, $p_i = \text{prob}_i - [c_i \times \text{slack}/c2]$, otherwise, $p_i = \text{prob}_i$.
$\text{alloc}_i = p_i \times c_i$.

Distribution of M dollars to $alloc_i$ reduces vulnerability by lowering fault probabilities on individual threats, but it does not minimize vulnerability over the entire fault tree, because the objective function derived in **Derivation 6.3** considers only fault tree threats. It ignores the double, triple, and other combination faults enumerated by the event tree. The amount of fault probability decrease is dictated by the fault tree inputs, in contrast to the optimal strategy, which is defined by all combination faults enumerated in the event tree.

Once again, assume the following inputs and the data provided by the shared bank account fault tree of Figure 6.7:

$$
\begin{aligned}
M &= \$100. \\
n &= 4. \\
prob_i &= 50\%, 80\%, 20\%, 80\%. \\
d_i &= \$0, \$20, \$0, \$20. \\
cost_i &= \$1, \$100, \$1, \$50. \\
c_i &= \$1/50, \$100/80, \$1/20, \$50/80 = 0.2, 1.25, 0.5, 0.625.
\end{aligned}
$$

To find p_i, and $alloc_i$, compute the following per the equations given above:

$$
\begin{aligned}
\text{Slack} &= \Sigma_{i=1}^{4}[prob_i \times c_i] - M \\
&= [50 \times 1/50 + 80 \times 100/80 + 20 \times 1/20 + 80 \times 50/80] - 100 \\
&= [1 + 100 + 1 + 50] - 100 = 52.
\end{aligned}
$$

$$
\begin{aligned}
c2 &= \Sigma_{i=1}^{n} c_i^2 = [0.02^2 + 1.25^2 + 0.05^2 + 0.625^2] = 1.956. \\
\text{Slack}/c2 &= 52/1.956 = 26.58. \\
\text{Slack} &> 0, \text{ so:} \\
p_1 &= prob_1 - [c_1 \times slack/c2] = 50.0 - [0.02 \times 26.58] = 49.5. \\
p_2 &= prob_2 - [c_2 \times slack/c2] = 80.0 - [1.25 \times 26.58] = 46.8. \\
p_3 &= prob_3 - [c_3 \times slack/c2] = 20.0 - [0.05 \times 26.58] = 18.7. \\
p_4 &= prob_4 - [c_4 \times slack/c2] = 80.0 - [0.625 \times 26.58] = 63.4.
\end{aligned}
$$

Now:

$$
\begin{aligned}
alloc_i &= p_i \times c_i \text{ for each fault, } i = 1 \text{ to } 4: \\
alloc_1 &= p_1 \times c_1 = 49.5 \times 0.02 = \$1 && \text{(H-Withdraw} = \$100). \\
alloc_2 &= p_2 \times c_2 = 46.8 \times 1.25 = \$58.5 && \text{(Balance} < 100). \\
alloc_3 &= p_3 \times c_3 = 18.7 \times 0.05 = \$0.94 && \text{(W-Withdraw} = \$50). \\
alloc_4 &= p_4 \times c_4 = 63.4 \times 0.625 = \$39.6, && \text{(Balance} < \$50).
\end{aligned}
$$

As p_i are the reductions, the difference ($prob_i - p_i$) is the amount of vulnerability remaining after the investment reduces each. Therefore, the allocation computed here leaves the following vulnerabilities:

$$p_1 = 50 - 49.5 \sim 0\% \quad \text{(H-Withdraw = \$100).}$$
$$p_2 = 80.0 - 46.8 \sim 33\% \quad \text{(Balance < \$100).}$$
$$p_3 = 20.0 - 18.7 \sim 0\% \quad \text{(W-Withdraw = \$50).}$$
$$p_4 = 80.0 - 63.4 \sim 17\% \quad \text{(Balance < \$50).}$$

The \$100 investment reduces sector vulnerability from 49.6% to approximately 0%, but there remain two nonzero fault probabilities in the allocation. How can the sector vulnerability be nearly eliminated while individual fault probabilities are nonzero? Once again, the AND logic in the fault tree prevents these nonzero vulnerabilities from contributing to the overall fault tree vulnerability. Even though the probability of (Balance < \$100) is 33%, the AND logic prevents this vulnerability from propagating up the tree to its root. Similarly, the probability of (Balance < \$50) is 17%, but this fault cannot propagate up the fault tree because it is ANDed with (W-Withdraw = \$50), which occurs with a probability of 0%.

Vulnerability reduction is also called *fault reduction*, because the objective is to reduce the probability of a fault. This is different than risk reduction, which aims to reduce financial risk. Although both strive to reduce the probability of faults, risk reduction does so to reduce the expected value of damage.

FINANCIAL RISK REDUCTION

Fault reduction is not the same as financial risk reduction. Risk reduction is influenced by the damage estimates, whereas fault reduction ignores damages. Use fault reduction to reduce the likelihood of a fault. Use risk reduction to reduce expected damages. From a policy point of view, fault reduction may be favored because it focuses on prevention of any fault, whereas risk reduction focuses on the damages associated with a fault. Fault reduction may be more important than risk reduction, especially if human life is involved.

Financial risk is defined as expected loss, measured in dollars, averaged over all faults. *Financial risk reduction*—abbreviated as *risk reduction*—is any preventive or protective measure that reduces financial loss, averaged across all faults. Risk reduction is different than fault reduction, because it considers damages as well as probabilities. As a consequence, risk reduction through an apportioned risk reduction strategy generally produces a different budget allocation than fault reduction.

As an example, consider two bridges A and B that are both subject to a 20% probability of failure due to an attack of some kind. Experts estimate the financial impact (physical, economic, loss of jobs, etc.) of destruction of bridge A as \$100 M and

TABLE 6.3. Comparison of Fault vs. Risk Reductions.

Fault#	Fault Reduction	Risk Reduction
1	$1.0	$1.0
2	$58.5	$58.0
3	$1.0	$1.0
4	$39.5	$40.0

bridge B as one billion dollars. Clearly the financial risk is greater with the more expensive bridge, even though the fault probabilities are the same. Thus, it makes more sense to allocate more money to protect the expensive bridge. Risk assessment of the two bridges depends on the financial damages. Vulnerability analysis depends only on the 20% fault probabilities.

In the shared bank account example, investing $100 toward fault reduction gives one investment strategy and risk reduction gives only a slightly different funding strategy (Table 6.3). This is because damage estimates are uniformly the same for husband and wife.

The following development differs slightly from the previous vulnerability reduction development, because of the inclusion of threat damages d_i into the equations. The damages may be considered as weights in the sum-of-squares formula used earlier. The addition of these weights modifies the resulting equations only slightly. The apportioned risk strategy problem can be solved, in closed form, as before (**Derivation 6.4**)[15]:

Given: $n =$ number of threats: $i = 1$ to n.

$M =$ the budget $= \Sigma_{i=1}^{n}[\text{prob}_i \times c_i]$.

$\text{prob}_i =$ probability of failure due to threat i.

$\text{cost}_i =$ the cost to reduce the probability of failure to zero.

$c_i = \text{cost}_i / \text{prob}_i =$ cost per percentage point.

$d_i =$ financial loss caused by failure i, i.e., the "risk."

Find: $r_i =$ the reduced probability of fault i occurring after the investment.

$\text{alloc}_i =$ dollar allocation to prevent fault due to threat i. This is computed. by multiplying $r(i) \times c_i$ for each threat i.

DERIVATION 6.4: RISK REDUCTION OPTIMIZATION

The derivation of risk reduction optimization is very similar to that of fault reduction with one major addition: d_i are added weights to the sum of squares formulation in the objective function RR.

[15]Of course we could also solve this problem using emergence, but whenever possible it is desirable to obtain a closed-form solution. Computers are fast and cheap, but humans are still more clever.

Let fault i occur with probability $prob_i$. Find the decrease r_i such that the sum of squares objective function is optimized. Use the following:

Given:

M = budget total.
n = number of faults
i = threat number; $i = 1$ to n.
$prob_i$ = likelihood that threat i occurs.
c_i = cost to reduce threat vulnerability i by 1% point.
d_i = damages incurred by threat i.

Solve for:

r_i = amount of reduction in risk i obtained by investment $alloc_i$.
$alloc_i$ = portion of budget M allocated to remedy threat i; $i = 1$ to n. This is computed by multiplying $c_i \times r_i$.

Step 1: Form an objective function RR:
Before allocation, suppose $r = \Sigma_{i=1}^{n} prob_i \times d_i$. After allocation, $r = \Sigma_{i=1}^{n}[prob_i - r_i] \times d_i$. Thus, the "before" and "after" differ only by the reduction r_i. We want to minimize the "after" risk without having plus and minus values of r cancel one another out, so we minimize of the sum of squares: $\Sigma_{i=1}^{n}[prob_i - r_i]^2 d_i$ instead of r.

Now, let the budget be fully allocated, which means $\Sigma_{i=1}^{n} alloc_i = \Sigma_{i=1}^{n} c_i \times r_i = M$. This places a constraint on the sum of squares above, so that the objective function can be written as a combination of sum of squares and budget constraint as follows:

$$RR = \Sigma_{i=1}^{n}[prob_i - r_i]^2 d_i + 2\lambda[M - \Sigma_{i=1}^{n} c_i \times r_i],$$

where λ is a Lagrange multiplier.
Step 2: Minimize RR to obtain λ and r(i):
Do this by setting the partial derivative of RR to zero with respect to λ and all n unknowns r_i and then by solving the resulting system of $n + 1$ simultaneous equations. Note: δ is the partial derivative operator.

I. $\delta RR/\delta\lambda = [M - \Sigma_{i=1}^{n} c_i \times r_i] = 0.$
II. $\delta RR/\delta r_i = -2[prob_i - r_i]d_i - 2\lambda \times c_i = 0.$
III. Solving II. for r_i: $r_i = prob_i + \lambda \times c_i/d_i$.

Now, substitute III into I and solve for λ:

$$M = \Sigma_{i=1}^{n} c_i \times [prob_i + \lambda \times c_i/d_i] = \Sigma_{i=1}^{n} c_i \times prob_i + \lambda\Sigma_{i=1}^{n} c_i^2/d_i.$$

Rearrange terms, and solve for λ:

$$\lambda = [M - \Sigma_{i=1}^{n} c_i \times prob_i]/\Sigma_{i=1}^{n} c_i^2/d_i.$$

With the formula for λ in hand, we can find r_i by substituting the expression above for λ into III to get:

$$\text{IV.}\quad r_i = \text{prob}_i - [c_i/d_i] \times [\{\Sigma_{i=1}^{n} c_i \times \text{prob}_i - M\}/\Sigma_{i=1}^{n} c_i^2/d_i].$$

Step 3: Consider cases:
Let Slack $= \Sigma_{i=1}^{n} c_i \times \text{prob}_i - M$ be the amount of money needed to reduce all fault probabilities to zero, minus the budget M. Slack represents the shortfall in funding. If this shortfall is greater than zero, it must be spread across n threats.

Case I: Slack less than or equal to zero: $r_i = \text{prob}_i$.

Case II: Slack is greater than zero (deficiency in funding):

$$r_i = \text{prob}_i - [c_i/d_i] \times [\text{Slack}/\Sigma_{i=1}^{n} c_i^2/d_i].$$

Step 4: Calculate the allocations a(i):

$$\text{alloc}_i = c_i \times r_i \text{ for all } i = 1 \text{ to n threats.}$$

Note: Objective equation RR does not constrain the results r_i to positive numbers. This limitation can be remedied by noting that negative probabilities are not realistic, and hence, the negative r_i's must be set to zero and the "excess" reallocated to other threats. This is done as follows:
Let $T = \Sigma_{i=1}^{n} \text{alloc}_i$ for only the a_i's that are positive. That is, ignore the negative allocations in summing T. Then adjust each remaining alloc_i as follows:

If alloc_i greater than zero, set alloc_i to **min**$\{M, \text{alloc}_i \times M/T\}$; otherwise set $\text{alloc}_i = 0$.

The probability reductions are adjusted:

$$r_i = \text{alloc}_i/c_i, \text{ for all } i = 1 \text{ to n.}$$

The allocation is obtained by the following (**Derivation 6.4**):

$$\text{Slack} = \Sigma_{i=1}^{n}[\text{prob}_i \times c_i] - M.$$

$$c2 = \Sigma_{i=1}^{n}[c_i^2/d_i]$$ Note: if $d_i = 0$, then c_i^2/d_i is assumed to be zero.,

$$\text{Slack2} = \text{Slack}/c2.$$

If Slack > 0, $r_i = \text{prob}_i - [c_i \times \text{Slack2}/d_i]$ Note: if $d_i = 0$, then $\text{Slack2}/d_i$ is zero.

Otherwise, $r_i = \text{prob}_i$.

$$\text{alloc}_i = r_i \times c_i \text{ for } i = 1 \text{ to n.}$$

Suppose we use the husband–wife shared bank account values as before. In addition, we must know the financial risk for each fault (Figure 6.7):

damage(1) = \$0	no damage due to (H-Withdraw = \$100).
damage(2) = \$20	overdraft penalty for (Balance < \$100).
damage(3) = \$0	no damage due to (W-Withdraw = \$50).
damage(4) = \$20	overdraft penalty for (Balance < \$50).

Then we get the results shown in Table 6.3 from the following:

$$M = \$100.$$
$$n = 4 \text{ faults.}$$
$$\text{prob}_i = 50\%, 80\%, 20\%, \text{and } 80\%.$$
$$\text{cost}_i = \$1, \$100, \$1, \$50, \text{respectively.}$$
$$c_i = 1/50, 100/80, 1/20, 50/80 = 0.02, 1.25, 0.05, 0.625.$$
$$\text{Slack} = \Sigma_{i=1}^{n}[\text{prob}_i \times c_i] - M = [1 + 100 + 1 + 50] - 100 = 52.$$
$$c2 = \Sigma_{i=1}^{n}[c_i^2/d_i] = [0.02^2/0 + 1.25^2/20 + 0.05^2/0 + 0.625^2/20] = 0.09766.$$
$$\text{Slack2} = \text{Slack}/c2 = 52/0.09766 = 532.48.$$

Because Slack and therefore Slack2 are positive, we use the formula:

$$r_i = \text{prob}_i - [c_i \times \text{Slack2}/d_i].$$
$$r_1 = 50.0 - [0.02 \times 532.48/0] = 50.0 - 0 = 50\%.$$
$$r_2 = 80.0 - [1.25 \times 532.48/20] = 80.0 - 33.3 = 46.7\%.$$
$$r_3 = 20.0 - [0.05 \times 532.48/0] = 20.0 - 0 = 20\%.$$
$$r_4 = 80.0 - [0.625 \times 532.48/20] = 80.0 - 16.6 = 63.4\%.$$

And to get the allocations, multiply probability reduction by cost per percentage point:

$$\text{alloc}_1 = r_1 \times c_1 = 50 \times 0.02 = \$1.$$
$$\text{alloc}_2 = r_2 \times c_2 = 46.7 \times 1.25 = \$58.4.$$
$$\text{alloc}_3 = r_3 \times c_3 = 20\% \times 0.05 = \$1.$$
$$\text{alloc}_4 = r_4 \times c_4 = 63.4\% \times 0.625 = \$39.6.$$

This allocation strategy adds up to \$100 and is very similar to the vulnerability reduction allocation because the damage estimates are the same for both husband and wife. For example, if the damage due to (Balance < \$50) was \$40 instead of \$20, the risk reduction allocation would have been \$1, \$54, \$1, and \$44, respectively.

The husband–wife shared bank account example, although somewhat contrived, illustrates the advantages and disadvantages of each strategy, as summarized below.

- Manual: Meets political and economic goals but is rarely optimal or even risk reducing.

- Ranked order: Most common method but does not guarantee the best allocation.
- Optimal: Guarantees the best allocation but often misses political or economic objectives.
- Apportioned: A blend of optimal and ranked order that often gives optimal solutions that are also satisfying because it spreads money around; most components get some amount of funding.

PROGRAM *FTplus*/FAULT TREE ANALYSIS

The calculations for fault and risk reduction are arduous and error-prone. Such calculations are best left for a computer. Fortunately they have been incorporated into the program *FTplus*.[16]

FTplus performs two major functions: (1) allows the user to draw a fault tree on the screen and compute the four allocation strategies described here and (2) automatically enumerates all outcomes of the event tree corresponding to the fault tree. Recall that the purpose of fault tree analysis in MBVA is to accurately model the microstructure of a critical node and to decide the best way to allocate funding to the components of the critical node.

FTplus displays both fault tree and event tree but in different display windows. Figure 6.8(a) shows the fault tree window, and Figure 6.8(b) shows the event tree window. The fault tree window shown in Figure 6.8(a) is where the user constructs a fault tree, selects an example by pulling down the EXAMPLES menu, selects and runs an allocation calculation, or switches between fault tree and event tree displays.

The FILE menu contains standard Open, Save, and SaveAs items. The EXAMPLES menu contains several example fault trees, including the examples in this book. The WINDOW menu contains FAULT TREE and EVENT TREE items, so you can switch between the two displays. The remaining buttons along the top of the window are used to edit an existing fault tree or to create a new one. In every case, a fault tree box is selected by clicking on it and then pressing one of these buttons. For example, to remove a box, select it, and then press the ERASE button.[17] To add a threat, select the box to connect the threat to, and press THREAT. EDIT is used to change values: Select an existing threat box, and then press EDIT.

The controls along the bottom of the display window are used to perform various resource allocation algorithms. After setting the appropriate radio buttons appearing on the left-hand side of this control object, pressing ALLOCATE causes the corresponding allocation algorithm to run. This button changes to PAUSE, which the user presses to stop the allocation program, perhaps to select another configuration of radio buttons. The SENSITIVITY button computes and displays a bar chart showing how calculated risk or vulnerability changes with changes in input

[16] *FTplus*—also known as FT.jar—runs on the Web as *FTplus* and on the desktop as *FT.jar*.

[17] Details of how to use *FTplus* and *FT.jar* are provided in the operations manual.

values, V (vulnerability), C (cost), and D (damage) for each threat. CLEAR erases the fault tree and starts the program over.

The two OBJECTIVE radio buttons appearing in the extreme lower left portion of the window are used to select the objective function: minimize the vulnerability, or minimize the risk. Selection of %VULNERABILITY REDUCTION means the program will allocate funding such that vulnerability is minimized. Selection of $RISK REDUCTION causes the program to allocate funds so that risk is minimized.

The DISPLAY radio buttons appearing next are used to set the display mode: $ALLOC V. THREAT computes the minimum objective function time and the $RISK V. $BUDGET button computes the objective function for various budget values ranging from zero to the maximum needed to reduce the objective function to zero. This setting is useful for determining how much risk reduction the user can afford.

The two STRATEGY columns contain radio buttons for selecting which allocation strategy to compute. MANUAL allows the user to enter allocations and then computes the risk as well as the event tree values corresponding to the manually entered allocations. RANKED implements the ranked order allocation strategy, and APPORTIONED implements the apportioned risk reduction strategy. MINIMAL computes the optimal risk reduction strategy allocation using the emergence organizing principle described earlier in this chapter. The organizing principle used to compute each allocation result is defined by the ***FTplus* Organizing Principle** given in the sidebar.

Each threat box in the fault tree is assigned values: v = fault probability, p = reduced fault probability after an allocation strategy has been applied, c = cost to eliminate the vulnerability, d = damages if an attack succeeds, and a = allocation to be computed.

FTplus takes budget $v = prob_i$, $c = cost_i$, and $d = damage_i$ as inputs and then computes the reductions p_i and allocations a_i for each fault from one to n. Editing the value stored in the root of the fault tree sets the budget: Select the root, press EDIT, and then enter the desired budget value. Similarly, selecting the threat node, pressing EDIT, and entering the values into the resulting data input dialog changes the values shown in the threat nodes.

Selecting EVENT TREE from the WINDOW menu switches from fault tree to the event tree display [Figure 6.8(b)]. Instead of drawing a tree with lines, a code is used to designate which event probability corresponds with a combination event (outcome). Below each outcome is a row of numbers designating which threats are included in the outcome. For example, a row of numbers such as "1 3 4" means this outcome is a combination of threats 1, 3, and 4.

Each outcome bar chart displays the probability of a fault before to the allocation (gray numbers and bars) and the current fault probability that is a result of allocations (the red numbers and bars). The outcome events are numbered from one to $2^n - 1$.

The number of outcome events grows exponentially as the number of threats increase, so the event tree window compresses the event tree when necessary.

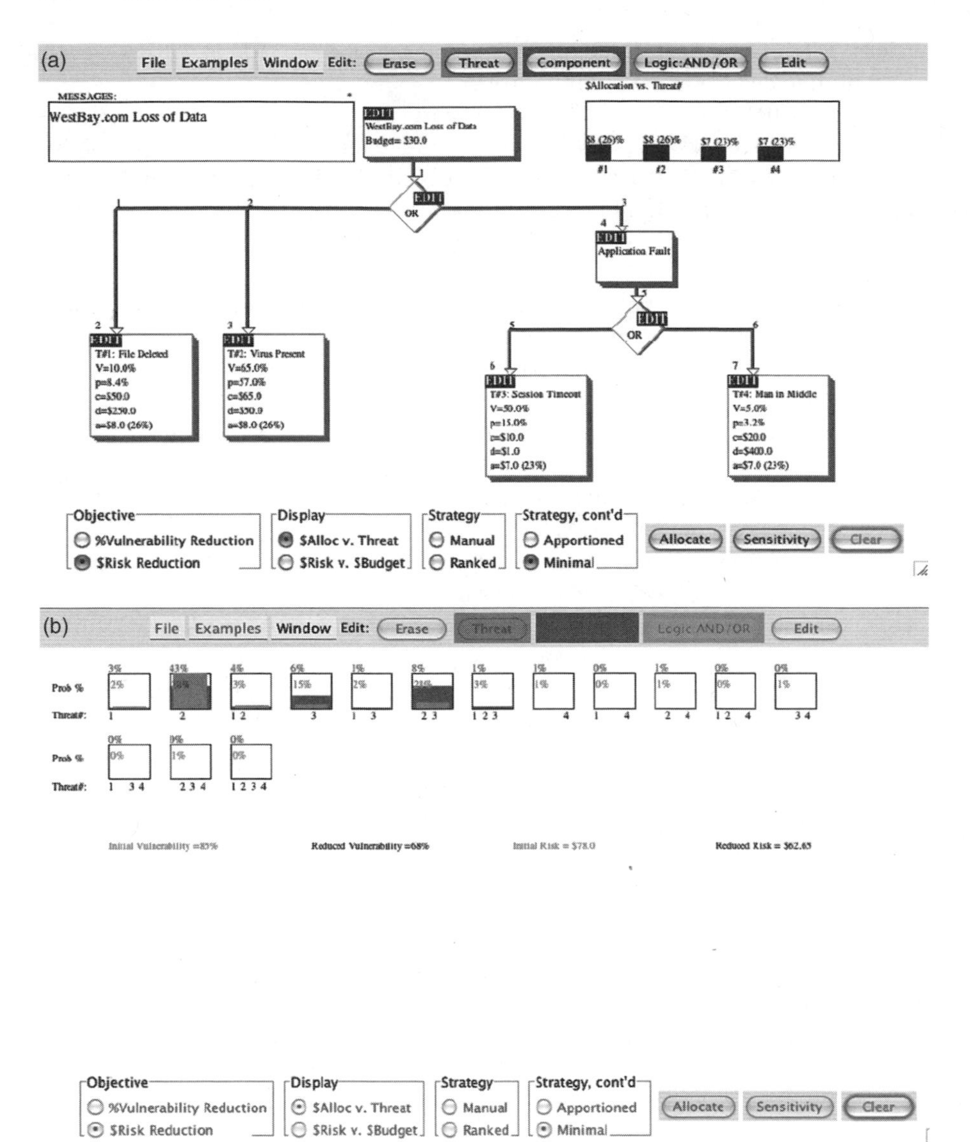

Figure 6.8. Windows of *FTplus*: (a) fault tree window and (b) event tree window.

First, only nonzero probability events are displayed, and then as the number of events grows even larger, the event tree is compressed into a text-only format. V = p (T: a b c) means the event occurs with probability p, and the combination threats are a, b, and c. The bar charts are replaced by a list of nonzero outcomes only.

For more details of how to run FTplus, consult the FTplus User's Guide provided online.

FTPLUS ORGANIZING PRINCIPLE

Given: Vulnerability V, Cost C, Damage D, and Allocation A.
Manual:
For each threat, $i = 1, 2, \ldots, n$, compute the cost c_i to reduce the fault probability V_i, by one percentage point, and then apply the allocation A_i to "buy down" the vulnerability:

1. $c_i = C_i/(100 \times V_i)$, where C_i is expressed as a percentage and V_i is the fault probability (vulnerability) associated with the threat.
2. Reduced vulnerability $v_i = \text{maximum } \{0, V_i - A_i/(100 \times c_i)\}$.
3. Reduced risk $r_i = v_i \times d_i$.

Ranked:
Given budget M: For each threat, $i = 1, 2, \ldots, n$, compute vulnerability/risk:

1. Rank order the threats with the highest vulnerability/risk at the top of the list.
2. Repeat from top of the list $j = \text{top}$, until M is zero:
   a. Compute the cost C_j to reduce the highest-ranked vulnerability/risk threat j to zero.
   b. If $M > C_j$, allocate C_j to threat j. Otherwise, allocate M to threat j.
   c. Subtract C_j from M.
   d. If M is zero, stop. Otherwise, repeat with next in list: $j <- (j - 1)$

Minimal (Optimal):

1. Repeat forever:
2. Randomly select a DONOR threat, and randomly select a RECIPIENT threat.
3. If $V_{RECIPIENT} > 0$, deduct \$1 allocation from DONOR and add \$1 allocation to RECIPIENT. Otherwise, go to step 2.
4. For the entire fault tree, compute the vulnerability, if the objective is to reduce fault probability; compute the risk, if the objective is to reduce risk.
5. If the allocation in step 3 reduces vulnerability or risk, resume with step 2; otherwise restore DONOR and RECIPIENT to original values, and resume with step 2.

Apportioned:

1. Apply the apportionment formula to the threat nodes of the fault tree, only. (The event tree combinations are ignored.)
2. Reallocate negative values as described in Derivations 6.3 and 6.4.

ANALYSIS

The field of vulnerability analysis and risk assessment is still young; no absolutely best strategy has emerged for properly allocating funds. MBVA is currently the most complete method known for screening of nodes to find the most important ones, modeling risks and vulnerabilities as fault trees, and allocating a budget such that worst-case or best-case objectives are met. MBVA establishes a primitive, yet mathematically rigorous, method of analysis. It is based on two fundamental principles that operate over different scales: network-wide availability analysis and individual critical node analysis.

Network-wide analysis examines the large-scale structure of a critical infrastructure sector and attempts to identify the vital components through a mathematically rigorous availability analysis. The goal of network-wide availability analysis is to identify the critical nodes by self-organization. The emergent answer is "discovered" by a simple organizing principle that takes points from donor nodes and links and gives them to recipient nodes and links. What emerges is a pattern: Vital nodes are preferred over less-vital nodes and links. The emergence validates the scale free network theory, which says the nodes with the highest degree are the most vital. But the network does not need to be scale free. Regardless of the critical infrastructure sector's structure, network-wide availability analysis finds the most important nodes (and links) as indicated by the allocation of points.

Network-wide analysis is a more powerful generalization of scale-free network analysis because it works on the structure and the damage values associated with networks. It takes availability points from low-degreed nodes and concentrates them around high-degreed or high-valued nodes. With little analytical foundation, network-wide availability analysis produces the best allocation of funding to nodes. It maximizes overall network security.

Network analysis exploits the structure of a network to reduce risk by optimal allocation. If random networks are at one end of the spectrum of the structure, scale free is at the other end and small world falls somewhere in between. In fact, most critical infrastructure sectors will contain some of both random and scale-free structure. But network analysis, and the algorithms provided by *NA.jar* and *NetworkAnalsyis.html*, works regardless of the amount of structure and the classification of the network. This software optimally allocates availability points to random, scale free, small world, and all networks in between.

The techniques described here suggest an even grander strategy of self-help; i.e., emergence shows how critical infrastructure owners can seek out greater availability through a barter system that takes from donors and gives to the most critical recipients. Any policy that can promote this form of self-help would set into motion an evolutionary process of infrastructure protection and automatic target hardening. Over time, such critical infrastructure networks would increase their reliability and resilience against attacks. It is not clear, however, what form such a policy might take. It would have to reverse century-long trends established by regulatory agencies and recent legislation that encourages competition rather than cooperation among players in the telecommunications, power, and energy sectors, for example.

The second big idea of this chapter is that the most rigorous methods of risk assessment are not particularly absolute. We have explored four strategies and found them to give compelling but different answers! Ranked order, apportioned, and optimal allocation strategies are convincing because they are rigorous, but not very satisfying, because they disagree with one another. This observation underscores the immaturity of this field.

Many vulnerability analysis tools use rank order allocation, but as we have shown, it is not guaranteed to produce optimal allocations.[18] In fact, rank order allocation is often based on a weighted sum of characteristics that have nothing to do with vulnerability or risk. For example, what is the vulnerability or risk associated with psychological damage? Such intangibles are often rated from 1 to 10 and added to other ratings that have equally noncompelling value.

Rank order allocation should be used carefully, because although it is the most popular technique, it is also the one least likely to be based on scientific evidence. (One could argue that the fault probabilities, costs, and damage values used in MBVA are equally suspect because we do not know how to estimate these quantities accurately.)

Apportioned allocation is more "democratic" but equally arbitrary. Its value lies in the balance it establishes between absolute minimization and "load balancing" of demand from various constituents of a critical infrastructure. Apportioned allocation is quantitative, but is it valid? The honest answer is "no," because it does not seek out and find the best allocation possible. But the practical answer is "maybe" because it spreads funding across the sector while attempting to minimize sector risk.

The optimal allocation organizing principle seeks out and finds the absolute minimum vulnerability or risk, given accurate cost and damage estimates. It does so by minimizing the mathematical function that represents vulnerability and risk. But it is brutal in its exactness, leaving some components unfunded. This is more of an artifact of the "sharp corners" in the equations than it is an artifact of the sector being modeled. Probabilities are exact numbers, and models based on them tend to be sharp-cornered. If it was possible to soften the model, perhaps the optimal risk reduction strategy would be more satisfying.[19]

Nonetheless, optimal risk reduction allocation is a benchmark for all others. It provides a way to compare the results of other strategies. In the final analysis, policy makers need to know what the best strategy is, regardless of its political or economic consequences.

[18]CARVER and Port Security Risk Assessment Tool (PSRAT) are two examples that use some kind of ranking to determine which vulnerability or risk is greatest.

[19]"Sharp corners" are a result of functions with $p_1(1 - p_2)(1 - p_3)$ terms, which when minimized tends to drive p_1 to zero and p_2, p_3 to one. This causes the volume defined by $p_1(1 - p_2)(1 - p_3)$ to have "many zeros"—all indistinguishable from one another.

EXERCISES

1. Which of the following is a major step in MBVA (select one)?
   a. Threat analysis—what are the threats?
   b. Gap analysis—what is the funding gap in the infrastructure strategy?
   c. Resource allocation.
   d. Rank vulnerabilities by cost and likelihood.
   e. Give vulnerabilities a score.
2. The goal of the network-wide investment strategy is to (select one):
   a. Identify the hubs.
   b. Use redundancy to eliminate risk.
   c. Find the scale-free power law.
   d. Allocate most of the money to links.
   e. Increase network availability.
3. The most common risk reduction strategy is (select one):
   a. Ranked order allocation
   b. Protect the critical nodes
   c. Protect the hubs
   d. Optimal allocation
   e. Physical security
4. Apportioned risk reduction ... (select one):
   a. Maximizes the difference between fault probability and reduced fault probability.
   b. Minimizes the difference between fault probability and reduced fault probability.
   c. Minimizes the difference between fault probability and risk.
   d. Maximizes the difference between fault probability and risk.
   e. Produces the same allocation for vulnerability and risk reduction.
5. Optimal risk reduction ... (select one):
   a. Maximizes the equation for risk.
   b. Minimizes financial risk due to damages.
   c. Minimizes the sum-of-squares.
   d. Gives the lowest cost allocation.
   e. Maximizes fault probability multiplied times damage.
6. Donors and recipients in organizing principles ... (select one):
   a. Minimize risk.
   b. Maximize availability.

c. Exchange one unit of something.
d. Solve unsolvable optimization problems.
e. Simulate nodes in a network.

7. Generally, the best funding strategy for a network of nodes and links is to . . . (select one):
   a. Fund the scale-free network hub.
   b. Fund the highest-degreed nodes.
   c. Fund the highest-valued nodes.
   d. Fund the highest-degreed and highest-valued nodes.
   e. Invest something in nodes and links or both.

8. Availability is defined as . . . (select one):
   a. The complement of fault probability.
   b. The time between failures.
   c. Reliability in networks.
   d. The optimal allocation of points in a network.
   e. The optimal allocation of points to nodes but not links.

9. [Advanced] Given the barbell vulnerability formula below, what is the network-wide availability of a network whereby all e links and n nodes are assigned an availability of 50% and the degree of all nodes is two?

   $V_{barell} = (1 - a_i a_{lk} a_j)$, where nodes i and j and link l_k form the barbell.

   a. 87.5%e
   b. 50%e
   c. 50%n
   d. 50%(e + n)
   e. 87.5%n

10. MBVA designates cost and the cost-per-percentage point as the cost of (select one):
    a. Checking account penalties.
    b. Removal of a vulnerability.
    c. Financial damage to a component.
    d. The cost to replace a component.
    e. Cashing a check.

11. MBVA designates damage as the cost of . . . (select one):
    a. Repairing a sector.
    b. Loss of people, equipment, and productivity.
    c. Capital expenditure.

d. Financial risk.
e. Bouncing a check in a shared account.

12. Use ranked order risk reduction allocation to allocate \$100 to a sector with two bridges A and B. Assume the probability of faults is 25% for destruction of A and 40% for destruction of B. A cost \$250 to protect, and B cost \$150. In the event of a successful attack on each bridge, A would suffer damages of \$500 and B would suffer damages of \$800. Assume an OR-tree model; what allocation results from a ranked order risk reduction allocation strategy?
    a. A: \$100, B: \$0
    b. A: \$0, B: \$100
    c. A: \$93, B: \$7
    d. A: \$7, B: \$93
    e. A: \$50, B: \$50

13. What is the optimal risk reduction allocation for the two-bridge sector described in problem 12?
    a. A: \$100, B: \$0
    b. A: \$0, B: \$100
    c. A: \$93, B: \$7
    d. A: \$7, B: \$93
    e. A: \$50, B: \$50

14. [Advanced] What is the value of R, for an OR-tree containing two threats, each with a fault probability of 50% and a cost of \$100, but destruction of one causes damages of \$1000 and the other \$500?

$$R = \Sigma_{j=0}{}^{k} FT(j)[\Sigma_{i=1}{}^{n}(d_i \times b_i(j))\Pi_{i=1}{}^{n}(1 - p_i)^{bi\wedge(j)} \times p_i^{bi(j)}]$$

    a. 75%
    b. 50%
    c. \$1000
    d. \$500
    e. \$750

15. [From Chapter 5] The city of San Lewis Rey has a budget of \$100 million and a system of bridges in their transportation infrastructure they want to protect against BOMB and SCADA attacks and POWER outages (because they are drawbridges that need power to operate). See Figure 5.12.

    In addition to the facts collected by the city engineers and given in Chapter 5, it was determined that the cost to protect each bridge and the damages that would result should a bridge be accidentally or purposefully destroyed are as follows:

Fault and Fault Probability Table.

Faults	Probability
POWER outage	5%
BOMB attack	25%
SCADA attack	10%

Threats, Costs, and Damages Table.

Threats	Cost	Damages
POWER	\$5M	\$5M
BOMB	\$100M	\$500M
SCADA	\$2M	\$1M

Given the fault tree of Figure 5.12, a budget of M = \$25 million, and the data given above, answer the following questions.

Q1: What is the financial risk of this sector before allocation of \$25 million?

a. \$18.4 million
b. \$500 million
c. \$506 million
d. \$107 million
e. \$0 million

Q2: What is the optimal risk reduction allocation of \$25 million?

a. Power: \$5 million, Bomb: \$10 million, SCADA: \$10 million
b. Power: \$5 million, Bomb: \$15 million, SCADA: \$5 million
c. Power: \$5 million, Bomb: \$18 million, SCADA: \$2 million
d. Power: \$0 million, Bomb: \$25 million, SCADA: \$0 million
e. Power: \$5 million, Bomb: \$100 million, SCADA: \$2 million

Q3: What is the reduced vulnerability due to each threat after optimal risk reduction allocation?

a. Power: 5%, Bomb: 25%, SCADA: 10%
b. Power: 5%, Bomb: 20%, SCADA: 10%
c. Power: 5%, Bomb: 21%, SCADA: 0%
d. Power: 1%, Bomb: 20%, SCADA: 0%
e. Power: 0%, Bomb: 21%, SCADA: 0%

Q4: According to the optimal risk reduction allocation, how much of the budget is allocated to prevent a combination BOMB + SCADA failure?

a. None
b. $52.25 M
c. $23.75 M
d. $2.375 M
e. $20 M

Q5: What is the difference between fault reduction allocation and risk reduction allocation in terms of the optimal allocation strategy?

a. They are the same.
b. Risk reduction favors combination outcomes.
c. Fault reduction favors combination outcomes.
d. $100 M is not a large enough budget to determine the difference.
e. None of the above.

Q6: San Lewis Rey discovers a budget shortage and cannot afford $25 million to secure the bridges. Instead, they must do what they can with $2 million. What is the optimal risk reduction allocation, now?

a. Power: $0 million, Bomb: $2 million, SCADA: $0 million
b. Power: $1 million, Bomb: $1 million, SCADA: $0 million
c. Power: $0 million, Bomb: $1 million, SCADA: $1 million
d. Power: $0 million, Bomb: $0 million, SCADA: $2 million
e. Power: $2 million, Bomb: $0 million, SCADA: $0 million

Q7: What is the ranked order risk reduction allocation given a reduced budget of $2 million?

a. Power: $0 million, Bomb: $2 million, SCADA: $0 million
b. Power: $1 million, Bomb: $1 million, SCADA: $0 million
c. Power: $0 million, Bomb: $1 million, SCADA: $1 million
d. Power: $0 million, Bomb: $0 million, SCADA: $2 million
e. Power: $2 million, Bomb: $0 million, SCADA: $0 million

Q8: What is the *least amount* of money San Lewis Rey must allocate (total budget) to produce a reduced financial risk of zero, given these cost and damage values?

a. $107 million
b. $0 million
c. $2 million
d. $7 million
e. $5 million

CHAPTER 7

Water

This chapter applies the tools described in Chapters 4–6 to the first Level 1 critical infrastructure—water. First we trace the evolution of water as a valuable resource protected by legislation and regulation to a critical infrastructure that supplies drinking water to 280 million Americans and is indirectly linked to food production (food/agriculture infrastructure) as well as industrial production capacity. Then we produce a model-based vulnerability analysis (MBVA) that zeros in on vulnerabilities of critical nodes for one of the nation's largest cities—the Hetch Hetchy water system that supplies water to the San Francisco Bay Area. This case study illustrates MBVA and shows how to optimally allocate $1.6 billion in water supply improvement funds for the preservation and protection of Hetch Hetchy.

In the first half of this chapter, we show how water supply legislation has evolved from a public health issue (biological contamination of drinking water) to an environmental protection issue (chemical and radiological contamination) and, finally, to a terrorism issue. The main contemporary concern is that terrorists might disrupt the supply of drinking water with a *denial of service* attack.

After reading the second half of this chapter, you should be able to perform a vulnerability analysis on any water supply system as well as justify how best to allocate resources for protecting the water infrastructure in your region or city. This is the first opportunity we have to apply the complete MBVA methodology to a real system.

The following major topics and concepts are described in detail:

1. *Purity vs. Terrorism*: Public health legislation at the turn of the 20th century was focused on water purity and the prevention of disease. The U.S. Public Health Service (PHS) was responsible for protecting drinking water in communities across the country. Today the U.S. Environmental Protection Agency (EPA) is responsible for protecting the water supply system from biological, chemical, and radiological contamination as well as countering denial of service attacks perpetrated by terrorists.
2. *Drinking vs. Agricultural and Industrial Uses of Water*: By far the preponderance of legislation and regulation of water has been aimed at drinking water,

Critical Infrastructure Protection in Homeland Security: Defending a Networked Nation, edited by Ted G. Lewis

and yet over 80% of the water supply is used for agricultural and industrial applications. Intelligent life on this planet depends on water, but so does civilized food production and industrial economy.

3. *SDWA of 1974*: The Safe Drinking Water Act (SDWA) of 1974 is the foundation on which modern water regulation is based. It also transferred responsibility from the U.S. Public Health Department to the EPA. Since 1974, the SDWA has been modified many times, but it still stands as the foundation for contemporary water safety.
4. *Bio-terrorism Act of 2002*: Title 4 of the *Public Health Security & Bioterror Response Act of 2002* extended the SDWA of 1974 to include a new threat: terrorism. It also directs water communities to perform vulnerability analysis for a new failure mode: denial of service (cutting off the supply of water entirely).
5. *Critical Node/Link Approach to Water*: We learn that the water sector lends itself to *scale-free network* analysis. We illustrate this phenomenon by analyzing Hetch Hetchy—the water supply system of the San Francisco Bay Area—and note that tunnels are the critical nodes (links) in Hetch Hetchy.
6. *MBVA*: The Hetch Hetchy case study verifies the usefulness and power of the MBVA technique. We show how to best allocate $1.6 billion to the protection of reservoirs, treatment plants, pipes and valves, and tunnels, against bomb, earthquake, and chemical/biological attacks.
7. *Interdependence*: Finally, we show how interdependencies among water, agriculture, transportation (airports), and power make water one of the primary (Level 1) critical infrastructures. It is myopic to restrict our thinking to drinking water. In fact, other sectors depend heavily on nondrinking as well as drinking water. The water sector may have evolved out of public health concerns, but today it is an essential component of economic vitality.

FROM GERMS TO TERRORISTS

Before the development of the germ theory by Louis Pasteur in the 1880s, water was simply a resource to be exploited for powering water wheels and quenching the thirst of humans, animals, and crops. But soon after Pasteur developed his theory, water became a recognized vector for the transmission of diseases. Dr. John Snow showed how cholera was transmitted from wells to homes via water in 1885.[1] And by 1914, the PHS began setting standards for the *bacteriological* quality of drinking water. But these standards had to be promulgated by water utilities that served sizeable communities. Utilities maximized profit, sometimes at the expense of water purity, and small utilities were not closely monitored. As a

[1]"25 Years of the Safe Drinking Water Act: History and Trends," EPA 816-R-99-007, December 1999, http://www.epa.gov/safewater/consumer/trendrpt.pdf.

consequence, it took the country decades to "purify" all water supplies consumed in the United States.

As more and more unhealthy substances were identified, they were added to the list of substances regulated by the maximum contamination level (MCL) standard. Bacteriological standards were revised in 1925, 1946, and 1962. In 1960, a PHS study showed that only 60% of drinking water met PHS purity standards. Consequently, a 1962 revision increased the number of substances falling within the regulations to 28 substances—the most rigorous standards until 1974. Table 7.1 lists these substances.

A 1972 PHS study of the Mississippi River found 36 *chemical* contaminants in drinking water *after* treatment plants had processed it. The treatment plants were not filtering out these hazardous chemicals, pollution, pesticides, and other chemical and radiological contaminants. Biological contamination was but one of many contaminants in the Mississippi River water supply. The 1972 study underscored the importance of filtering out nonbiological contamination and led to the creation of the modern foundation of water legislation—the *Safe Drinking Water Act of 1974*.

The SDWA establishes the foundation of modern regulations for protecting the purity of water *and water systems*. Because the list of contaminants had grown to include chemicals and other hazardous materials, the responsibility for protecting drinking water and associated processing systems was transferred to the EPA. This foundational act was revised in 1986 and in 1996 and 2002, but it remains the bedrock of water legislation.

The concept of water as a critical resource expanded once again as acts of terrorism multiplied during the 1990s. On May 22, 1998, President Clinton signed Presidential Decision Directive 63 (PDD-63), which identified, among other sectors, *drinking* water as one of America's critical infrastructures. People cannot survive long without water. Terrorists merely need to deny water service for a few days or week to cause major disruptions in the health, environment, and commerce of the country.

TABLE 7.1. These Contaminants are Regulated per the 1962 Public Health Service Standards.[1]

Alkyl Benzene Sulfonate (ABS)	Lead
Arsenic	Manganese
Barium	Nitrate
Beta and photon emitters	Phenols
Cadmium	Radium-226
Carbon Chloroform Extract (CCE)	Selenium
Chloride	Silver
Chromium	Strontium-90
Color	Sulfate
Copper	Threshold Odor Number
Cyanide	Total Coliform
Fluoride	Total Dissolved Solids
Gross alpha emitters	Turbidity
Iron	Zinc

Water "purity" legislation evolved from biological, to environmental, and then to denial of service "contamination." PDD-63 identified the issue, but the Bio-Terrorism Act added terrorism to the list of contaminants. The *Public Health Security and Bio-terrorism and Response Act of 2002* was signed into law by President George W. Bush on June 12, 2002. It was the most significant event affecting water security since the SDWA of 1974. Title IV of this act addresses the water sector and provides several penalties for perpetrators of attacks on water systems.

Shortly after the Bio-terrorism Act of 2002 was signed, the EPA completed the first *classified* Baseline Threat Report describing likely modes of terrorist attack and outlining the parameters for vulnerability assessments by *community* water systems. This report remains classified. One can only speculate that the threats identified by the EPA report are similar to the ones identified in the case study described later in this chapter.

In December 2002, the EPA provided funds to the American Water Works Association (AWWA)—a professional society for water system professionals—to form the Water Information Sharing and Analysis Center (WaterISAC) as prescribed by the National Strategy for Critical Infrastructure Protection. The WaterISAC is a consortium of professional associations and vendors focused on the promotion of water works safety and security. It brings together the private and public sectors to implement the strategies of the SDWA and its descendants. It also provides training and education to its members in subjects such as vulnerability analysis and risk assessment.

FOUNDATIONS: SDWA OF 1974

The SDWA assigns responsibility for water safety to the EPA. But the focus before 1974 was on biological purity. After 1974, the focus expanded to chemical, biological, and radiological purity. The shift from biological contamination to environmental pollutants signaled a phase shift in public policy regarding water. One more shift occured in 2002 when bioterrorism was added to the SDWA foundation.

Minor modifications in 1996 broadened the scope of responsibility of the EPA, (Table 7.2). The EPA is now responsible for entire water supply systems, not just drinking water coming from household taps. This means the EPA is responsible for protecting the entire system, including water from rivers, lakes, pipes, and treatment plants. It includes protection against physical, biological, chemical, radiological, and cyber-threats.

But the regulatory power of the EPA does not extend to all water systems. For example, it does not regulate private wells serving 25 or fewer consumers. And public or *community water* systems must serve at least 3300 consumers to fall within the EPA's jurisdiction. In addition, the stringency of the law increases as the size of the water supply system increases. The EPA distinguishes *community* water supply systems from *private drinking* water systems and other systems such as agricultural and industrial water supply systems.

The EPA does not assume full responsibility for enforcing the SWDA. Instead its strategy is to partner with states, tribes, and private utilities. It aims to regulate and fund local enforcement of the *National Primary Drinking Water Regulations*, which

TABLE 7.2. The 1996 SWDA Amendments Require EPA to Enforce the Following Mandates.[2]

Consumer Confidence Reports

All community water systems must prepare and distribute an annual report about the water they provide, including information on detected contaminants, possible health effects, and the water's source.

Cost–Benefit Analysis

The EPA must conduct a thorough cost–benefit analysis for every new standard to determine whether the benefits of a drinking water standard justify the costs.

Drinking Water State Revolving Fund

States can use this fund to help water systems make infrastructure or management improvements or to help systems assess and protect their source water.

Microbial Contaminants and Disinfection Byproducts

The EPA is required to strengthen protection for microbial contaminants, including *cryptosporidium*, while strengthening control over the byproducts of chemical disinfection. Two new drinking water rules in November 1998 addressed these issues; others will follow.

Operator Certification

Water system operators must be certified to ensure that systems are operated safely. The EPA issued guidelines in February 1999 specifying minimum standards for the certification and recertification of the operators of community and nontransient, noncommunity water systems.

Public Information and Consultation

The SDWA emphasizes that consumers have a right to know what is in their drinking water, where it comes from, how it is treated, and how to help protect it. The EPA distributes public information materials (through its Safe Drinking Water Hotline, Safe water website, and Water Resource Center) and holds public meetings, working with states, tribes, water systems, and environmental and civic groups to encourage public involvement.

Small Water Systems

Small water systems are given special consideration and resources under the SDWA to make sure they have the managerial, financial, and technical ability to comply with drinking water standards.

Source Water Assessment Programs

Every state must conduct an assessment of its sources of drinking water (rivers, lakes, reservoirs, springs, and ground-water wells) to identify significant potential sources of contamination and to determine how susceptible the sources are to these threats.

[2]Understanding the Safe Drinking Water Act, December 1999, EPA 810-F-99-008. The EPA's Office of Ground Water and Drinking Water, website: http://www.epa.gov/safewater/.

define enforceable MCLs for particular contaminants in drinking water (Table 7.1). It specifies that certain proven methods of decontamination be used to treat water to remove contaminants. It also sets standards for drinking water communities according to the number of people served by the system.

MORE EXTENSIONS: THE BIO-TERRORISM ACT OF 2002[3]

Title IV of the Bio-terrorism Act of 2002 extends the SDWA to cover terrorism and modern asymmetric threats such as supervisory control and data acquisition (SCADA) attacks. SCADA is described in Chapter 8. SCADA is the name given to the computer and digital network infrastructure that supports the surveillance and operation of water, power, and energy sectors.

The 2002 Act recommends hardening of targets by adding intruder detection equipment, installing fences, gating, lighting, locks, tamper-proof hydrants, and making improvements to SCADA. It provides funds for training in operations and the handling of chemicals. It requires that water works employees and contractors submit to security screening and provides penalties for breach of confidentiality.

Some highlights of the act are as follows:

- Provides up to $160M in FY02 and "such sums as may be necessary" in FY03–FY05 to (1) perform physical and SCADA vulnerability analysis of all systems with 3300 or more consumers according to the following timetable:
  - March 2003 for communities of 100,000 or more.
  - December 2003 for communities of 50,000 to 100,000.
  - June 2004 for communities of 3300 to 50,000 consumers.
- Restricts who has access to vulnerability assessment information, and specifies penalties of up to 1 year in prison for anyone who "recklessly reveals such assessments."
- Grants up to $5M for small communities (less than 3300 consumers).
- Requires all communities to develop an Emergency Response Plan to "obviate or significantly lesson impact of terrorist attacks."
- Provides up to $15M in FY02 and "such sums as may be necessary" in FY03–FY05 to:
  - Work with the Center for Disease Control (CDC) to "prevent, detect, and respond" to chemical, biological, and radioactive contamination of water.
  - Review methods by which terrorists can disrupt supply or safety.
  - Review means of providing alternative supply in event of disruption.
  - Create a water information sharing and analysis center (WaterISAC).
- Amend the SDWA to extend wording about *water safety* to include wording about disruption of services by *terrorists.*

[3]http://www.fda.gov/oc/bioterrorism/PL107-188.html#title4.

THE WATERISAC[4]

The WaterISAC (http://www.WaterISAC.org) is a consortium of private water works companies. It was developed with the advice of utility representatives under a grant from the EPA to the Association of Metropolitan Water Agencies (AMWA). AMWA is the lead agency.

The WaterISAC Board of Managers comprises water utility managers appointed by the national drinking water and wastewater organizations below. There are also two at-large seats, filled by the Board of Managers:

American Water Works Association
Association of Metropolitan Sewerage Agencies
Association of Metropolitan Water Agencies
AWWA Research Foundation
National Association of Water Companies
National Rural Water Association
Water Environment Federation
Water Environment Research Foundation

The WaterISAC is a bridge between the public and the private sectors operating within the water sector. It has established the following goals and provides the following products for its members:

- Alerts on potential terrorist activity.
- Aggregation of information on water security from federal homeland security, intelligence, law enforcement, public health, and environment agencies.
- Maintain databases of chemical, biological, and radiological agents.
- Identify physical vulnerabilities and security solutions.
- Provide its members with notification of cyber-vulnerabilities and technical fixes.
- Perform research and publish reports and other information.
- Provide a secure means for reporting security incidents.
- Recommend/provide vulnerability assessment tools and resources.
- Provide emergency preparedness and response resources.
- Provide secure electronic bulletin boards and chat rooms on security topics.
- Summarize open-source security information.

IS WATER FOR DRINKING?

After a century of focusing on biological, then environmental, and now terrorist threats to the water supply, the EPA is working with states, tribes, drinking water

[4]http://www.WaterISAC.org, 1620 I Street, NW, Suite 500, Washington, D.C. 20006.

and wastewater utilities (water utilities), and other partners to enhance the security of water, water works, sources of water, and wastewater utilities. It has set the following objectives for itself:

1. The EPA will work with the states, tribes, drinking water and wastewater utilities (water utilities), and other partners to enhance the security of water and wastewater utilities.
2. The EPA will work with the states, tribes, and other partners to enhance security in the chemical and oil industry.
3. The EPA will work with other federal agencies, the building industry, and other partners to help reduce the vulnerability of indoor environments in buildings to chemical, biological, and radiological (CBR) incidents.
4. The EPA will help to ensure that critical environmental threat monitoring information and technologies are available to the private sector, federal counterparts, and state and local governments to assist in threat detection.
5. The EPA will be an active participant in national security and homeland security efforts pertaining to food, transportation, and energy.
6. The EPA will manage its federal, civil, and criminal enforcement programs to meet our homeland security, counter-terrorism, and anti-terrorism responsibilities under Presidential Decision Directives (PDDs) 39, 62, and 63 and environmental civil and criminal statutes.

But the national strategy as implemented by the EPA addresses only a portion of the problem. In California, for example, 80% of the water managed via supply systems, treatment plants, aqueducts, and regulated utilities goes to agriculture, not drinking water. And this does not address the needs of industry. Water is needed to process silicon into computer chips in the $370 billion semiconductor industry. Without water, Silicon Valley would shrivel up as quickly as the Central Valley (major agricultural area of California). In addition, major hydroelectric power plants depend on the abundance of water to generate power for the San Francisco International Airport, for example.

Thus, the question is, "should the water sector be extended beyond drinking water?" Agricultural and industrial uses of water have become as important to national security as drinking water, so why not incorporate these interdependencies? These questions suggest that the concept of denial of service (DoS)—as applied to all water supply systems—is a major vulnerability to public health, agriculture, and the industrial base. Water affects at least 3 of the 11 critical infrastructures defined in the national strategy.

The following case study illustrates the interdependency of water with the economy and food supply of the San Francisco Bay Area. It underscores the vulnerability of a water supply that quenches the thirst of a major metropolitan area—one that serves the famous Silicon Valley, perhaps America's most powerful generator of economic power and is a "cousin" to the California Aqueduct system that supports one of the largest and most productive agricultural regions of the United States.

THE CASE OF HETCH HETCHY

San Francisco is well acquainted with disaster. The 8.3-magnitude earthquake and subsequent fires of 1906 are still remembered by the city. In more modern times, it suffered heavy damage in the 1989 Loma Prieta Earthquake. These natural disasters have forced San Francisco to constantly hone the skills of its firefighters and emergency response personnel to deal with disaster:

> At 5:04 P.M., Tuesday, October 17, 1989, as over 62,000 fans filled Candlestick Park for the third game of the World Series and the San Francisco Bay Area commute moved into its heaviest flow, a Richter magnitude 7.1 earthquake struck. It was an emergency planner's worst-case scenario. The 20-second earthquake was centered about 60 miles south of San Francisco, and was felt as far away as San Diego and western Nevada. Scientists had predicted an earthquake would hit on this section of the San Andreas Fault and considered it one of the Bay Area's most dangerous stretches of the fault.
>
> Over 62 people died, a remarkably low number given the [rush hour] time and size of the earthquake. Most casualties were caused by the collapse of the Cypress Street section. At least 3,700 people were reported injured and over 12,000 were displaced. Over 18,000 homes were damaged and 963 were destroyed. Over 2,500 other buildings were damaged and 147 were destroyed.
>
> Damage and business interruption estimates reached as high as $10 billion, with direct damage estimated at $6.8 billion. $2 billion of that amount is for San Francisco alone and Santa Cruz officials estimated that damage to that county will top $1 billion.[5]

The water supply, however, was minimally impacted by the 1989 disaster. City workers sampled the quality of the water the next day and noted many breaks in lines but no major disruptions in the availability of drinking water. The eight hills throughout the city lost power, and firefighters were forced to pump water from the Bay to put out fires, but the city's 4-day supply of water remained intact.

> The greatest damage to the water system consisted of approximately 150 main breaks and service line leaks. Of the 102 main breaks, over 90 percent were in the Marina, Islais Creek and South of Market infirm areas. The significant loss of service occurred in the Marina area, where 67 main breaks and numerous service line leaks caused loss of pressure.[6]

[5]The October 17, 1989 Loma Prieta Earthquake, http://www.sfmuseum.net/alm/quakes3.html#1989.
[6]Memorandum to Tom Elzey, PUC General Manager from Art Jensen, Acting General Manager, November 21, 1989. Museum of the City of San Francisco.

The damage was minor when considering the size and complexity of the city's water system. Twelve gatemen run the whole system. "The system" contains over 1300 miles of pipeline connecting 8000 hydrants and 45,000 valves. It delivers 80 million gallons per day to 770,000 city dwellers. The San Francisco Public Utilities Commission (SFPUC), which bills to 160,000 meters, sells the surplus to another 1.6 million suburban users around the Bay Area.[7]

The major lesson learned from 1989 water supply damage was to buy more backup power systems. This disaster tested the plumbing and purity of water, not its availability. The consequences of an extended DoS event have yet to be tested. San Francisco was lucky because water kept flowing into the city from 175 miles away. The Hetch Hetchy valley and reservoir located in Yosemite National Park supplies most of the city's water. What happens if this huge water resource dries up? This is the case of Hetch Hetchy, which experienced a minor disruption in 2002.

MODEL-BASED VULNERABILITY

In November 2002, San Francisco voters approved legislation to finance the largest renovation of a water delivery system in San Francisco history. The $3.6 billion capital program funded 77 projects to repair, replace, and seismically upgrade the water system's aging pipelines, tunnels, reservoirs, and dams. In this analysis, we look at the first stage of this decade-long renovation—a $1.6 billion investment designed to harden the drinking water supply.

We have $1.6 billion to spend on Hetch Hetchy. But $1.6 billion is not a large amount to protect such a huge system from biological, chemical, radiological, and terrorist attacks. In addition, Hetch Hetchy is a vast system of reservoirs, treatment plants, pipes, and tunnels. It is too large to completely protect every component, every mile, and every pump. The size of Hetch Hetchy demands that we decide what is most important and what is critical.

Fortunately, we have devised a method—MBVA—that separates critical nodes from less important nodes, identifies the vulnerabilities of greatest magnitude, and provides several methods of allocating the $1.6 billion in remedies. We can use MBVA to identify critical nodes and explore investment strategies that provide the best approach to protecting the water supply. The steps of MBVA are as follows:

1. Take inventory: Identify major components of Hetch Hetchy.
2. Identify critical nodes in Hetch Hetchy—focus on these.
3. Build model of Hetch Hetchy—Fault tree.
4. Analyze Hetch Hetchy for vulnerabilities—Event tree.
5. Suggest optimal resource allocation—protect Hetch Hetchy.

[7] http://Sfwater.org.

CRITICAL NODE ANALYSIS

San Francisco maintains six municipal wells and 980 acres of lakes and land, but the bulk (65% or more) of its water comes from the pure lakes, reservoirs, and streams of the Hetch Hetchy. Hetch Hetchy consists of 14 reservoirs, 22 pumping stations, several tunnels, and several treatment plants, filtration plants, and storage temples. This system delivers 400 million gallons of drinking water per day to 2.4 million customers. It is so big and complex that the first step is to identify the major components of the Hetch Hetchy network from the map of Figure 7.1.

The components of Figure 7.1 are represented as nodes and links in the network of Figure 7.2. Nodes represent treatment plants, reservoirs, lakes, tunnels, storage facilities, filtration plants, and powerhouses, and links represent river flows and pipelines. The network represents the flow of water from the lakes and reservoirs in Yosemite—on the right of Figure 7.2—to the city of San Francisco designated as node SFO—on the left of Figure 7.2. Although tunnels are link-like parts of the water flow, they have been designated nodes in the network because they are important components, as we shall see.

This network also represents several powerhouses as nodes. Hetch Hetchy provides electrical power to San Francisco and the International Airport located 12 miles south of the city. Power is used to run the treatment plants and pumping stations within the system. Hydroelectric generators along the Hetch Hetchy system power most utilities for the city of San Francisco. Not only is power and water interdependent in this case, but also Hetch Hetchy is dependent on itself for power!

The main power transmission line can be seen in Figure 7.1, but we ignore it for the purposes of this analysis.[8]

It is not always obvious what to designate as a link or node. For example, Figure 7.2 contains three *Merge* nodes—places where pipelines come together or join. These are treated as nodes, but we could have treated them as links. In addition, storage facilities are represented as nodes, even though they could be ignored. For example, Sunol Temple is a storage node that resides near the San Antonio Reservoir and appears in the network as a node with a single link.

The next step in our analysis of the water network is to determine its macroscale structure. This tells us something about the architecture of the network. Is it scale free? Is it a small-world network? Can we identify the nodes that are most vital to the operation of this network? Maybe this network is random. Regardless of the exact mathematical structure of the water supply network, we want to find the hub or hubs of the network, because the critical nodes of any network are its hubs.

Recall that we identify network hubs by counting the number of links emanating from each node, converting these counts into percentages, and then plotting the percentages on a histogram. The vertical axis of this histogram is the percentage of nodes with a certain number of links, and the horizontal axis is the corresponding

[8]Another network could be constructed that models the interdependencies of water, power, and agriculture, but we resist the temptation here.

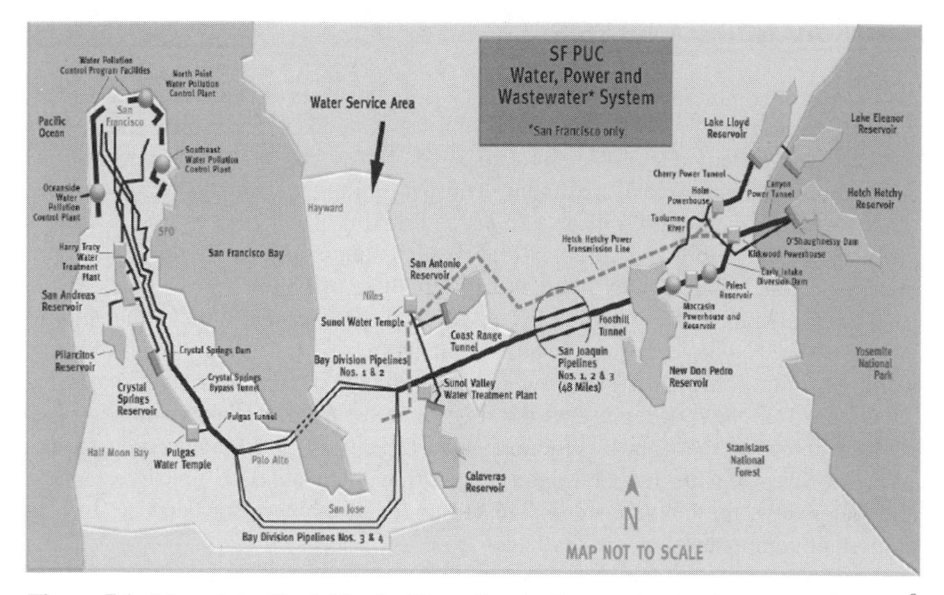

Figure 7.1. Map of the Hetch Hetchy Water Supply System showing how expansive it is.[9]

degree of these nodes. If the histogram declines according to a power law, the network is scale free. Otherwise, it is not. Even if the histogram is not shaped like a pure power law, it may tell us that the network is a small world, or that it has rare nodes with many links, and many nodes with a small number of links. In any case, the objective is to find the rare nodes with many links, because destruction of high-degreed nodes has the greatest impact on the network.

The results of network analysis are shown in Figure 7.3. The histogram on the right was obtained by counting the links emanating from each node of Figure 7.2. The percentage of nodes with one link is 20%; the percentage with two links is 52%; three links is 20%, and with four links, the percentage is 8%.

Is this a scale-free network? No, because the histogram is not shaped like the "mathematically exact" network on the left-hand side of Figure 7.3. But if we ignore the nodes with one link and fit a power law to the abbreviated histogram on the right, we get the histogram on the left. The best fit is in fact a power law with the exponent equal to approximately 1.5. In other words, the Hetch Hetchy network of nodes with only two, three, and four links is scale free.[10]

From this analysis we can identify the hubs in the Hetch Hetchy Network as the nodes with four links, i.e., the New Don Pedro Reservoir and the Crystal Springs Tunnel. The New Don Pedro Reservoir is a critical node near the source of water,

[9]"Hetch Hetchy Water and the Bay Area Economy," Bay Area Economic Forum, October 2002, http://www.bayeconfor.org/baefinfrastruct.htm.

[10]Strictly speaking, the exponent should lie between 2 and 3, but we do not need to be mathematically rigorous. Our purpose is to identify the most connected nodes.

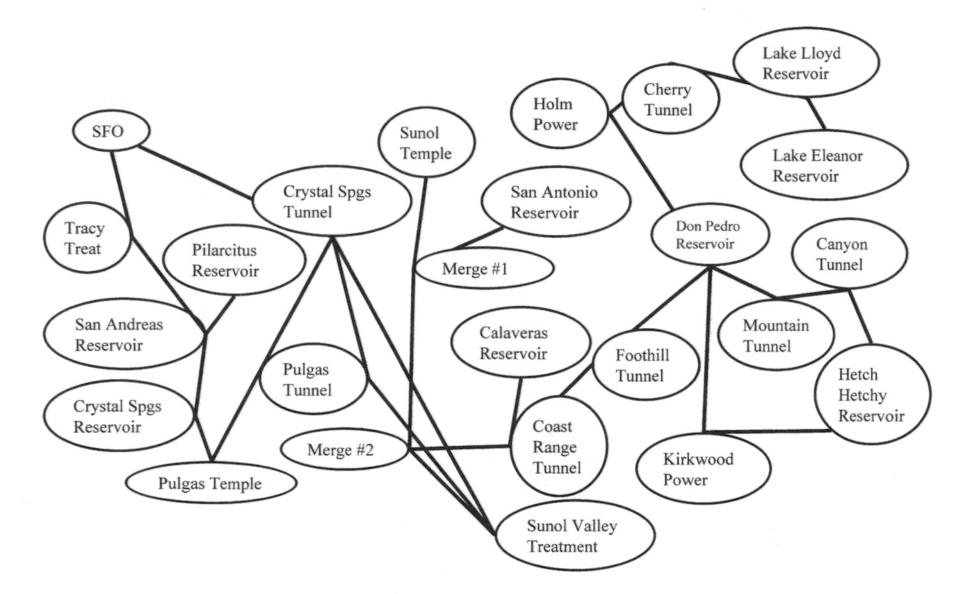

Figure 7.2. The Hetch Hetchy network is an abstraction of the components shown in Figure 7.1. Nodes represent treatment plants, tunnels, lakes, and reservoirs. Links represent the connection among the nodes, e.g., pipes and water flows.

and the Crystal Springs Tunnel is a critical node located near the down-stream end—near San Francisco.

Clearly, these two components are important. Before the network analysis, it was not so clear that these two nodes are critical. But in hindsight, it is obvious that

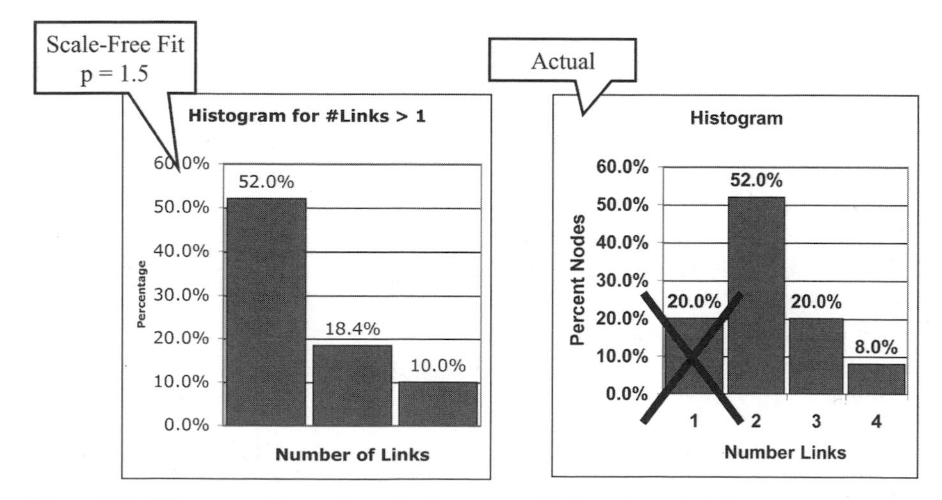

Figure 7.3. Histogram of a mathematically exact scale-free network (left) and the histogram of the Hetch Hetchy Network of Figure 7.2 (right).

failure in either of these nodes turns off most of the water flowing into San Francisco and the surrounding suburbs.

But these are not the only important nodes in this network. If funding was available, we could protect other nodes as well, but financial surpluses are rare. If we cannot afford to protect everything, we should at least protect these critical nodes. This is the strategy we employ here—first protect the nodes with the highest number of links. Only after these critical nodes have been protected do we turn attention to secondary "targets."

HETCH HETCHY FAULT TREE MODEL

Vulnerability is quantified in terms of fault probabilities—the probability that a certain attack is successful. In practice, fault probabilities are obtained from observations of the probability of failure gathered from over a century of earthquakes in the Bay Area. That is, the probability of failures in pumps, storage tanks, and treatment plants is low. However, the likelihood of damage to pipes and tunnels in the area is much higher and more significant, as we shall soon see. To simplify the analysis, assume the following scale for estimating the probability of a successful attack:

Never Happen	0%
Low	20%
Medium	50%
High	80%
Certain to happen	100%

Figure 7.4 shows a fault tree model for the critical nodes of this water sector, along with estimates of their vulnerability and damages. The Don Pedro Reservoir is considered vulnerable to a Chem/Bio attack, with a probability of success equal to 20%. The Crystal Springs Tunnel is vulnerable to a Bomb or Quake attack with a probability of 84%, and the pipelines connecting everything together are considered vulnerable with a fault probability of 50%.

Note that this is an OR-tree, because all logic blocks are OR gates. Recall that OR-trees are highly vulnerable, because any single, double, triple, and higher combination events lead to sector failure. That is, any one of the faults can occur, or any combination, resulting in a fault in the whole tree. For example, the probability of a successful attack on the critical tunnel node is computed using the OR logic formula:

$$\begin{aligned}\text{Prob(Tunnel Fault)} &= 1 - (1 - \text{Prob(Bomb)}) \times (1 - \text{Prob(Quake)}) \\ &= 1 - (1 - 0.20) \times (1 - 0.80) = 1 - (0.8)(0.2) \\ &= 1 - 0.16 = 84\%.\end{aligned}$$

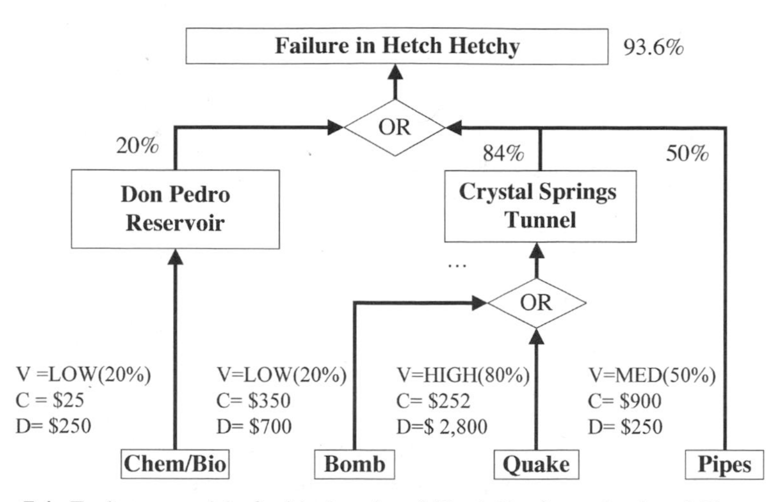

Figure 7.4. Fault tree model of critical nodes of Hetch Hetchy and vulnerability, cost, and damage estimates.

Chem/Bio-Threat to Reservoirs

How were these vulnerabilities determined, and how does an analyst obtain the other estimates shown in Figure 7.4? It is always difficult to estimate damages incurred by an event that has yet to take place, but one common technique is to look at similar incidents that have taken place. Another approach is to develop highly sophisticated models of the infrastructure and its impact on the community. In this section we examine both. We can learn something from the largest-ever cryptosporidium outbreak that contaminated drinking water in Milwaukee, Wisconsin. And we can learn even more about the impact of damage to the water supply system of this region of the country from a study performed by the Bay Area Economic Forum.[11] First, what is the nature of the Chem/Bio threat? A North Carolina report on the 1993 cryptosporidium mishap in Milwaukee, Wisconsin suggests that Chem/Bio attacks are real, but perhaps not as devastating to the availability of drinking water as we might think:

> Cryptosporidium came to national attention in 1993 in Milwaukee, Wisconsin, where 400,000 people were sickened. The protozoan was traced to a water filtration plant that served a portion of Milwaukee with drinking water. An investigation found there was a strong likelihood the organism passed through the filtration process and entered the water supply distribution system. The actual origin of this organism has been speculated to have come from animal operations located in the tributaries of Milwaukee River. These tributaries drain directly into Lake Michigan, just north of where the water intake is located.[12]

[11] "Hetch Hetchy Water and the Bay Area Economy," Bay Area Economic Forum, October 2002, http://www.bayeconfor.org/baefinfrastruct.htm.

[12] Robert E. Holman, "Cryptosporidium: A Drinking Water Supply Problem," Water Resources Research Institute of the University of North Carolina. Special Report No. 12, November 1993.

Milwaukee's cryptosporidium outbreak caused damages of \$75–118 million.[13] Although this was—and still is—the largest known biological incident to affect drinking water in the United States, the per-capita damages were modest—approximately \$80 per person for medical treatment and \$160 per person for loss of productivity. Therefore, we should use conservative damage estimates when modeling the effects of a biological or chemical attack. What happens when this model is applied to Hetch Hetchy?

Assume the per-capita damage caused by a Chem/Bio attack on the Don Pedro Reservoir is \$200. There are 2.4 million people in the Bay Area, so if 50% of the population is contaminated, the damage would be 1.2 million times \$200 or \$240 million. Furthermore, suppose another \$10 million is needed to clean up and repair the reservoir. This brings the total damage estimate to \$250 million—an estimate we used in the fault tree of Figure 7.4.

Chem/Bio attacks on large bodies of water are not easy to do. First, it takes a large amount of contamination because the natural tendency of nature is to break down contaminates, which dilutes their effectiveness. In addition, the EPA has done a good job of regulating large water systems so that their treatment plants are equipped with chemical and biological detection and purification equipment. For these reasons, the probability of a successful Chem/Bio attack on Don Pedro Reservoir is LOW (20%). This number is also added to the fault tree.

Finally, we come to the cost of preventing this attack altogether. The equipment for detection and contamination removal is not cheap, but it is inexpensive relative to other threats. The filtration equipment at or downstream of Don Pedro Reservoir can be installed for tens of millions rather than billions of dollars. Physical barriers to restrict access to the reservoir are not practical, because this is a major recreational resource in Central California. But they are not needed, if treatment plants are properly equipped and operated. Therefore, the cost to protect this component against this threat is a modest \$25 million—10% of the \$250 million damage estimate.

To summarize, the foregoing analysis provides concrete inputs to the fault tree model, as follows:

$$\text{Damage, } d = \$250 \text{ million.}$$
$$\text{Cost, } c = \$25 \text{ million.}$$
$$\text{Vulnerability, } V = \text{LOW, which is, numerically, } 20\%.$$

Earthquake Threat to Crystal Springs Tunnel

Now turn attention to the threats arrayed against Crystal Springs Tunnel. These are Quake, which represents a natural disaster, and Bomb, which represents a manmade

[13]Phaedra S. Corso, Michael H. Kramer, Kathleen A. Blair, David G. Addiss, Jeffrey P. Davis, and Anne C. Haddix, "Cost of Illness in the 1993 Waterborne Cryptosporidium Outbreak, Milwaukee, Wisconsin," *Emerging Infectious Diseases*, vol. 9, no. 4, April 2003.

attack on the tunnel. A nonprofit industrial organization called the Bay Area Economic Forum (BAEF) does studies to support the economic well-being of the San Francisco Bay Area. In October 2002, the BAEF released a report entitled: "Hetch Hetchy Water and the Bay Area Economy." This report estimated the impact of a 7.9-magnitude earthquake on the Bay Area, providing a sound basis for the estimated cost and damages expected of a major earthquake in this area of the country.

The BAEF report makes an impression: A 7.9 earthquake along the Hayward Fault would produce a loss in productivity and physical damage of $17 billion. Similarly, the combined economic and infrastructure damage caused by an earthquake along the San Andreas Fault would exceed $28 billion!

The area covered by Figure 7.4 is perhaps one tenth of this amount, so an estimate of $2.8 billion is reasonable. This estimate covers the tunnel and pipelines leading up to it. Business losses are one thing, and physical damages are another. For the San Andreas Fault area, approximately one half of the damages are attributed to economic losses, and the other half are attributed to physical damages. This estimate does not include psychological damages caused by fear and panic.

The BAEF report estimates the cost to harden the entire Hetch Hetchy system against earthquakes as $3.6 billion. But once again, Figure 7.4 models less than 10% of the entire system, or $360 million. Another data point provided by the BAEF report can be used here: The BAEF recommends that $144 million be invested to harden the Irvington Tunnel, which is located near the San Andreas Fault. Figure 7.4 estimates the cost to harden Crystal Springs Tunnel by averaging the two estimates above: one half of $144 million plus $360 million, or $252 million.

Lessons learned in 1906 and 1989 taught the city that earthquakes are a major threat to water, power, and transportation. In fact, it ranks higher in probability of occurring than all other faults. So, Quake vulnerability is considered HIGH, which has a numerical value of 80%. Is this reasonable? Note that Quake is the highest-valued fault probability in Figure 7.4, followed by destruction of the pipes. Chem/Bio and Bomb attacks are the lowest-valued threat. In this region of the country, it is entirely rational to assume that earthquakes are four times more likely to damage the Crystal Springs Tunnel than a Chem/Bio or Bomb attack, and almost 50% more likely to happen than pipeline damage.

Threat of Pipe Failure

In November 2002, a leak in the main pipeline connecting Hetch Hetchy with the Bay Area treatment and distribution network cut the water supply to San Francisco in half. It stopped 210 to 240 million gallons per day from flowing to the Bay Area, but for only a few days. Fortunately, there is a 4-day supply of water "in the system," which buffered the effects of this accident.

But burst pipelines can disrupt the supply, even if for only a short time. Hence the pipeline component is highly vulnerable but a relatively low-risk component of the sector. Therefore, the Pipe vulnerability is rated MEDIUM, or likely to happen with a probability of 50%.

Various estimates for repairing pipe leaks such as the one that occurred in 2002 put the bill at around $6 million. But this was for a single location. Hetch Hetchy contains 150 miles of pipeline, so it is appropriate to multiply $6 million times 150 to obtain the cost estimate associated with this threat. This is how the Pipe cost estimate of $900 million used in Figure 7.4 was obtained.

Bomb Threat

The November 2002 pipeline failure highlights the criticality of Hetch Hetchy to small incidents with big consequences. It made people think what might be the impact of a more serious, widespread incident, such as an earthquake. Experts claimed that a major quake could leave the city without water for 2 months! But what about a much smaller manmade attack? What would a strategically placed bomb do to the city's water supply?

During the El Nino winter of 1996–1997, a landslide occurred on the northeast hillside above Polhemus Road in San Mateo County, which damaged homes and blocked Polhemus Road. The landslide temporarily buried an important water transmission pipeline in the Hetch Hetchy system called the Crystal Springs Bypass Pipeline. The 96-inch pipeline runs nearly parallel to and south of Polhemus Road and transports an average of 90 million gallons of drinking water per day from the Sierra Nevada Mountains to communities in San Francisco and on the Peninsula, including San Mateo.

Ninety million gallons per day is 25% of the daily flow of Hetch Hetchy. The implication is that a bomb attack could easily deny consumer access to 25% of the total water supply, or 25% of the impact caused by threat Quake. Thus, damages from such an attack can be reasonably estimated to be 25% of $2800 or $700 million, and the cost to protect the critical node against a bomb is estimated to be 25% of $252, or $63 million.

But a manmade attack using a bomb is less likely than an earthquake or a break in a pipeline. In fact, 25% of Quake vulnerability yields a Bomb vulnerability rating of LOW, which is numerically 20%.

Risk Analysis

Before continuing, we should ask whether the model developed and captured in the form of the fault tree of Figure 7.4 is realistic and useful. Does it represent the sector, and is it useful for performing the risk analysis we seek?

Note that the damage done by all of these threats leads to highly different repair delays. And repair delays have a lot to do with productivity losses. For example, the BAEF report estimates that repairing a pipeline takes 20 times as long as repairing a pumping station. Tunnels can take up to 30 times as much time. In this sense, the model does not directly capture these delays, but productivity damages are incorporated in the model. The fault tree lacks the expressive power to model time delays, but it does have the expressive power to model productivity losses.

The next question is, "how do we know if the estimates derived here are correct?" Once again, we can check these results by comparing them with the estimates given by others. If we add all of the damage estimates, for single and multiple threats, the total is approximately \$32.2 billion—about the same number obtained by the BAEF. Similarly, the cost to harden against the four threats described here is between \$1.24 billion and \$9.9 billion, which straddles the \$3.6 billion recommendation of the BAEF. But this analysis focuses on the two most critical nodes, whereas the BAEF study covered the entire sector. So Figure 7.4 is perhaps overly pessimistic.

According to MBVA, the next step is to construct the event tree corresponding to the fault tree of Figure 7.4. Figure 7.5 shows this tree, which identifies all possible combination faults. Clearly, Quake and the combination of a Quake and Pipe fault are the highest-ranking vulnerabilities. They account for over 50% of the sector's vulnerability, which is 93.6%. But vulnerability is not the same as risk. To compute risk, we must multiply vulnerability by damages. The results of computing risk for each outcome and summing over the entire event tree of Figure 7.5 are summarized in Table 7.3. The overall sector vulnerability is 93.6%, and the sector risk is \$2.568 billion. Also note in Table 7.3 the bill for hardening everything: \$9.92 billion. Suppose the city has only \$1 billion to do the best it can? What is the best way to invest \$1000 million to secure Hetch Hetchy's critical nodes?

ALLOCATION OF RESOURCES

The next question is, "how do we allocate resources to harden the targets identified by this analysis?" According to the MBVA technique, we need to plug the numbers

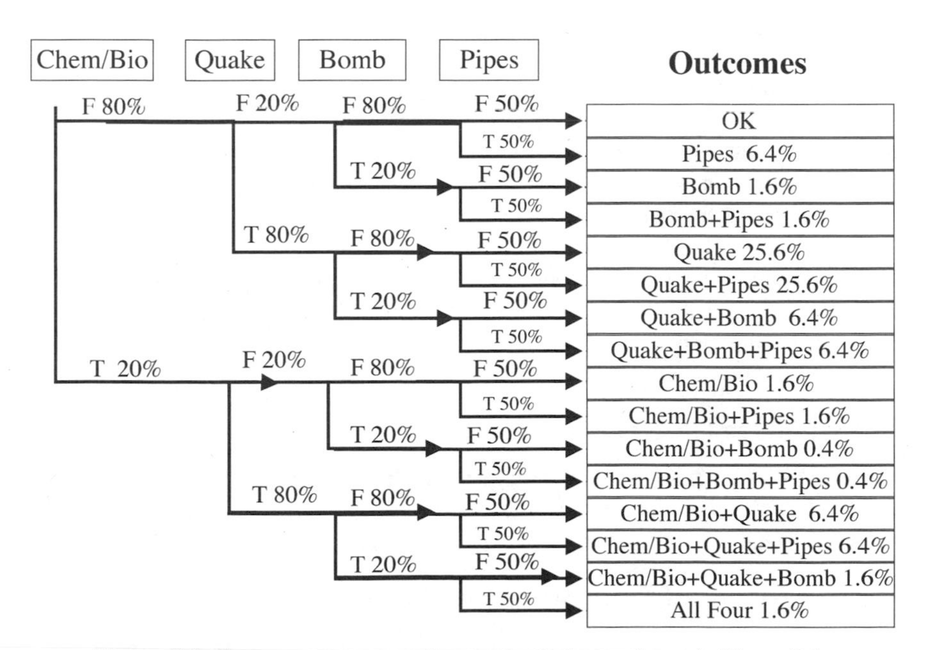

Figure 7.5. Event-tree corresponding to the fault tree shown in Figure 7.4.

TABLE 7.3. Summary of Event Tree Analysis of Figure 7.5.

Fault	Vulnerability	Probability (V)	Cost (C)	Damage (D)	$Risk
1. Chem/Bio	LOW	1.6%	$25	$250	$4
2. Quake	HIGH	25.6%	$252	$2,800	$717
3. Bomb	LOW	1.6%	$63	$700	$11
4. Pipes	MEDIUM	6.4%	$900	$250	$16
5. Chem/Bio + Quake		6.4%	$277	$3,250	$208
6. Chem/Bio + Bomb		0.4%	$88	$950	$4
7. Chem/Bio + Pipes		1.6%	$925	$500	$8
8. Quake + Bomb		6.4%	$315	$3,500	$224
9. Quake + Bomb + Pipes		6.4%	$1,215	$3,750	$240
10. Quake + Pipes		25.6%	$1,152	$3,050	$781
11. Bomb + Pipes		1.6%	$963	$950	$15
12. Chem/Bio + Quake + Bomb		1.6%	$340	$3,750	$60
13. Chem/Bio + Quake + Pipes		6.4%	$1,177	$3,300	$211
14. Chem/Bio + Bomb + Pipes		0.4%	$988	$1,200	$5
15. Chem/Bio + Quake + Bomb + Pipes		1.6%	$1,240	$4,000	$64
TOTALS = Sum Probabilities =		**93.6%**	$9,920	$32,200	$2,568

given by Figure 7.4 into *FTplus* software to compute the allocations that satisfy the policy makers' investment strategy objectives.[14] In Chapter 6, we developed four strategies: manual, ranked order, apportioned, and optimal. In this section we compute all four and argue that the "best" strategy is to minimize risk by allocating the $1000 million budget to completely harden against Chem/Bio, Bomb, and Quake events, at the peril of the pipeline vulnerability.

Figures 7.6 and 7.7 show results from using the program *FTplus* to construct the fault tree of Figure 7.4 and compute the allocation obtained by running all four allocation strategies. Table 7.4 summarizes the results. What do these numbers mean to the security of Hetch Hetchy?

Manual Allocation Strategy

Suppose we allocate an equal amount of money to each threat. There are four threats and $1000 million to invest, so each threat is allotted $250 million. This strategy guarantees that all threats are treated equally, and so the money is spread across

[14]Program *FTplus*, a.k.a. *FT.jar*, is available on the disk accompanying this book and on the course website.

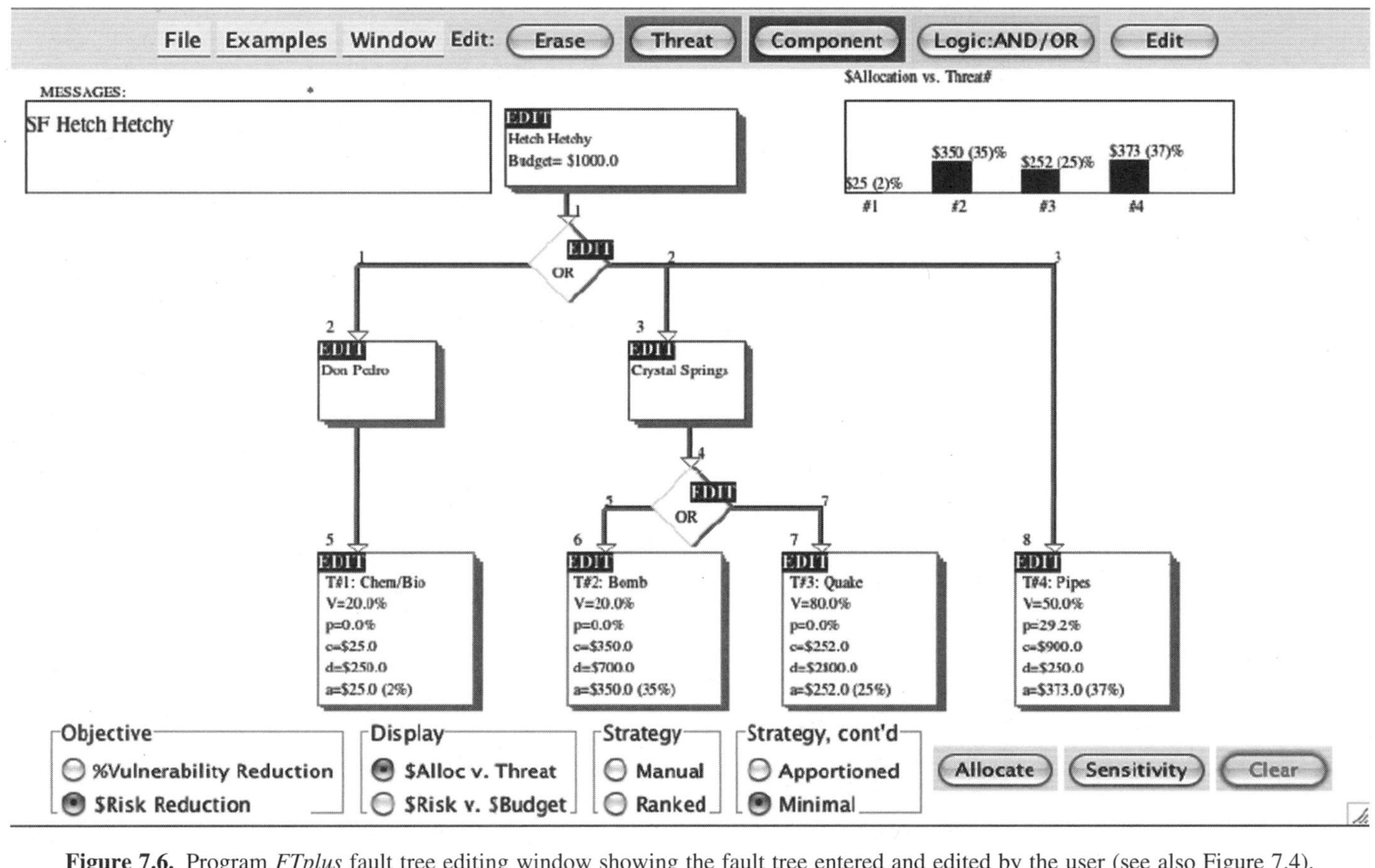

Figure 7.6. Program *FTplus* fault tree editing window showing the fault tree entered and edited by the user (see also Figure 7.4).

Summary of Results

#	Threat	Vulnerability	Manual%	$Manual	Ranked%	$Ranked	Apportioned%	$Apportioned	Minimal%	$Minimal
Fault	Fault	Fault	Fault	Fault	Fault	Fault	Fault	Fault	Fault	Fault
1	Chem/Bio	20.0	0.0	250.0	20.0	0.0	1.2	23.71	0.0	25.0
2	Bomb	20.0	5.71	250.0	20.0	0.0	14.37	98.50	13.65	111.0
3	Quake	80.0	0.63	250.0	0.0	252.0	2.58	243.85	0.0	252.0
4	Pipes	50.0	36.11	250.0	8.44	748.0	14.78	633.92	15.99	612.0
Risk	Risk	Risk	Risk	Risk	Risk	Risk	Risk	Risk	Risk	Risk
1	Chem/Bio	20.0	0.0	250.0	20.0	0.0	1.51	23.11	0.0	25.0
2	Bomb	20.0	5.71	250.0	0.0	350.0	7.55	217.74	0.0	350.0
3	Quake	80.0	0.63	250.0	0.0	252.0	0.34	250.92	0.0	252.0
4	Pipes	50.0	36.11	250.0	27.88	398.0	21.76	508.21	29.27	373.0

Figure 7.7. Program *FTplus* summary window showing the results from manual, ranked order, apportioned, and optimal allocations.

all critical nodes and their threats. As a consequence, the Chem/Bio vulnerability is driven to zero, and the vulnerability due to QUAKE is nearly eliminated. But the BOMB threat probability remains at 6%, and the PIPE vulnerability remains at 36%. This strategy, although appearing to treat all threats equally, actually favors Chem/Bio and QUAKE risk reduction over the others. The result is a lingering sector vulnerability of 40% and a sector risk of $149 million. Although this is a major improvement over the initial sector risk of $2.56 billion, it is far from the best strategy.

Ranked Order Allocation Strategy

Ranked order allocation funds the highest vulnerability threat, in the case of vulnerability reduction, and the highest risk threat, in the case of risk reduction. When there is enough funding, the threats with the highest valued risk are funded in rank order. When applied to the Hetch Hetchy water supply, the strategy fully funds Quake and Bomb threats, reducing their fault probabilities to zero and then uses the remaining funds to partially fund the next highest risk threat: pipes. And then the funds are depleted, so the Chem/Bio vulnerability is not funded; therefore, it remains.

Ranked order allocation reduces the sector vulnerability to 42%, which is the highest of all the strategies. Chem/Bio remains 20% vulnerable, and pipes remain 28% vulnerable, and the sector risk is reduced from $2.56 billion to

TABLE 7.4. Summary of Allocation Results Obtained from *FTplus*.

Threat	Manual	Ranked	Apportioned	Optimal
Chem/Bio	$250	$0	$23	$26
Quake	$250	$350	$218	$350
Bomb	$250	$252	$252	$252
Pipes	$250	$398	$507	$372
Sector V =	40%	42%	29%	29%
Sector Risk =	$149	$111	$121	$62

$111 million. Therefore, ranked order allocation produces the worst vulnerability result, but the second-best risk reduction result. This is an example of the difference between vulnerability and risk. This strategy also shows how important it is to decide what is important—risk reduction or fault probability (vulnerability) reduction.

Apportioned Allocation Strategy

The apportioned allocation strategy spreads money across the threats proportional to risk: $23 million to Chem/Bio, $218 million to Quake, $252 million to Bomb, and $507 million to Pipes. This reduces vulnerability to the minimum possible with $1000 million, but it is next to the worst risk reduction of the four strategies. Threat vulnerabilities are reduced to 2% for Chem/Bio, 8% for Bomb, and 22% for Pipes. Interestingly, this strategy reduces threats from earthquakes to zero, as does the other three strategies, except for the manual allocation. Therefore, this strategy strikes a compromise between numerically superior vulnerability reduction and an allocation that should make everyone happy with their funding levels.

Apportioned allocation does what it says: It spreads the money across all threats using a weighted sum of squares metric. It ignores the event tree outcomes and optimizes only on the threats shown in the fault tree. As a consequence, apportioned risk reduction is 100% higher than the optimal allocation strategy. This is because the optimal allocation strategy minimizes the risk (vulnerability) across the entire event tree.

Optimal (Minimal) Allocation Strategy

Finally, Table 7.4 shows the results for optimal allocation. This allocation is similar to the apportioned allocation strategy, but it guarantees the absolute minimal risk. The Chem/Bio threat receives $26 million, Quake gets $350 million, Bomb funding is $252 million, and the remaining $372 is spent on hardening Pipes. This allocation produces the smallest risk value possible: $62 million. It also produces a reduced vulnerability of 29%—the same as the apportioned risk reduction allocation strategy.

Optimal allocation can be used to achieve a minimum vulnerability allocation or a minimum risk allocation. Only the minimum risk allocation is shown in Table 7.4. As it turns out, the allocation is the same in both cases—fault reduction and risk reduction. This allocation minimizes both vulnerability and risk for the sector.

Risk can be reduced 98%—from $2.56 billion to $62 million—with an investment of $1 billion. This is made possible by careful analysis and optimal allocation. In particular, the worst allocation strategy shown in Table 7.4 is to evenly divide the $1 billion across all four threats. Apportioned allocation is second from the bottom, but it does a good job of reducing vulnerability while mitigating risk. Ranked order allocation produces a slight improvement over apportioned, but it leaves the sector highly vulnerable. But notice that ranked order eliminates the two highest-risk threats—Bomb and Quake.

Manual strategies should rarely be used. Ranked order allocation should be used only when the worst-case threat is to be removed. Optimal allocation always provides a benchmark to measure all others against, and apportioned allocation seeks a "middle-of-the-road" solution. In this analysis, apportioned and optimal are suitable solutions to the question of "how should resources be allocated to protect this sector?"

ANALYSIS

What can we learn about the water sector from this chapter? First, interest in protecting water has evolved during the past century from a resource that was threatened by germs to one that must now be protected from disease, environmental pollution, and terrorist attacks. Along the way, the responsibility for water has shifted from the PHS to the EPA. Might its next transition be to the Department of Homeland Security? This is an unlikely shift, because water is one of the few critical infrastructures that is truly local. It is not regulated by interstate commerce, and water security is not enhanced by access to national intelligence. This is in sharp contrast to other sectors described in this book. Therefore, water is somewhat unique.

The strategy of the EPA is to partner with community water systems and local governments. In an act of pure federalism, the EPA has chosen to share responsibility across federal, state, and local levels as well as the private–public divide. But terrorism is a radically different kind of threat than disease and acts of God. Terrorism is a cunning threat that adapts and counters static attempts to foil it. It seems that an intelligence and law enforcement arm of the EPA will be needed to work with state and local law enforcement groups. For example, who is going to stand guard around the tunnels of Hetch Hetchy?

By far the preponderance of legislation and regulation of water has been aimed at drinking water, and yet over 80% of the water supply is used for agricultural and industrial applications. Although nobody advocates that drinking water is no longer essential, the economic impact of denying water to agriculture and factories surrounding the San Francisco Bay Area is staggering. Is the next shift of emphasis in policy concerning this sector going to be a greater emphasis on nondrinking water? Or will agriculture and industry have to provide its own backup supply as it does now with backup power?

The Bio-terrorism Act of 2002 provides penalties for the misuse of information about the water supply. Will this legislation make it less likely that interagency information sharing will take place? Does it introduce a barrier to coordination? The WaterISAC is a public (and open) organization, not a military or governmental organization that has highly structured security regulations on dissemination of information.

This chapter shows that the water sector can be analyzed as a *scale-free network*. The application of the MBVA method revealed a tunnel and a reservoir as critical nodes in Hetch Hetchy. Awareness of tunnels as critical may be a surprising result, because much of the emphasis on water purity has been focused on treatment plants and reservoirs. Although these are important, it is highly unlikely that treatment plants and reservoirs are critical nodes in general. Analysis of Hetch Hetchy shows that water purity can be maintained at low cost. But physical assets such as tunnels pose a much greater risk because their destruction is easy and the time to repair them is high. A collapsed tunnel could lead to a long period of denial of service. For example, the BAEF report estimates that repairing a Hetch Hetchy tunnel can take up to 30 times as much time as any of the other components. The system has about 4 days of supply, but a tunnel takes up to 60 days to repair!

Finally, we showed how interdependencies among water, agriculture, transportation (airports), and power make water one of the primary (Level 1) critical infrastructures. Other sectors depend heavily on water, even though the water sector has traditionally been concerned with drinking water. This simple view of the water sector is destined to change over time just as the story of water has changed since the early 1900s. Terrorism, and its corresponding asymmetric challenges, will perhaps drive the evolution of water supply policy to its next level.

One final point merits mentioning: *cascade failure*. The experience of November 2002 whereby the Hetch Hetchy pipeline break led to a stuck gate is instructive. We will see this effect again in several other sectors. This is known as a *cascade failure*—a series of failures that are tripped by a seemingly modest fault. It is often the case that major disasters start with a small failure that propagates to greater failures, which in turn lead to a major system collapse.

Cascade failures have been observed in some of the most costly and dramatic disasters, such as the sinking of the Titanic on April 15, 1912, the Apollo 13 fire on March 16, 1970; the largest sea port disaster in the United States at Texas City, Texas, on April 16, 1947 (killing almost 500 people); the partial nuclear reactor meltdown at Three Mile Island on March 28, 1979; the Union Carbide chemical spill that killed 7000 people in Bhopal, India, on December 3, 1984; and the Northeastern power grid blackout of 2003.[15] Cascade failures are not modeled by the technique described here. What cascade failures lurk in today's water systems?

EXERCISES

1. Which of the following is the primary issue in protecting water? (Select only one):
   a. Terrorism
   b. Drinking water supplies
   c. Agricultural water supplies
   d. Industrial water supplies
   e. All of the above

[15]J. R. Chiles, "Inviting Disaster: Lessons From the Edge of Technology," *HarperBusiness*, 2001, 2002. http://www.invitingdisaster.com.

2. When did the responsibility for the safety of water transfer from the PHS to the EPA?
   a. 1914
   b. 1962
   c. 1974
   d. 2002
   e. 2003
3. Why did regulation of water move from PHS to EPA?
   a. Emphasis shifted from biological to environmental contamination.
   b. Emphasis shifted to terrorism.
   c. The EPA had more money.
   d. PHS was abolished.
   e. Emphasis shifted from chemical to environmental contamination.
4. The Bio-terrorism Act of 2002 extends the SDWA of 1974 as follows:
   a. Includes acts of terrorism.
   b. Requires vulnerability assessments.
   c. Establishes the WaterISAC.
   d. Specifies prison term penalties.
   e. All of the above.
5. Which of the following is the critical node in the San Francisco County Water System (Hetch Hetchy) as determined by network analysis?
   a. New Don Pedro Reservoir and Crystal Springs Tunnel
   b. The Treatment Plants
   c. Hetch Hetchy And Lake Lloyd Reservoirs
   d. Sunol Valley Treatment and Kirkwood Power Plants
   e. Merge #2 and Merge #3
6. If fault A occurs with a probability of 50% and fault B occurs with a probability of 33%, what is the probability of (A OR B) occurring?
   a. 83%
   b. 67%
   c. 50%
   d. 33%
   e. 17%
7. According to the analysis shown here, which vulnerability is of highest concern in the Hetch Hetchy case study?
   a. Tunnels
   b. Chem/Bio
   c. SCADA
   d. Power outage
   e. Storage tank failure

8. In general, fault probability reduction differs from financial risk reduction in the MBVA method of resource allocation. Why?
   a. The cost to make the improvement differs in each allocation.
   b. Input to MBVA is sensitive to errors.
   c. MBVA does not include psychological costs.
   d. Hetch Hetchy's most serious vulnerability is treatment plants.
   e. The damage estimate differs from vulnerability to vulnerability.
9. The MBVA result obtained by the Hetch Hetchy case study says apportioned and optimal risk reduction yield the same vulnerability reduction. So why is optimal risk reduction considered better?
   a. It produces the greatest amount of risk reduction.
   b. MBVA contradicts itself.
   c. Both are correct: Apportioned and optimal produce the same result.
   d. Tunnels and reservoirs are not threats, but only vulnerabilities.
   e. I do not know why.
10. Earthquake experience has shown that the water supply is most vulnerable to:
    a. Broken pipes
    b. Contamination of the lakes and reservoirs
    c. Collateral fires and explosions
    d. Collapsing tunnels
    e. Collapsing freeways
11. The main lesson learned from San Francisco earthquakes is:
    a. Earthquake-resistent pipes are needed.
    b. Drinking water is no longer pottable.
    c. Collapsing tunnels block the flow of water.
    d. 80% of the water is used for agriculture.
    e. Backup power is essential.
12. The largest water supply contamination disaster in the United States was:
    a. Hetch Hetchy, November 2002
    b. Milwaukee, Wisconsin's cryptosporidium contamination in 1993
    c. The Loma Prieta Earthquake in the San Francisco Bay Area
    d. Hurricane Fran in 1996
    e. Hurricane Dennis in 1999
13. In terms of time delays caused by the time to repair a water sector component, the Hetch Hetchy water supply is most vulnerable to:
    a. Broken pipes
    b. Collapsing tunnels
    c. Collapsing freeways
    d. Broken pumps and gates
    e. Insufficient budget

14. Which of the following is the least interdependent with the water sector? (Select only one):
    a. Transportation
    b. Power
    c. Agriculture
    d. Silicon Valley Industry
    e. Public Health
15. The water supply system of San Lewis Rey is very similar to the Hetch Hetchy water supply network, but it is somewhat simpler (see Figure 7.8). What is the best way for San Lewis Rey planners to invest $10 million to harden this network? Use the following input data:

    Threats and their vulnerabilities, costs to fix, and damages:
    Reservoirs and Lakes:
    Cryptosporidium: V = 5%; C = $50 million; D = $250 million.
    SCADA malfunctioning Valve: V = 10%; C = $20 million; D = $40 million.
    Treatment:
    Bomb: V = 5%; C = $100 million; D = $400 million.
    SCADA malfunctioning Valve: V = 10%; C = $20 million; D = $5 million.
    Tunnels, Storage, and Storage Temples:
    Bomb: V = 15%; C = $15 million; D = $1 million.

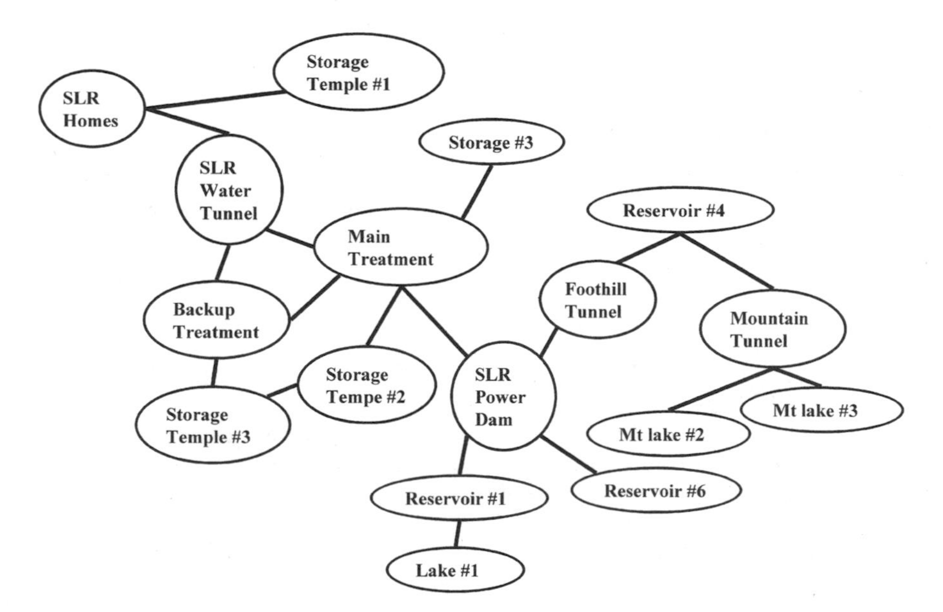

Figure 7.8. San Lewis Rey water supply network.

Power Dam:

Bomb: V = 10%; C = $50 million; D = $500 million.

SCADA control attack: V = 20%; C = $40 million/%; D = $50 million.

a. Which asset is the most critical node?
   i. Main Treatment
   ii. SLR Power Dam
   iii. Mountain Tunnel
   iv. SLR Water Tunnel
   v. Lake #1

b. Show the fault tree for the critical node. What type of fault tree is it?
   i. OR-tree
   ii. AND-tree
   iii. AND-OR-tree
   iv. OR-AND-tree
   v. Event tree

c. What is the optimal allocation of $10 million to reduce financial risk in the critical node?
   i. $10 million to BOMB; $0 million to SCADA
   ii. $5 million to BOMB; $5 million to SCADA
   iii. $1 million to BOMB; $9 million to SCADA
   iv. $9 million to BOMB; $1 million to SCADA
   v. $0 million to BOMB; $10 million to SCADA

d. What is the sector risk after the allocation?
   i. $10 million
   ii. $0 million
   iii. $18.5 million
   iv. $20.5 million
   v. $5 million

e. What is the sector vulnerability after investment?
   i. 10%
   ii. 14.05%
   iii. 18.5%
   iv. 15%
   v. 7.5%

f. How much money does it take to reduce risk to zero?
   i. $405 million
   ii. $100 million

iii. $20 million
iv. $5 million
v. $120 million

Sidebar 7.1. Historical timeline for the evolution of water safety and prevention of terrorist attacks on drinking water.

1880s—Louis Pasteur develops germ theory and notes that water is a vector.
1885—Dr. John Snow proves cholera was transmitted by drinking water.
1914—PHS sets standards for the *bacteriological* quality of drinking water.
1925, 1946, and 1962—PHS revises standards. The 1962 revision called for the regulation of 28 substances and established the most rigorous standards until 1974.
1960—PHS study shows that only 60% of drinking water met PHS standards.
1972—PHS study of Mississippi River reveals 36 *chemicals* contaminating drinking water processed by treatment plants.
1974—SDWA establishes foundation of modern regulations for protecting the purity of water and water systems. Enforcement transferred to EPA.
1986, 1996—Revisions to SDWA-1974.
1993—Cryptosporidium outbreak in Wisconsin kills over 50 people and infects 400,000 consumers of public water.
May 22, 1998—President Clinton signed Presidential Decision Directive 63 identifying drinking water as one of America's critical infrastructures.
June 12, 2002—President Bush signs into law the Public Health Security and Bio-terrorism and Response Act of 2002.
August 1, 2002—EPA completes the *classified* Baseline Threat Report describing likely modes of terrorist attack and outlining the parameters for vulnerability assessments by *community* water systems.
December 2002—WaterISAC becomes operational.
March 31, 2003—Water systems serving more than 100,000 people submit vulnerability assessments to the EPA
December 31, 2003—Water systems serving between 50,000 and 100,000 people are required to submit vulnerability assessments to the EPA.
June 30, 2004—Water systems serving between 3,300 and 50,000 people are required to submit vulnerability assessments to the EPA.

CHAPTER 8

SCADA

SCADA is an acronym for *supervisory control and data acquisition*. SCADA systems are composed of computers, networks, and sensors used to control industrial processes by sensing and collecting data from the running process, analyzing that data to determine how best to control it, and then sending signals back through a network to adjust or optimize the process. Several definitions exist. We chose the following, because it is operational and descriptive:

> An industrial measurement and control system consisting of a central host or master (usually called a master station, *master terminal unit* or MTU); one or more field data gathering and control units or remotes (usually called remote stations, *remote terminal units*, or RTU's); and a collection of standard and/or custom software used to monitor and control remotely located field data elements. Contemporary SCADA systems exhibit predominantly open-loop control characteristics and utilize predominantly long distance communications, although some elements of closed-loop control and/or short distance communications may also be present.[1]

The terms "SCADA" and "distributed control systems" (DCS) are often used interchangeably, but SCADA is usually reserved for systems that are geographically dispersed. SCADA systems typically rely on communication networks to connect remote terminal units (RTUs) to the MTU. The DCS, energy management system (EMS), and programmed logic controller (PLC) are various kinds of control systems that are similar to SCADA systems described here.

However, according to industry experts, SCADA security is a discipline unto itself and requires special considerations. "SCADA is only one type of control system. The terms SCADA and DCS are not, and should not be used interchangeably."[2] For our purposes, the distinction among the various kinds of control systems is not necessary because all kinds of systems are used in controlling critical infrastructures. Any control systems used in any critical infrastructure system that may render the sector vulnerable will be of interest in this book.

[1]http://www.sss-mag.com/glossary/page4.html.
[2]Personal communication with Joe Weiss, April 2005.

Critical Infrastructure Protection in Homeland Security: Defending a Networked Nation,
edited by Ted G. Lewis

A thorough understanding of control systems is necessary because automation supports much of modern technological society. They run major portions of the transportation, energy, power, and water sectors as well as most manufacturing processes. If you have ever ridden in a subway, train, or automobile, your safety has been in the electronic hands of a SCADA system.

In this chapter, you will learn the following general concepts:

1. *SCADA, DCS, and Other Control Systems Are Pervasive*: Automation of critical infrastructure sector processes found in water works, power, and transportation systems continues to increase, placing control in the hands of a machine, with human oversight. This trend will continue, making the study of control system security more relevant over time.
2. *Responsibility Is Scattered*: The National Institute of Standards and Technology (NIST) and the National Security Agency (NSA) created the National Information Assurance Partnership (NIAP) to set standards and promulgate best practices, but there is no SCADA ISAC. Rather, industry is working with NIST within the Process Controls Security Requirements Forum (PCSRF).
3. *SCADA Is Vulnerable to Cyber-Intrusion*: Because the components of SCADA systems are selected on the basis of cost, their security has historically been sacrificed to lower cost and reduce their consumption of power. The result: Most SCADA systems are unprotected and so need to be hardened against cyber-attack.
4. *SCADA Policies Should Focus on Information Assurance*: Because SCADA is generally vulnerable to asymmetric cyber-intrusion, the focus of SCADA policy should be on cyber-security and policies that reinforce best IT security practices.
5. *Redundancy Works*: Although duplication of equipment is expensive, SCADA systems (and IT systems in general) can be physically protected using redundancy of computers, communications, and facilities as illustrated by the case study in this chapter. However, redundancy can become an added vulnerability for cyber-events, especially if redundant computers are connected to the same network.

WHAT IS SCADA?

Figure 8.1 shows a simplified representation of a typical SCADA system. The system is operated through one or more operation control centers (OCCs) containing computers, networks, and databases. The SCADA System Database, for example, stores the state of the entire system—the condition of all sensors, valves, switches, and so on. Sensors and control actuators are represented as RTUs. The state of RTUs is stored in the database and viewed through an operator user interface, typically computer monitors, big-screen displays, and switches and dials mounted on a wall.

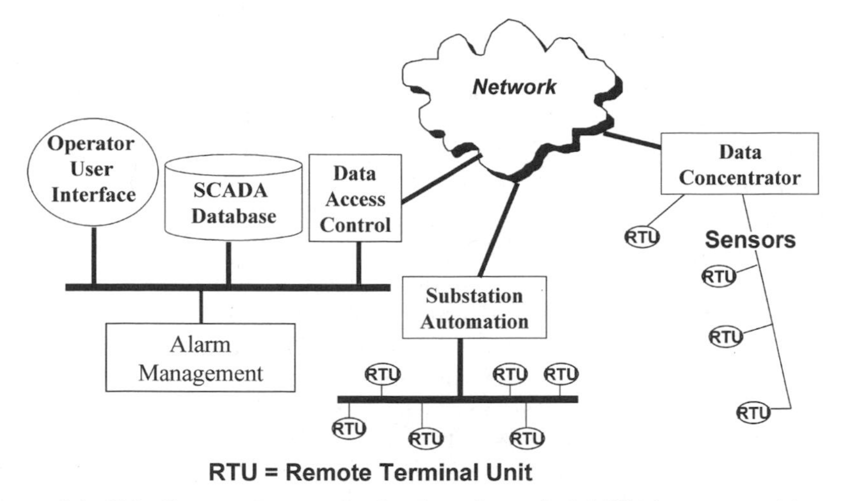

Figure 8.1. This diagram shows a simple view of a typical SCADA system and its components consisting of computers, networks, databases, RTUs, and software.

SCADA is a kind of remote control system, meaning the valves, gates, switches, thermostats, and regulators being controlled are many miles away from the OCC. The RTUs located close to the devices being controlled report back to the OCC through a network. The RTUs can produce data—the data acquisition part of SCADA—or accept commands from the OCC to open/close a valve in a water pipe, or report leaks in an oil or gas pipeline.

An alarm management system also runs off the database. It constantly evaluates the state of the system by processing records in the database and monitoring the data streaming in from the network. If a certain reading is out of bounds or exceeds a threshold, an alarm is tripped, alerting the human operators. For example, in a power generation plant control system, an alarm may sound when temperatures exceed a certain threshold or sensors attached to the power grid detect a failure in a power line. In an oil pipeline system, sensors may collect data regarding leakage and report the location to an alarm management system so that repairs can be ordered. A transportation control system such as found in a subway or light rail train might report dangerous conditions to the alarm management system so that collisions are avoided.

Data are managed in a hierarchical fashion in most SCADA systems. The raw data collected by an RTU are aggregated at the RTU itself and then passed to a substation where they are summarized or aggregated some more, and then transmitted to one or more operation control centers where they are analyzed and summarized. SCADA is a kind of distributed computing system that collects data from geographically distributed sensors and delivers it to one or more processing servers. Both RTUs and MTUs can be distributed, especially if redundant computers are employed to increase reliability and security. This idea will be illustrated in a case study presented later in this chapter.

WHO IS IN CHARGE?

PDD-63 does not specifically reference SCADA as a critical infrastructure, nor does it name SCADA as a component of other infrastructure sectors. Rather, cyber- and physical security are given equal weight. According to PDD-63, "Critical infrastructures are those physical and cyber-based systems essential to the minimum operations of the economy and government." The Homeland Security Act of 2002 (H.R. 5005) assigned responsibility for information security to the Under Secretary for Information Analysis and Infrastructure Protection, which in turn created the National Cyber Security Division (NCSD) to address cyber-security within critical infrastructure systems. In addition, the Department of Homeland Security's (DHS's) Science and Technology Division conducts research in SCADA security as it pertains to infrastructure protection, with one exception: The Department of Energy (DOE) is responsible for SCADA in electric power.

Section 225 of the Homeland Security Act of 2002 (CYBER SECURITY ENHANCEMENT ACT OF 2002) specifies penalties for cyber-crime (up to 20 years for doing harm, and life imprisonment for attacks that result in death), requires the DHS Under Secretary to report on cyber-crimes to Congress, and outlaws Internet advertising of devices that may be used in cyber-attacks.

The Homeland Security Act falls short on details for preventing attacks on computer systems—whether they are SCADA or standard information technology systems used by government, business, or consumers. The responsibility for establishing standards and guidance has been delegated to a combination of agencies and industrial groups. The NIST and the NSA have partnered to fill in the details concerning information security. This partnership is called the NIAP.

According to NIST, the NIAP is a partnership between NIST and the NSA:

> The National Information Assurance Partnership (NIAP) is a U.S. Government initiative designed to meet the security testing, evaluation, and assessment needs of both information technology (IT) producers and consumers. NIAP is collaboration between the National Institute of Standards and Technology (NIST) and the National Security Agency (NSA) in fulfilling their respective responsibilities under the Computer Security Act of 1987. The partnership, originated in 1997, combines the extensive security experience of both agencies to promote the development of technically sound security requirements for IT products and systems and appropriate metrics for evaluating those products and systems. The long-term goal of NIAP is to help increase the level of trust consumers have in their information systems and networks through the use of cost-effective security testing, evaluation, and assessment programs. NIAP continues to build important relationships with government agencies and industry in a variety of areas to help meet current and future IT security challenges affecting the nation's critical information infrastructure.[3]

[3]http://niap.nist.gov/.

The NIAP has further delegated responsibility for working with the private sector to the PCSRF. The members of this public–private organization are EPRI, American Gas Association (AGA), Association of Metropolitan Water Agencies (AMWA), and the Society of Instrumentation, Systems, and Automation (ISA). Government participation comes from the NSA, DOE, and NIST. The so-called *Common Criteria for Information Technology Security Evaluation*, also known as the ISO/IEC 15408 standard, is being used to document the results of the NIAP effort.

There is no SCADA ISAC. However, there is an IT-ISAC that encompasses all information assurance issues.[4] Major digital network and computer security companies are members of the IT-ISAC: Computer Sciences Corporation, Symantec Corporation, General Dynamics, Cisco, Computer Associates, Hewlett-Packard, International Business Machines, Microsoft Corporation, Oracle Corporation, and RSA Security, Inc. Note the conspicuous absence of control system vendors and SCADA end-users in the IT_ISAC.

The NIAP/PCSRF initiative serves to set standards for control systems. The IT-ISAC provides linkages among the private sector companies who have a vested interest in making their network and computer products secure. But these governmental and industrial groups do not specifically address the processing needs of vertical sector components such as water treatment plants, power generation plants, and traffic control networks. These verticals are served by their own ISACs or various governmental agencies, but the vertical ISACs typically lack SCADA expertise. For example, the DOE, which is responsible for power and energy infrastructure protection, has its own SCADA initiative. The DOE makes its own recommendations apart from the IT-ISAC, NIAP, and the power industry (see Table 8.1).

From the foregoing, we can see that responsibility for SCADA and control system security is scattered across governmental agencies and commercial groups. Multiple initiatives by government agencies overlap and often duplicate one another, i.e., DHS, NIST, NSA, and DOE all seem to play similar roles. Because SCADA cuts across various infrastructure sectors, the private industrial groups also overlap and support dual programs. For example, the various ISACs such as the IT-ISAC, WaterISAC, and Electric Power ISAC (EP-ISAC) perform similar functions when it comes to SCADA.

SCADA standardization efforts are spread across commercial and nonprofit organizations as well as governmental partnerships like NIAP. In addition, 3C, World Internet Society",4>W^{3}C, World Internet Society, and the IEEE promote their own information technology standards that may or may not address control system security. This adds to the confusion on where to go for authoritative information. Who is in charge? Many private and public groups claim responsibility for SCADA. But like SCADA, the "command and control" of SCADA protection is spread far and wide. It is everywhere.

[4]http://www.it-isac.org/.

SCADA EVERYWHERE

Control systems such as SCADA are used in almost every kind of industry. Application is not limited to critical infrastructure sector processes. Widespread adoption is driven by efficiencies and economies. Automation reduces labor costs and decreases reaction time. But SCADA is fundamentally at risk because it is an information technology that has ignored security for decades. Most SCADA networks are as open as the telephone system and as vulnerable as a telephone line.

Below is a sampling of applications where SCADA reduces costs and decreases reaction time, by automating various industrial processes:

Food manufacturing
Pharmaceuticals manufacturing
Discrete parts manufacturing
Environmental controls monitoring
Auto manufacturing
Railways/transit operations
Monitor and control mail sorting
Lock and gate security
Money production
Naval ship onboard monitoring
Power generation DCS
Transmission grid management
Power distribution DCS
Automatic metering
Oil refinery control
Oil pipeline management
Gas production
Gas pipeline management
Gas distribution
Gas supply management
Automatic metering
Clean water treatment
Wastewater treatment
Water supply management
Dams/aqueducts/spillways
Transportation control—subways, trams, and people movers at airports
Highway monitoring and control
Automation of bridge controls

Figure 8.2. The 132-mile North-to-South Pacific Pipeline delivers crude oil from the oil fields of Bakersfield, California, to refineries on the coast next to Los Angeles.

The pervasiveness and extent of SCADA applications are staggering. For example, the flow of electric power through 672,000 circuit miles of overhead high-voltage transmission lines is governed by independent control systems that run unattended and parallel to the lines. Eighty percent of the nation's power is generated by 270 utilities. Each utility can contain up to 50,000 data collection points. In addition, the major DCS-controlled power generation plants are connected to over 3000 public and private electric utilities and rural cooperatives that make up the electric power grid. The market for power plant control was $1.5 billion and growing at about 6% per year in 2003.[5]

The United States currently uses 3250 billion-kilowatt hours of electricity, annually. A large part of this is generated by consuming 94 quadrillion BTUs of energy piped through 409,637 miles of interstate pipelines. Most of these are monitored and controlled by SCADA.

For example consider the Pacific Pipeline, which originates near the oil fields of Bakersfield, and runs 132 miles to energy-hungry southern California (see Figure 8.2). This plumbing transports 130,000 barrels/day of heavy crude oil from Bakersfield in the north to the Los Angeles refinery district located on the Pacific Coast. A parallel fiber optic network also runs the length of the pipeline so that SCADA computers can scan the entire length of the pipeline four times per second. The computers are looking for pipeline leaks that could lead to breaks. This pipeline cross several earthquake faults between Bakersfield and Los Angeles.

According to the Newton-Evans Research Company, 75% of the world's gas and oil pipelines of 25 km or more in length are monitored and controlled by SCADA systems.[6] Spending on these SCADA systems exceeds $200 million annually and

[5]ARC Advisory Group, http://www.ARCweb.com.
[6]Newton-Evans Research, Baltimore, MD, http://www.newton-evans.com.

is growing 30% per year. SCADA reduces the operational costs of gas and oil delivery by automating surveillance and emergency management. As we shall see, it also opens the door for asymmetric attacks on the power and energy delivery system.

SCADA VULNERABILITY ANALYSIS

If SCADA networks and industrial control systems were as simple as the schematic in Figure 8.1 suggests, the vulnerabilities would be limited—perhaps even inconsequential. But in reality, SCADA networks are intertwined with corporate networks, vendor connections, business partner connections, related websites, accounting and business process applications, and corporate databases. In practice, most SCADA systems live in a messy world of interdependent information systems (see Figure 8.3). This is where vulnerabilities can be found.

Access to SCADA networks has steadily grown as productivity needs have increased, the number of business partners has grown, and the ease of networking has prompted public utilities, energy companies, and power operators to connect everything to everything else. Communication has improved efficiency and lowered cost, but it has also opened SCADA to network intrusion. It has added more vulnerability to the infrastructures it was designed to enhance.

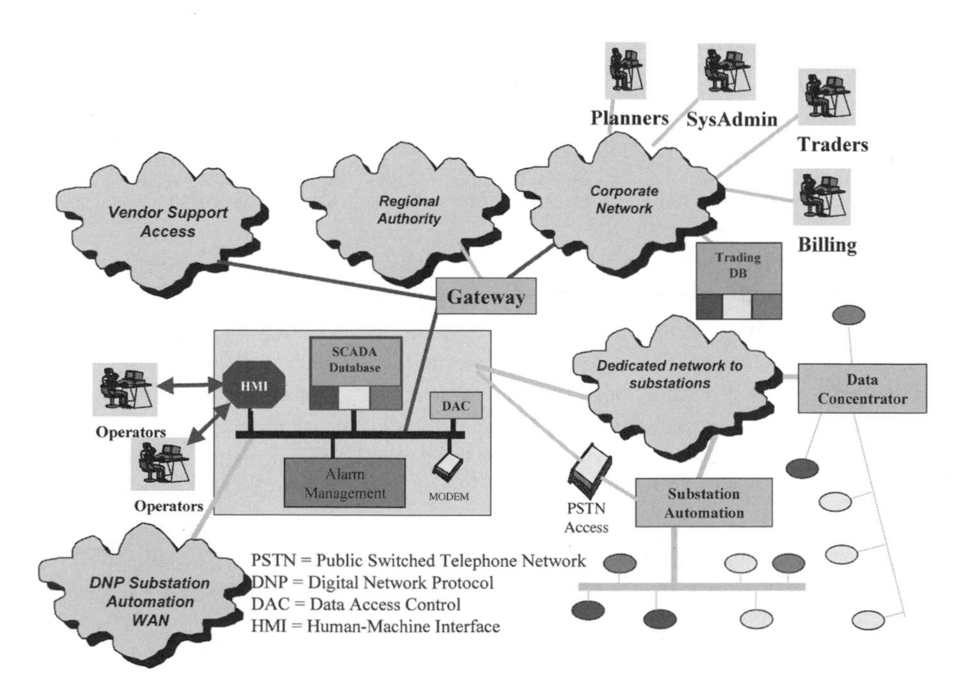

Figure 8.3. Most SCADA systems are open to access by several partners and vendors.

To make matters worse, most devices in SCADA networks are low-cost and low-powered—optimized to be deployed by the tens of thousands. The RTUs are often inexpensive microcomputers with limited memory. They are not designed to support impenetrable security. For example, they usually do not support difficult to crack cryptography or employ expensive firewall equipment that can block unauthorized access. Many RTUs are accessible over a simple dial-up telephone—an access method that can be used by anyone from anywhere in the world.

SCADA networks employ nearly every form of communication from Internet, Public Switched Telephone Network (PSTN), Advanced Digital Network (ADN) (a form of PSTN similar to DSL), digital radio, digital satellite, and Wi-Fi wireless. All of these methods of communication have well-known security weaknesses. All are vulnerable to attack, and all are connected to the inner workings of the critical infrastructures they monitor.

The vulnerabilities of SCADA include the vulnerabilities of general information systems, plus additional SCADA-specific vulnerabilities:

1. Policy issues—Have best practices been put into place?
2. Business process problems—Are there vulnerabilities in the process?
3. System vulnerabilities—Are there vulnerabilities in the design of the system?
4. Open connectivity—Are there too many access routes that are unprotected?
5. Weak identification and authentication—Do users frequently change passwords?
6. Reliance on vulnerable technology—Is the technology vulnerable?
7. Protection out-paced by threat—Has your anti-virus software and patches been updated?
8. Few security features built into technology—Is your equipment out of date?
9. In addition, SCADA and control systems generally must run 24 hours per day, every day, without failure, and be impervious to unauthorized access to PLCs and RTUs in the field.

Perhaps the biggest security hole in SCADA systems is traced to its openness and connectivity with related internal business systems and external partners as illustrated in Figure 8.3. This openness has its advantages: Business processes are made more efficient, and the resources needed to run SCADA systems can be shared with other IT functions. In addition, the skills needed to maintain SCADA are not altogether different than general IT support skills.

The concentration of IT assets and streamlined networking of SCADA with other IT processes has its downside: It leaves SCADA vulnerable to denial of service attacks, Internet viruses, and malicious software. This is a familiar story; economic and competitive forces make it attractive for businesses to connect SCADA with everything else in the enterprise, and yet this is exactly the wrong thing to do if security is paramount. In addition, typical information technology workers may lack the specialized knowledge needed to secure SCADA components.

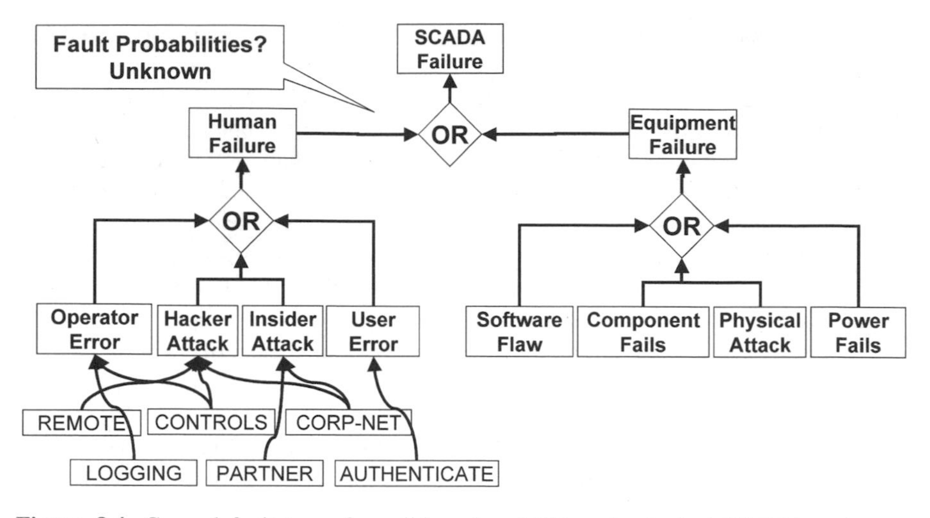

Figure 8.4. General fault tree of possible vulnerabilities of a typical SCADA system as described in Figure 8.3.

Figure 8.4 shows a partially completed fault tree of vulnerabilities that are attributed to linkage to corporate networks and access by business partners. On the Human Failure side of the fault tree, the major threats are Operator Error, Hacker Attack, Insider Attack, and User Errors that lead to intrusion. On the Equipment Failure side of the fault tree are Software Flaws, Component Failures, Physical Attack, and Power Failure threats.

Only the detail under *Human Failure* is described here. The threats described under Human Failure are the most common weaknesses found in typical SCADA systems (see Figure 8.3).

Remote Access (REMOTE): Almost all SCADA systems allow dial-up connections via old-fashioned (but inexpensive) modems. In many cases, the dial-up connection is not protected and allows anyone to directly access the RTUs or SCADA database, or both.

System Controls Lacking or Nonexistent (CONTROLS): The security of many SCADA systems have simply been overlooked or eliminated to save money. They are not protected against cyber-attack because the necessary control software has not been implemented. Encryption, for example, introduces additional overhead and adds to the cost of a SCADA component.

Weaknesses in Corporate Networks (CORP-NET): Many SCADA systems are linked to corporate networks through a shared network link, a computer that is connected to both, or indirectly through dial-up communication lines. Thus, if the corporate network is penetrated, the SCADA network is indirectly penetrated.

No Logging (LOGGING): System operators typically keep operational logs of events that have taken place during each work shift. The same logic applies to the SCADA system. Each data access—what kind of access was made and by whom—should be logged in a file. Unauthorized access should be denied, and the unauthorized attempt should be logged. Failure to keep logs and control access is like leaving the front door of your home open to burglars.

Vendors and Partners (PARTNER): The so-called "perimeter" of a SCADA network is expanded because the access points have been extended by allowing more and more users to connect and access the databases maintained by SCADA. Although this improves the efficiency of business processes, it also increases vulnerability. Business partners and vendors should be required to follow the same authentication and security procedures as employees.

Individual User Authentication Rarely Enforced (AUTHENTICATE): Perhaps the most common vulnerability comes from the users. Passwords are the first line of defense, and yet most users either do not use passwords, or they do not change them frequently enough to ward off password crackers. Passwords must be managed just like the keys to your car or house.

These (and other) threats are exploited through operator errors, hacker attacks, insider attacks, and user errors. They may lead to system failure as shown in Figure 8.4. Exact vulnerabilities (fault probabilities) for each of these generic threats are not generally known, nor are the financial damages resulting from a successful attack. This makes it difficult to estimate financial risk. The following example illustrates the difficulty of making these estimates.

CASE STUDY: SFPUC SCADA UPGRADE

We used the Hetch Hetchy water supply network in Chapter 7 to illustrate vulnerability analysis on the San Francisco Public Utilities Commission (SFPUC) water supply system. We can do the same for the SFPUC water supply SCADA system that underwent a major upgrade in 1999–2000 to harden it against natural and manmade disasters.

In November 2002, San Francisco voters approved legislation to finance the largest renovation of a water delivery system in San Francisco history. The $3.6 billion capital program contained 77 projects to repair, replace, and seismically upgrade the water system's aging pipelines and tunnels, reservoirs, and dams. The first phase of the massive renovation amounted to $1.6 billion. In addition, a $10.5 million upgrade of the SFPUC water SCADA system was approved and implemented as described here. The purpose of this case study is to illustrate how SCADA systems like the SFPUC SCADA can be hardened against attacks on its information technology infrastructure.

Redundancy as a Preventive Mechanism

The SFPUC water SCADA network reduced its vulnerability using *redundant components*. As this example illustrates, duplication of communication links, OCCs, and computer equipment goes a long way toward hardening SCADA. Redundancy may be one of the most effective methods of protecting other critical infrastructures.

Redundancy dramatically reduces the probability of a fault because the probability of individual component failures are *multiplied* rather than added together. Hence, if an individual component fails with probability of 10%, two identical components fail with a probability of 10% times 10% or 1%. Three identical components fail with a probability of 10% times 10% times 10% or 0.1%. This dramatic reduction in vulnerability is more than a mathematical fact—it works in practice!

This is exactly how the SFPUC SCADA system was hardened. Triple redundancy means each critical component of the SCADA system is duplicated three times. The resulting fault probability is decreased by three orders of magnitude. So instead of a 1% probability of failure, a triple redundant system has an extremely small 1-part-per-million failure probability (1% × 1% × 1% is 0.000001, or 1 part in 1 million).

The upgraded SFPUC water SCADA network is shown in Figure 8.5. The nodes of this network represent OCCs, RTUs (78 RTUs in the entire system are

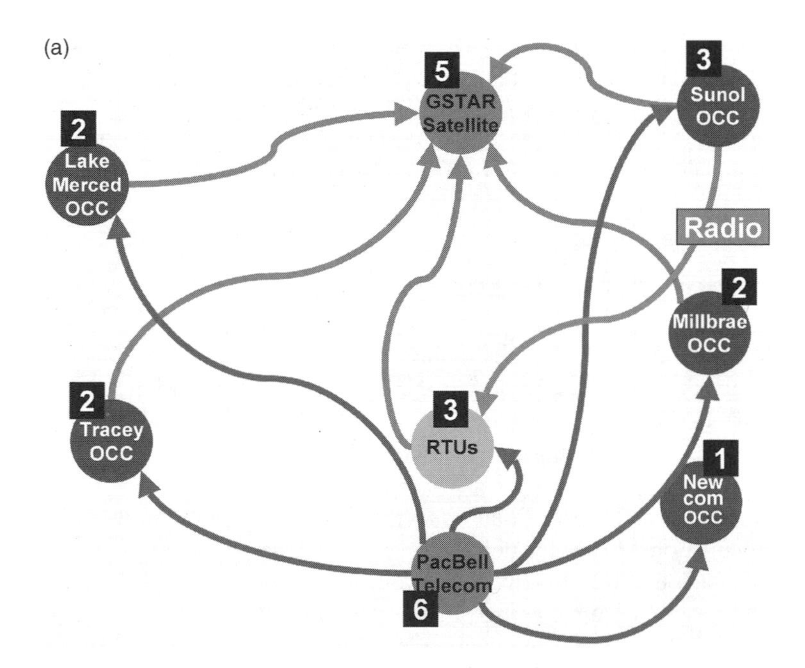

Figure 8.5. SFPUC water SCADA network showing the major network nodes and their connections via the SCADA communication paths. The numbers next to each node indicate the number of links. (a) Original SCADA network with communication paths as links. (b) SCADA network after removal of PacBell node. (c) SCADA network after removal of PacBell and GSTAR satellite nodes.

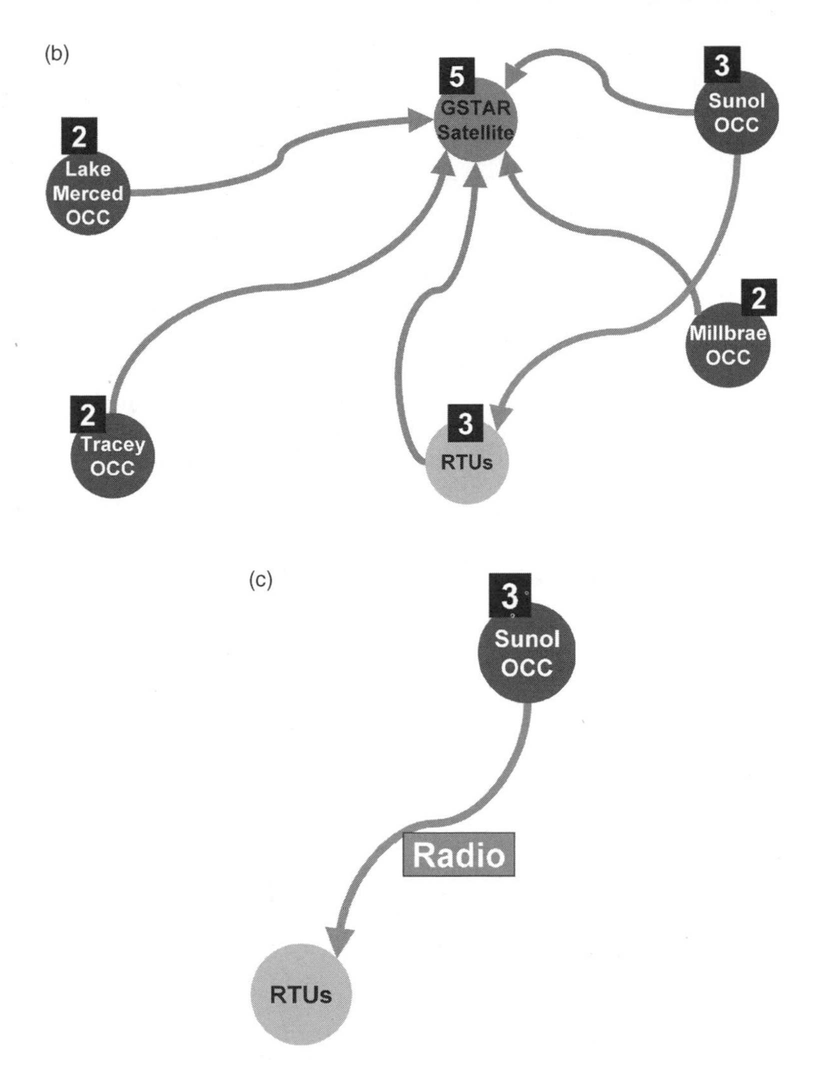

Figure 8.5. *Continued.*

represented by one node in Figure 8.5), and the major communication services—satellite, digital radio, and telephone lines.

Redundancy exists in the form of three OCCs—one each at Tracey, Lake Merced, and Sunol. Data are distributed to all three OCCs simultaneously. There are redundant servers and multiple workstations inside each OCC. The servers work from the same SCADA database so there is always a backup server in case the primary server fails. Furthermore, failure in one OCC does not lead to overall failure, because operations are transferred to another OCC.

The communication links in this SCADA network are also redundant. Multiple communication paths to the 78 RTUs in the field are implemented by two

PacBell telephone-wired networks (ADN and PSTN). In addition, there are several wireless links—a digital radio link, UHF radio link, and satellite links—to each OCC. If one path fails, there are two other alternatives.

Each node of Figure 8.5 is labeled with a number indicating how many links connect the node to other nodes. A scale-free network analysis of Figure 8.5 reveals hubs, as we expect from a scale-free network (see Figure 8.6). But the shape of the histogram for this network is only a very rough approximation to the power law of a scale-free network. If we ignore nodes with single links, $p = 1.5$ is the best fitting power law exponent. Note that there is only one node with the highest frequency of links.

The most significant hubs are the nodes with six and five links, respectively. These are the PacBell node with six links and the GSTAR satellite node with five links. Because of its unique radio link, the Sunol OCC node is the third-most critical node in the SCADA network. As we shall see, failure of the PacBell, GSTAR, and Sunol hubs overwhelms the triple redundant network and causes the entire SCADA network to fail. But this is highly unlikely, because all three must fail at once.

At first glance, it seems ridiculous to suggest that a satellite in space or the entire PacBell network system might be vulnerable to an attack. But the threat comes from a cyber-attack, not a physical attack. Cyber-attacks can disrupt satellites 23,000 miles in space as easily as placing a roadblock across an interstate highway. Similarly, the PacBell network does not need to be physically destroyed to render it useless. Cyber-attacks are extremely asymmetric because a single vandal or terrorist can launch a major denial of service attack that reduces the capacity or functionality of an entire network.

Cyber-attacks are doubly threatening because of their epidemic-like behavior. Once infected by a computer virus, the SCADA network serves as a kind of "disease vector" carrying the virus to other parts of the network. Infection of one OCC can lead to infection of the other two. Similarly, infection of the one communication path can spread throughout the entire telephone, satellite, or radio system. This SCADA network may be a cascade network as described in Chapter 4.

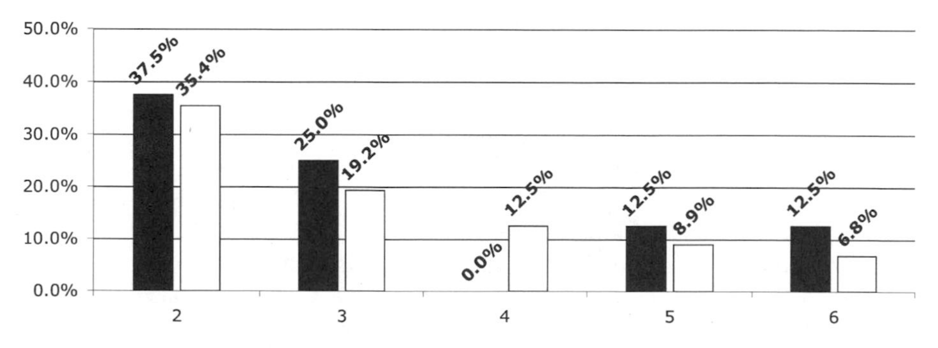

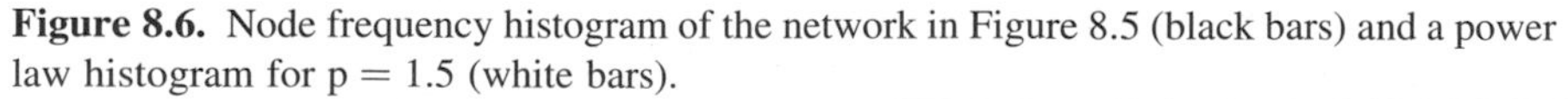

Figure 8.6. Node frequency histogram of the network in Figure 8.5 (black bars) and a power law histogram for $p = 1.5$ (white bars).

The question naturally arises, "what happens if these nodes are disabled?" The PacBell node has six links, so what happens if it is disabled? Figure 8.5(b) shows the result. If the wired network paths provided by PacBell fail, the field RTUs are still connected via the GSTAR satellite path, so the SCADA network continues to operate. Similarly, if the GSTAR hub goes down, the PacBell hub continues to connect the OCCs to the RTUs.

What happens if both PacBell and GSTAR nodes are disabled? Now two of the three redundant systems are disabled. Figure 8.5(c) shows this. The redundant OCC at Sunol takes over, and the redundant radio links continue to operate. Hence, the SCADA network can withstand disruption of PSTN, ADN, and VSAT links. Triple redundancy keeps the network operable. (Performance may be degraded, but the system still works.)

To have any major impact at all, an attacker must knock out multiple nodes. A cyber-attack needs to disable the GSTAR, PacBell, and Sunol nodes. A physical attack on the Tracey, Lake Merced, and Sunol OCCs has a lesser impact. To disconnect the SCADA network from its RTUs, a physical attack would have to disable four OCCs: Tracey, Lake Merced, Sunol, and Millbrae. It is highly unlikely that such a carefully coordinated physical attack would succeed. This is the value of redundancy in any system.

Before Redundancy Was Added

To dramatize the effects of redundancy, suppose the SFPUC SCADA network has no redundancy as shown in Figure 8.7. This system has only one communication infrastructure shown as the PacBell telecommunications network and only one OCC shown as the Tracey OCC. Clearly, a failure in PacBell would disconnect the entire network, and a failure in Tracey OCC would result in loss of control for the entire system. Hence, the PacBell node is critical, as is the Tracey OCC.

As prescribed by Chapter 4, critical nodes are the components we care most about and, hence, invest in them the most. Suppose this network is hardened using an investment strategy suggested by MBVA, assuming a budget of $10.5 million and improvement costs and damages as shown in Figure 8.8. In this analysis, we have assumed the vulnerabilities due to cyber-attack are MEDIUM for Tracey OCC and LOW for the PacBell hub. Using the standard fault probabilities, in the table below, this gives 50% and 20%, respectively, for OCC and communication faults:

Vulnerability	Fault Probability
Certain	100%
High	80%
Medium	50%
Low	20%
Impossible	0%

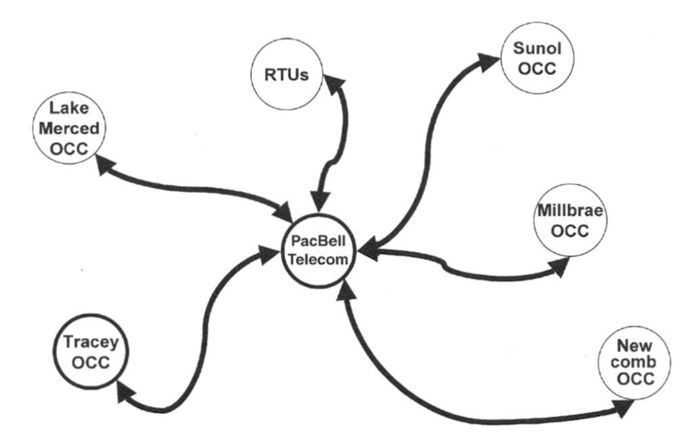

Figure 8.7. SFPUC SCADA network before upgrade—no redundancy.

Figure 8.8 combines the fault tree with the event tree for the vulnerabilities assumed on each critical node (Tracey OCC and PacBell Comm) revealed in Figure 8.7. Thus, the COMM-only fault probability is 10%, the OCC-only fault probability is 40%, and both faults are likely to occur with a fault probability of 10%. Sector vulnerability is 60%, and sector risk is $110 million.

Figure 8.8 and Table 8.1 includes the costs and damages for each component—$10 million and $200 million for OCC and $5 million and $50 million for the PacBell node. Recall the budget is $10.5 million, which is not enough to cover the $15 million needed to reduce all vulnerabilities to zero. Therefore, an allocation strategy is needed to decide how much to allocate to each components. These

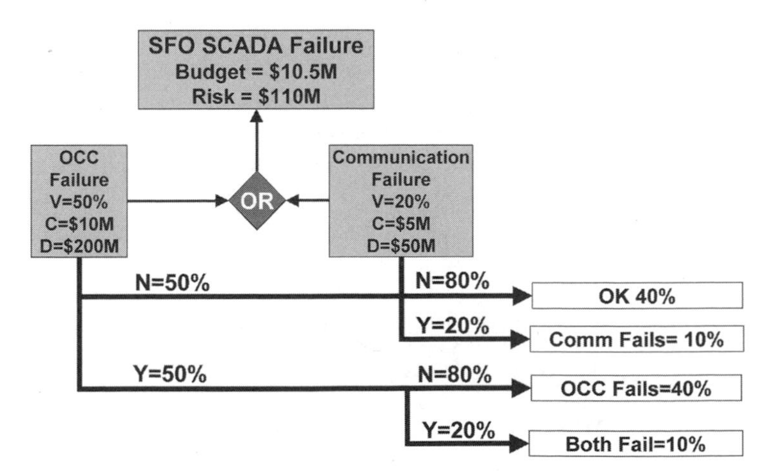

Figure 8.8. Combined fault and event tree for nonredundant SCADA network.

TABLE 8.1. Results of Apportioned Risk Reduction Allocation for the SCADA Fault Tree of Figure 8.8.

Threat	OCC	COMM
Vulnerability	50%	20%
Cost	$10	$5
Damage	$200	$50
Vulnerability Allocation	$8.20	$2.30
Reduced Vulnerability	8.7%	10.9%
Risk Allocation	$9.38	$1.12
Reduced Vulnerability	3.1%	15.5%

numbers were entered into program *FTplus* to obtain the apportioned allocation results shown in Table 8.1.[7]

Apportioned vulnerability reduction—reducing the fault probabilities—leads to an investment strategy that allocates 78% of the budget in OCC fault reduction—$8.2 million versus $2.3 million on the communication fault probability reduction. This allocation reduces the fault probability of OCC failure to 8.7% and the probability of COMM failure from 20% to 10.9%. Sector vulnerability drops from 60% to approximately 19% (single and double faults add up to 18.79%).

Apportioned risk reduction is similar, but it allocates even more to the OCC threat. In Table 8.1, the allocation strategy for reducing risk means investing heavily in OCC protection ($9.38 million) and only $1.12 million in protecting communication links. The reason is clear: The damages resulting from an OCC failure are four times that of a COMM failure—$200 million versus $50 million. Reducing risk through this allocation strategy reduces the fault probability of an OCC failure to 3.1%, and it reduces the COMM fault probability to 15.5%, which when combined with the double combination fault (OCC and COMM, both), results in a sector vulnerability of 18.1%, and the risk to $14 million—down from $110 million. Hence, the $10.5 million investment removed $96 million in risk from this sector.

Which strategy is best? If the goal is to reduce the probability of any failure in any component, select the vulnerability reduction strategy. If it is to limit the financial damage resulting from an attack, then use the risk reduction strategy. In the end, there is only a slight difference between vulnerability and risk reduction, principally because the probability of an OCC failure and the damages it produces are both higher than those of the COMM threat.

Adding Redundancy

In this section, we study the effects of adding redundancy to the SFPUC SCADA network of Figure 8.7 and repeat the analysis to compare the two results. In the end, the new and improved redundant SFPUC SCADA network is extremely

[7]Only the apportioned risk reduction strategy is reported here. The others are left as an exercise for the reader.

secure because redundancy is a bargain compared with target hardening. Risk drops to an extremely low level, because fault probabilities are multiplied instead of added together. The resulting redundant network, redundant computers, and redundant OCCs are shown in Figure 8.5(a), and the corresponding fault tree is shown in Figure 8.9(a).

Figure 8.9(a) models redundancy with AND logic. Recall that AND logic reduces vulnerability in the tree by multiplying fault probabilities. Thus, the fault probability of an OCC failure declines from 50% to 6.25%, and communication failure declines from 20% to 0.8%. The overall sector vulnerability declines

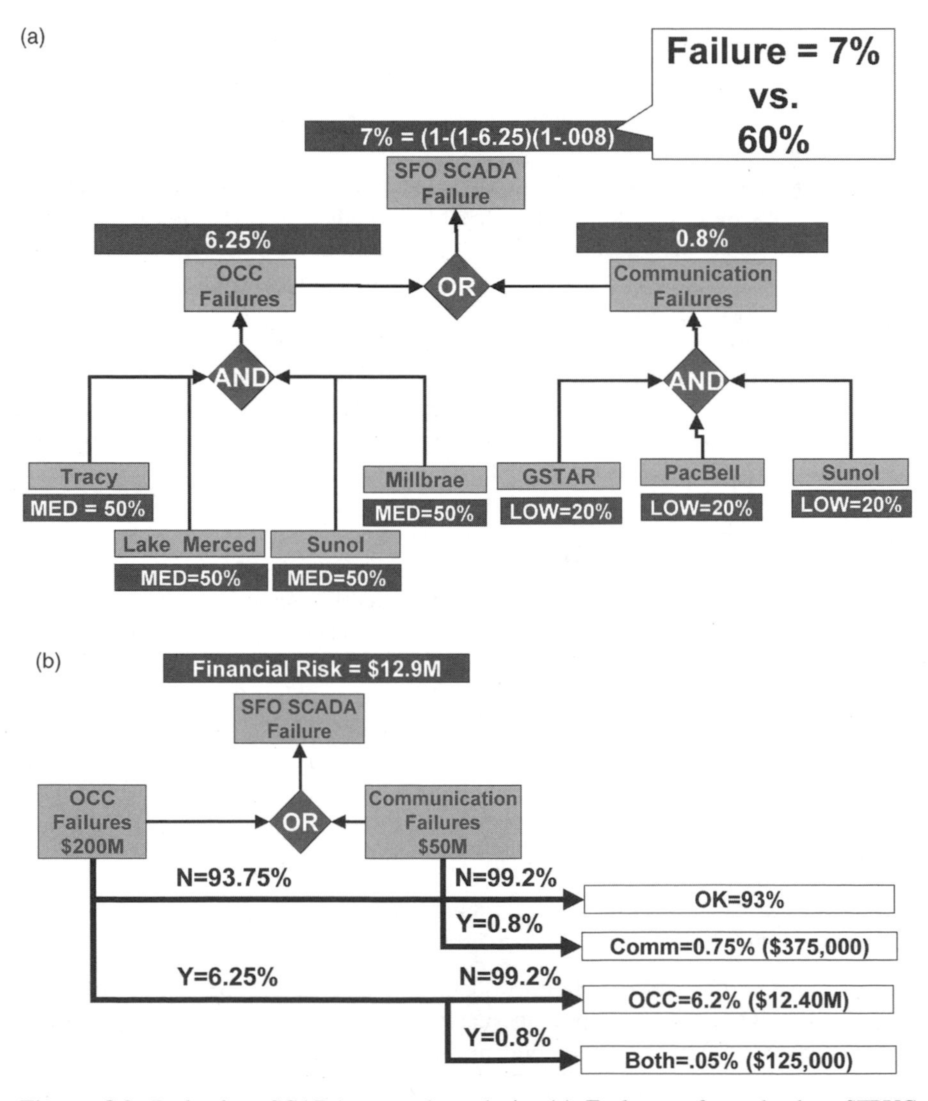

Figure 8.9. Redundant SCADA network analysis. (a) Fault tree for redundant SFPUC SCADA network showing assumed component vulnerabilities and sector vulnerability. (b) Event tree for redundant network showing damages and financial risk.

from 60% to 7%—nearly an order of magnitude difference! This is the power of redundancy.

Figure 8.9(b) shows the resulting event tree and damages assumed for Figure 8.8. Instead of $110 million, the risk of this redundant network is $12.9 million—an order of magnitude reduction. Assuming the redundancy can be purchased for $10.5 million, we can compare the results of adding redundancy with the prior analysis. Redundancy produces a lower risk than risk reduction allocation—$12.9 million versus $14 million. Why the difference?

Table 8.2 summarizes reduction in vulnerabilities as a result of resource allocation—using apportioned risk reduction—compared with the reductions resulting from triple redundancy. In round numbers, apportioned allocation "buys down" the faults to 3%, 16%, and less than 1% (assumed to be zero), whereas triple and quadruple redundancy reduces faults to 0.75%, 6.2%, and 0.05%, respectively. The total fault probability remaining after nonredundant risk reduction is 19%, whereas the total remaining after adding redundancy is 7%. In other words, target hardening with redundancy is twice as effective as target hardening without redundancy. Redundancy is a cost-effective strategy; no wonder it is often used to prevent failures in complex systems.

But there is a downside to redundancy in any computer system: It may exacerbate the cyber-threat. If two or more redundant computers are connected to the same network, a computer worm or virus might contaminate all redundant systems. This may lead to a false sense of security, when in fact, all redundant systems fail at once. Therefore, it is important to carefully implement redundancy, especially when defending against cyber-attacks.

ANALYSIS

What does the SFPUC water SCADA example tell us about SCADA in general? After making major investments in the SFPUC water SCADA, what were the benefits? In summary, this case illustrates the following:

- Cyber-intrusion is a major risk to control systems. Protect SCADA by isolating it from the corporate network, encrypting its data, and enforcing validated passwords. Isolation may be implemented through physical separation or through

TABLE 8.2. Comparison of Results for Apportioned Risk Reduction Allocation and Risk Reduction by Redundancy.

Threat	Original Vulnerability	After Risk Allocation	Redundancy Probability
Communication	10.0%	3.0%	0.75%
OCC	40.0%	16.0%	6.20%
Both	10.0%	less than 1%	0.05%
		19.0%	**7.0%**

firewall machinery that logically separates corporate networks from control system networks.[8]

- Redundancy can reduce fault probabilities by an order of magnitude. Wherever a critical node is found, duplicate its function with double- or triple-redundant "backups." Because redundancy multiplies fault probabilities, instead of being additive, it is an effective means of protecting any system. But redundancy can also give operators a false sense of security, especially when it comes to cyber-security. Computer viruses and worms can simultaneously infect all redundant computers, rendering them all unusable.
- Communication security is as important as OCC security. Treatment plants, pipelines, and control centers are important, but do not overlook the network that connects them together. Careful implementation of user authentication, network isolation, and redundant communication links (as illustrated by the SFPUC example) can reduce vulnerabilities.
- The SFPUC water SCADA system is highly secure, because of triple redundancy in servers, communication links, and OCCs. This greatly improved system is an example of how to correctly increase SCADA security and harden typical water SCADA networks against natural and manmade attacks. But cyber-attacks may still succeed against such a redundant system unless they are properly isolated and administered.

ANALYSIS

SCADA is found in almost every industrial process and most critical infrastructures. It is pervasive. The trend is for even more automation because it lowers costs and increases speed and efficiency. The sectors most impacted are Level 1 infrastructures—water, power, energy, and telecommunications. But SCADA and related control systems are employed in other critical infrastructures such as transportation, food, and agriculture.

Attacks on infrastructures that use any automated control system can be extremely asymmetric. As the study of the SFPUC water SCADA system illustrated, an inexpensive cyber-attack on a SCADA network can bring down the entire network. Fortunately, the likelihood of a successful attack on the triple-redundant SFPUC network is extremely low. Furthermore, it is not clear that damages would be very great—not nearly as great as the estimates used here.[9] A successful attack on a water SCADA may lead to contamination or destruction of equipment; there is no recorded cases of SCADA attacks leading to mass casualties.

Responsibility for SCADA security is scattered across governmental and commercial bureaucracies. This is unfortunate, but understandable, as SCADA applications are scattered across hundreds of industries. There is no SCADA ISAC

[8]A firewall is a computer that filters or blocks input and output data ports, thus restricting user access to a network or subnetwork.

[9]In general we do not know the extent of damage that might be inflicted by a successful cyber-attack on some SCADA system.

because SCADA and control system security is vertical industry-specific. According to Weiss, "end-users won't share critical information with an ISAC that could act as a policeman and also with an organization they don't know or trust."[10]

The SFPUC water SCADA case study points out how important it is to build in duplication of services. Redundancy works. But redundancy is expensive, and it may not prevent a cyber-attack. So it is important to allocate resources carefully. The analysis showed that communication links and OCCs are the critical nodes of the SFPUC SCADA system. In general, many expensive assets will be concentrated in OCCs. Hence, they need to be hardened. Additionally, the ease of executing cyber-attacks on communication links means we cannot ignore the communication "nodes." Even satellites in space are vulnerable to cyber-attacks. The asymmetric

TABLE 8.3. DOE's 21 Steps to SCADA Security.[11]

1. Identify all connections to SCADA networks.
2. Disconnect unnecessary connections to the SCADA network.
3. Evaluate and strengthen the security of any remaining connections to the SCADA network.
4. Harden SCADA networks by removing or disabling unnecessary services.
5. Do not rely on proprietary protocols to protect your system.
6. Implement the security features provided by device and system vendors.
7. Establish strong controls over any medium that is used as a backdoor into the SCADA network.
8. Implement internal and external intrusion detection systems and establish 24-hour-a-day incident monitoring.
9. Perform technical audits of SCADA devices and networks, and any other connected networks, to identify security concerns.
10. Conduct physical security surveys, and assess all remote sites connected to the SCADA network to evaluate their security.
11. Establish SCADA "Red Teams" to identify and evaluate possible attack scenarios.
12. Clearly define cyber-security roles, responsibilities, and authorities for managers, system administrators, and users.
13. Document network architecture, and identify systems that serve critical functions or contain sensitive information that requires additional levels of protection.
14. Establish a rigorous, ongoing risk management process.
15. Establish a network protection strategy based on the principle of defense-in-depth.
16. Clearly identify cyber-security requirements.
17. Establish effective configuration management processes.
18. Conduct routine self-assessments.
19. Establish system backups and disaster recovery plans.
20. Senior organizational leadership should establish expectations for cyber-security performance and hold individuals accountable for their performance.
21. Establish policies and conduct training to minimize the likelihood that organizational personnel will inadvertently disclose sensitive information regarding SCADA system design, operations, or security controls.

[10]Personal communication with Joe Weiss, April 2005.

[11]http://www.utc.org/?v2_group = 0&p = 3629.

cyber-attack scenario is countered by implementing triple-redundant links, OCCs, and highly secures computers. This case study is a good lesson in how to harden such targets, but we postponed the deeper discussion of cyber-security until the final three chapters.

At this time in history, damages done by cyber-terrorists have been minor. There have been no deaths, and the cost to the economy has been relatively low. SCADA components are vulnerable, but damages are so low that SCADA security has not gained much attention as a major threat. Will this change if a major event occurs? It is important for us to differentiate between vulnerabilities and risk. A very low-value threat associated with a low fault probability is of little interest. Interest may still be low if a system is highly vulnerable, but financial risk is extremely low. When both vulnerability and risk are high, we should be extremely interested in target hardening. Redundancy is an effective means of preventing highly likely, high-risk events.

Even though no SCADA attack has gained much public attention, the potential for major damage to SCADA networks and indirectly to the economy still remains. Thus, SCADA and other control system policies should focus on hardening of targets against cyber-intrusion. The DOE has provided 21 steps for protecting SCADA in the power sector (see Table 8.3). These steps are general enough to apply to all SCADA systems. They are policies that every security-conscious organization should follow.

EXERCISES

1. What is SCADA (select only one)?
   a. Secure communications for data analysis
   b. Secure communications for data acquisition
   c. Supervisory control and data analysis
   d. Supervisory control and data acquisition
   e. Supervisory control and distributed analysis
2. Which of the following are *not* components of SCADA (select all that apply)?
   a. Sensors
   b. Controls
   c. Flood gates
   d. System database
   e. Data concentrator
3. Why is encryption *not* used more often (always) in SCADA communications (select only one)?
   a. Keys are not long enough.
   b. Encryption introduces overhead and cost.
   c. Analog communications cannot be encrypted.

   d. Cellular modems can be war-dialed.
   e. There is nothing secret in most water supply systems.

4. SCADA networks should be protected by (select all that apply)?
   a. Isolation from corporate network
   b. Increasing latency
   c. Encryption of data
   d. Validation of passwords
   e. Governmental regulation of the OCCs
5. Who is the lead federal department responsible for protecting critical infrastructure from SCADA attack (select only one)?
   a. Department of Homeland Security
   b. EPA
   c. Dept of Energy
   d. Dept of Treasury
   e. NIAP
6. What would be the effect of changing all AND logic into OR logic in the fault tree of Figure 8.9?
   a. The number of event tree outcomes expands from 4 to 128.
   b. The sector fault probability changes from 60% to 96.8%.
   c. The vulnerability of an OCC failure changes from 6.25% to 93.75%.
   d. All of a, b, and c.
   e. None of a, b, and c.
7. What ISAC has responsibility for SCADA security?
   a. NIAC
   b. DHS
   c. IT-ISAC
   d. WaterISAC
   e. SCADA-ISAC
8. What is the argument against redundancy?
   a. It does not always work.
   b. It adds to the cost of a system.
   c. It reduces fault probabilities.
   d. It reduces financial risk.
   e. Systems still fail.
9. In the SFPUC case study, why did redundancy produce a lower sector risk, overall, than apportioned risk reduction allocation?
   a. Apportioned reduces financial loss—redundancy lowers fault probabilities.
   b. Apportioned does not work correctly.

c. Redundancy is a more cost-effective way to reduce faults because of the multiplicative effect of redundancy versus the additive effect of funding.
d. Fault trees are not perfect.
e. All of the above.

10. Why is cyber-intrusion such a major threat to SCADA?
    a. SCADA systems are notoriously vulnerable to cyber-attacks.
    b. Cyber-SCADA attacks have historically been disastrous.
    c. Cyber-SCADA attacks have killed people.
    d. Cyber-SCADA protection is expensive.
    e. Scientific studies have concluded that these are the worst vulnerabilities.

11. The gas and oil pipeline system of San Lewis Rey is monitored and controlled by a SCADA network much like the Pacific Pipeline system described in Figure 8.2 (see Figure 8.10). A fiber optic network is attached to the pipes and substations along with RTUs that measure temperature and pressure along the pipeline. An OCC at node SLR-13 controls values and flows of gas and oil through the entire network. The OCC is connected to all RTUs through the fiber optical cable that runs along the pipes—no wireless or satellite networks supplement this connection.

    Perform a network-wide analysis of this network, and identify where to allocate 1000 availability points such that this network is as secure as possible.

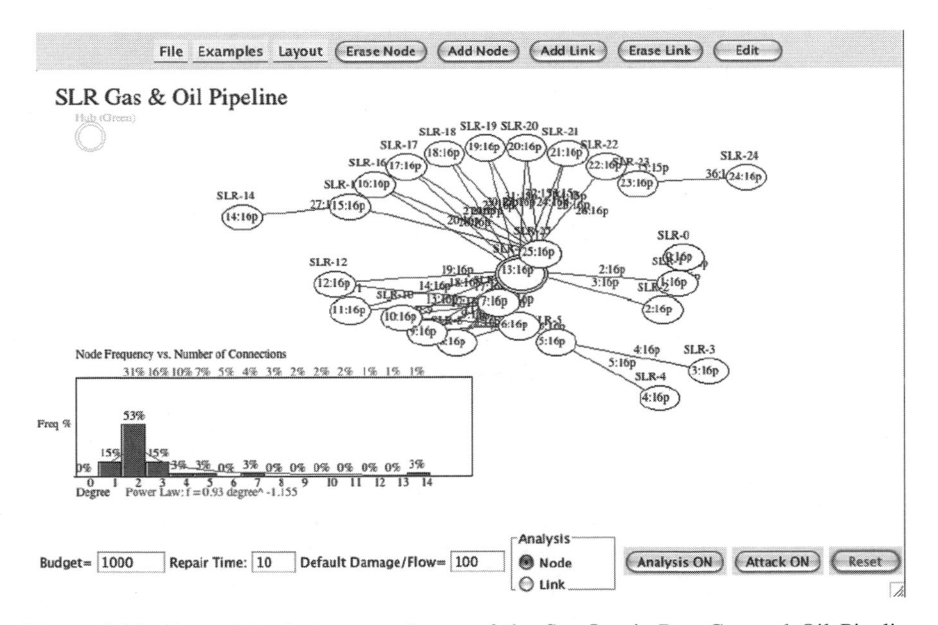

Figure 8.10. *NetworkAnalysis* screen image of the San Lewis Rey Gas and Oil Pipeline SCADA Network showing the results of network-wide allocation.

Q1: Is this network scale-free?

a. Yes, with p between 2 and 3.

b. No, because p is less than 1.

c. Yes, but with p between 1 and 2.

d. No, because there are no hubs.

e. No, because it is a small-world network.

Q2: Given 1000 points to allocate, what is the risk after allocation?

a. 9332

b. 6000

c. 5000

d. 6500

e. 5069

Q3: Given 1000 points to allocate, how many nodes receive a full 100 points?

a. 8

b. 7

c. 5

d. 4

e. None

Q4: Which node is the hub?

a. SLR-25

b. SLR-13

c. SLR-25 and SLR-13

d. SLR-7

e. None

Q5: Given 1000 points to allocate, and assuming all link and node damages equal 100, how many links receive points after network-wide allocation?

a. All

b. None

c. One

d. Two

e. 10%

Q6: Given 200 points to allocate, and assuming all links and node damages equal 100, network-wide allocation gives all the points to nodes . . .?

a. Nodes #6, 7, and 13

b. Nodes #7, 13, and 25

c. Nodes #13 and 25

d. Nodes #5, 6, 7, and 13

e. The hub, node #13, gets all points

12. Using the fault tree of Figure 8.8 to model node SLR-13 of the San Lewis Rey pipeline system described in problem 11 and Figure 8.10, and the following inputs to *FTplus*, answer the questions below. Assume a budget of $15.

Threat	Vulnerability	Cost	Damage
OCC	20%	$20	$100
Comm	80%	$30	$10

Q1: What is the allocation of $15 to each threat using the ranked allocation strategy for fault reduction?

a. Fault Reduction Allocation: $0 to OCC, $15 to Comm
b. Fault Reduction Allocation: $13 to OCC, $2 to Comm
c. Fault Reduction Allocation: $15 to OCC, $0 to Comm
d. Fault Reduction Allocation: $5.5 to OCC, $9.5 to Comm
e. Fault Reduction Allocation: $9.5 to OCC, $5.5 to Comm

Q2: What is the allocation of $15 to each threat using the minimal (optimal) allocation strategy for fault reduction and then for risk reduction?

a. Fault: $15 to OCC, $0 to Comm. Risk: $0 to OCC, $15 to Comm
b. Fault: $10 to OCC, $5 to Comm. Risk: $10 to OCC, $5 to Comm
c. Fault: $5 to OCC, $10 to Comm. Risk: $10 to OCC, $5 to Comm
d. Fault: $10 to OCC, $5 to Comm. Risk: $5 to OCC, $10 to Comm
e. Fault: $0 to OCC, $15 to Comm. Risk: $15 to OCC, $0 to Comm

Q3: What is the allocation of $15 to each threat using the apportioned allocation strategy for risk reduction?

a. Risk: $0 to OCC, $15 to Comm
b. Risk: $15 to OCC, $0 to Comm
c. Risk: $5 to OCC, $10 to Comm
d. Risk: $5.5 to OCC, $9.5 to Comm
e. Risk: $9.5 to OCC, $5.5 to Comm

Q4: Assuming a budget of $15, which objective is met (and is the same allocation of $15) for all allocation strategies described in this problem?

a. Apportioned risk and fault reductions
b. Minimal risk and fault reductions
c. Ranked risk and fault reductions
d. Fault reduction for all allocation strategies
e. Risk reduction for all allocation strategies

CHAPTER 9

Power

In 2000, the National Academy of Engineering named modern power grids—those vast electrical power generation, transmission, and distribution networks that span the country—the top engineering technology of the 20th century. In the Academy's opinion, the power grid surpassed invention of the automobile, airplane, delivery of safe and abundant water, and electronics as the most important engineering accomplishment. Electrical power is what makes modern society tick. It is essential. So it comes as no surprise that the grid is one of the fundamental infrastructures of the United States. It is a Level 1 critical infrastructure.

In this chapter you will learn the following concepts and be able to apply them to the challenge of electrical power grid vulnerability analysis:

1. Historically, the components of power—generation, transmission and distribution, consumption, and SCADA control—have been *vertically integrated.* Since 1992, the components have been decoupled through deregulation legislation, which has introduced *economic vulnerabilities* into the grid.
2. The power grid has been, and continues to be, shaped by a combination of *governmental regulation and the laws of physics*—these two do not always work together.
3. *There is no shortage of power*, but there is a shortage of distribution capacity. This shortage aggravates the problem of protection, because a small "spot shortage" can have widespread damaging effects on major regions of the country.
4. The "Architecture" of the grid is that of a *small-world network*—clustered nodes connected to other clustered nodes through a combination of many short and a few long links. This leads to the notion that the bigger the grid, the harder it can fall.
5. Because of the small-world architecture of the grid, and the laws of physics, the grid is *vulnerable to cascade effects* that can sweep through power grid interconnects like a contagion sweeps through human populations.
6. The *greatest vulnerabilities exist "in the middle,"* that is, in the transmission and distribution layer of the power grid. Faults occur and propagate

Critical Infrastructure Protection in Homeland Security: Defending a Networked Nation,
edited by Ted G. Lewis

throughout major portions of the country, because of critical links, insufficient distribution capability, and cascade failures.

7. SCADA systems are an essential component of the deregulated grid; hence, the *grid is vulnerable to SCADA attacks*. Called energy management systems (EMS), the hardware and software of EMS are susceptible to the same vulnerabilities as other information technology systems.
8. No single power generator is critical—the largest source of power provides less than 1% of the national capacity. It is a myth that the most vital components of the nation's power sector are power plants. This points once again to the "middle" of the grid as the most likely place for failures to occur.

FROM DEATH RAYS TO VERTICAL INTEGRATION

Electrical power has historical roots in the famous Pearl Street, New York utility created by Thomas Edison in the 1880s. This first utility supplied direct current (DC) electrical power to 59 Manhattan customers. Edison was convinced that DC was the best way to deliver electricity, but Serbian immigrant Nikola Tesla had a better idea: alternating current (AC). Tesla was Edison's rival in all things having to do with harnessing the power of the electron. He is the father of all modern electric power generation technology (generators), distribution (transmission lines and substations), and appliances (motors).

A titanic power struggle between Tesla and Edison ensued over the advantages of AC versus DC. When Tesla sold his patent rights to George Westinghouse, Edison's feud shifted from Tesla to Westinghouse. Edison derided AC. At one point he used the electric chair to show that AC was unfit for consumers. Tesla countered with daring demonstrations of his own:

> Tesla gave exhibitions in his laboratory in which he lighted lamps without wires by allowing electricity to flow through his body, to allay fears of alternating current. He was often invited to lecture at home and abroad. The Tesla coil, which he invented in 1891, is widely used today in radio and television sets and other electronic equipment. That year also marked the date of Tesla's United States citizenship.
>
> Westinghouse used Tesla's system to light the World's Columbian Exposition at Chicago in 1893. His success was a factor in winning him the contract to install the first power machinery at Niagara Falls, which bore Tesla's name and patent numbers. The project carried power to Buffalo by 1896.
>
> In 1898 Tesla announced his invention of a teleautomatic boat guided by remote control. When skepticism was voiced, Tesla proved his claims for it before a crowd in Madison Square Garden.
>
> In Colorado Springs, Colo., where he stayed from May 1899 until early 1900, Tesla made what he regarded as his most important discovery—terrestrial stationary waves. By this discovery he proved that the Earth could be used as a conductor and would be as responsive as a tuning fork to electrical vibrations of a certain frequency.

> He also lighted 200 lamps without wires from a distance of 25 miles (40 kilometres) and created man-made lightning, producing flashes measuring 135 feet (41 metres). At one time he was certain he had received signals from another planet in his Colorado laboratory, a claim that was met with derision in some scientific journals.
>
> Tesla was a godsend to reporters who sought sensational copy but a problem to editors who were uncertain how seriously his futuristic prophecies should be regarded. Caustic criticism greeted his speculations concerning communication with other planets, his assertions that he could split the Earth like an apple, and his claim of having invented a death ray capable of destroying 10,000 airplanes at a distance of 250 miles (400 kilometres).[1]

Eventually, the Tesla–Westinghouse approach won out and established AC as the standard technology for power generation and distribution. AC could be transmitted over longer distances than DC, easily powered motors used in factories and homes, and could be voltage-stepped up/down to accommodate different needs for a diverse consumer.

By 1896 the Tesla–Westinghouse collaboration resulted in hydroelectric power generation at Niagara Falls and AC transmission to Buffalo 20 miles away. Edison's DC power networks were limited to one mile. The world had entered the modern era of electrical power generation, but the business of electrification had yet to be invented.

In 1920, the precursor to the modern power regulator, the Federal Power Commission (FPC) was set up by Congress to coordinate hydroelectric projects. The FPC would grow over the decades and eventually become the Federal Energy Regulatory Commission (FERC), with a budget exceeding $200 million and vast regulatory powers over natural gas and electrical power. But in the 1920s, electrical power generation, transmission, and distribution was owned by large interstate holding companies that optimized the flow of power from fuels such as coal or hydroelectric generators. They exercised control of their regions of the country by vertically integrating all aspects of production, distribution, and marketing. These vertical monopolies standardized on 60-Hz (cycles per second) and 240/120-volt current, but they were stove-piped islands when it came to interoperability. Two AC signals have to be synchronized as shown in Figure 9.1(c), before they can be combined across vertical monopolies. Synchronization would remain a technical challenge into the 21st century.

Figure 9.1(a) shows what two AC signals look like when they are identical except for a *phase shift*. One signal's voltage level lags the other. Figure 9.1(b) shows two identical power signals except that one oscillates faster than the other. The net effect of phase or frequency differences is a distorted signal, which is bad for motors and appliances that use the power. Clean power results when both signals oscillate at the same frequency and same phase as in Figure 9.1(c). Thus, it is important that signals be in phase and at the same frequency when merged onto the same transmission lines, transformers, and other electrical machinery.

[1] http://www.neuronet.pitt.edu/~bogdan/tesla/bio.htm.

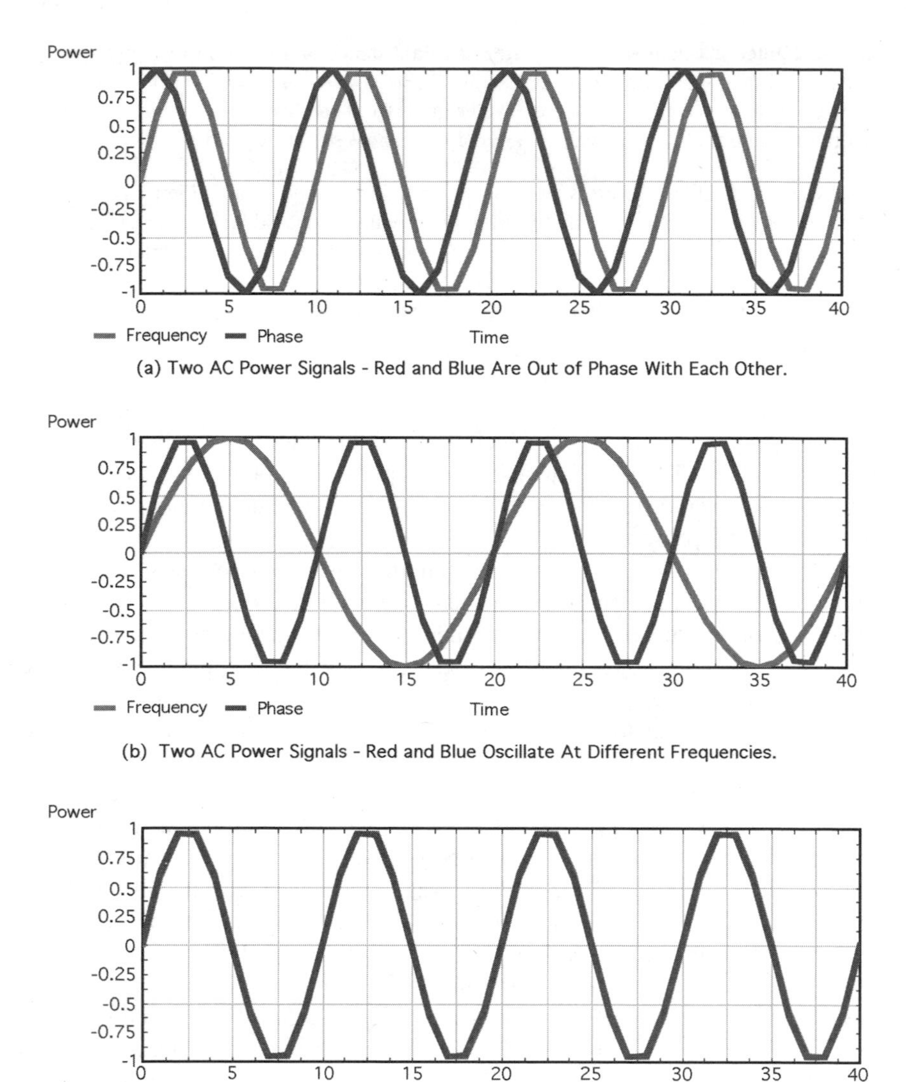

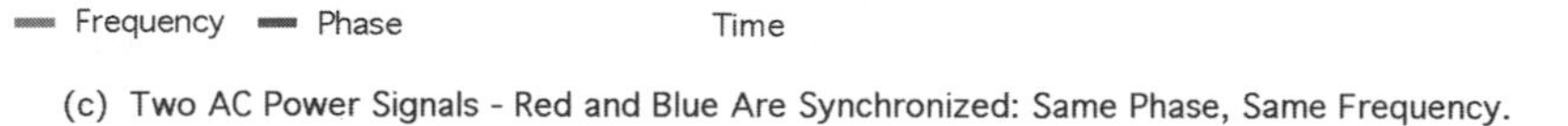
(c) Two AC Power Signals - Red and Blue Are Synchronized: Same Phase, Same Frequency.

Figure 9.1. AC power must be synchronized, which means two power signals must be at the same frequency and same phase.

Standardization and synchronization was needed before privately held vertical monopolies could interoperate. *Universal access*—the ability for anyone in the United States to get electrical power service—had not yet arrived. It would require interoperability, a technical capability that was lacking among the local monopolies.

The Federal Power Act of 1920, the Natural Gas Act of 1938, and the Public Utility Holding Company Act (PUHCA) of 1935 changed the landscape by empowering the FPC to regulate the sale and transportation of natural gas and electricity across state borders. Together, these laws define power and energy transmission as interstate commerce, which is the exclusive purview of the Legislative Branch of Government. Thus, a state could not directly regulate that commerce, but Congress could. The PUHCA shaped the electric power industry until 1992.

A series of legal modifications to the PUHCA expanded the power of Congress to regulate power and energy companies. For example, the Natural Gas Act was amended in 1940 to charge the FPC with the responsibility for certifying and regulating natural gas facilities—going beyond simply regulating the sale of power across interstate boundaries.

The Northeast Blackout of 1965 highlighted the vulnerability of the vertically integrated power grid. As local holding companies were encouraged to interoperate and borrow power from one another to accommodate surges in demand, they also became more fragile. A loss of capacity in one region could lead to a series of failures that could collapse entire regions. Thus, the cascade failure was born, forcing a shift in federal regulatory legislation from pure regulation and universal access to safety and reliability.

The first prerequisite for prevention of cascade failures is that the power grid must be extremely reliable. Even a relatively insignificant component such as a power line must not fail. Thus, the North American Electric Reliability Council (NERC) was formed shortly after the blackout in 1965. The NERC is a not-for-profit company formed to promote the reliability of bulk electric systems that serve North America.[2]

The energy crisis of the 1970s brought fuel price inflation, conservation, and a growing concern for the environment. Congress began to shift its emphasis once again from reliability to clean and inexpensive power. The Public Utilities Regulatory Policies Act (PURPA) was enacted in 1978 to promote conservation of energy. But it had an important side effect: It opened the vertically integrated monopolies to competitors. PURPA required the electric utilities to buy power from "qualified facilities" (QFs). Thus was born the non-utility generator (NUG) and independent power producer (IPP). This side effect would be expanded in 1992 when the vertical monopolies were broken up by deregulation legislation.

In 1977, Congress transferred the powers of FPC into FERC—an independent agency that regulates the interstate transmission of natural gas, oil, and electricity. The FERC maintained the shape of the electrical power sector during the 1980s

[2]"NERC's members are the ten Regional Reliability Councils whose members come from all segments of the electric industry: investor-owned utilities; federal power agencies; rural electric cooperatives; state, municipal and provincial utilities; independent power producers; power marketers; and end-use customers. These entities account for virtually all the electricity supplied in the United States, Canada, and a portion of Baja California Norte, Mexico." http://www.nerc.com.

and early 1990s. As part of its responsibility, the FERC[3]:

- Regulates the transmission and sale of natural gas for resale in interstate commerce.
- Regulates the transmission of oil by pipeline in interstate commerce.
- Regulates the transmission and wholesale sales of electricity in interstate commerce.
- Licenses and inspects private, municipal, and state hydroelectric projects.
- Approves the siting of and abandonment of interstate natural gas facilities, including pipelines, storage, and liquefied natural gas.
- Oversees environmental matters related to natural gas and hydroelectricity projects and major electricity policy initiatives.
- Administers accounting and financial reporting regulations and conduct of regulated companies.

The FERC interprets and implements regulatory statutes that grant an exclusive franchise to electric utilities in exchange for low-cost universal access by all consumers. Regulation forced the monopolies into a "cost plus" business model rather than a model that encouraged innovation and expansion of power options. This resulted in highly efficient, reliable, and environmentally sensitive power, at the expense of technological advancement.

The era of regulated, vertical monopolies as represented by layers in Figure 9.2 came to an end in 1992 with the enactment of the Energy Policy Act (EPACT). EPACT dramatically changed the emphasis once again. In addition to retaining clean, environmentally safe, and reliable power, Congress now required utilities to provide "nondiscriminatory" transmission access to the transmission and

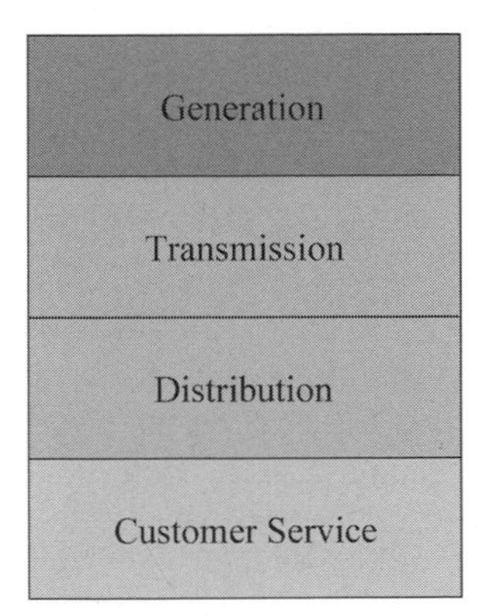

Figure 9.2. From 1935 to 1992, utilities held a regional monopoly over generation, transmission, and distribution of electrical power.

[3]http://www.ferc.gov/.

distribution layers shown in Figure 9.2. Any *qualifying facility* (QF) could use any part of the power grid to deliver its power to consumers. In fact, the old monopolies were required to buy from their competitors when the price was right. EPACT was intended to create competitive electricity markets that would lower consumer prices and increase supply. Unfortunately, EPACT plunged the grid into chaos. According to one industry expert, the modern deregulated power industry is like a gasoline industry that fixes the price of oil at $30 per barrel but allows the retail price of gasoline to go to $450 per gallon!

> A particularly extreme example of the new sensitivity of prices occurred during the latter part of June 1998. For several days, spot-market prices for electricity in the Midwest experienced almost unheard-of volatility, soaring from typical values of about $25 per megawatt-hour (2.5 cents per kilowatt-hour) up to $7,500 per megawatt-hour ($7.50 per kilowatt-hour). Because the affected utilities were selling the power to their customers at fixed rates of less than 10 cents per kilowatt-hour, they lost a lot of money very quickly.
>
> The run-up in prices was so staggering that it might take an everyday analogy to appreciate it. In the 1970s, drivers howled when the price of gasoline tripled. Imagine your consternation if, one day, you pulled into a gas station and discovered the price had increased three hundredfold, from $1.50 per gallon to $450 per gallon.
>
> Most of us would look for alternative transportation. But with electricity you do not have options. With no way to store it, the affected utilities had a choice of either paying the going rate, or pulling the plug on their customers on the hottest day of the year. The total additional charges incurred by the utilities as a result of the price spike were estimated to be $500 million.[4]

As we shall see, this peculiar mixture of physics and economics will lead to vulnerabilities in the grid that must be considered when establishing policies for protection of this very critical infrastructure. In particular, the grid has been made more vulnerable at the point in history when it should be made less vulnerable. Economics has been given precedence over security. Deregulation encourages competition, but it discourages investment in the grid. This is an example of the "Tragedy of the Commons"—a phenomenon well known in other industries[5]:

> The tragedy of the commons develops in this way. Picture a pasture open to all. As a rational being, each herdsman seeks to maximize his gain. Explicitly or implicitly, more or less consciously, he asks, 'What is the utility to me of adding one more animal to my herd?' This utility has one negative and one positive component. 1) The positive component is a function of the increment of one animal. Since the herdsman receives all the proceeds from the sale of the additional animal, the positive utility is nearly 1. 2) The negative component is a function of the additional

[4]T. Overbye, "Reengineering the Electric Grid," *American Scientist*, 88, 3, p. 220, May–June 2000.
[5]W. F. Lloyd, *Two Lectures on the Checks to Population*, Oxford Univ. Press, Oxford, England, 1833, reprinted (in part) in *Population, Evolution, and Birth Control*, G. Hardin, Ed., Freeman, San Francisco, 1964, p. 37.

> overgrazing created by one more animal. Since, however, the effects of overgrazing are shared by all the herdsmen, the negative utility for any particular decision-making herdsman is only a fraction of 1. Adding together the component partial utilities, the rational herdsman concludes that the only sensible course for him to pursue is to add another animal to his herd. And another; and another.... But this is the conclusion reached by each and every rational herdsman sharing a commons. Therein is the tragedy. Each man is locked into a system that compels him to increase his herd without limit—in a world that is limited.[6]

The benefit of the shared grazing pasture of the commons can be formalized in mathematical terms as the sum of each herdsman's gain (n animals), minus the herdsman's cost (C/n, where n is the number of grazing animals owned by the herdsman and C is the total cost of the commons, which is shared among the herdsmen). In other words, each herdsman is motivated to increase n because he or she will gain (n − C/n) in personal wealth as a consequence. Assuming C is constant, each herdsman can maximize profit by increasing n without bound, because (n − C/n) gets larger and larger as n increases. The best strategy of an individual herdsman is to increase n, but at some point, the capacity of the shared pasture is exceeded. When this happens, all herdsmen lose, because the pasture is depleted of grass and all animals die.

Any infrastructure sector could suffer the fate of the tragedy of the commons. The shared transmission lines in the electrical power grid, shared pipelines in the gas and oil sectors, and the shared long-distance lines of the telecommunications industry are in danger of falling into this category, if not protected by regulatory policies that encourage investment in the shared assets of each.

OUT OF ORDERS 888 AND 889 COMES CHAOS

The Energy Policy Act of 1992 unleashed forces whose effects have yet to be fully realized. At its core, EPACT opened up the formerly closed transmission and distribution grid to all comers (FERC Order 889). The power companies of the vertically integrated era are now required to buy power from qualified facilities and allow competitors to use their transmission and distribution lines. But they can only charge consumers a usage fee set by state regulators—not them. Retail prices are fixed, whereas wholesale prices are allowed to float. The new grid is a competitive marketplace—almost. Floating wholesale prices can be inflated to the advantage of the seller, but each state sets retail prices as low as possible for political reasons. This has created chaotic economic shockwaves in states like California where power brokers have been allowed to "game the system" through predatory pricing contracts. Enron is perhaps the most notorious example of this practice.

The modern deregulated grid is still regulated, but not for the purpose of monopolies. Rather, it is regulated for the purpose of encouraging innovation

[6]G. Hardin, "The Tragedy of the Commons," *SCIENCE*, Vol. 162, 13, December 1968, http://www.sciencemag.org.

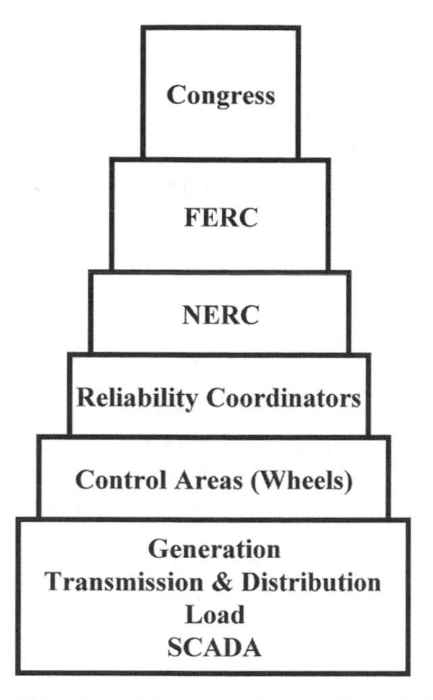

Figure 9.3. The power grid is shaped by many layers of regulation: congress, the FERC, NERC, reliability coordinators, control areas, and finally the laws of physics.

through competition. Still, it is a regulated industry with layers of regulators as shown in Figure 9.3. By Order 888, the FERC created independent system operators (ISOs) that essentially replaced the monopolistic utilities with nonprofit "broker" companies. ISOs are where buyers meet sellers. According to Overby[4]:

> In a bid to ensure open and fair access by all to the transmission system, in Order 888 FERC envisioned the establishment of several region wide entities known as ISOs, or Independent System Operators. The purpose of the ISO is to replace the local utility's operation of the grid by a private, not-for-profit organization with no financial interest in the economic performance of any market players. In short, the job of the ISO is to keep the lights on, staying independent of and therefore impartial to the market players. As of the end of 1999 ISOs were operating the electrical grid in California, New England, New York, Texas and the coordinated power market known as PJM (Pennsylvania–New Jersey–Maryland).

ISOs buy and sell electricity into the layered grid shown in Figure 9.3. Under EPACT, the responsibilities of an ISO are to:

- Control the transmission system.
- Maintain system reliability.

- Provide ancillary services such as system and voltage control, regulation, spinning reserve, supplemental operating reserve, and energy imbalance.
- Administer transmission tariff.
- Manage transmission constraints.
- Provide transmission system information (OASIS).
- Operate a power exchange (optional).

Sometimes the ISO separates the buying and selling activity from the regulation and reliability activities. In this case, they set up a separate power exchange. These are trading centers where utilities and other electricity suppliers submit price and quantity bids to buy and sell energy or services. Enron Online was one such exchange. It bought power on contract and resold its contracts to utility companies like PG&E in California. At one time, Enron Online had cornered enough of the California market that it could charge whatever it wanted. This led to the California energy crisis in the late 1990s.

The FERC requires an ISO to monitor its energy market for manipulation or abuses by the participants. This requirement covers both the power exchange (auction-based) market and the bilateral transactions in the region (wheeling). An ISO's authority to take corrective action when market abuses are identified depends on the nature of the abuse. In the case of abuses by Enron in 2002, the Department of Justice—not the ISO—pursued malfeasance charges against Enron executives.

Congress legislates and the FERC regulates through cooperation with the NERC. The NERC has divided the United States and Canada into geographical areas called *reliability coordinators*. Each reliability coordinator oversees the operation of a number of *control areas* sometimes called "wheels." Buying and selling across control areas is called "wheeling" in the terminology of grid operators. The 17 reliability coordinators and control areas of North America are shown in Figure 9.4. Reliability coordinators operate from one or more control areas, which in turn monitor and adjust the flow of electrons throughout their region of responsibility.

The economics of the deregulated grid often conflict with the laws of physics because:

- Electrons cannot be easily stored or inventoried; hence, spot markets can be volatile, thus encouraging gaming of the system.
- The grid cannot easily redirect power to where it is needed; this foils demand-and-supply economics with both short-term and long-term implications.
- It is difficult to quantify the exact amount of power available at any point in time, which introduces human errors in the process of stabilizing the grid.
- A certain portion of the gird is "down" at any point in time because of maintenance, which makes it difficult for operators to estimate transmission and distribution capacity.

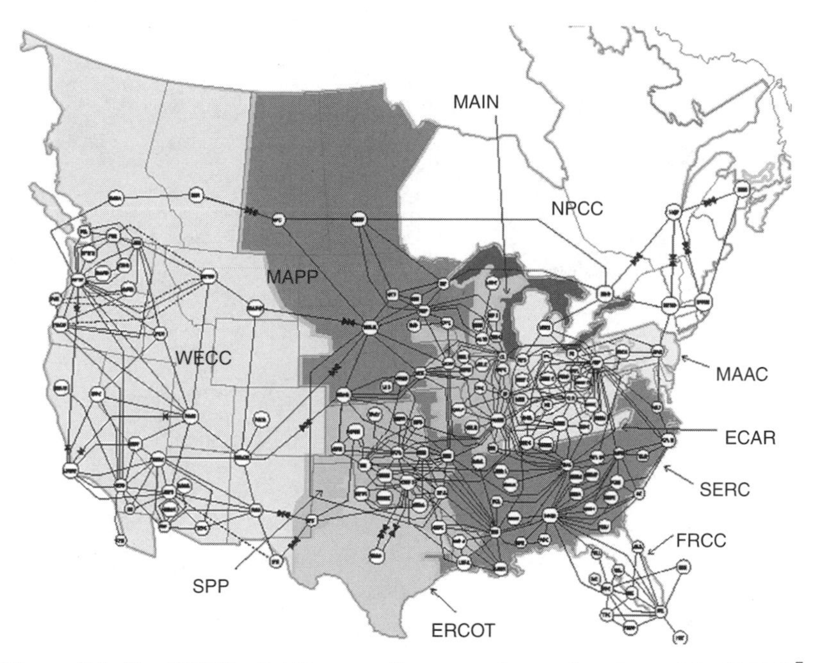

Figure 9.4. The NERC reliability coordinators and control regions of the NERC.[7]

ENERGY SECTOR ISAC[8]

The Electricity Sector ISAC (ES-ISAC) should not be confused with EISAC, the Energy ISAC that deals with oil and natural gas information sharing. The ES-ISAC is run by the NERC and serves the electricity sector. It provides sharing among its electric sector members, federal government, and other critical infrastructure industries. Specifically, the mission of the ES-ISAC is to collect and analyze security data and disseminate its analysis and warnings to its members, the FBI, and the Department of Homeland Security (DHS).

The ES-ISAC recommends and publishes the Department of Energy vulnerability assessment methodology called the Vulnerability Assessment Survey (VAS) program. The VAS is a high-level prescription for what needs to be assessed, along with recommendations on which grid components are the most important to evaluate.

THE GRID

The North American Electric Grid (grid) is one of the largest and most complex manmade objects ever created. It consists of four large 60-Hz AC synchronous

[7]U.S.–Canada Power System Outage Task Force, "Interim Report: Causes of the August 14th Blackout in the United States and Canada," November 2003.
[8]http://www.esisac.com.

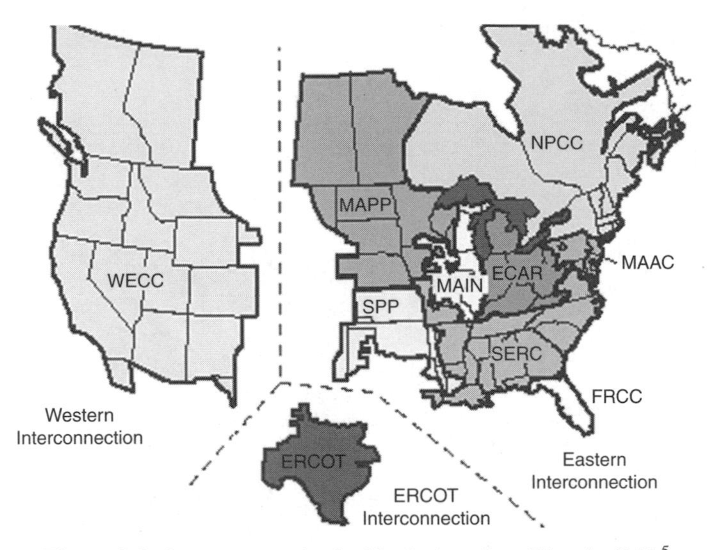

Figure 9.5. Interconnects in the North American Electric Grid.[5]

subsystems called the Eastern Interconnect, Western Interconnect (WSCC), Texas (ERCOT), and Quebec Interconnect. Figure 9.5 shows the four interconnects plus some sub-divisions of each.[9]

The Eastern Interconnect has about 670,000 MW of capacity and a maximum demand of about 580,000 MW. The Western Interconnect has about 166,000 MW of capacity and a maximum demand of about 135,000 MW. ERCOT has 69,000 MW of capacity and a maximum demand of 57,000 MW. Thus, there is approximately 15% more generation capacity than demand at peak levels. The North American Electric Grid has sufficient power, but it lacks the transmission and distribution capacity needed to meet surge demand. This is a consequence of the historical development of vertical monopolies and the regulatory policies of Congress. It is also the grid's major weakness.

Theoretically the grid can move power from one place to another to meet demand. For example, peak power consumption in the Eastern Interconnect occurs 3 hours before peak demand in the Western Interconnect simply because of time zones. In addition, weather conditions ameliorate the demand for power. During the winter, Los Angeles sends power to heat homes in the Northwest, and during the summer, Bonneville Power transmits power to southern California to run air conditioners.

But this is theory. In reality, the grid is not robust enough to transmit power to where it is needed most. Instead, the grid has to be constantly monitored to meet demand and guard against cascading events such as tripped lines or power plants

[9]To see an animation of real-time flow of electricity in the Eastern Power Grid, visit http://powerworld.com/Java/Eastern.

that are taken offline for maintenance. This challenge is mediated by SCADA/EMS at all levels throughout the grid.

The grid is made up of four major components: SCADA/EMS, generation, transmission and distribution, and consumer load. The last three are managed by various SCADA/EMS systems. Figure 9.6 illustrates this as a unified system of components called the grid:

1. *Generation*—source of electric energy: Coal provides fuel for over half of the U.S. electric power generators. There are more than 10,000 different generating units with a total capacity of about 800,000 MW in the United States. The largest generation plant is Grand Coulee Dam, WA (7000 MW from hydro), and the next largest are Polo Verde, AZ (3700 MW from nuclear), W.A. Parish, TX (3600 MW of coal), and Scherer, GA (3400 MW from coal). Generation is fueled 56% by coal, 21% by nuclear, 9.6% by natural gas, 9.5% by hydroelectric, and 3.4% by petroleum. Most hydroelectric generators are in the east and west, most nuclear generators are in the midwest and east, and thermal electric generation plants are spread throughout the United States.
2. *Transmission and Distribution*—the substations, transformers, and wires that carry the power from generation to load. There are more than 150,000 miles of high-voltage transmission lines in the United States. High-voltage lines operate at voltages up to 765 kV (kilovolt), with many 500-kV, 345-kV and 230-kV lines. Higher voltage lines typically consist of three wires attached to poles and towers by large conspicuous insulators. They are easy to identify from a passing automobile, bus, or train. Generally, they are in the open and unprotected. When a transmission line becomes too hot or shorts, it is said to have "tripped." Perhaps the most common fault in the grid stems from tripped lines. Often a line is overloaded in an attempt to shift power to where it is needed. The line heats up, sags, and touches a tree or the ground. Contact causes the circuit to short into the ground, and the line has to be shut down.

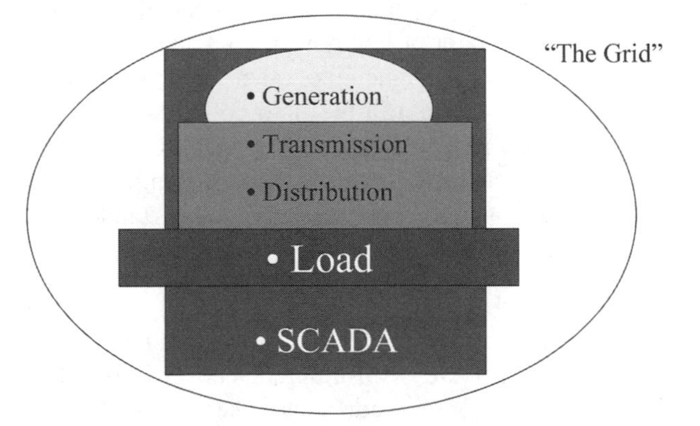

Figure 9.6. Grid components are generators, transmission and distribution, load, and SCADA/EMS control systems.

Thus, a series of cascade failures can begin with a tripped high-power line. The greatest vulnerabilities of the grid are in the middle of the grid—its transmission and distribution network. The state of the transmission and distribution network is maintained by regional ISOs and the *Open Access Same-Time Information System* (OASIS) database. OASIS is an Internet-based database used by transmission providers and customers. It provides capacity reservation information, transmission prices, and ancillary services.

3. *Load*—consumers are in complete control of demand; utilities must supply enough power to meet the load *at all times*. Total peak demand is about 710,000 MW, but the peaks occur at different times in different regions. In addition, demand can make dramatic swings—from 20,000 MW to 35,000 MW over a 1-week period, and as much as 8000 MW to 20,000 MW on an hourly basis. This means the SCADA/EMS system must be highly responsive, and the operators must be alert. Gas-fired peaker plants are commonly used to meet surges in demand, but it may not be possible to distribute the additional capacity to where it is needed, because of inadequate transmission and distribution capacity. Hence, there is no shortage of power, but there is a shortage of transmission and distribution capacity. This, and the wild swings in demand, is the major reason for blackouts.
4. *SCADA and Other Control Systems*—the control of all components of the grid. This component includes EMS and power plant automation hardware and software. The main measure of how well the grid is doing is called the *area control error* (ACE). It is the difference between the actual flow of electricity into an area and the scheduled flow. Ideally, the ACE should always be zero, but due to changing load conditions, ACE varies. Most wheels use *automatic generation control* (AGC) to adjust ACE. The goal of AGC is to keep ACE close to zero. Loss of a generator, transmission line, tower, or transformer can cause abrupt changes in ACE. It can take many minutes for AGC to rectify the loss and bring ACE back to zero. This is done by a complicated series of steps involving simulation of the intended change (say to increase the power from a generator or buy power from an adjacent qualified facility). Power control systems work much like other sector's SCADA systems. Many remote terminal units (RTUs) in the field collect data and control switches. The RTU data goes into a database, where EMS software calculates the next setting of the switches. And like other control systems, the control network sometimes hangs from the same towers and poles as the power lines themselves.

In conclusion, the grid is most vulnerable in the middle—the transmission and distribution layer—because there is insufficient capacity to deliver all available power generated by the major interconnects. The system operators must chase ACE to meet unpredictable demand. This means either producing more power near to the load or buying power from other parts of the grid. Under deregulation, they are encouraged to buy and sell from each other to drive ACE to zero. They are also required to allow competitors to use the old vertical monopoly's transmission and distribution layer. Buying and selling, or "wheeling," opens the grid to economic and technical vulnerabilities—a kind of double jeopardy for this sector.

PROGRAM *PowerGridSim*

The foregoing ideas have been turned into a simulation called *PowerGridSim*. We can use it to study the deregulated market economics of the grid and to gain insight into the complexity of ACE. *PowerGridSim* greatly simplifies the real power grid by representing major components as follows: Source nodes represent generators; destination nodes represent loads (consumers), and links represent the transmission lines that connect generators to loads.

Figure 9.7 shows the output display of *PowerGridSim*. The network of nodes and links is displayed in the upper portion of the display, and a bar chart of load and generator power levels is shown on the right side. A time-varying line graph of power level versus time is shown in the window on the left. Clicking on the GRAF-ON/GRAF-OF button opens this window.

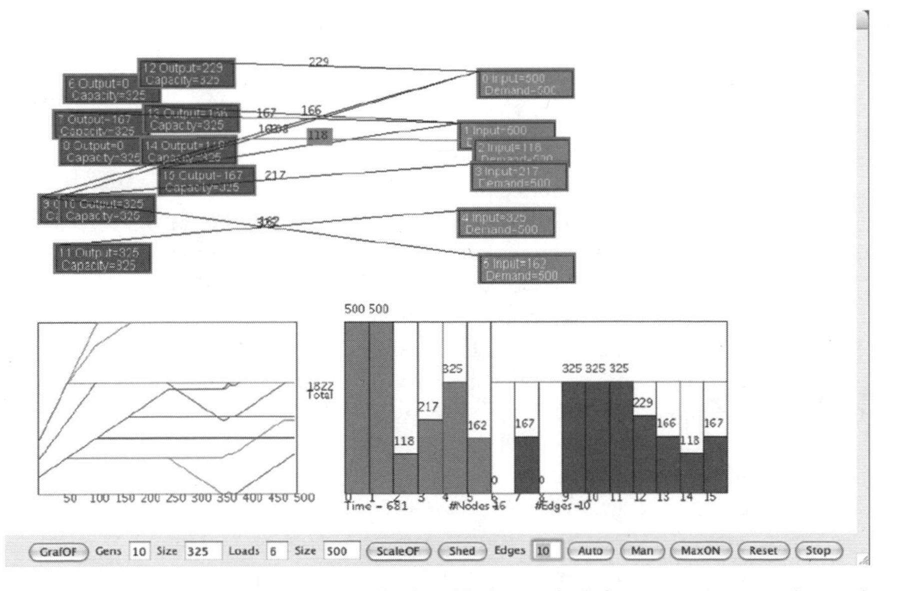

Figure 9.7. The *PowerGridSim* output display: Nodes on the left are generators, nodes on the right are loads, and links in between are transmission lines. The amount of electric power—measured at each generator and load—is shown as a bar graph on the right, and as a time-varying line graph on the left.

All control buttons and input values are placed along the bottom of the display. To start the simulator, click GO, and to automatically populate the simulation with nodes and links, click AUTO. The MAN button allows you to enter new nodes or change the values on any node that is selected by clicking on it, and then clicking on AUTO. In this way, you can customize the simulator.

Selecting a node (generator or load) followed by clicking on SHED causes the node to be taken offline. It ceases to produce electrical power if it is a generator node and ceases to consume power if it is a load. SHEDding a generator or load causes the network to adapt to a new situation, just like a real power grid. You will notice how power output and consumption begins to oscillate as the grid adapts. Power oscillation after a disturbance is one of the primary difficulties in controlling a grid.

PowerGridSim uses a simple organizing principle to simulate an ISO marketplace. When the user clicks the SCALE-ON button, each load simply tries to find more power by dropping one source and picking up another source. Each load acts on its own, but what emerges is an optimal allocation, often reaching the goal of maximizing the total power delivered to all loads. Here is how the simulator approximates the market.

Organizing Principle:

1. Drop weakest supplier (old link).
2. Randomly select replacement supplier (new link).
3. Measure total power supplied over all loads.
4. Repeat 1 through 3, forever.

The question is, "does this simple organizing principle—exercised in isolation by each load – produce the maximum output over all loads?" To find out, launch the simulator and let it run for several minutes.[10]

By clicking SCALE-ON, *PowerGridSim* tries to find the best power generation to load allocation possible through the process of emergence. But the random selection of a new supplier of power continues forever, according to the organizing principle above. This may cause the simulation to oscillate around suboptimal allocations. Therefore, clicking the MAX-ON button causes *PowerGridSim* to maintain the maximum level of power on each generator as it reaches its maximum output level. Does this guarantee the best allocation over all time?[11]

Even this simple model illustrates the real-world grid. Here is what can be learned from *PowerGridSim*:

- Maximum generator output is not used, leading to suboptimal power utilization.
- Loads are not provided with 100% of their request, even though the generators generate sufficient power.

[10]Not always: It depends on the number and connectivity of the transmission links!

[11]Sometimes SCALE-ON must take from one generator to give to another, which may not be possible when MAX-ON is set.

- Buying and selling introduces fluctuations in power.
- Constraints on the number and placement of links leads to suboptimal utilization of available power, simply because power cannot flow from generator to load.

PowerGridSim shows that the ISO market mechanism underlying the power grid is suboptimal, because it does not guarantee 100% utilization of available power. An inadequate number and placement of links that connect generators to loads cause this. In other words, the grid's weakest links are in the "middle."

PRELUDE TO VULNERABILITY ANALYSIS

The grid distributes power through a network of transmission lines, towers, substations, and transformers delivering power to where it is needed—within limits. For example, New York City wakes up an hour before Chicago, so the demand on the grid peaks an hour earlier in New York City than Chicago. The operators of the grid reduce power going to New York and increase power going to Chicago as the day progresses. In this way, the power grid adapts to the load. The grid is a network, similar to the Internet in some ways, and very different in other ways.

Power grid networks seem to be random. That is, they seem to have no obvious structure. But regulation, economics of generation, and technical limitations have all contributed to shaping the grid. It turns out the grid is an example of a *small-world* network.[12] Recall that a small world is a network that has clusters of highly connected nodes with "long-distance" links that connect all nodes through a relatively small number of connections.

A small-world network can be created from a random network by replacing randomly selected links with shortest-distance links. This was illustrated in Chapter 4 using the program *ShortestLink*. In the case of the power grid, the substitution of random links with shortest links creates regional control areas—neighborhoods that emerge as "critical" to the overall operation of the grid. This is the first observation that leads to the best strategy for resource allocation in the MBVA method to follow.

What are the implications of a power grid that is essentially a small-world network? To understand why such a network is vulnerable to market forces as well as manmade attacks, consider the following experiment. Let each power generator have a maximum generation capacity and each consumer (load) a constant demand for power. In between the generators and the loads, we will assume an arbitrary network of substations that switch power from generators to consumers (loads). Thus, we have a network consisting of generator nodes, consumer nodes, and substation nodes. It has links representing transmission lines that connect generators to substations and substations to consumers. Admittedly this is an oversimplification of

[12]S. H. Strogatz, "Exploring Complex Networks," *NATURE*, vol. 410, March 2001, http://www.nature.com.

a large-scale power grid such as the Western Interconnect, but the simplification will serve to illustrate several important concepts.

Power lines have varying capacities. The higher the voltage, the more efficient it is to transmit power. So generators deliver power to large-capacity, long-haul transmission lines (733,000 V, for example), which in turn deliver power to substations. The substations step the voltage down, say to 230,000 V, and then transmit to other substations, which do the same. Finally, when the electricity reaches your back yard, it is reduced to 240/120 V. This is the idea behind the grid—use high-voltage lines to move power over long distances, and small voltage lines to move power around the consumer's home, factory, and so on.

Now assume that the network of generators, substations, and consumer loads is randomly connected by a fixed number of links. The links can be changed to accommodate consumer demand, but initially they are arbitrarily connected to the nodes in our imaginary random network.

The power grid is run like a marketplace (which it is). Whenever a consumer needs more power, the link delivering the least amount of power is dropped in favor of any other link that can supply power. That is, we randomly select a new supplier by dropping a weak link and replacing it with a randomly chosen link. We do this over and over. If we always replace a weaker link with a link that delivers more power, we eventually reach a point of maximum output. This experiment is identical to the organizing principle in *PowerGridSim*. But what if we modify the organizing principle of *PowerGridSim* so that the resulting network automatically identifies the best place to spend limited funds to harden the grid against attack? What organizing principle causes a random network to emerge as a hardened network? This is the task performed by the simulation program *RNet*.

PROGRAM *RNet*: AVAILABILITY BELONGS IN THE MIDDLE

Suppose each power generator and substation is given the same amount of money to improve component *availability*, i.e:, to increase the reliability of generator or substation.[13] Availability (reliability) of 100% means the node is completely protected, whereas an availability of 0% means the node is completely offline. If availability is 50%, the node is available only one half of the time. What should be our policy for allocating availability—and therefore funding—to the components of the grid?

We can use people, money, or "points" as a measure of availability hardening. Allocation of a certain number of points is equivalent to allocation of a certain amount of money to the infrastructure. Availability points are expressed as a percentage, ranging from 0% to 100% at each component of the grid. The program *RNet*, described here, uses the same concepts described in Chapter 6 and the program *NetworkAnalysis*.

[13]Chapter 6 defines availability as $(1 - p)$, where p is the fault probability associated with a vulnerability.

The *available power* delivered to the consumer by the grid is computed by multiplying availability percentage times the power generated or passed through the transmission and distribution layer in the "middle" of the grid. This is called the *service* of each node or link, and the sum of services across all loads is called the *service level* of the grid. For each node and link in the grid, service is the product of availability times capacity of node or link. Therefore, the service level for a certain grid is the sum of the services delivered to each load. If a generator produces 100 MW of power at maximum capacity, and availability is 50%, then only one half of the maximum power generated can be delivered over a period of time. Its service is 50 MW. Further suppose that this generator is connected to a single load by a link with 50 MW of transmission capacity, but only 10% availability. The service level of this grid is 10% times 50 MW or 5 MW. In general, the grid service level is the sum of services reaching all loads. As availability goes up, the grid service level also rises. Hence, the objective of any grid is to maximize service level.

Given a certain number of total availability points, R, what is the best way to spread them around a grid? In other words, how should R be allocated to nodes and links such that service level is maximized? Figure 9.8 shows the display panel of the program *RNet*, which does the labor of computing how best to allocate availability points to a grid consisting of power generators, substations, loads, and transmission lines.

The *RNet* simulator is a version of the program *NetworkAnalysis* that has been modified to simulate generators, substations, and loads.[14] They work much the same way. First, it generates a random power grid from inputs such as the number of generators (RED nodes), loads (BLUE nodes), and substations in between (BLACK nodes). Also, the user may specify the maximum number of transmission lines connecting generators to substations and substations to loads. Finally, the total number of availability points is entered, and when GO is clicked, a random grid is generated. Initially, the availability points are evenly spread across the generators and substations. Hence, the generators and substations are equally available.

RNet's job is to determine how best to allocate an insufficient number of availability points to the grid such that the service level delivered to consumers (loads) is maximized. That is, we want to distribute as much flow of electricity as possible, under conditions of arbitrary attacks on the components of the grid. But we can afford only a limited amount of availability, because funding is limited.

Here is an organizing principle that finds the best allocation of availability points. Remove a small amount of availability from a randomly selected component (generator, substation), and give it to any other randomly selected component. If this increases the service level of the entire network, keep going. If not, backup and restore the availability points to their previous levels. Continue this process

[14]The program *RNet* is available from the disk or the website.

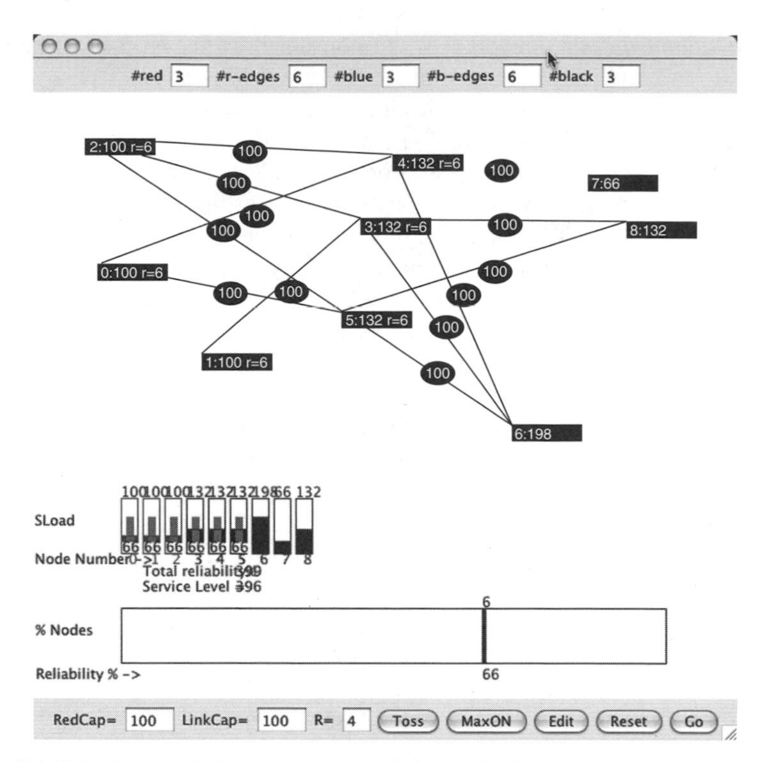

Figure 9.8. Display panel of program *RNet*. Nodes on the left are generators; the middle are substations, and the right are loads. A bar graph shows how much power is flowing through each node and how many availability points are allocated to each node.

repeatedly until the total service level is no longer increased, regardless of reallocation of availability to components.

Clicking on the MAX-ON button of *RNet* starts this process. As the organizing principle repeats, the bar graph on the left side of the display shows two results: the power at each node, and the number of availability points allocated to each node. At first, these numbers will change dramatically. After some time, the number of availability points allocated to each node will begin to gravitate toward the middle—the substations. In fact, points will be taken away from power generators and given to substations, thus maximizing the amount of power delivered to consumers.

Figure 9.9 illustrates the result of this experiment on a simple grid consisting of two generators, three substations, and one load. Initially, there is 300% availability to share among five components, so each component is allocated 60%. This produces a service level of 18 MV, even though the two generators produce a total of 30 MV. The service level is a function of the connectivity (links) as well as the availability of nodes. If there is not enough connectivity to supply maximum power to the load, then it does not matter how much power the power generators produce.

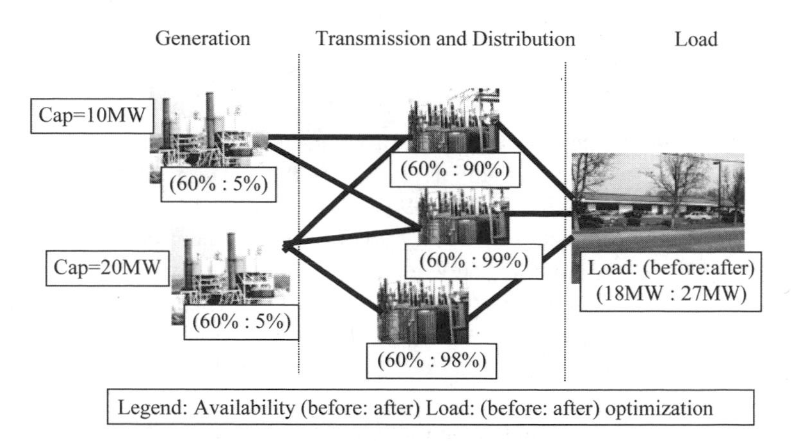

Figure 9.9. Before and after allocation of availability in a simple grid.

After performing the organizing principle many times, an optimal allocation emerges from *RNet*. The two generators are allocated 5% each, and the three substations are allocated 90%, 99%, and 98%, respectively. In other words, most availability is allocated to the middle layer. Substations must be more available than power generators!

This is an important and counter-intuitive result, because it says it is more important to harden the middle of the grid than the power generators. The reason: There is sufficient power generation, even if one should fail, but there is insufficient transmission and distribution capacity in the grid. Substations are more critical than generators. Links are more important than generator nodes.

In fact, the importance of substations and transmission lines was underscored in 1996 when the Western Interconnect was disrupted by a failure in a single link connecting Oregon and California. This small fault in a minor component of the Western Power Grid spread throughout the 11 Western states, pulling down the entire grid. Barabasi describes this failure in network terms:

> On a day of record temperatures, at 15:42:37 on August 10, 1996, the Allison-Keeler line in Oregon expanded and sagged close to a tree. There was a huge flash and the 1,300-megawatt line went dead. Because electricity cannot be stored, this enormous amount of power had to be suddenly shifted to neighboring lines. The shift took place automatically, funneling the current over to lower-voltage lines of 115 and 232 kilovolts, east of the Cascade Mountains. These power lines were not designed, however, to carry this excess power for an extended time. Loaded up to 115% of their thermal ratings, they too failed. A relay broke down in the 115-kilovolt line, and the excess current overheated the overloaded Ross-Lexington line, causing it too to drop into a tree. From this moment things could only keep deteriorating. Thirteen generators at the McNary Dam malfunctioned, causing power and voltage oscillations, effectively separating the North-South Pacific Intertie near the California-Oregon border. This shattered the Western Interconnected Network into

isolated pieces, creating a blackout in eleven U.S. states and two Canadian provinces.[15]

The 1996 power outage in the Western 11 states and Canada was prophetic. A similar failure on a relatively minor portion of the Eastern Grid led to massive outages in 2003. Knowledge of the 1996 outage was not sufficient to prevent the 2003 blackout. This is because we still do not have adequate tools to perform vulnerability analysis on complex networks such as the Eastern Interchange. Even the insight gained by *RNet* is insufficient to completely understand the power grid of the United States, because *RNet* does not take into consideration the dynamic (time-varying) behavior of electrical power generation, distribution, and consumption.

The foregoing has obscured some important policy-level insights that should be summarized here:

1. The grid is a small-world network, not a scale-free network. Its network structure is characterized by clusters of densely linked nodes with a small number of "long-distance" links connecting to other, more distant clusters.[16] Figure 9.4 illustrates this concept; the Eastern Interconnect has many clusters that are easily identified by simply looking at a map.
2. Failures can propagate throughout a small-world network; thus, the grid is prone to cascade failures as exemplified by the massive blackouts of 1996 and 2003. In the following analysis, we will trace the steps of the cascade failure that happened in August 2003.
3. Optimal allocation of funding to increase availability favors nodes in the middle—the transmission and distribution nodes of the grid. The best critical infrastructure protection strategy is to invest in these nodes, because they connect generators to loads. Without them, power never reaches the consumer.
4. Because the grid is a small-world network, our vulnerability analysis must be modified to identify *critical clusters* instead of single nodes; steps #1 and #2 of MBVA must identify highly linked clusters by analyzing clusters of linked nodes in the grid. As it turns out, these are self-evident, because the clusters formed by control areas or wheels are easily identified from maps. Maps such as the one shown in Figure 9.4 easily expose critical clusters when network analysis is applied.
5. Historical data exist for making reasonable assumptions regarding fault probability, cost, and damage estimates—the three quantities needed by MBVA.

[15]A.-L. Barabasi, "Linked: How Everything Is Connected To Everything Else and What It Means for Business, Science, and Everyday Life," *PLUME*, p. 119, 2003.
[16]D. J. Watts and S. H. Strogatz, "Collective dynamics of 'small-world' networks," *NATURE*, 393, June 1998.

THREAT ANALYSIS

One way to identify threats to the grid is to create "red team" scenarios by pretending to be the attacker. *Red teams* are attackers, and *blue teams* are defenders. Maj. Warren Aronson, U.S. Army, and Maj. Tom Arnold, U.S. Marine Corps, prepared the following four attack scenarios while playing the role of red team.[17] They focused on fuel supply, transformers, SCADA, transmission substations including towers, and power generators. These were chosen because they are the most likely asymmetric attack targets; they cost little to attack, and yet they can create enormous damage and destabilize the grid. If the red team can create an unstable grid, argued the red team, NERC rules will direct operators to propagate the instability across the entire control area, and perhaps create a blackout across the entire interconnection.

Attack Scenario 1: Disruption of Fuel Supply to Electric Generation Plants

The process of supplying electricity begins with the transportation of power plant fuel by water, rail, road, or pipe to power generation plants. The largest source of North American electricity comes from coal-fired, thermal-generating plants. These plants have historically maintained a reserve of 60 to 90 days supply of coal near each generator complex. However, the variance in seasonal demand for coal and the increasing dependence on "just-in-time" shipments to reduce storage costs and environmental impact has reduced local inventories significantly. Gas and fuel oil-fired plants generally have little, if any, fuel storage capacity onsite.

A red team might disable, or at least significantly degrade, a major portion of regional power generation by attacking key components in the transportation system fueling an electric generation facility. A specific example of this attack might involve the Powder River Basin in Wyoming, which provides 56% of the coal supplied to the United States. Only three railroad lines service the region, carrying 305 million tons of coal annually to generation plants in more than a dozen states. Moving the same volume of coal by truck—currently the only alternative to rail—is both prohibitively expensive and restricted by available trucks and drivers, which currently support other consumers. The destruction of an important bridge, like the triple bridge over Antelope Creek, would stop coal transport on one of two primary lines feeding rail hubs for distribution to multiple states. Destruction of one to three similar targets immediately before peak periods of seasonal electricity demand could disable much of the country's generation capacity for periods of weeks to months.

[17]CS 3660 projects, Summer 2002.

Attack Scenario 2: Destruction of Step-Up/Step-Down Transformers

Transformers are the key links between generation and transmission substations as well as between transmission and distribution subsystems. Most transformers are mechanically simple, consisting of wound copper coils encased in tanks of oil. The oil cools the coils to prevent the high-voltage current from melting the copper, breaking the wire and opening the transformer circuit. Step-up transformers servicing generation plants are very large and heavy, with some weighing hundreds of tons. The size of the devices makes movement from the manufacturer to installation locations slow. In addition to their size, most are custom made and cost in excess of one million dollars; as a consequence, neither electricity producers nor transformer manufacturers maintain a large inventory. Replacement times can vary from months to over a year. Step-down transformers can be equally difficult to replace and represent choke points for electric flow.

A devastating attack against step-up and step-down transformers is relatively simple. A single person can accomplish outright destruction quickly and inexpensively by planting explosives or driving a vehicle or material handling equipment into the side of a transformer. An even easier attack may be possible without entering the facility; puncturing the side of a transformer with a weapon like a rifle would cause coolant oil to leak, resulting in overheating before the attack is detected. Although heat sensors might shut down the transformer before fatal overheating, the loss of oil would temporarily stop the flow of electricity while the substation was shut down and the transformer was isolated and repaired.

Attack Scenario 3: Disruption of Communications in SCADA

SCADA control complexes provide constant monitoring and adjustment to all subsystems of the electrical power system. Electric utility companies recognize the importance of these sites and have taken measures to protect them from physical attack. They are normally well protected and located in hardened structures, often behind layered security or below ground. Attackers may be insiders or launch a direct assault on the facilities. However, recruiting existing employees sympathetic to the attacker's cause or placing a team member in a trusted position in such a facility requires total faith in that individual and may take considerable time. Direct assault against a facility requires information on facility configuration, extended surveillance to discover security procedures, well-trained assault forces, overt action, and relative strength favoring the attacker, which is not typical of an asymmetric attack.

The weakest points in a control system are usually the communications networks that link complexes. Although these links normally have some form of redundancy, they are still susceptible to attack. Some components of the communication system will likely be exposed to observation and thus vulnerable to physical attack; some examples include telephone wires strung on poles and externally mounted antennas essential to radio system operations. Physically protected components may be vulnerable to stealthy attack using directed energy weapons or other forms of electronic warfare. Other attacks of the cyber-variety could target the published protocols used by SCADA systems. Regardless of the method chosen, the goal of these attacks is to both seize control of a system and cause operations to occur outside of safe operating parameters, destabilizing the ACE or disrupting recovery efforts after other events.

Attack Scenario 4: Creation of a Cascading Transmission Failure

In accordance with NERC rules, generators and switching circuits are designed to automatically go offline when they operate outside of safe operating ranges. Control circuits usually make less dramatic adjustments before the system exceeds these limits, but "fail safe" devices will shut down even in the absence of external commands. When multiple failures occur nearly simultaneously, it is possible that the cumulative effect is a spike, which impacts the entire system.

By design, if a major transmission line or substation fails, current surges and voltage transients will trip circuit breakers. In reaction to the circuit breakers being tripped, the protective circuits at the affected generation plants will shut down the turbines to prevent them from over-speeding due to the loss of load. As the load increases on the remaining generators, the rotation of the generator turbines decreases. As the turbine generators slow down, the power company must start to shut off some customers or bring more power online. If more power is not added or the load decreased quickly enough, other generators will start to shutdown automatically. This will continue the cascading effect by putting even a greater load on the remaining generators, which will likewise start to shut down. Preventing a cascading transmission failure is dependent on quick and effective action and therefore a responsive SCADA/EMS system.

FAULT TREE AND EVENT MATRIX

The foregoing red team analysis can be captured in a fault tree as shown in Figure 9.10. Any one of the six faults leads to failure of the entire grid; thus, the fault tree consists of OR logic. There are six threats, which lead to $2^6 = 64$ outcomes in a fully populated event tree. The full event tree would contain single, double, triple, and so on, up to 6-tuple combination outcomes. The corresponding fault probabilities would diminish until the likelihoods of triple, quadruple, or 6-tuple events are extremely unlikely. Therefore, the following event matrix analysis is designed to reduce the 64 combinations to 21 single and double faults by ignoring the rare triple combinations—from 3 up to the single 6-tuple event.[18]

Figure 9.11 lists all 21 single and double faults predicted by the formula for single and double faults. The fault probability of the entire fault tree will be higher than the

[18]This analysis follows from the event matrix discussion in Chapter 5.

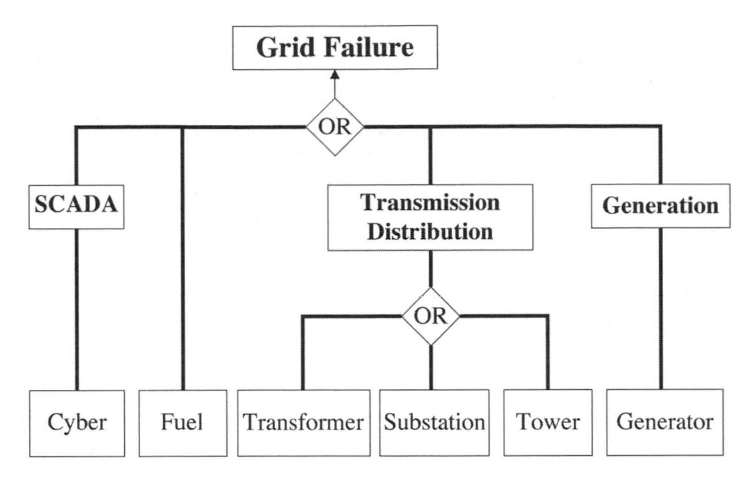

Figure 9.10. The fault tree model of the red team scenarios contains six threats.

event matrix, because combinations have been ignored in the matrix.[19] A two-dimensional matrix can hold single and double faults, and a three-dimensional matrix can hold single, double, and triple faults. A six-dimensional matrix would be needed to hold all faults generated by the fault tree of Figure 9.10.

	Cyber	Fuel	TRansformer	Substation	TOwer	Generator
Cyber	C	C, F	C, TR	C, S	C, TO	C, G
Fuel		F	F, TR	F, S	F, TO	F, G
TRansformer			TR	TR, S	TR, TO	TR, G
Substation				S	S, TO	S, G
TOwer					TO	TO, G
Generator						TO

Figure 9.11. Event matrix of red team threats for single and double faults. C = Cyber, F = Fuel, TR = Transformer, S = Substation, TO = Tower, G = Generator.

[19]The "missing" outcomes account for another 10% vulnerability.

Risk AFTER investment:

	Cyber	Fuel	TRansformer	Substation	TOwer	Generator
	C = 100 D = 1,000	C = 1,000 D = 2,000	C = 10 D = 10	C = 100 D = 100	C = 100 D = 1	C = 10 D = 500
Cyber		\$150 (1.4%)				
Fuel		\$267 (4.7%)	\$84 (1.5%)	\$82 (1.5%)	\$71 (1.5%)	\$123 (1.4%)
TRansformer				None	None	
Substation					None	
TOwer					None	
Generator						

Figure 9.12. A summary of apportioned risk reduction allocation of \$1000 million to the fault matrix of Figure 9.11. C = Cost, D = Damages, and V = Vulnerability = LOW = 20% for all threats. Each cell in the matrix contains the amount of investment and fault probability after the investment.

RESOURCE ALLOCATION

Suppose all vulnerabilities are LOW (20%), and costs and damages for each threat are shown at the top of Figure 9.12. For example, damage due to a cyber-incident is estimated at \$1000 million, but it cost \$100 million to protect against; the damage due to a generator fault is \$500 million, but it cost \$10 million to prevent. Assuming vulnerabilities are the same, $V = 20\%$, then single faults occur with fault probability $V(1 - V)^5$ and double faults occur with probability $V^2(1 - V)^4$.[20] Substitution of $V = 0.2$ into these results in single fault vulnerability of 6.55% and double fault vulnerability of 1.63%.

Figure 9.12 summarizes the result of treating each single and double outcome as a fault, and then allocating funds to each according to the apportioned risk reduction allocation method. For example, the row in Figure 9.12 representing the FUEL threat contains allocations for single event FUEL and double events FUEL AND TRANSFORMER, FUEL AND SUBSTATION, FUEL AND TOWER, and FUEL AND GENERATOR.

The amount of each investment, and the corresponding reduced fault probability, are given by the entries in the matrix of Figure 9.12. It is clear that the extremely

[20]$\text{Prob}(N,K) = V^K(1 - V)^{(N-K)}$, where N = number of threats and K = order of combination fault: 1, 2, 3 . . . N.

high damages caused by intervention in the fuel supply to generators leads to the highest investment: $267 million to harden FUEL supplies. This reduces the probability of this fault from 6.55% to 4.7%.

The second highest investment goes to a double fault: Cyber and Fuel. In this case, our allocation strategy recommends an investment of $150 million, which reduces vulnerability from 1.63% to 1.4%. Similar reductions, as shown in Figure 9.12, lower the overall risk from $540 million to $484 million, or 21%. Over 80% of the $1 billion budget is allocated to one third of the threats. In other words, although apportioned allocation tries to spread the investment across most threats, in this case, most of the funding goes to protect the power grid against fuel shortages and cyber-attacks.

In summary, the red team exercise and event matrix analysis suggests the following strategy to protect this power grid. Assuming a budget of $1000 million:

1. Fuel supply vulnerability is the top single fault and should get $267 million.
2. (Cyber and Fuel) is second and should get $150 million.
3. Eighty percent of the budget should be invested to reduce vulnerability in 33% of the threats.

How good is the event matrix approximation? Table 9.1 shown below summarizes the differences, using results obtained from program *FTplus* and the event tree analysis. In general terms all methods rank the threats in approximately the same order from highest to lowest priority, as follows:

1. Fuel
2. Cyber
3. Generator
4. Substation and transformer

But the simplification provided by the event matrix shows significant approximation errors. For example, the event matrix allocation yields a sector vulnerability

TABLE 9.1. A Comparison of Results of the Event Matrix vs. Fault Tree Analysis for the Power Grid Model Described in Figures 9.10–9.12.

	Event Matrix	Minimum Risk	Ranked Risk	Apportioned Risk
Vulnerability	64%	50%	55%	33%
Risk	$424	$66	$103	$127
$Cyber		$100	$100	$82
$Fuel		$890	$894	$820
$Transformer				$8
$Substation				$82
$Generator		$10	$6	$8

of 64%, whereas the risk minimization vulnerability is 50%. The event matrix grossly over-estimates the reduced risk—stating it as over $400 million—versus a minimum risk value of $66 million when running *FTplus*.

It seems that the event matrix shortcut should be used only when the fault probabilities are below 20% and that the analyst is mainly concerned with estimating the rank order of vulnerabilities, and not the precise allocation needed to reduce risk.

Regardless of the method used to obtain the investment strategy, this analysis makes important suggestions about how to generally harden the electrical power grid:

- Backup fuel supplies, and secure alternative fuel sources.
- Review SCADA and other control system protocols and fail-safe procedures to reduce likelihood of cascading blackouts.
- Install redundant communications for SCADA and other control system networks.
- Install physical security at key generators.
- Harden or shield transformer substations.
- Review SCADA and other control system cyber-security policies and procedures.
- Standardize and keep spare transformers in stock.

VULNERABILITY ANALYSIS OF THE EASTERN INTERCONNECT

Would the foregoing vulnerability analysis and implementation of the corresponding prevention measures have avoided the 2003 blackout that put 50 million people in the dark? Perhaps it would not, because fuel shortages and transformer faults were not the cause of the 2003 blackout. But a careful MBVA study of the Eastern Grid would have identified the region of the country where the 2003 blackout originated.

Suppose we apply the MBVA method to the case of the Eastern Interconnect, which is located in the eastern part of the map shown in Figure 9.4. Applying the network analysis suggested by step one and two of MBVA to the Eastern Interconnect subset in Figure 9.4 yields the histogram shown in Figure 9.13.[21] Clearly, this distribution does not obey a power law, and so it is not scale free. But the rare nodes (with 16 or more links, say) made evident by Figure 9.13 are the centers of clusters or neighborhoods predicted by a small-world model. They are the locus of nodes that are the most critical components of the grid. Critical nodes can be identified in Figure 9.13.

In this case, critical node clusters are actually *control areas*, as follows:

[21]Actually, the results in Figure 9.13 were obtained from a magnified version of Figure 9.4.

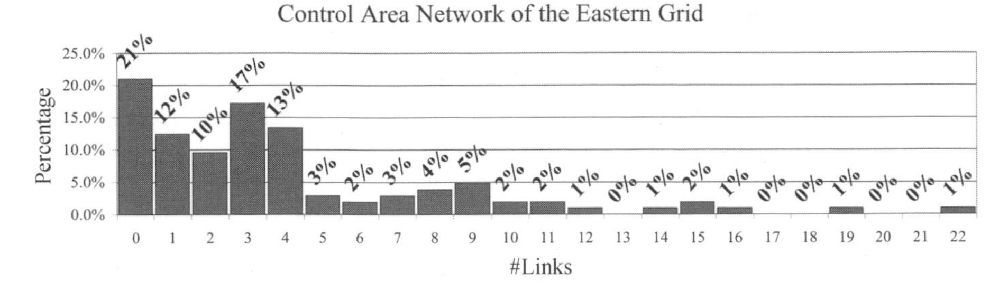

Figure 9.13. Histogram of control area links of the Eastern Interconnect.

Control Area	Number of Links	Reliability Coordinator
AEP (Ohio)	23	MISO
AMRN (S. Illinois)	20	MISO
SOCO (Alabama)	17	TVA
CSWS (Oklahoma)	16	SPP
TVA (Tennessee)	16	TVA

The top nodes, and hence the most critical nodes are AEP, AMRN, and SOCO, the most highly connected control areas. These nodes are critical because of their connectivity but also because according to NERC regulations—Operating Policy 5: Emergency Operations—each control area "shall establish a program of manual and automatic load shedding" if synchronization problems originate that "could result in an uncontrolled failure of components of the interconnection." The NERC says each control area "shall communicate its current and future status to neighboring systems, control areas, or pools throughout the interconnection" whenever voltage levels threaten the reliability of the interconnection.

These two rules can lead to *cascade* failure, because they facilitate the propagation of faults from one operator to another. This is not entirely the fault of the NERC, but rather the result of physical limitations on the transmission and distribution network in the middle—electrical power can neither be stored nor easily diverted to any load in the grid. Once the cascading begins, it is difficult, but not impossible, to stop. The solution is to have effective SCADA/EMS software that quickly identifies departures from ACE and deftly reallocates electrons in a stable manner. Thus, SCADA/EMS is a critical component in the operation of the grid.

CASE STUDY: THE EASTERN GRID BLACKOUT OF 2003[22,23]

The infamous Eastern Interconnection blackout of August 14, 2003 cut off power to millions of Americans in eight states and millions of Canadians in one province. Lasting 2 days, the blackout shed 12% of NERC capacity. The economic costs,

[22]"Interim Report: Causes of the August 14th Blackout in the United States and Canada." November 2003, U.S.–Canada Power System Outage Task Force.

based on loss of electricity sales to consumers, range from $7 billion to $10 billion. The insurance industry lost $3 billion. Compare this with the damages and cleanup costs of TMI-2 (Three Mile Island Nuclear Power Plant #2) that melted down in 1979. TMI-2 incurred damages of $973 million—one tenth the damage done by the Northeastern Grid failure of 2003.

According to the final report issued after the 2003 blackout, "The initiating events of the blackout involved two control areas—FirstEnergy (FE) and American Electric Power (AEP)—and their respective reliability coordinators, MISO and PJM."[23] AEP (American Electric Power, Inc.) is a control area within the MISO (Midwest Independent System Operator) reliability coordinator's area of responsibility. MISO is a network of linked brokers, control areas, and SCADA systems that includes AEP and FE (FirstEnergy, Inc.) as shown in Figure 9.14.

AEP of Columbus, OH, owns and operates more than 42,000 MW of generating capacity in the United States and in some international markets.[24] It is the largest electricity generator in the United States. AEP is also one of the largest electric utilities in the country, with almost five million customers linked to its 11-state electricity transmission and distribution grid.

FirstEnergy Corp. of Akron, OH, is the fourth-largest investor-owned electric system in the United States. Its seven electric utility operating companies serve 4.3 million customers within 92,400 square kilometers of Ohio, Pennsylvania, and New Jersey.[25] It also provides natural gas service to approximately 150,000 customers in the Midwest.

The following sequence of events are broken down in stages so you can follow what happened and how cascade failures start out small and seemingly insignificant and grow, eventually overwhelming the entire grid. The sequence of events leading to the outage is documented in greater detail in the Interim and Final reports produced by the U.S.–Canada Power System Outage Task Force.

What Happened on August 14th[26]

- FirstEnergy's control-room alarm system was not working, which meant operators did not know transmission lines had gone down, did not take any action to keep the problem from spreading, and did not alert anyone else in a timely fashion.
- MISO's tools for analyzing the system were also malfunctioning, and its reliability coordinators were using outdated data for monitoring, all of which kept MISO from noticing what was happening with FirstEnergy in time to avert the cascading.
- MISO and PJM Interconnection, the neighboring reliability coordinator, had no procedures to coordinate their reactions to transmission problems.

[23]"Final Report on the August 14, 2003 Blackout in the United States and Canada: Causes and Recommendations," April 2004. U.S.–Canada Power System Outage Task Force, p. 12 (dated January 12, 2004). https://reports.energy.gov.

[24]http://www.aep.com.

[25]http://www.firstenergycorp.com.

[26]http://www.spectrum.ieee.org/WEBONLY/special/aug03/tline.html.

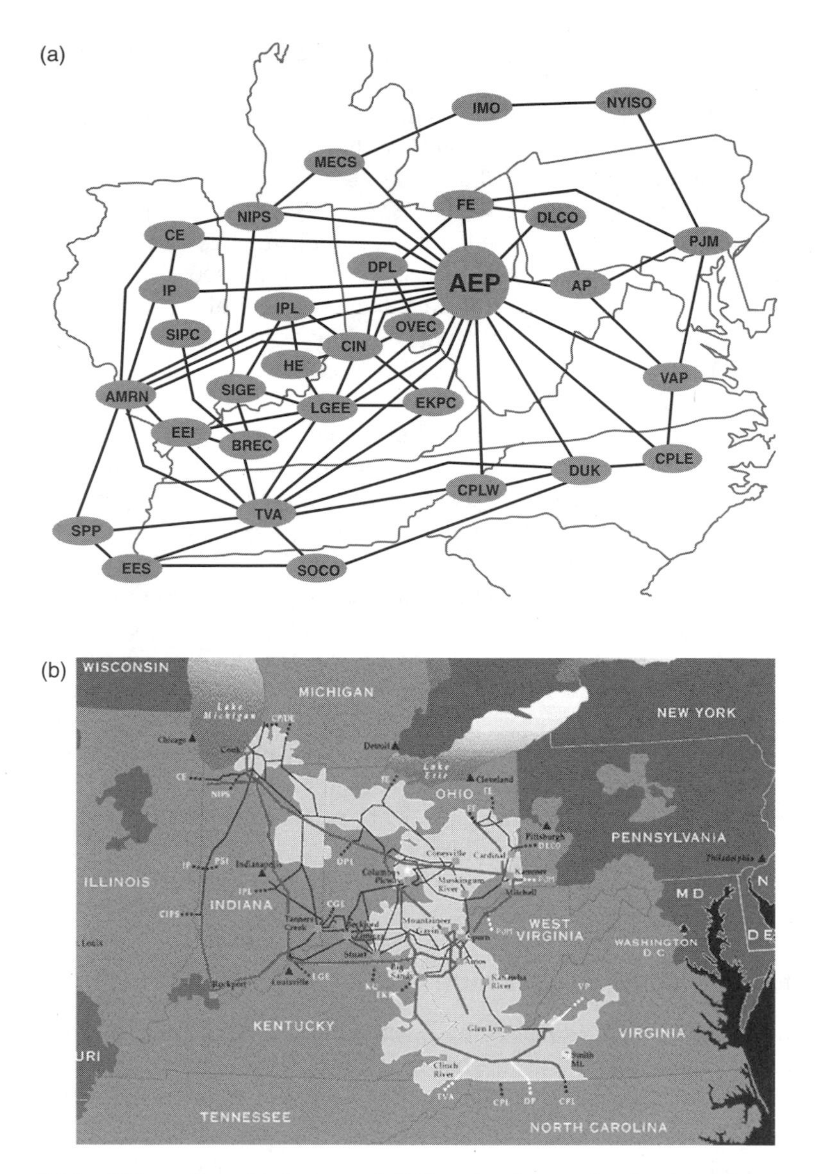

Figure 9.14. AEP small-world cluster and AEP regional map showing its regional area of responsibility. (a) AEP is a small-world neighborhood. (b) Geographical location of AEP and its control area.

These failures started out small, with an error in control software and a shorting of an obscure power line in Ohio.

Phase I: Power degradation seems like a routing fault that could have been corrected immediately, had the control software been working. But this small fault was not the end of the story. It would gather momentum like a snowball and lead to other, bigger faults.

All times are in Eastern Daylight Savings Time, August 14, 2003. AEP refers to American Electricity Power, and FE refers to FirstEnergy, Inc. MISO is the Midwest Independent System Operator. PJM is the Pennsylvania–Jersey–Maryland ISO.

Phase I: Power Degradation

12:15 EDT: MISO SCADA/EMS state estimator software has high error—it is turned off.
13:31 EDT: Eastlake Unit #5 generation tripped in Ohio.
14:02 EDT: Stuart-Atlanta 345-kV line tripped in Ohio due to contact with a tree.

Phase II: Computer Failure

14:14: FE SCADA/EMS alarm software fails.
14:20: FE SCADA RTUs fail.
14:27: Star-South Canton 345-kV line tripped.
14:32: AEP called FE regarding Star-South Canton line.
14:41: FE transfers software applications to backup computer.
14:54: FE backup computer failed.

Phase III: Cascade Line Failures Begin

15:05: Harding–Chamberlin 345-kV line overheats, shorts with tree.
15:31: MISO called PJM to confirm Stuart-Atlanta line was out.
15:32: Hanna-Juniper 345-kV overheated, sags, and shorts out.
15:35: AEP unaware of Hanna-Juniper failure.
15:36: MISO unaware of Hanna-Juniper failure.
15:41: Star-South Canton tripped, closed, tripped again, unknown to AEP and PJM.

Phase IV: Cascading Collapses Transmission

15:39 to 15:58: Seven 138-kV lines trip.
15:59: Loss of the West Akron bus causes five more 138-kV lines to trip.
16:00 to 16:08: Four more 138-kV lines trip, Sammis-Star 345-kV line overheats and trips.

Phases 5, 6, and 7

16:10 to 16:12: Transmission lines disconnect and form isolated islands in Northeast U.S. and Canada.

When it was all over, 263 of the 531 generators were shut down in the United States and Canada. The cascade that began in MISO spread to other reliability coordinators: Quebec, Ontario; New England, New York; and PJM. It may never be known with certainty why a major portion of the Eastern Grid "cascaded," but the 2003 blackout was at least aggravated by errors in the control system software, inadequate controls to recover from tripped power lines, and operator errors.

The entire Northeastern United States and parts of Canada faded to black when a relatively small fault snowballed into a medium-sized fault, and then into a major fault. This was an accident, but it suggests a more serious conclusion—the 2003 blackout would have been relatively easy as a premeditated attack. Such an attack would not require much more than tripping a few high-voltage lines or disabling the power control systems. However, the terrorist would have to have a high degree of engineering knowledge.

This blackout qualified as a "1-in-100-years" event. It was "large" by several metrics—it covered a large geographical area, it affected a large population, and its economic impact was large. In addition, it shared many of the common faults observed in power outages over the past century:

- Power lines making contact with trees and shorting
- Underestimation of generator output
- Inability of operators to visualize the entire system
- Failure to ensure operation within safe limits
- Lack of coordination
- Ineffective communication
- Lack of "safety nets"
- Inadequate training of operators

ANALYSIS

Theoretically, a large power grid can shift power from one end of the country to the other because it can be extremely adaptable to changing demand and localized faults. Even if the largest dozen or so centralized power plants fail, power can theoretically be transferred from somewhere else. The largest dozen power plants supply less than 5% of national power demand, and there is a 15% surplus. In addition, different regions of the grid reach peak load at different times, so when demand peaks in one part of the grid, the demand can be satisfied by a valley in another part.

So, the larger the grid, the more adaptable it is—theoretically. But in practice, the grid is too complex to guarantee isolation of faults (versus cascading), and vertical integration over the past century has led to regional interoperability problems. Simply put, it is still not possible for the grid to adapt to demand on a national scale because there is insufficient capacity in the transmission and distribution network. To make matters worse, the SCADA/EMS systems are not sophisticated enough to properly automate the regulation of ACE. Although there is no shortage of power, there is a shortage of distribution capability, SCADA sophistication, and trained operators.

The 1992 EPACT was aimed at decoupling the layers of the old vertical monopolies, but at the present time, this has increased, rather than decreased, the vulnerability of the grid. In addition to economic vulnerability (the "gaming of prices" by predator utility brokers like Enron), the network is vulnerable to technical vulnerabilities (SCADA software errors, complex interdependencies that are not fully understood). EPACT has deregulated the generation and load components, but left the middle component on its own in a world that views the transmission and distribution component "someone else's problem." There is no money to be made from the middle. Thus, economic forces are working against protection of the most vulnerable part of the grid.

The grid may simply be too big and complex to fully control. In fact, the grid may not be entirely necessary. In 1902 there were 50,000 independent, isolated power generating plants in the United States and only 3624 central power plants.[27] The grid started out as a decentralized, distributed generation network. Immediately after World War I, the price of coal soared and urbanization favored centralization of generation. Technology advanced rapidly during the 1920s, which lowered the cost of building centralized plants. In addition, their owners drove the independents out of the market by lowering monthly bills to consumers and by diversifying applications to make up for the loss of revenue during nonpeak periods of the day. The vertically integrated and centralized power companies sold their nonpeak power to electrified train systems (subways and commuter trains), factories, and large building owners to power elevators. Thus, centralized generation won out, and today we have a grid that has formed a small-world network. We have demonstrated that small-world networks consist of vulnerable clusters such as AEP that can bring down the sizeable regions they serve.

But the grid does not have to remain the way it is today. If the grid was redesigned and regulatory legislation was to favor *distributed generation* (wind, solar, and fuel cell generators at factories, shopping malls, and neighborhoods), the grid would be made almost invincible because it would truly be adaptable. In distributed generation systems, most of the time most of the power comes from only a few yards away. Solar generators do not produce during the night, and wind power does not produce during calm weather, so the grid would still be needed. But it would be needed less of the time, and when it fails, the local generation facility would provide enough power to keep critical services like hospitals operating. Most importantly, the lack of sufficient transmission and distribution capacity would be less critical. Distributed generation may turn out to be cheaper than hardening the grid with its complexity and vast expanse of transmission and generation network.

This leaves SCADA/EMS as the vulnerability of greatest concern. And the cyber-threat to power is real. The SQLSlammer worm penetrated a private computer network at Ohio's Davis-Besse nuclear power plant in January 2003. It disabled a safety monitoring system for nearly 5 hours and shut down a critical control network after moving from a corporate network, through a remote computer onto the local area network that was connected to the control center network. Slammer might have affected critical control systems at Davis-Besse. As it turned

[27]A. Friedlander, "Power and Light: Electricity in the U.S. Energy Infrastructure 1870–1940," Corporation for National Research Initiatives, 1996, p. 51. request@cnri.reston.va.us.

out, the affected systems were used to monitor, not control, the reactor. The safety of Davis-Besse was not jeopardized.

However, by 2005, more than 60 cyber-security events have impacted power control systems, including three nuclear plants.[28] This number is likely to grow as the Internet becomes intertwined with non-Internet control networks. Unfortunately, SCADA/EMS components—computers, networks, and software—will remain complex and unreliable for a long time because securing an information system is well known to be problematic. Thus far, it has been impossible to build software that is guaranteed to be bug-free. These software flaws lead to networks becoming disconnected, data being lost, and computers being disabled. As long as software is flawed, there will be faults in industrial control systems such as SCADA and EMS. This "cyber-security problem" is addressed in the final two chapters that deal with cyber-threats and cyber-security.

EXERCISES

1. Why are there high- and low-voltage lines?
   a. Cities need more power than farms.
   b. Electrons travel farther on high voltage.
   c. Electricity travels more efficiently at high voltage.
   d. Electrons travel faster at high voltage.
   e. Electricity has lower resistance at low voltage.
2. AC won over DC because AC:
   a. Works better in radios and TVs.
   b. Is an international standard.
   c. Operates at 60 cycles per second, which is compatible with clocks.
   d. Can be transmitted at high voltages.
   e. Can be switched and easily stepped down in voltage.
3. Before it was called the FERC, it was called:
   a. FPC
   b. CIA
   c. NERC
   d. MISO
   e. NCS
4. The Federal Power Act of 1935 established federal regulatory control over power because:
   a. It was the right thing to do.
   b. Rural areas needed power too.

[28]Personal communication with Joe Weiss.

c. The Great Depression was in full effect.
d. Congress wanted to establish power over power.
e. Interstate commerce allows the federal government to step in.

5. NERC and load sharing through wheeling was established soon after:
   a. Enactment of the Federal Power Act of 1920.
   b. The Northeast Blackout of 1965.
   c. Soon after the problem of synchronization was solved in the 1970s.
   d. Enactment of PURPA in 1978.
   e. Soon after deregulation in 1992.

6. Electrical power was deregulated by enactment of:
   a. EPACT
   b. PURPA in 1992
   c. Establishment of ISO's in 1992
   d. Tragedy of the Commons Act of 1992
   e. ISOs

7. ISOs were authorized by FERC to:
   a. Monitor the operators.
   b. Look for abuses by participants.
   c. Run power exchange markets.
   d. Maintain their independence.
   e. All of the above.

8. The ES-ISAC is:
   a. The same as EISAC.
   b. Run by NERC.
   c. Run by the FERC.
   d. Run by an ISO.
   e. Run by the Department of Homeland Security.

9. Which one of the following is *NOT* a power grid within NERC?
   a. ERCOT
   b. Western Interconnect
   c. Quebec Interconnect
   d. Midwestern Interconnect
   e. Eastern Interconnect

10. Which of the following is *NOT* a component of the electrical power system of the United States (select only one)?
    a. Dams
    b. Power plants

c. Transmission grids
d. Distribution networks
e. SCADA

11. Why is the electrical power system particularly *immune* to disasters, accidents, and attacks (check all that apply)?
    a. The grid is large and distributed like a mesh graph.
    b. Transformers are difficult to damage.
    c. Utility companies rapidly buy and sell power to each other.
    d. Power is more efficiently transmitted at high voltage.
    e. The ACE/AGC mechanism.

12. Why is the electrical power grid particularly *vulnerable* to disasters, accidents, and attacks (check all that apply)?
    a. Most power comes from a few central power plants.
    b. Coal fuel supplies in the United States depend on critical railroad assets.
    c. Large transformers are difficult to replace.
    d. SCADA systems are generally open and thus vulnerable.
    e. Hydroelectric dams are vulnerable.

13. In the United States, power, transmission grids, and fuel comes from (check all that apply):
    a. International sources
    b. Public and private entities (public and private ownership)
    c. SCADA systems
    d. Interdependent suppliers within the United States
    e. Middle Eastern countries

14. In the United States, over 50% of electrical power is generated by:
    a. Burning coal
    b. Burning natural gas
    c. Burning foreign oil
    d. Burning the midnight oil
    e. Nuclear fusion

15. The most likely asymmetric attack on power generation is (select only one):
    a. Bombing of Grand Coolee Dam
    b. Attacking a nuclear power plant
    c. Coordinated attack on substations using *PowerGridSim*
    d. Cyber-attack on SCADA systems that control power
    e. Bombing of Hoover Dam on the Colorado River

16. The most likely asymmetric attack on the transmission component of power is (select only one):
    a. Towers carrying high-voltage power
    b. Interstate tie lines
    c. Local distribution network transformers
    d. Transmission lines supplying power to major areas such as Chicago
    e. Bombing of the load component
17. The most vulnerable (and most damaging if attacked) part of the transmission component of the power system is (select only one)?
    a. Hard to replace, major transformers
    b. Three-wire power lines
    c. Local distribution substations
    d. Deregulation of the utilities
    e. Transmission poles
18. How is the power grid different than the Internet (select one)?
    a. The grid has no routers and switches.
    b. The grid sends packets of electricity.
    c. The Internet sends packets of information.
    d. Packets have source and destination addresses.
    e. Power grid circuits can send electrons to specific loads.
19. Why does ACE rarely equal to zero?
    a. The load is constantly changing.
    b. Generators generate unpredictable output.
    c. SCADA/EMS software often fails.
    d. The weather is constantly changing.
    e. The grid is too big and complicated to understand.
20. Why is the grid vulnerable in the middle?
    a. There is insufficient transmission and distribution capacity.
    b. Transformers are critical and unprotected.
    c. Substations are unreliable.
    d. Everything depends on generators.
    e. Fuel is in short supply.
21. Using the program *Rtree*, repeat the *availability allocation* experiment with a grid that is shaped like a tree: One generator is at the root with transmission lines connecting it to two substations, A and B. Then, A is connected to two other substations, A1 and A2, and B is connected to two more substations, B1 and B2. A1, A2, B1, and B2 are connected to a single load. Each transmission line has a maximum capacity of 500 MW. Assume there is enough funding to

provide 500% availability for the seven nodes (generator plus six substations), what is the optimal allocation of availability to each substation?

a. 100% goes to the generator, and the remaining 400% is spread evenly.
b. 100% goes to the generator, and 100% to A, 100% each to A1, A2.
c. 50% to generator, 300% to one substation, 150 to the other substation.
d. 100% to generator, 100% to B, 100% to B1, B2.
e. 100% to generator, 100% to B, 100% to A, 50% each to B1, B2. A1, A2.

22. The network-wide allocation of 1000 availability points to the network of Figure 9.14 (AEP cluster) yields 100 points each to the following nodes, assuming damages are all 100 (nodes and links):
    a. AEP, AMRN, SOCO, CSWS, TVA
    b. AEP, DUK, TVA, LGEE, CIN, DREC, AMRN
    c. MISO, TVA, SPP, JPM
    d. MISO, TVA, SPP
    e. AEP, TVA, LGEE, FE, CIN, AMRN

23. Figures 9.15 and 9.16 show a fault/event tree and area control network for the power grid that serves San Lewis Rey. Find the critical node(s) in this grid and then perform an MBVA study that results in a strategy for how to strengthen the grid. Use all strategies provided by the program *FTplus* to find allocations of $200 million to reduce risk. Use the inputs below and

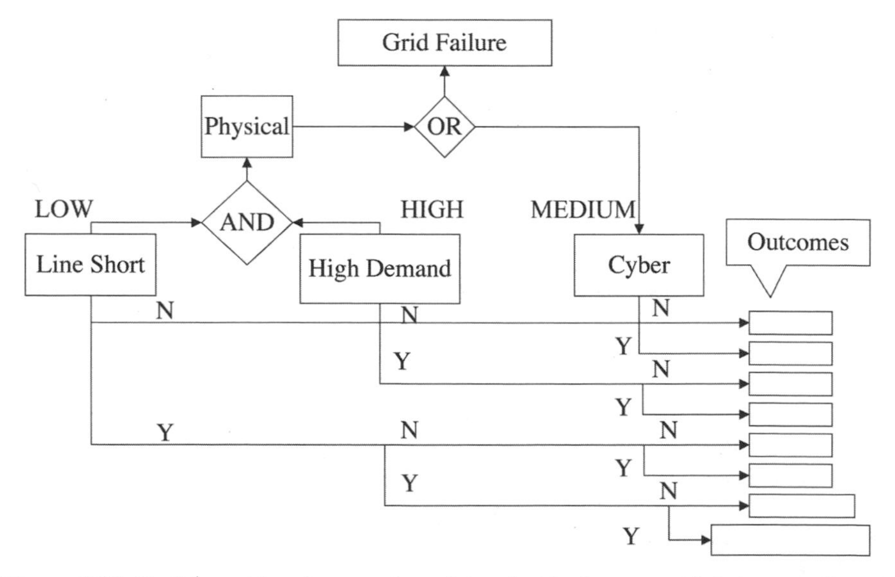

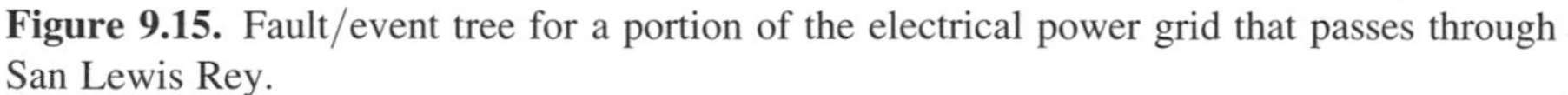

Figure 9.15. Fault/event tree for a portion of the electrical power grid that passes through San Lewis Rey.

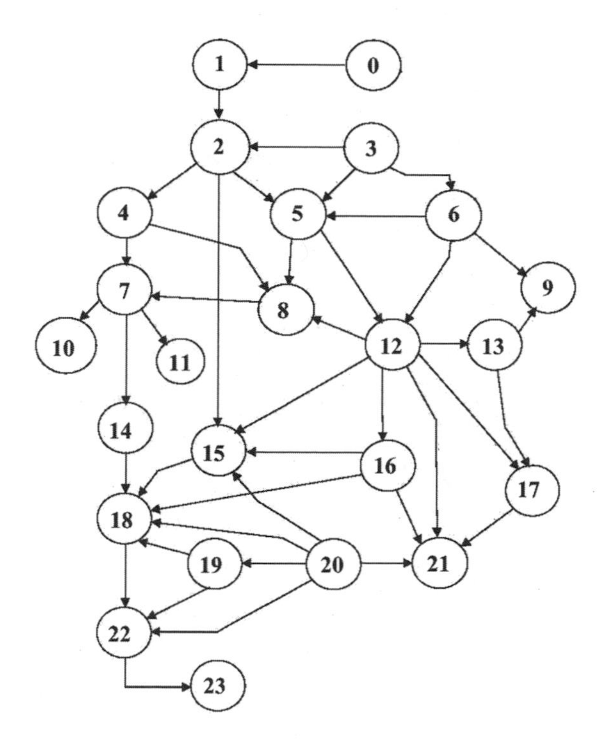

Figure 9.16. Area control network for San Lewis Rey electric power.

Figure 9.15. Why is it *not* possible for all fault combinations to trip a grid failure?

Inputs to SLR MBVA Study.

Assume	Line Shorts	High Demand	Cyber-Attack
Vulnerability(i)	LOW	HIGH	MEDIUM
Cost(i)	\$300 M	\$150 M	\$200 M
Damages(i)	\$100 M	\$250 M	\$500 M

CHAPTER 10

Energy

Energy—nuclear, chemical, and biological—propels everything. It runs power plants, heats our homes, powers our cars, and air conditions our offices. Without energy, the telephone and Internet would not work, and the modern conveniences we take for granted would vanish.

Most energy consumed in the United States comes from petroleum, coal, and natural gas (NG). Figure 10.1 shows the past, present, and future consumption of energy broken out into the different sources. Currently, burning NG generates less than 20% of electrical power, and NG accounted for only 25% of energy consumption in 2003, but the country is quickly transitioning to even greater dependence on NG. For instance, 88% of new power plants constructed today are NG-powered.[1]

> Homes, businesses, industries, and electric power generators are projected to increase their combined consumption of natural gas 54 percent by 2025.[2]

NG burns cleaner than coal. In addition, although the United States is heavily dependent on foreign imports for its petroleum, America produces 85% of its own NG. Most of the remaining 15% comes from Canada. So NG has geopolitical advantages over petroleum and environmental advantages over coal. Even so, petroleum remains the number one energy consumed in the United States, followed by NG. Because NG is rapidly expanding into the power generation industry, it is reasonable to assume that gas and oil will remain the most consumed energy through 2025 (see Figure 10.1). Hence, study of the energy sector is principally an investigation into the petroleum and NG supply lines that power America.

This chapter examines the critical nodes of the petroleum and NG supply chains[3] that deliver much of the energy consumed by the United States. We illustrate how this supply chain works through three case studies: the Southern California

[1]http://www.eei.org/industry_issues/energy_infrastructure/natural_gas/index.htm.
[2]http://www.eei.org/industry_issues/energy_infrastructure/natural_gas/index.htm.
[3]According to Dr. Warren H. Hausman of Stanford University, "The term supply chain refers to the entire network of companies that work together to design, produce, deliver, and service products." http://www.supplychainonline.com/cgi-local/preview/SCM101/1.html.

Critical Infrastructure Protection in Homeland Security: Defending a Networked Nation,
edited by Ted G. Lewis

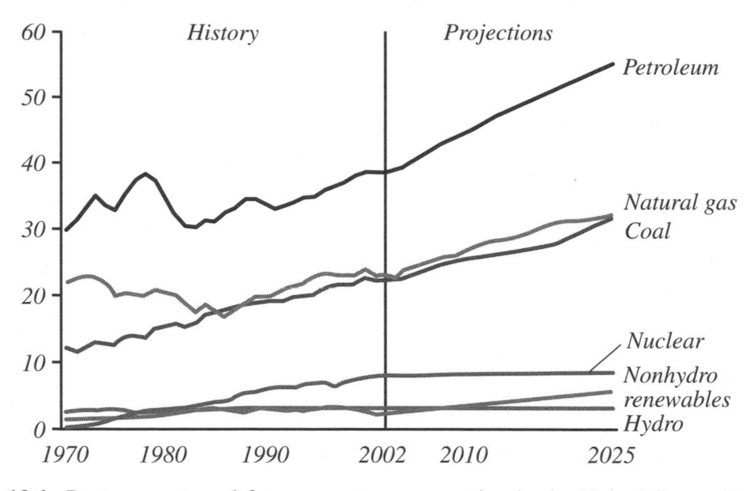

Figure 10.1. Past, present, and future energy consumption in the United States: Petroleum and NG will remain as the most consumed energy for the foreseeable future.[4]

(Kinder-Morgan) refined petroleum pipeline, a Northeast refined petroleum pipeline cluster, and a critical node located in one Atlantic Coast storage terminal.

In this chapter you will learn the following concepts and be able to apply them to the analysis of America's NG and oil supply chain:

1. Most U.S. energy comes from vast oil and natural gas (NG) supply chain networks that are probably small-world networks.
2. Major components of oil and NG supply chains are wellheads, refineries, transmission pipelines, storage, distribution pipelines, and SCADA.
3. Oil and NG supply chains are regulated much like the electric power grid, and like the power grid, are undergoing radical transformation due to deregulation of the energy supply chain.
4. Oil and NG supply chain networks are characterized by heavy concentrations—clusters—of refineries, transmission links, and storage terminals. They are also heavily interdependent with transportation and power (for running pumps and SCADA).
5. Major vulnerabilities exist in the oil and NG supply chains because of clustering of refineries, clusters of transmission pipelines, clusters of storage facilities, and vulnerable SCADA networks.
6. Oil and NG vulnerabilities are concentrated in four major components of the energy supply chain: refineries, large transmission pipelines, large centralized storage facilities, and, to a lesser degree, SCADA.

[4]http://www.eia.doe.gov/oiaf/aeo/index.html.

7. Critical nodes of the refinery component of the supply chain are vulnerable to physical attack because they produce a large percentage of output and they are clustered along the Gulf of Mexico.
8. Critical nodes in the transmission pipeline component of the supply chain are large-volume clusters that carry oil and natural gas to key areas of the country such as New York and Los Angeles.
9. Critical nodes in the storage component of the supply chain are large-capacity clusters located in key transportation nodes such as Perth Amboy, NJ.
10. SCADA is important, but less critical because disruption of control has less impact than destruction of the physical components of the supply chain.

THE RISE OF OIL AND AUTOMOBILES[5]

When Edwin Drake discovered petroleum in Titusville, PA, in 1859, the market for oil was in making kerosene used in lighting. Pennsylvania soon dominated the kerosene lamp oil business, holding 80% market share. But then two important things happened: (1) Edison perfected the electric light bulb, and (2) Henry Ford revolutionized transportation with the mass-produced gasoline-powered automobile. The Edison electric light bulb was better and cheaper than a kerosene lamp, and the automobile was a better form of transportation than the horse and buggy. Both transformed the fledgling oil industry from a niche business into a mainstream consumer products economy. What was not so evident in the 1860s was the extent to which oil would become dependent on transportation, and vice versa.

The oil business quickly became intertwined with transportation. Although oil was plentiful in Pennsylvania, the kerosene and gasoline consuming population was miles away in New York. Getting the product to market became the tail that wagged the dog. Crude oil was poured into wooden whiskey barrels and carried by wagon from the wellhead to the train station where it was transported to Northeast refineries. Then it was once again distributed by horse drawn wagons. Hence, moving oil became as profitable as the oil itself.

The Teamsters controlled the supply lines from wellhead to consumer. And like any monopoly, they began charging high prices for moving crude to refinery and refined kerosene to consumer. The cost to move a whiskey barrel of oil 5 miles to a rail station was greater than the railroads charged to move the same barrel from Pennsylvania to New York City. An alternative to this chokehold had to be found. Thus was born the first pipeline in 1863. Pipelines were soon, and still are today, the most economical way to move petroleum from wellhead to refinery, and from refinery to consumer.

The famous Tidewater pipeline, opened in 1879 to bypass even more costly middlemen, was the first *trunk* line—major energy transmission pipeline. It was half-owned by John D. Rockefeller who secured control of the oil supply chain

[5]http://www.pipeline101.com/History/index.html.

to control the oil market. Thus, monopoly power passed from the Teamsters to Rockefeller. He rapidly expanded the pipeline network to Buffalo, Cleveland, and New York City.

The business model for petroleum and natural gas was well established when Texas oil was discovered. Extract raw crude from the ground, send it to the North-eastern refineries by transmission pipeline, and then distribute the refined product to large metropolitan centers in the East. Thus, a profitable energy business was linked to control of a very long transportation network that often exceeded 10,000 miles in length. The barrier to entry was significant—how many competitors could afford to build such a pipeline? It soon became a business—Standard Oil—run by monopolist Rockefeller.

Standard Oil was not the only monopoly held by a single powerful industrialist in the late 1800s. But it was one of the most persistent. The Sherman Anti-trust Act was passed in 1890, making certain practices of a monopoly illegal. The Hepburn Act of 1905 declared trans-border transmission lines a form of interstate commerce and hence subject to regulation by the Federal Government. But it was not until 1912 that Standard Oil was forced to break up after a lengthy and colorful struggle involving President Theodore Roosevelt and powerful industrialists of the Gilded Age. In the end, seven regional companies replaced Rockefeller's Standard Oil Company and the petroleum industry continued to grow at a rapid pace. After World War I, pipeline networks were serving much of the nation, rising to 115,000 miles in the 1920s. Today they exceed 200,000 miles.

NG, and its more compact form, liquid natural gas (LNG), are also transported largely by pipeline. Pipeline middlemen controlled the market and hence the price of getting NG to consumers. So in 1938 Congress passed the Natural Gas Act (NGA), which allowed the Federal Power Commission (FPC), forerunner of the FERC, to set the prices charged by interstate pipelines but not the prices charged by producers.[6] In 1940, the NGA was amended to add regulation of the NG facilities themselves to the FPC's responsibilities.

In the famous *Phillips Decision* of 1954, the U.S. Supreme Court determined that the NGA covered wellhead prices as well. The FPC was now responsible for regulating the prices along much of the supply chain. The Court sought to protect consumers from "exploitation at the hands of natural gas companies."[7] The government sought to control the source, refining, and distribution of gas and oil for the benefit of taxpayers and voters.

Oil production in the United States was outstripped by demand during the 1950s and 1960s, which led to an increasing dependence on imported oil. The Colonial

[6]There seems to be a pattern here: Federal government regulators attempted to put limits on the prices that an electric utility may charge consumers but allowed the wholesale prices of power to float. This backfired in states such as California because of shortages in the capacity of the power grid to distribute power to consumers. In the case of NG, regulation had to be expanded to cover the entire supply chain. Is this the eventual fate of electric power grid operators?

[7]*Phillips Petroleum Co. v. Wisconsin*, 1954. This Supreme Court decision resulted in an expansion of the FPC's jurisdiction, and in the aftermath of the decision, NG applications under the NGA exploded, far exceeding the volume of electrical and hydroelectric regulation handled by the FPC.

Pipeline was the largest privately financed project in history in 1968 when it was constructed to deliver oil products from the Gulf of Mexico states to the Northeastern United States. The 800-mile Trans-Alaskan pipeline delivered 2 million barrels per day when it opened in 1977. It delivers 1 million barrels today. Thus, increasing demand and decreasing domestic supply pushed importation of oil ever upward.

By 2003, the United States was consuming 20 million barrels of oil *per day*, 68% of it delivered by pipeline networks and only 27% by boat. The remainder was moved by truck and rail. Most NG and oil is moved from wellhead to refinery to distributor, and then to consumer, by pipe (see Figure 10.2). This is the most economical way to provide such an enormous quantity of product over such long distances to so many consumers. But it is also the source of vulnerabilities, as we shall see later in this chapter.

Notice the network clustering in Figure 10.2. Are these pipelines yet another example of small-world networks? Whether they are scale-free or small-world networks, they exhibit a high degree of clustering as can be seen in Figure 10.2. We will

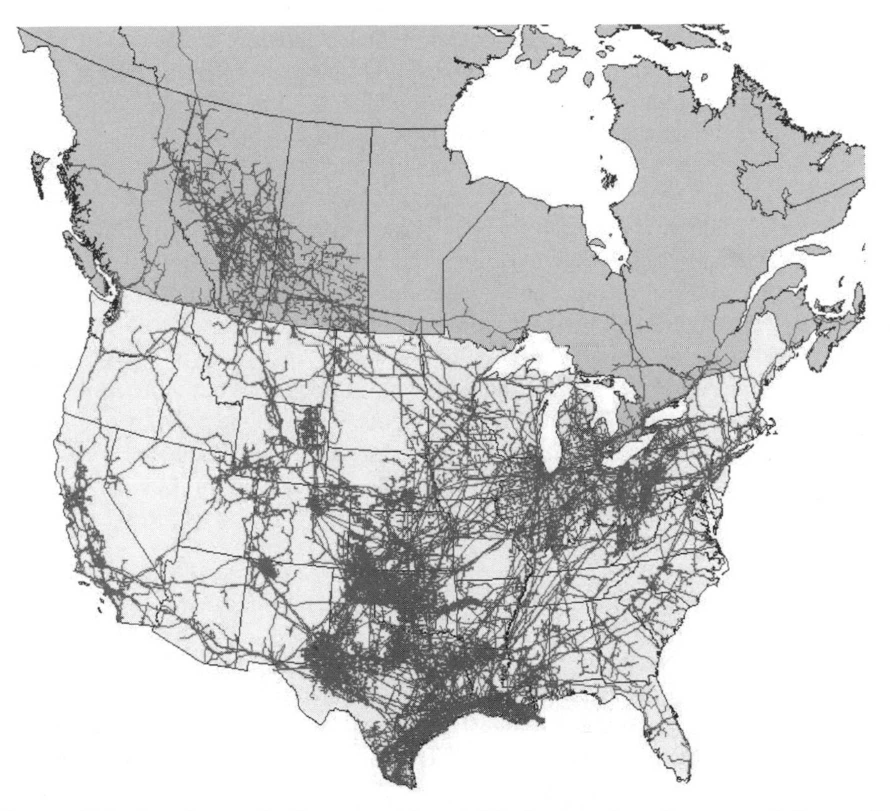

Figure 10.2. Petroleum pipeline networks of North America. Courtesy of Pennwell MapSearch: http://www.mapsearch.com/digital_products.cfm.

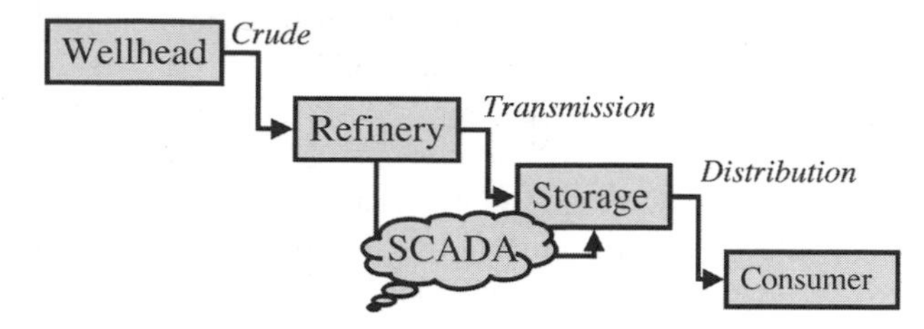

Figure 10.3. Simplified supply chain of the NG and petroleum energy supply.

show that clustering is indeed the main source of vulnerability in the supply chains of NG and oil.

THE SUPPLY CHAIN

Supply chain management (SCM) is the management of all steps in the delivery of a product or service to consumers. The NG and oil industry is principally an SCM industry with the added burden of governmental regulation. It consists of exploration, drilling, operation of crude pipelines, and operation of refineries for the production of fuel, plastics, and so on. Trunk line or "transmission pipes" are the arterials that deliver refined products such as gasoline and aviation fuel to terminals located around the country. Distribution refers to the sale and delivery of these products to consumers from storage terminals.

A greatly simplified supply chain is shown in Figure 10.3. Unrefined product from a wellhead is transported to a refinery by a pipeline system or boat. The refinery converts the crude into refined products, which in turn are transmitted over long distance to terminals, where they are stored. Then a distribution network of pipelines, trucks, and so on delivers the product to consumers.

The vast miles of pipeline are monitored by various SCADA systems that report anomalies such as leaks and broken components. An example of a pipeline SCADA network is given in the chapter on SCADA—the crude pipeline that delivers oil from the fields around Bakersfield, California, to the refineries in the Los Angeles area.

PADDs

For purposes of reporting, the supply chain is divided into five regions called *Petroleum Administration for Defense Districts* (PADDs). They were created during World War II when gasoline was rationed and are still used today for tracking supplies. Broadly speaking, the five PADDs cover the East Coast (PADD1), the Midwest (PADD2), the Gulf Coast (PADD3), the Rocky Mountain Region (PADD4), and the West Coast (PADD5). Supply chain data can be obtained from the U.S. Department of Energy's Energy Information Administration, which collects and publishes oil supply data by PADD (http://www.eia.doe.gov).

For example, in 2001, the percentage of oil produced and imported by each PADD was:

PADD #	Production	Importation
1. East Coast	~0%	~100%
2. Midwest	10%	90%
3. Gulf Coast	90%	10%
4. Rocky Mountain	~100%	~0%
5. West Coast	45%	55%

Figure 10.4 shows the crude oil supply and the refined product networks that link together the various components of the supply chain. The PADDs are also shown as shaded states. Note that crude mainly flows from Canada or the Gulf Coast, and refined product flows mostly to the East Coast and Midwest.

Refineries

The United States has more refining capacity (20%) than any nation in the world and refines 96% of all petroleum products it uses. Refineries are distillation factories. That is, they take in crude oil and distill it into various petroleum products according to their specific gravity (density). Lighter molecules percolate to the top of the distillery, and heavier particles stay near the bottom.

A 42-gallon barrel of crude produces 44 gallons of refined products, because of gains in the distillation process. Less than one half is turned into gasoline. Assuming your automobile tank holds 20 gallons, each of the 140 million registered automobiles and trucks in the United States consumes 140 million barrels of oil just to "fill 'er up!" (Recall that 2001 consumption was 840 million gallons per day in the United States. Filling up all registered vehicles consumes at least one sixth of total consumption for 1 day.)

The products obtained from a barrel of oil are summarized below along with the amount of each:

Gallons	Product
19.5	Gasoline
9.2	Heating oil
4.1	Jet fuel
11.2	Asphalt, feedstocks, lubricants, kerosene, etc.
44	TOTAL

In 2003 there were 152 refineries in 32 states. The top 10 refineries produced almost 20% of the total, and the top 2 (Baytown, TX, and Baton Rouge, LA) produced over 5% of the national supply. Refineries are highly concentrated along the Gulf Coast. In fact, most of the high-volume refineries are clustered along the

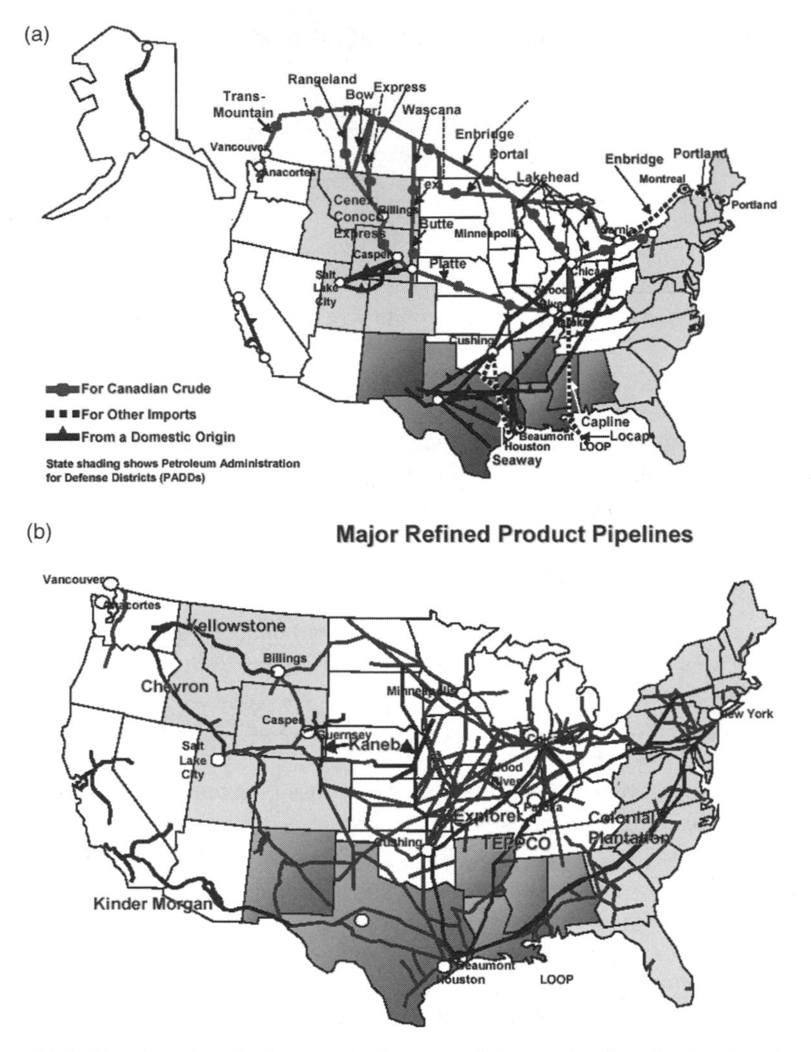

Figure 10.4. Map showing the largest pipelines supplying crude oil and refined products to PADDs in the U.S. market.[8] (a) Crude oil pipelines. (b) Refined oil pipelines.

coastline between Galveston, TX, and Baton Rouge, LA. This geographic region poses a major vulnerability in the refinery component of the energy supply chain. This vulnerability was demonstrated in 2005 when Hurricane Katrina slammed into the Gulf Coast and damaged the nation's energy supply chain. Without power, these major refineries were unable to produce NG and oil products essential to the rest of the country.

[8]C. J. Trench (ctrench@concentric.net), "How Pipelines Make the Oil Market Work—Their Networks, Operation and Regulation," A memorandum prepared for the Association of Oil Pipe Lines and the American Petroleum Institute's Pipeline Committee, December 2001, pp. 8–9. http://www.aopl.org.

Transmission

Large pumps and compressors move billions of gallons of product along pipelines each year. It is not uncommon for a pumping station to be powered by a 4000-horsepower diesel or electric motor, for example. Although these pumps and compressors are backed up with auxiliary power, it shows how dependent the energy sector is on the power sector. Ironically, without power, energy cannot be moved along its supply chain. And without energy, power cannot be generated.

In addition, operators depend on SCADA to monitor the pipeline and its contents, looking for leaks and other anomalies. SCADA is less critical, but without it, operators are blind. Writing for the Allegro Energy Group, Cheryl Trench describes how pipeline SCADA works:

> Pipeline employees using computers remotely control the pumps and other aspects of pipeline operations. Pipeline control rooms utilize Supervisory Control And Data Acquisition (SCADA) systems that return real-time information about the rate of flow, the pressure, the speed and other characteristics. Both computers and trained operators evaluate the information continuously. Most pipelines are operated and monitored 365 days a year, 24 hours per day. In addition, instruments return real-time information about certain specifications of the product being shipped—the specific gravity, the flash point and the density, for example—information that are important to product quality maintenance. Oil moves through pipelines at speeds of 3 to 8 miles per hour. Pipeline transport speed is dependent upon the diameter of the pipe, the pressure under which the oil is being transported, and other factors such as the topography of the terrain and the viscosity of the oil being transported. At 3–8 mph it takes 14 to 22 days to move oil from Houston, Texas to New York City.[9]

Figure 10.5 illustrates how petroleum products are "sequenced" in a pipeline. Like packets of data on the Internet, multiple segments travel on the same pipe. For example, a segment of kerosene is transported along with a segment of gasoline. This leads to *transmixing*—the unintentional mixing of products, which often requires some reprocessing at a terminal. But this is an extremely efficient way to move different products over the same (expensive) network. Allegro Energy Group president, Cheryl Trench, describes how sequencing works (Colonial Pipeline is the largest oil pipeline in the United States):

> Pipeline operators establish the batch schedules well in advance. A shipper desiring to move product from the Gulf Coast to New York Harbor knows months ahead the dates on which Colonial will be injecting heating oil, for instance, into the line from a given location. On a trunk line, a shipper must normally "nominate" volumes—ask for space on the line—on a monthly schedule . . . As common carriers, oil pipelines cannot refuse space to any shipper that meets their published conditions of service. If shippers nominate more volumes than the line can carry, the pipeline operator allocates space

[9]C. J. Trench (ctrench@concentric.net), "How Pipelines Make the Oil Market Work—Their Networks, Operation and Regulation," A memorandum prepared for the Association of Oil Pipe Lines and the American Petroleum Institute's Pipeline Committee, December 2001, http://www.aopl.org.

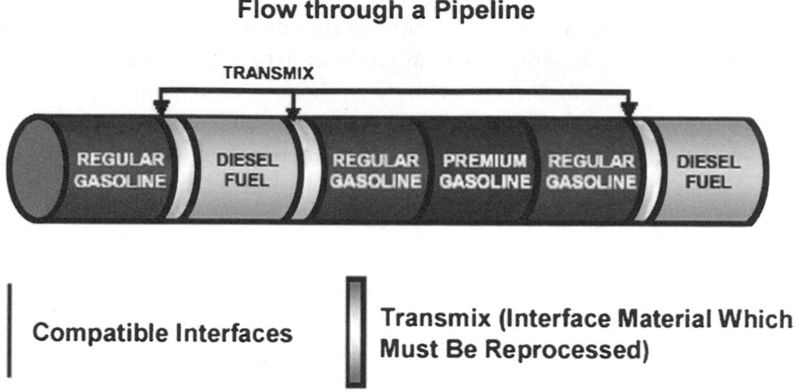

Figure 10.5. Pipelines are "multiplexed" by sequencing different products on the same pipeline.

in a non-discriminatory manner, usually on a *pro rata* basis. This is often referred to in the industry as "apportionment."[10]

Transport4

In August 1999, several major pipeline companies formed a joint venture to create Transport4 (T4)—an SCM website dedicated to scheduling product sequences through the major pipelines. Buckeye Pipe Line Company, Colonial Pipeline, Explorer Pipeline, and TEPPCO Pipeline opened up T4 to connect any pipeline carrier to any customer. Today, 80% of all product is scheduled via T4.

Storage

Products transmitted through major pipeline systems like the 5500-mile Transcontinental Pipeline (Transco) that delivers 95 million gallons per day to the East Coast (PADD1) end up in storage farms where millions of gallons of heating oil, aviation fuel, and gasoline are stored before being distributed to consumers. These tanks are large, conspicuous, and because of the *law of increasing returns*, clustered. As a consequence, they are highly vulnerable.

Natural Gas Supply Chains

The natural gas network architecture is much like the petroleum network architecture. It too is highly clustered and is most likely a small-world network. In some

[10]Ibid, pp. 15–16.

ways, it is even more critical because NG currently provides 25% of U.S. energy, and the percentage is expected to rise. Thus, its criticality is growing.

The NG supply chain is even more expansive than the petroleum supply chain. It consists of 280,000 miles of transmission pipeline and 1.4 million miles of distribution pipeline. The NG network is expanding at a rate of 14%/decade. Over the next 20 years, the United States will need 255,000 more miles of pipeline.[11] The National Petroleum Council estimates that utilities will spend $5 billion/year to build out this capacity.

SCADA

SCADA systems are designed to keep pipeline systems safe. They do this by monitoring pumps, valves, pressure, density, and temperature of the contents of the pipeline. An alarm sounds when one or more of the measurements go out of bounds, so operators can shut down pumps and compressors. Operators must consider the local terrain, the product that is inside a pipe, and numerous physical characteristics of the pipeline. If safety limits are exceeded, a SCADA system can automatically shut down a pipeline within minutes.

As we shall see, SCADA plays a relatively minor role in vulnerability analysis because a SCADA shutdown may cause loss of revenue, but not loss of life. Also, SCADA vulnerabilities can be mitigated for a relatively modest investment compared with the investment required to replace a refinery, storage tank, or section of pipeline.

REGULATORY STRUCTURE

The regulatory structure of the NG and petroleum supply chain shapes the industry in much the same ways it shapes the power industry. Congress delegated oversight of NG, oil, and electric power to the FERC as shown in Figure 10.6. The FERC must work with the Office of Pipeline Safety (OPS) positioned within the Department of Transportation, the Environmental Protection Agency (EPA), and the Department of Energy (DOE). The Atomic Energy Commission (AEC) is responsible for oversight of nuclear power plants. Why all of these agencies?

Gas and oil pipeline companies are considered *common carriers*, which means they must operate as all interstate commerce companies: They must provide nondiscriminatory access to their networks. But the energy supply chain overlaps other domains because of its impact on the environment and dependence on transportation. Operational dependencies lead to complex regulatory controls. The NG and oil supply chains cross many regulatory boundaries as well as geographical boundaries.

The Natural Gas Pipeline Safety Act (NGPSA) of 1968 authorized the Department of Transportation (DOT) to regulate pipeline transportation of various gases,

[11]National Petroleum Council, Meeting the Challenge of the Nation's Growing Natural Gas Demand, December 1999.

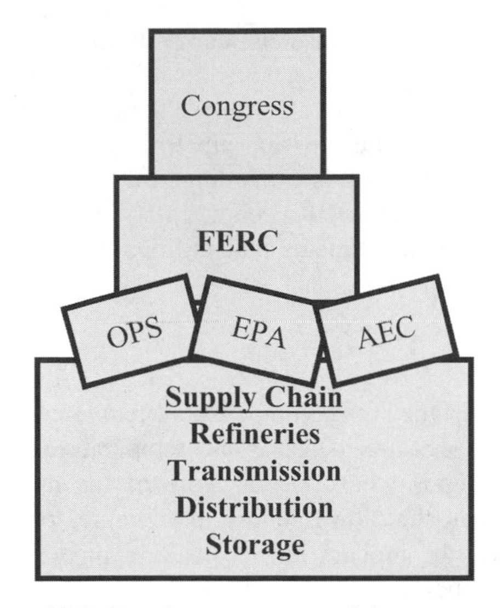

Figure 10.6. Regulatory structure of the energy sector.

including NG and LNG. As a consequence, DOT created OPS, but the emphasis was on safety, not regulation. The National Environmental Policy Act (NEPA) of 1969 made the FPC responsible for reporting environmental impacts associated with the construction of interstate NG facilities. And then the task of coordinating federal efforts to cope with electricity shortages was taken from the FPC and given to the Office of Emergency Preparedness in 1970. To complicate matters even more, the FPC was converted into the FERC in 1977 and the National Energy Act of 1978 that includes the Public Utility Regulatory Policies Act (PURPA) required gradual deregulation of NG. In 1979, Congress passed the Hazardous Liquid Pipeline Safety Act (HLPSA), which authorized the DOT to regulate pipeline transportation of hazardous liquids.

The Energy Policy Act of 1992 required the FERC to foster competition in the wholesale energy markets through open access to transmission facilities. By 1996, the FERC issued a series of orders forcing common carrier companies to carry electricity, NG, or petroleum products from a variety of competing suppliers. This re-regulation of the energy sector was euphemistically called "deregulation."

By 2003 the regulatory structure was cloudy at best. The FERC handled regulation, the DOT/OPS handled pipeline safety, and the EPA handled hazardous materials. The DOE provided information on the energy sector through its Energy Information Agency (EIA) at http://www.eia.doe.gov. HSPD-7, issued by President Bush in December 2002, distributes responsibility for the energy sector across DOE, DHS, DOT, and EPA. Today, they all have a role in making energy safe.

The FERC sets prices and practices of interstate pipeline companies and guarantees equal access to pipes by shippers. It also establishes "reasonable rates" for transportation via pipelines. These charges typically add a penny or two to the price of a gallon of gasoline, for example. But the FERC is not responsible for safety, oil spills, or the construction of the energy supply chain. Safety is regulated by the OPS. Oil spills are the responsibility of the EPA (Hazardous Material Spills). And responsibility for nuclear energy falls on the DOE and the AEC.

For example, when a pipeline ruptured and spilled 250,000 gallons of gasoline into a creek in Bellingham, WA, in June 1999, the OPS stepped in. This accident killed three people and injured eight. Several buildings were damaged, and the banks of the creek were destroyed along a 1.5-mile section.

Is there any difference between NG and oil when it comes to regulation? The FERC is empowered to grant permission for anyone to construct and operate interstate pipelines, interstate storage facilities for NG, or LNG plants. It also handles requests by anyone who wants to abandon facilities when, for example, pipelines get old and need to be upgraded or replaced with a new pipeline. But the FERC does not regulate local distribution pipeline companies. These are regulated by state authorities. Where does the FERC's power end and the EPA's begin? Where does federal regulation end and state regulation begin? In the case of electric power, rates are set by each state. In the case of gas and oil, the FERC sets rates. In the case of a terrorist attack, the Department of Homeland Security must deal with at least two other departments and one agency (the DOE, DOT, and EPA).

THE ENERGY ISAC

The mission of the Energy ISAC (E-ISAC) is to provide threat and warning information to member companies in the energy sector. It is funded by the Federal Government and run by a third-party company from an undisclosed location. Its members are the companies that produce NG, oil, coal, and so on. The E-ISAC is not the same as the Electricity Sector ISAC (ES-ISAC).

According to http://www.energyisac.com, the E-ISAC provides its members with:

- Information on threats and vulnerabilities (physical, cyber-security, and interdependencies).
- How to respond to threats (both physical and information security tips).
- A forum for members to communicate best practices in a secure environment.

VULNERABILITY ANALYSIS

In this section we evaluate the energy sector by applying model-based vulnerability analysis (MBVA) to the refinery, transmission, and storage portions of the oil and NG supply chains described in Figure 10.7. Refineries, major transmission pipelines, and major storage facilities seem to be the most critical components, because of their concentration and capacities. In addition, many of these critical components are

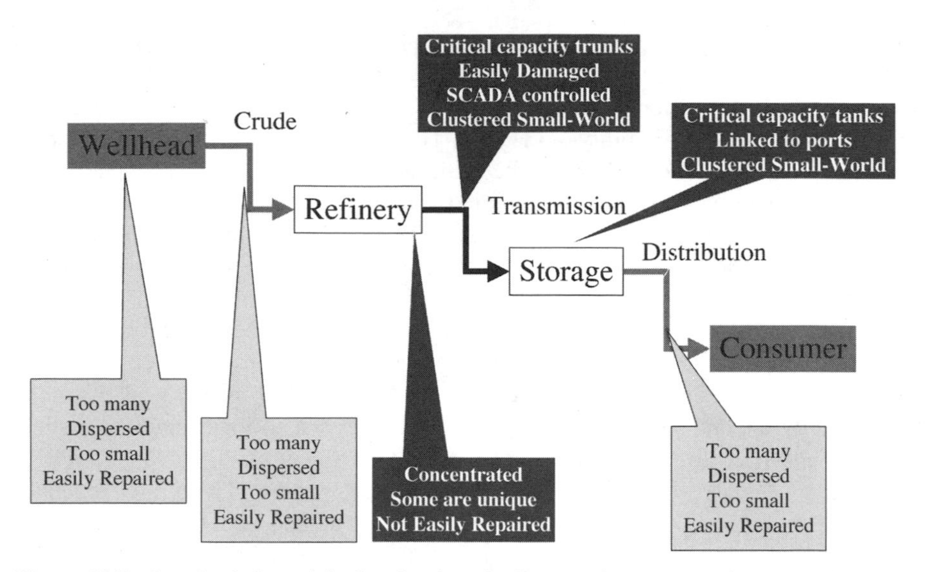

Figure 10.7. Supply chain revisited: refineries, pipelines, and storage are the most vulnerable components of the NG and oil supply chain network.

wide open; they are easily accessed and therefore vulnerable to symmetric and asymmetric attacks.

Because of the obvious concentration of assets and open access, we should focus attention on refineries, pipelines, and storage facility components of the energy supply chain. The following analysis shows how MBVA is applied to the energy sector, principally through four case studies: concentration of refineries along the Gulf Coast; vulnerability of the Los Angeles Petroleum Transmission Pipeline operated by Kinder-Morgan; vulnerability of the NG pipeline cluster located in a corner of Fairfax County of Virginia, near Washington, D.C.; and the vulnerability of the major storage terminal located in New York harbor. These are perhaps the most critical nodes of the U.S. energy sector.

Concentration of Refineries

The refinery component of the oil supply chain is characterized by a cluster of large-volume producers located very close to one another. The Gulf Coast area south of Houston, TX, stretching east to Lake Charles, LA, and then to Baton Rouge, LA, constitutes a critical node near the beginning of the petroleum supply chain.

Five of the top ten refineries fall into this geographic cluster. Which of the following locations contain these five?

1. Baytown, TX
2. Baton Rouge, LA

3. Texas City, TX
4. Whiting, IN
5. Beaumont, TX
6. Deer Park, TX
7. Philadelphia, PA
8. Pascagoula, MS
9. Lake Charles, LA
10. Wood River, IL

The top ten refineries produce 19.4% of all refined petroleum products in the United States. The five largest refineries located along the critical Gulf Coast region produce 11% of the total U.S. supply. Figure 10.8 lists the top five, how many barrels of product they produce each day, and where they are located.

The largest refinery in the United States belongs to ExxonMobil. It is located in Baytown, TX (in the Galveston Bay south of Houston) and is surrounded by smaller refineries. It is the largest of many refineries in the Galveston Bay cluster. It produces aviation fuels, lubricants, base stocks, chemicals, and marine fuels and lubricants.

In 1900, Galveston was the site of the deadliest hurricane in U.S. history. Over 8000 people lost their lives. Then in 1947, the largest port disaster in U.S. history wiped out Texas City. A ship containing fertilizer caught on fire and spread throughout the port and much of the town. Over the years, the worst refinery disasters have followed large clusters like Texas City: Whiting, IN; Texas City, Pasadena, and Amarillo, TX; Baton Rouge, LA; Romeoville, IL; and Avon and Torrence, CA. The bottom line is that refineries are highly vulnerable, and concentrations such as the Galveston Bay cluster are both highly critical to the oil supply chain and vulnerable to damage.

Figure 10.9 summarizes refinery vulnerability in the form of the familiar fault-tree. Refineries can be shut down because of lack of power, insufficient supply of crude, or fire damage. Power outages may last for only a few hours. Destruction of crude oil pipelines can deny service to a refinery for perhaps days. An explosion or fire can cause longer term damages, perhaps for months. Thus, the "cost" of each incident varies and the probability of each is unknown with any certainty.

The cost of closing a refinery depends on the volume of output (lost revenues) and the size of the refinery. Refinery replacement can cost more than $1 billion, and the

Rank#	1	2	3	5	9
Corporation	EXXON MOBIL	EXXON MOBIL	BP PLC	EXXON MOBIL	PDV AMERICA
Refiner	ExxonMobil	ExxonMobil	BP Products	ExxonMobil	Citgo Petroleum
Location	Baytown, TX	Baton Rouge, LA	Texas City, TX	Beaumont, TX	Lake Charles, LA
#Barrels/day	523,000	491,500	437,000	348,500	324,300
Market %	**11.0%**				

Figure 10.8. Rankings, names, and locations of the most productive refineries in the Gulf Coast Region. Total output is 11% of national refined product (2.2 Mmbl/day of 20 Mmbl/day).

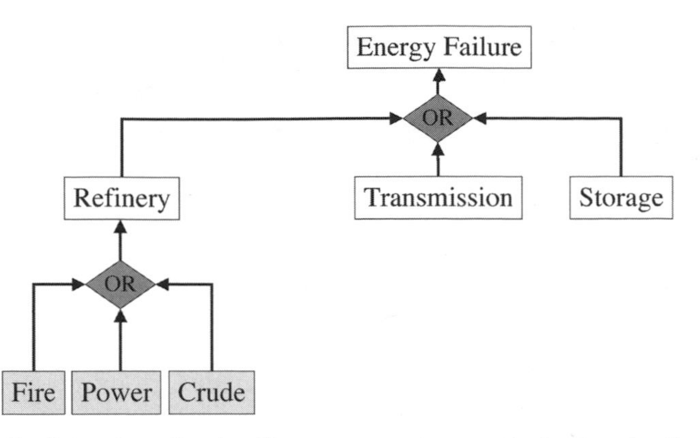

Figure 10.9. Fault-tree for refineries. Fires, power outages, or lack of crude oil can result in shutting down a refinery. Failure in a refinery, transmission pipeline, or storage terminal can result in an energy failure.

loss of production (500,000 barrels/day) can have severe implications on revenues as well as shortages that lead to price increases at the gasoline station.

Case 1: Gulf of Mexico Oil Field Network

PADD 3 (Gulf of Mexico) oil fields form a major network of refineries, pipelines, and a major import port called the Louisiana Offshore Oil Port (LOOP). In the foregoing introduction, refineries were critical, because the largest in the nation reside within this network. In addition, LOOP accounts for approximately 13% of the nation's total import of crude. Taken together, the PADD 3 network shown in Figure 10.10(a) is vital to the energy sector.

Network analysis of the Gulf Oil Field network shows that it has a scale-free structure [see Figure 10.10(b)]. In fact, this network obeys a near-perfect power law with $p = 1.179$. What are the critical nodes of this vast network? Obviously, the hubs are critical simply because of their high degree of connectivity. These are located at HOUMA, Clovelly, and under the sea directly south of Lafayette and New Orleans.

Assuming all damage values are equal ($d = 1$), a budget of 100 availability points is allocated by *NetworkAnalysis* as follows:

Node#	Name of Node	Number of Availability Points
67	Houma	36
70	Clovelly	43
34	EE-331 (under sea)	21%

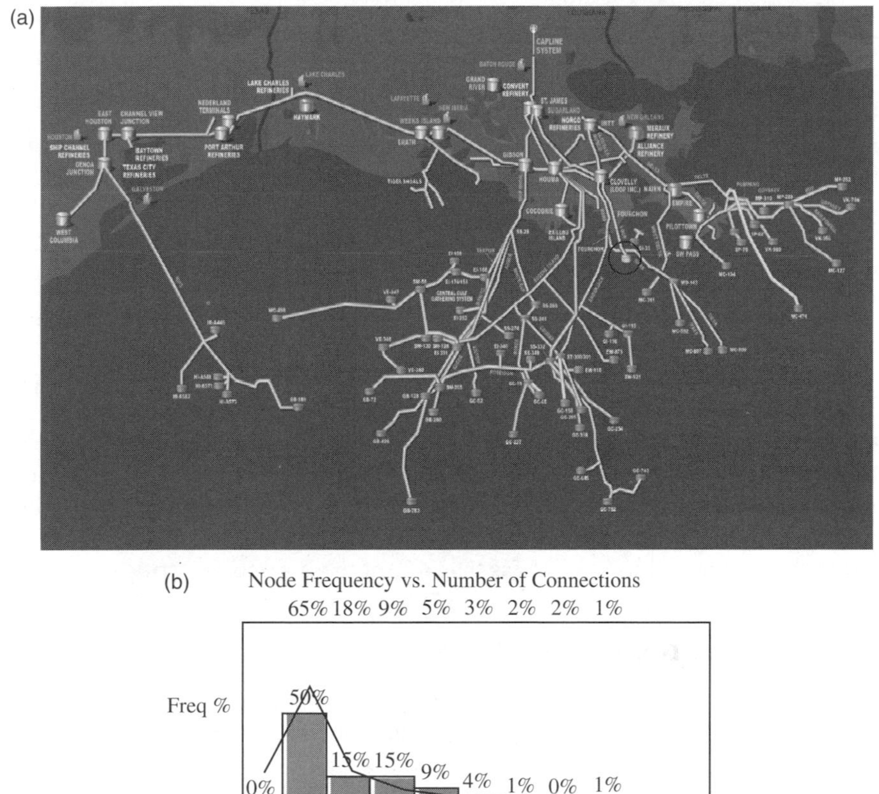

Figure 10.10. Gulf of Mexico oil fields and refineries. (a) Network of Gulf of Mexico oil fields and refineries. LOOP is circled.[11] (b) Node frequency histogram and power law ($p = 1.79$) fit for the network of (a).

This allocation makes sense because Houma is the network hub with the most connections, and Clovelly and EE-331 rank as the next-most connected hubs. *NetworkAnalysis* simply allocated the most points to the most highly connected nodes, i.e., the nodes with the highest degrees.

But this analysis assumes that all nodes are equal. If we give refineries and the offshore port LOOP damage values proportional to their value, the allocation will change. Suppose all refineries are assigned a damage value of 3, because the largest refineries produce approximately 3% of the national total, and let LOOP be assigned a damage value of 13, because it is the port of entry for 13% of all

[12]http://www.shellpipeline.com/cd_maps/SPL403_D_gc_crude_f.pdf.

imported crude. All other nodes and links are assumed to have a damage value of 1. The results of this weighted network are:

Node or Link	Availability Points
Link 88: (LOOP-> Clovelly)	57
Node 69: (Norco Refinery)	25
Node 18: (Lake Charles Refinery)	13
Node 71: (LOOP)	3
Node 69: (Houma)	2

This result also makes sense because refineries and the LOOP node are responsible for sizeable quantities of product. They are the largest sources of flow within the network. Therefore, when damage values are assigned to critical components such as refineries and ports, the Gulf Oil Field Network reveals the critical nodes: They are major refineries and the LOOP node and, to a lesser extent, the hubs identified by the scale-free network analysis. There are over 100 nodes and another 100 or so links in this network. Less than 4% of these assets are critical.

Case 2: Critical Transmission Nodes in Southern California

Now, suppose we examine an energy sector transmission network that serves the West Coast.[13] The petroleum supply chain that provides fuel to 8 million people in Los Angeles, San Diego, Las Vegas, and Phoenix, is shown in Figure 10.11.[14] The pipeline link to San Diego supplies the U.S. Navy at the San Diego port. Hence, it is important to both civilians and the military.

The network shown in Figure 10.11 is owned and operated by Kinder-Morgan, Inc., and is sometimes referred to as the Kinder-Morgan pipeline. The Kinder-Morgan network is shown in Figure 10.11 along with the name and degree of each node. This network is scale free, with the power law exponent equal to 1.1. What are the critical nodes? The three nodes with the highest number of links are Colton, Niland, and Watson (because it is the source node). These nodes are critical, because damage to any one disrupts flow to at least four major metropolitan areas.

Where should a defender invest limited resources to get the best protection? Suppose the damage value of all nodes and links is set to 50, except for Watson, which is set to 100, because it is the source for the entire network flow. If an allocation budget of 100 availability points is all the resource we have, then all 100 points should go to protect Watson. But if the budget is 200 points, Watson, Colton, and Niland should all receive points. As it turns out, allocation of 100 points to Watson and the remaining 100 points to Niland and Colton yield the

[13]A red team consisting of two U.S. Navy officers, Charles Hurst and John Cardillo, and two civilian analysts, Bill Brickner and Kristen Pearson, did this analysis.

[14]Professors Brown and Carlyle, Operations Analysis, Naval Postgraduate School, Fall 2003.

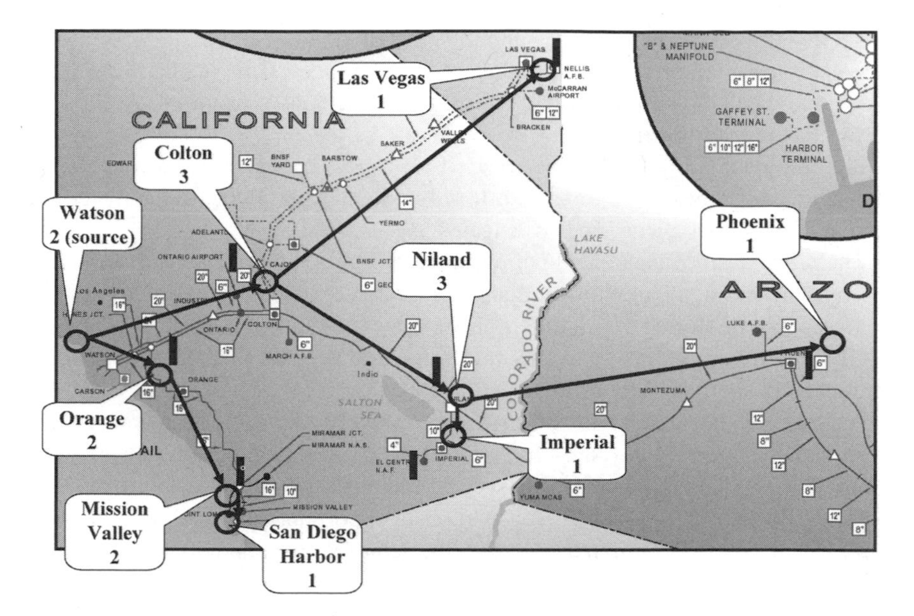

Figure 10.11. Network analysis of the Southern California Transmission Pipeline Network.

same minimum risk regardless of how many points Niland and Colton receive! This is known as a saddle point—a "flat surface" in the minimum risk plane where any distribution of points to Niland and Colton yields the same risk.

This network has a tipping point, however. If the damage value of Niland is changed to 51 instead of 50, Niland will receive all remaining 100 points. Conversely, if the damage value of Colton is changed to 51 instead of 50, it will receive all 100 remaining points. Therefore, risk is minimized when 100 points are allocated to Watson and 100 points are divided between Niland and Colton, in any proportion.

Now for the fault tree analysis of these three critical nodes: Because Kinder-Morgan is the only network for fuel transmission within the region, alternative transportation methods are neither competitive nor viable—there is simply too much product that must be delivered. More than 12.2 million barrels per month of refined petroleum products, primarily unleaded gasoline, flows outbound, from the Watson facility.

The network's fragility and the region's dependence on the pipeline's supply were highlighted in August 2003 when a break occurred in a very small section of the eastern portion of the network near Phoenix, AZ. The break disrupted 30% of the Phoenix area's supply, resulting in shortages that caused a dramatic spike in gasoline prices. Some motorists paid as much as \$4 per gallon for gasoline.[15] Repair took 29 days because of an antiquated pipeline. The immediate economic impact to the region was severe and lasted for nearly a month. The effects were

[15]At the time, the average pump price of gasoline was \$1.50/gallon.

also felt well beyond Phoenix, as supplies were diverted from coastal California to Arizona in an attempt to alleviate some of the shortages. This incident resulted in higher fuel costs up and down the California coast, thus expanding the economic impact far beyond Phoenix.

If Watson should fail, all flows cease. If Colton is damaged, Las Vegas, Niland, and indirectly, Phoenix are impacted. If only Niland fails, Phoenix is affected, but not Las Vegas, San Diego, or Los Angeles.

The refinery at Watson Terminal on South Wilmington Avenue, Carson, CA, is the most critical node, obviously. Spanning 19.3 acres, with 15 petroleum product tanks capable of storing 823,850 barrels of product, Watson accepts product from smaller refineries in the area and transmits it throughout southern California, Arizona, and southern Nevada (CalNev line).

Assume that the vulnerability of these three critical nodes is HIGH (80%). Furthermore, assume numerical values for prevention costs and damages are shown in Figure 10.12. What is the best way to spend $100 million on this supply chain? Using *FTplus* and the risk reduction strategies ranked, apportioned, and minimal the same result is yielded: $100 million should be allocated to Watson and $0 to the other two critical nodes. But using fault reduction strategies, we get a more interesting result:

Fault Reduction Allocations.

Strategy	Watson	Colton	Niland
Ranked	$77	$15	$8
Apportioned	$0	$64	$36
Minimal	$0	$50	$50

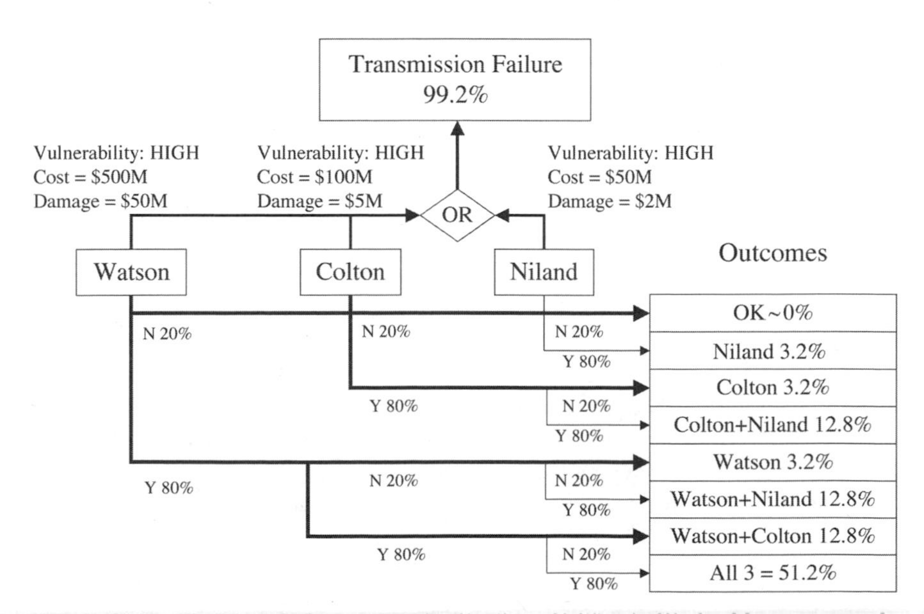

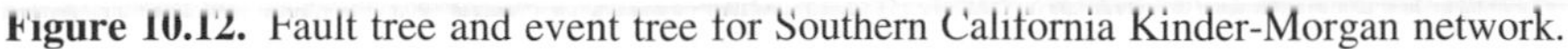
Figure 10.12. Fault tree and event tree for Southern California Kinder-Morgan network.

Why is the fault reduction allocation much different than the allocation for risk reduction? Take a close look at the fault tree in Figure 10.12 again. The most likely event is a combination fault where *all three critical nodes are attacked simultaneously*. Fault reduction allocation ignores the damage values and uses probabilities only. And probability calculations can yield nonintuitive results. Remember that the vulnerability of each node is the same for all nodes: 80%. The only variable in fault reduction is cost.

Ranked allocation attempts to reduce the highest vulnerability first. As the cost to harden Watson is the highest, it is also the most vulnerable. Therefore, $77 million is allocated to Watson. Colton is twice as costly to harden as Niland ($100 versus $50 million), so two thirds of ($100 − $77) = $23 million goes to Colton ($15 million) and $8 million goes to Niland.

Minimal (optimal) allocation attempts to drive the mathematical value of vulnerability to zero, which is done in fault minimization by driving fault probabilities down. The "least expensive" fault probabilities are Colton ($100/80% = $1.25/percentage point) and Niland ($50/80% = $0.625/percentage point). Watson fault reduction cost $500/80% or $6.25/percentage point to reduce. Therefore, minimal allocation "buys down" Colton and Niland vulnerability.

Apportioned allocation attempts to spread funding across all vulnerabilities by reducing the squared error difference between "before and after" fault probabilities. When the objective is to reduce fault, and not risk, only the fault probabilities are considered by *FTplus*. Note that the cost per percentage point of Colton ($1.25) is twice that of Niland ($0.625), and so it makes sense that the apportioned strategy allocates twice as much funding to Colton as Niland, because it cost twice as much to reduce the vulnerability of Colton than to reduce the vulnerability of Niland.

Note: The minimal (optimal) allocation strategy is the only one discussed here that considers triple combination event probabilities. Ranked and apportioned process the fault tree threat numbers only. But minimal computes the vulnerability across all single, double, and triple combinations of events.

Case 3: The NG Pipeline Cluster in Virginia

In this case study, we show that multiple NG pipelines in close proximity to one another pose a major vulnerability to the supply of NG to millions of people living in the East. Clustering of pipelines in the same geographical location may reduce operational and supply costs, but it also increases the likelihood of attack because they become attractive asymmetric targets. An inexpensive attack on such a cluster would render massive destruction to the energy economy.

Most NG comes from the Gulf Coast and Canada and heads toward the East Coast where the large metropolitan populations consume large quantities of NG to heat homes (80%) and generate electrical power (20%). The largest of these transmission networks is the Transcontinental Pipeline, also known as Transco. Transco runs from Houston to the New York harbor terminal (see Figure 10.13).

Figure 10.14 lists the top seven NG pipeline networks along with their capacities and pipeline lengths. All pipelines are over 10,000 miles long, and taken together,

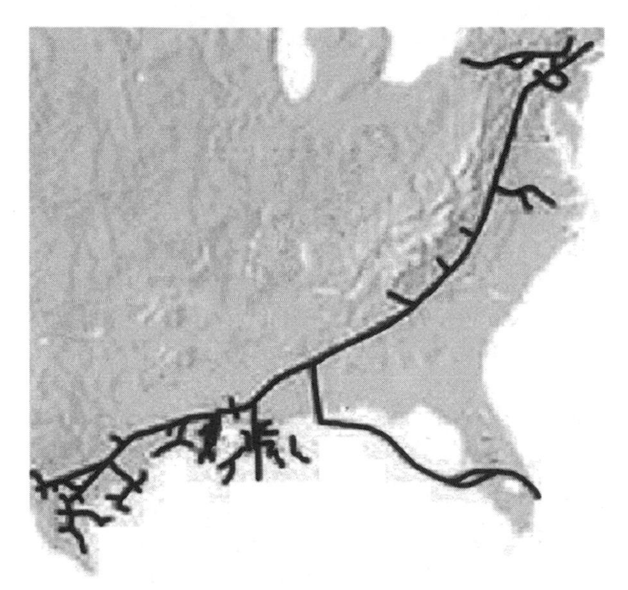

Figure 10.13. Map of the 10,600-mile Transco Pipeline. Courtesy of http://www.tgpl.twc.com/htm/inside.htm#Pipeline%20Profile.

they account for 35% of all NG consumption in the United States. The Transco pipeline ships 7300 MMbl/day (million barrels/day), and the longest, Tennessee Pipeline (14,700 miles), pumps nearly as much (7200 MMbl/day). These seven networks account for 79,500 of the 212,000 miles (38%) of pipeline and collectively pump 46.8 MMbl/day of the nation's 133-MMbl/day consumption.

Three networks listed in Figure 10.14 are of particular interest because they provide most of the energy consumed by the populous PADD1 region (Eastern United States). Number one Transco, number three Columbia, and number six Dominion form a cluster that is highly vulnerable to attack. Together they pump 20,913 MMbl/day of NG into the northern half of PADD1 (16% of national supply). But what makes these three particularly interesting is that they come geographically close to one another in northern Virginia, near Washington, D.C. For simplicity, let us call this the *Cove Point Intersection*. It is a highly desirable target because this is where these three large-volume pipelines come close to one another.

Rank	1	2	3	4	5	6	7
Name	Transcontinental	Columbia	Tennessee	ANR	Texas Eastern	Dominion	El Paso
Owner	Williams	NiSource	El Paso	El Paso	Duke	Dominion	El Paso
Capacity -MMbl/day	7,362	7,276	7,271	6,667	6,438	6,275	4,882
Length -miles	10,636	11,215	14,761	10,600	12,118	10,000	10,200

Figure 10.14. Top seven NG pipeline networks in the United States.

Cove Point Intersection

West of Washington, D.C., across the Potomac River in the Virginia counties of Fairfax, Prince William, and Loudoun, lies the nexus of three major pipelines: Transco, Columbia, and Dominion (a.k.a. CNG). Dominion runs through Leesburg Station; Transco runs through Loudoun Station, Nokesville, and Dranesville. Pleasant Valley forms the center of a network cluster that channels 16% of the nation's supply of NG to major markets to the north (see Figure 10.15).

Cove Point was recently reopened as a port of entry for even more NG in the form of LNG. Cove Point LNG was created to provide transmission capability from Cove Point to Pleasant Valley. The Cove Point LNG pipeline interconnects the Dominion, Transco, and Columbia lines, thus linking together the pipelines making up the critical Cove Point Intersection. Disruption of this geographic cluster would unplug 20,000 MMbl/day of LNG and NG that supplies New York and points north.

How vulnerable is Cove Point Intersection? There have been no major incidents involving these three pipeline companies, but there have been several accidents in the same neighborhood. Colonial's petroleum products transmission pipeline runs through the Cove Point Intersection, and Colonial Oil Pipeline, Inc., has the dubious distinction of paying the largest civil penalty in EPA history for seven spills over a period of years.

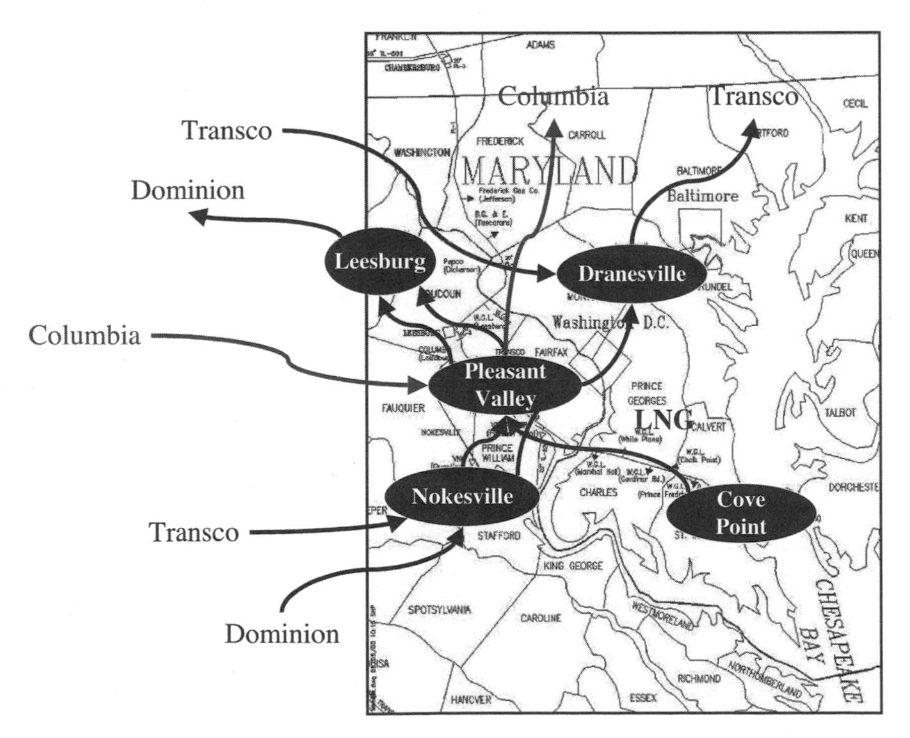

Figure 10.15. The Cove Point Intersection cluster depicted as a network.

Colonial Oil Pipeline experienced a spill March 28, 1993, when a rupture occurred in one of its oil pipelines in Fairfax County, which is near Pleasant Valley. The Colonial pipeline released over 400,000 gallons of oil and polluted 9 miles of the Sugarland Run Creek and seeped into the Potomac River. The EPA and Colonial have a litigious relationship:

> The Department of Justice and the Environmental Protection Agency today announced a settlement with Colonial Pipeline Company, resolving charges that the company violated the Clean Water Act on seven recent occasions by spilling 1.45 million gallons of oil from its 5,500 mile pipeline in five states. Under the consent decree, Colonial will upgrade environmental protection on the pipeline at an estimated cost of at least $30 million, and pay $34 million, the largest civil penalty a company has paid in EPA history.
>
> Atlanta-based Colonial is the largest-volume pipeline transporter of refined petroleum products in the world, moving an average of 83 million gallons of petroleum products each day through an underground pipeline that stretches from Port Arthur, Texas, to Linden, N.J., passing through Louisiana, Mississippi, Alabama, Georgia, Tennessee, South Carolina, North Carolina, Virginia, District of Columbia, Maryland, and Pennsylvania.[16]

Oil spill incidents like the seven Colonial accidents are not uncommon. In fact, thousands happen each year even though most pipelines are underground.

> Since 1986, just in the U.S., there have been 3140 incidents, 1407 injuries, and 322 deaths from natural gas pipelines alone.[17]

Major Causes of Pipeline Faults in 2002.

Cause	Fault%	Damage%
Corrosion	24.50%	31.60%
Excavation	15.40%	13.30%
Weld/Joint	7.70%	33.40%
Pump	9.10%	3.80%
Vandalism	1.40%	0.20%
Totals	**58.10%**	**82.30%**

In 2002, 143 pipeline incidents cost $33 million and caused one fatality. The most common fault is corrosion, followed by accidental excavation[18]:

Pipeline leaks and spills are relatively easy to fix even though there have been land-based spills larger than the ocean-based Exxon Valdez environmental disaster. The public at large is relatively ignorant of land-based spills, because they are

[16] http://www.usdoj.gov/, April 1, 2003.
[17] http://www.pipelinesafetyfoundation.org/.
[18] http://www.ops.dot.gov.

quickly (and quietly) repaired. But nobody knows what damage terrorists can do. We can only claim that it is relatively inexpensive and easy to do physical harm to the major transmission networks.

We conclude that critical nodes in the transmission pipeline component of the LNG and NG supply chain are large-volume clusters that carry NG to key areas of the country such as New York. These transmission lines are often colocated in clusters that offer attractive targets.

Case 4: Colonial Pipeline and Linden Station Storage Facilities

The petroleum supply chain may not be as long, but it is even larger and more concentrated than the NG supply chain. In this section we examine the largest-capacity petroleum transmission network in the world, and its termination point in New York harbor. Colonial Pipeline transmits 95 MMbl/day of petroleum products to customers in PADD1, eventually ending up in Perth Amboy, NJ, where it is stored before being distributed by Buckeye Pipeline. The high concentration of vast supplies in storage tanks in and around Linden Station provides an eye-popping target. Colonial and Linden Station are critical link and critical node, respectively, in this supply chain.

Colonial Pipeline is big: 5500 miles long (from Texas to New York); it delivers an average of 95 million gallons/day of gasoline, kerosene, home heating oil, diesel fuels, and national defense fuels to shipper terminals in 12 states and the District of Columbia. It has a 20% market share of the national supply.

In 1996, Colonial Pipeline Co. delivered 820.1 million gallons of jet fuel directly to airports in PADD1: Dulles International Airport, Baltimore-Washington International Airport, Nashville Metropolitan Airport, Charlotte-Douglas Airport, Raleigh-Durham Airport, Greensboro Triad Airport, and Hartsfield Atlanta International Airport.

Colonial is owned by a joint venture among several major energy companies:

Koch Capital Investments Co.	25.27%
HUTTS LLC	23.44%
Shell Pipeline Co. LP	16.12%
CITGO Pipeline Investment Co.	15.8%
Phillips Petroleum International Investment Co.	8.02%
Conoco Pipe Line Co.	8.53%
Marathon Oil Co.	2.82%

Even more interesting than the enormous volume of products delivered by Colonial is the terminal where Colonial and others store their products before distribution to consumers. Colonial and other common carriers terminate at Perth Amboy, NJ, in the New York harbor. This storage terminal contains a cluster of storage terminals

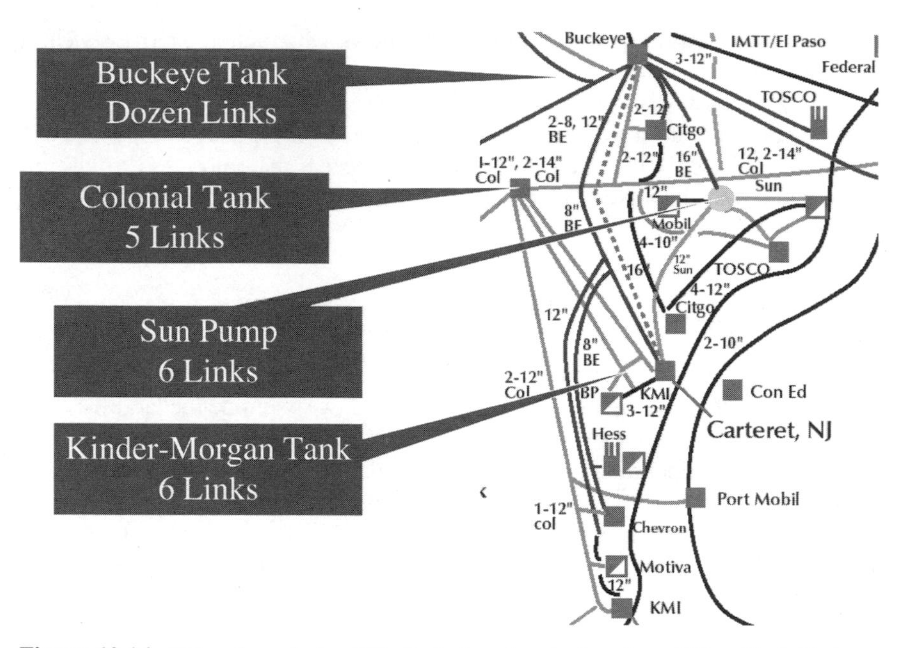

Figure 10.16. Network analysis of Perth Amboy terminal region of New York Harbor.

that are obvious targets. Although we are focused on the Linden Station in this case, note the numerous other terminals in Figures 10.16 and 10.17. There are hundreds of storage tanks located in close proximity to one another. This storage cluster concentrates vast storage tanks and hence is a high-value target.

Figure 10.16 summarizes a partial network analysis of this cluster.[19] Figure 10.17 is a closer view of the Linden Station area, which is typical of the clustering of storage and distribution points in this supply chain. Colonial products are stored near Linden Station and then distributed by Buckeye Pipeline, as shown in Figure 10.16. The implications for terrorism are obvious.

How realistic is this assessment? On January 2, 1990, a fatigue crack in an Exxon pipeline at Linden Station ruptured and spilled an estimated 567,000 gallons of fuel oil into the Arthur Kill waterway between New Jersey and Staten Island. The spill caused extensive environmental damage. Moreover, Exxon did not immediately detect the spill because operating staff had disabled the leak detection system, so pumping continued for 9 hours after the pipeline sprung the leak. Only when Exxon conducted a pressure test, pumping more oil into the water, did a Coast Guard team in a small boat see oil bubbling to the surface, and determined that the pipeline was the source. Exxon was ordered to shut down the pipeline.

The foregoing suggests that it is not only feasible for major oil incidents to happen at large terminals such as this, but they already have. Thus, the storage facility is yet another vulnerability in the supply chain. In fact, storage tank incidents are second

[19]A complete network analysis is left for the reader.

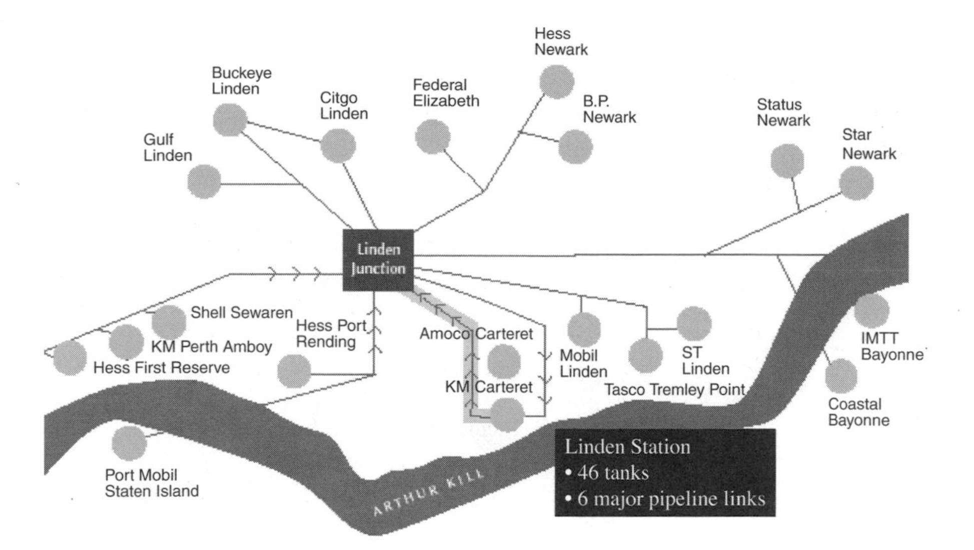

Figure 10.17. Close-up map of Linden Station area of terminal region.

only to pipeline leaks in terms of fault probabilities and loss of oil. In 1996, the second-most significant spillage (after pipelines) was caused by storage tanks.[20]

Oil Spills in 1996.

Incident	#Incidents	Gallons (Millions)
Tanker	5	0.3
Barges/Boats	12	1.3
Storage Tanks	**34**	**1.7**
Pipelines	40	3.3
Truck/Rail	3	0.06
Other	3	0.1

The bottom line on storage tanks: Critical nodes in the storage component of the supply chain are large-capacity clusters located in key transportation nodes such as Perth Amboy, NJ.

Figure 10.18 completes the fault tree for this sector. Note that loss of power and SCADA control rank lower as a threat than spills and disruptions of the supply chain. This is because most pipeline networks incorporate backup power and the damage—both in terms of duration and cost—caused by SCADA faults are minor compared with physical damage to refineries, pipelines, and storage terminals. SCADA is important but less critical because disruption of control would have less impact than destruction of the physical components of the supply chain.

[20]http://www.pipelinesafetyfoundation.org.

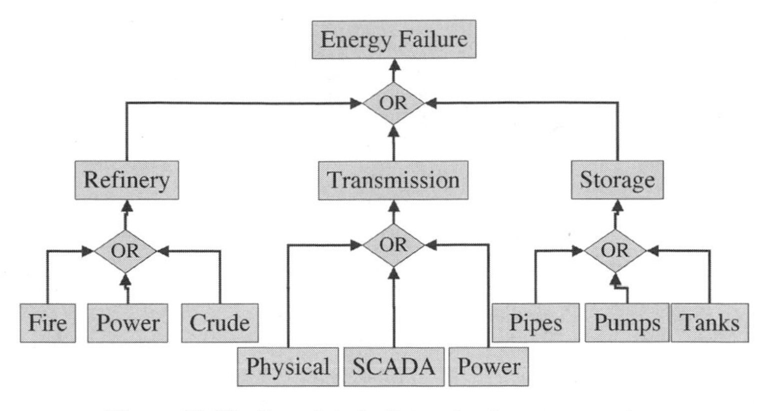

Figure 10.18. Complete fault tree for the energy sector.

ANALYSIS

The energy supply chain is vast and complex. It extends all the way from foreign wells in Saudi Arabia where 25% of the world's supply of crude oil is located to the homes and offices of every American. These vast oil and NG supply chain networks are probably small-world networks, although we have not attempted to prove this here. However, one thing is obvious: These supply chains are characterized by clustering of refineries, pipelines, and storage terminals.

Oil and NG supply chains are regulated much like the electric power grid and are undergoing radical transformation due to deregulation of the energy supply chain. The deregulation of pipelines, for example, means that products from competing companies must be allowed to travel over the same transmission pipeline. There is no economic incentive to build redundant pipelines. Existing pipelines are long and capable of hauling enormous quantities of energy. Thus, they are subject to the laws of *increasing returns*, which means there are only a few "hub carriers" like Colonial Pipeline, Transcontinental Pipeline, Dominion Pipeline, and Colombia Pipeline. Concentration of these "common carriers" translates into easy targets. We have no choice but to protect these assets, because building redundant capacity is not economically feasible. Thus, oil and NG supply chain networks will continue to be composed of clusters of refineries, transmission, and storage clusters. The question is, "how shall we harden these targets?"

This sector is also heavily interdependent with transportation and power. Historically, transportation of crude oil, refined oil products, and NG has shaped regulation as well as the business model for energy producers. Without access to markets far away from the wellhead, producers and refineries cannot get their product to market. Therefore, the regulatory power of gas and oil safety is vested in the Department of Transportation's OPS. But regulatory power is also vested in the DOE, EPA, and FERC.

When the world changed on September 11, 2001, the rules and regulations based on fair market competition and safety suddenly changed. But the regulating agencies have not. Today, responsibility for protecting the energy sector is shared with the Department of Homeland Security, Department of Transportation (OPS), DOE, FERC, and EPA. (Two months after September 11—December 2001—the DOE established the Office of Energy Assurance to work with the DHS. Other departments and agencies have established similar offices, leading to more fragmentation.)

The Federal Government generally recognizes the importance of the nation's pipelines, but how should this network be protected? The answer is not so clear, because the network is large and vast. It is not likely to be enhanced in any significant way for decades. According to the 2001 National Energy Policy:

> There are over two million miles of oil pipelines in the United States and they are the principal mode for transporting oil and petroleum products. Virtually all natural gas in the United States is moved via pipeline. Pipelines are less flexible than other forms of transport, because they are fixed assets that cannot easily be adjusted to changes in supply and demand. Once built, they are an efficient way to move products. A modest sized pipeline carries the equivalent of 720 tanker truckloads a day – the equivalent of a truckload leaving every two minutes, 24 hours a day, 7 days a week.[21]

How to protect this asset? The answer is vexing. One complicating factor is that the energy supply chain involves a wide array of technologies. Critical nodes in the supply chain are large-capacity clusters of refineries, pipelines, and storage tanks. To understand all of these clusters, one must understand environmental engineering, chemistry, fluid dynamics, SCADA hardware and software, mechanical engineering, and several other disciplines. This knowledge spans many governmental agencies. Lack of sector expertise is one of the fundamental challenges of critical infrastructure protection.

One can question the analysis made here by pointing to several weaknesses. The first critique might be that the pipelines and storage tanks are easily punctured, thus causing spills, but that these spills have historically been handled without causing a major loss of life. In fact, repairs seem to be routine and immediate. With a few notable exceptions, pipelines are repaired in a matter of hours or days.

The same can be said of energy SCADA. A modest investment in cyber-security and hardening of critical communication network operation centers can remedy weak SCADA systems. And the cost of this hardening is not huge. Are we making too much out of SCADA vulnerability in this sector?

So why be concerned? Historically, accidental incidents have taken only a few lives and cost a mere $33 million per year. Cunning terrorists have not perpetrated them. The question we should ask is, "given that it is rather easy to damage the energy supply chain, what damage might a clever and malicious attacker do?" It may be time to change strategies.

[21]http://www.whitehouse.gov/news/releases/2001/06/energyinit.html.

EXERCISES

1. Most energy consumed in the United States comes from:
   a. Natural gas
   b. Petroleum
   c. Coal
   d. Hydroelectric
   e. Wind and solar
2. Gas and oil pipeline safety is monitored by:
   a. FERC
   b. NERC
   c. EPA
   d. OPS
   e. DHS
3. Gas and oil supply chains are regulated by:
   a. FERC
   b. NERC
   c. EPA
   d. OPS
   e. DHS
4. The largest transmission pipeline, and largest privately financed project in the United States at the time, is:
   a. Colonial Pipeline
   b. Transco Pipeline
   c. Kinder-Morgan Pipeline
   d. East Texas Gas Pipeline
   e. Dominion Pipeline
5. Most pipelines are monitored and controlled by:
   a. OPS
   b. SCADA
   c. E-ISAC
   d. FERC
   e. DHS
6. How many PADDs are there in the energy supply chain of the United States?
   a. 5
   b. 2
   c. 3
   d. 1
   e. 50—one per state

7. In the United States, the top 10 refineries produce:
   a. Nearly all of the gas and oil.
   b. 20% of all petroleum products.
   c. 5% of all petroleum products.
   d. Over half of all petroleum products.
   e. Two thirds of all petroleum products.
8. To carry out its role as energy regulator, the FERC must work with:
   a. OPS
   b. EPA
   c. DOE
   d. E-ISAC
   e. All of the above
9. The energy sector was deregulated by:
   a. Energy Policy Act of 1992
   b. PURPA in 1978
   c. HLPSA of 1979
   d. FERC Order 888 in 1988
   e. DHS in 2003
10. Who is authorized to regulate local gas and oil distribution?
    a. FERC
    b. NERC
    c. State authorities
    d. DOE
    e. DHS
11. The most critical refinery node in the Northeastern petroleum supply chain is:
    a. Lake Charles, LA
    b. Baton Rouge, LA
    c. Wood River, IL
    d. Galveston Bay, TX
    e. Philadelphia, PA
12. The most critical node in the Southern California petroleum supply chain is:
    a. SoCal Pipeline
    b. Kinder-Morgan Line
    c. East Texas Pipeline
    d. Watson Terminal
    e. Phoenix, AZ

13. The largest NG transmission pipeline for the PADD1 area (Northeastern United States) is:
    a. Transco
    b. Columbia
    c. Dominion
    d. Tennessee Pipeline
    e. Colonial Pipeline
14. The largest petroleum (gasoline, kerosene, home heating oil) pipeline in the United States is:
    a. Transco
    b. Columbia
    c. Dominion
    d. Tennessee Pipeline
    e. Colonial Pipeline
15. The largest civil penalty levied against any company, up to that time, for pipeline leakage was against:
    a. Colonial
    b. Transco
    c. Columbia
    d. Tennessee
    e. Kinder-Morgan
16. The most common cause of pipeline faults in recent years has been:
    a. Terrorism
    b. Accidental excavation
    c. Corrosion
    d. SCADA failure
    e. Pump failure/power outage
17. The most common cause of oil spills from the petroleum supply chain is:
    a. Oil tankers
    b. Truck and rail incidents
    c. Storage and pipeline incidents
    d. The Exxon Valdez incident
    e. The Three-Mile Island incident
18. What percentage of the U.S. NG supply comes from foreign NG sources?
    a. 15%
    b. 90%
    c. 50%
    d. 0%
    e. 100%

19. Why is an attack on SCADA not considered a major threat?
    a. SCADA faults have been minor compared with physical damages, thus far.
    b. The energy sector uses very little SCADA.
    c. SCADA does not cost much to harden.
    d. Few operators understand SCADA.
    e. Nobody has been killed by SCADA attacks.
20. Why is the Kinder-Morgan pipeline most vulnerable to the triple event outcome?
    a. Watson is the source or entry point to the pipeline.
    b. Fault probabilities are all the same—80%.
    c. Fault probabilities are high; thus, their product is also high.
    d. This is an artifact of OR-tree fault trees.
    e. Apportioned risk reduction was used to allocate $100 million.
21. Perform a network-wide allocation experiment on the Kinder-Morgan pipeline (Figure 10.11) network assuming all node and link damages are 50, except the Watson node, which is 100, and a budget of 150 availability points. What is the allocation that minimizes network-wide risk?
    a. 66% of the points go to Watson; 34% to Colton.
    b. 66% go to Watson; 17% to Colton; 17% to Niland.
    c. 66% go to Watson; 10% to Colton; 24% to Niland.
    d. 66% go to Watson; 24% to Colton; 10% to Niland.
    e. All of the above.
22. (Discussion) Is the oil or NG transmission system a scale-free network? A small-world network? How can you determine this?
23. (Discussion) Is OPS, instead of DHS, responsible for counter-terrorism against pipeline systems? What is the role of OPS relative to terrorism?
24. (Discussion) How is the gas and oil supply chain like the power grid? How is it different?
25. (Discussion) Are oil refineries overly concentrated? Why? What is the alternative?
26. (Discussion) What kind of weapon would it take to disable a major storage terminal for 30 days or more?
27. (Discussion) Why are pipeline leaks—and attacks on pipelines—not typically considered major threats?

CHAPTER 11

Telecommunications

Telecommunications embraces all forms of electronic communications—communication satellite networks, landline telephone communications, and wireless cellular communications involving voice, data, and Internet. However, this chapter does not discuss broadcast communications such as radio, television, geographical position systems (GPS) navigation, and LORAN. Rather, this chapter focuses on the primary means of two-way personal communications: telephones and the networks that underlie telephony.

This chapter describes how the telecommunications sector is structured, how it works, what its vulnerabilities might be, and what strategies apply to prevention of disruption of the service. We discuss the following concepts:

1. Like the power and energy sectors, the telecommunications sector is shaped by its transition from vertical monopoly to re-regulated competitive oligopoly, which has weakened the sector and made it economically vulnerable.
2. Unlike the power and energy sectors, there is no clear-cut unity of the government's roles and responsibilities in the telecommunications sector: It is regulated by at least three agencies: the Federal Communications Commission (FCC's) Network Reliability and Interoperability Council (NRIC), Department of Commerce's National Telecommunications and Information Administration (NTIA), and Department of Homeland Security's (DHS's) National Communications System (NCS). In addition, the President's National Security Telecommunications Advisory Committee (NSTAC) has influence on the sector through its direct link to the Executive Branch of the Federal Government.
3. The three major telecommunication network infrastructure components are landlines, cellular, and extraterrestrial networks (communication satellites). These three have unique vulnerabilities as well as providing strength to the sector because they provide a level of redundancy.
4. Critical nodes are clustered around major metropolitan areas linked by the top 30 landline routes: Chicago, Atlanta, San Francisco, Dallas-Fort Worth,

Critical Infrastructure Protection in Homeland Security: Defending a Networked Nation, edited by Ted G. Lewis

Washington-Baltimore, New York, and Los Angeles. These are the most vulnerable geographical nodes in the sector.

5. The telecommunications sector is somewhat redundant (and therefore more available) because it consists of three overlapping systems: landlines, cellular, and extraterrestrial. But these three are also interdependent because they are connected through a system of gateways. Thus, a failure in one may lead to unexpected consequences in another.
6. The most critical nodes within the most critical geographical areas are (a) telecom hotels, (b) points of presence (POPS) and gateways, and (c) land earth stations (LES) that link communication satellites to terrestrial communication networks.
7. The most credible threats are cyber-attacks and high power microwave (HPM) attacks, followed by physical attacks on critical nodes: telecom hotels, POPS, and LESs. Telecom hotels are by far the most vulnerable components of the sector.
8. The national strategy for protecting the telecommunications sector must focus resources on protecting telecom hotels, and critical gateways and links and address asymmetric threats such as those posed by cyber-and HPM attacks. These critical nodes are where we should invest in target hardening.

ORIGINS

The modern age of communications is rapidly transitioning from analog (information is encoded in a continuous signal) to digital (information is encoded as a stream of digits—zeros and ones). As a consequence, we think of analog as an old technology and digital as a new technology. But digital communications is far older than analog. The telegraph machine (1837–1873) was the first *digital* communication system because it represented information as a series of digits just as modern digital systems do. Digital telegraphy was such a huge success that Western Union became the first telecommunications monopoly by 1866.

Telegraphy had one major drawback: It required a trained operator to translate the digital data into words and the reverse—words into digital code. This limited its usefulness as a consumer product. What people really wanted was a talking telegraph machine.

Sound is analog. Sound waves travel through the air as a continuous wave form. Thus, it seems only logical that a talking telegraph should encode sound (voice) as an analog signal—a continuous wave form. If only the energy of sound could be converted into electrical energy, transmitted as an electrical analog signal, and then converted back to analog sound at the other end, the telegraph could talk. Thus, the idea of a telephone was born.

Alexander Graham Bell demonstrated the first operating telephone in 1876. Bell combined his knowledge of speech therapy with recent theories of electricity postulated by his contemporaries to create the first telephony device to win a U.S. patent.

> Alexander Graham Bell (1847–1922) was a contemporary of James Clerk Maxwell, and, like Maxwell, was born in Scotland and educated in England. But while Maxwell was primarily a mathematical theoretician, Bell liked to make things. For two generations, his family had been leading authorities in elocution and speech, and, after the Bells moved to America in 1870, Graham Bell himself set up his own school in Boston for training teachers of the deaf, and then he became a professor of speech and vocal physiology at Boston University, where he specialized in teaching deaf-mutes to talk.
>
> In an age when the likes of Michael Faraday and Maxwell were beginning to plumb the secrets of an electronic universe, it wasn't much of a stretch for Bell to move from the mechanics of speech into the electrics of speech. As a speech therapist, he was, of course, dealing with waves, sound waves, and their reception. These waves did not travel very far, or very fast. But, he reasoned, if he could convert the sound of human speech into electrical oscillations, those sounds could be sent over wires, and then converted back into sound waves at the receiving end.
>
> He theorized that if he could find a way to make an electric current vary in intensity precisely as air varies in density during the production of sound, he could transmit speech. He had moved from speech therapy into one of the great human inventions of all time, one that would bring people together and shrink the planet itself.[1]

While on his honeymoon in England, Bell demonstrated his invention to Queen Victoria and received an order to put in a private line between Osborne House, on the Isle of Wight, and Buckingham Palace. By 1878, there were 5600 telephones in the United States. By 1882, there were 109,000, and by his death in 1922, there were 14 million telephones in the United States. Bell's patents expired in 1894, but by 1897, he had moved on to the study of aeronautics.

Bell, his father-in-law Gardiner Hubbard, along with Thomas Sanders formed Bell Telephone Company in 1877. They established their first telephone exchange network in New Haven, CT (21 telephones and 8 lines), and began expanding it outward—initially to Chicago, and eventually to San Francisco by 1915. Growth was rapid because Bell Telephone licensed its patents to others, thus attracting investments in local exchanges and "telephone companies." Soon, Bell Telephone Company became American Bell Telephone Company, suggesting it had become a national enterprise.

Licensing revenue allowed American Bell to buy controlling interest in Western Electric in 1882. Western Electric put American Bell into the equipment manufacturing business. Licensing and equipment manufacturing soon led to network system building. American Telephone and Telegraph Company (AT&T) was incorporated as a subsidiary of American Bell Company in 1885 for the sole purpose of building long-distance networks. In 1899, AT&T reorganized as an intellectual property (IP) holding company.

[1]http://www.acmi.net.au/AIC/BELL_BIO.html.

From 1898 to 1924, the telecommunications industry was engaged in a "telecommunications war" because of competition and rapid technological change in the industry. For example, the automated exchange and self-dial telephone invented by Almon Strowger made Bell's equipment obsolete. Thus, competition from upstarts fragmented the industry. By 1903, there were 2 million telephones from independent companies versus 1,278,000 from Bell. In addition, Bell Telephone had developed a reputation for high prices and poor service.

AT&T fell on hard times as the Bell System faltered and bankers began to take over. Floundering under Graham Bell's leadership and pressures from independents, J. P. Morgan gained control of the company and installed Theodore Vail as president in 1907. The rescue of the Bell system also marked the beginning of its downfall as an unregulated company, because Morgan's monopolistic consolidation of the independents would lead to the regulation of AT&T by Congress:

> In 1907 Theodore Vail returned to the AT&T as president, pressured by none other than J.P. Morgan himself, who had gained financial control of the Bell System. A true robber baron, Morgan thought he could turn the Bell System into America's only telephone company. To that end he began buying independents by the dozen, adding them to Bell's existing regional telephone companies. AT&T management re-organized the regional holding companies in 1911, a structure that held up over the next seventy years. But Morgan wasn't finished yet. He also worked on buying all of Western Union, acquiring 30% of its stock in 1909, culminating that action by installing Vail as its president. For his part, Vail thought telephone service was a natural monopoly, much as gas or electric service. But he also knew times were changing and that the present system couldn't continue.[2]

Under Vail's leadership and Morgan's pressure to consolidate, AT&T became a vertically integrated monopoly by 1911. Soon, the Department of Justice sued AT&T in 1913, claiming it had violated the Sherman Anti-trust Act. The lawsuit resulted in restricting, but not stopping, AT&T. After 1913, the Kingsbury Commitment stopped AT&T from buying independents without Department of Justice permission, required AT&T to interoperate with independents, and forced AT&T to divest itself of Western Electric. But by 1924, AT&T owned 223 of 234 independents! Vail and Morgan believed in monopolies:

> For much of its history, AT&T and its Bell System functioned as a legally sanctioned, regulated monopoly. The fundamental principle, formulated by AT&T president Theodore Vail in 1907, was that the telephone by the nature of its technology would operate most efficiently as a monopoly providing universal service. Vail wrote in that year's AT&T Annual Report that government regulation, 'provided it is independent, intelligent, considerate, thorough and just,' was an appropriate and acceptable substitute for the competitive marketplace.
>
> The United States government accepted this principle, initially in a 1913 agreement known as the Kingsbury Commitment. As part of this agreement, AT&T agreed to

[2]http://www-t.zhwin.ch/it/su/Block3/TelephoneHistory/history2.htm.

> connect non-competing independent telephone companies to its network and divest its controlling interest in Western Union telegraph. At several later points, as political philosophy evolved, federal administrations investigated the telephone monopoly in light of general antitrust law and alleged company abuses. One notable result was an anti-trust suit filed in 1949, which led in 1956 to a consent decree signed by AT&T and Department of Justice, and filed in court, whereby AT&T agreed to restrict its activities to the regulated business of the national telephone system and government work.[3]

Remnants of the early telecommunications wars remain today. Local exchanges operate in restricted regions called local access transport areas (LATAs), and the local telephone company is called a local exchange carrier (LEC). Before re-regulation of the industry in 1996, it was illegal for LECs to cross LATAs without permission from the FCC. This hampered adoption of new technology because LECs were monopolies within their LATAs, and there was only one long-distance company—AT&T.

Today, the old Bell system companies are called competitive local exchange carriers (CLECs), and the long-distance companies—including the Bell Long Lines system—are called interexchange carriers (IECs). Thus, we have a system that is shaped by years of regulation and natural monopoly.

Telecommunications was declared a natural monopoly from 1934 to 1996. But the industry did not stand still for 32 years. Rather, it went through a long period of divestiture. This long and winding road began with the Telecommunications Act of 1934. Congress asserted its control over broadcast and telecommunications companies and established the FCC as regulator of airwaves and all things having to do with telecommunications. It declared the electromagnetic spectrum public, not private, property. Electromagnetic spectrum governs radio and television broadcasts. It governs digital and analog forms of communication.[4] For example, commercial broadcasters must obtain licenses from the U.S. Government, and the government requires that broadcasters serve the public interest by donating a certain amount of broadcast time to public announcements. The FCC regulates telephone services, both wired landlines and wireless cellular networks, as well.

One of the hallmarks of this period was the achievement of *universal access*. Operating as a regulated monopoly, AT&T was able to serve 99% of the population regardless of where people lived. Rural as well as densely settled metropolitan areas received telephone service under the 1934 law.

But in 1974, the Department of Justice began taking a long series of steps leading to divestiture and re-regulation of the natural monopoly set up by the 1934 law. A long drawn out lawsuit from 1974 to 1984 led to the breakup of AT&T in 1984. In a profound decision, the 22 wholly owned Bell Operating Companies were separated from AT&T. The resulting 7 regional "Baby Bells" became CLECs and no longer operated as protected monopolies.

[3] http://www.att.com/history/history3.html.

[4] It may seem strange then that the Federal Government does NOT regulate the Internet, which is digital communication technology in the most precise definition of the phrase!

The next major step in divestiture came in 1996 with the Telecommunications Act of 1996. This law replaced the 1934 law. Its impact is still rippling through the industry today.

The 1996 law re-regulates the industry by forcing the IECs to rent their networks to LECs and vice versa. It unbundled services provided by LECs, long-distance carriers, and cable TV operators. For example, it requires line sharing, especially digital subscriber line or broadband (xDSL), across all carriers. The idea is to open up long distance to local telephone companies, and conversely to allow long-distance companies access to the "last mile," i.e., the homes and offices of consumers. But it still limits ownership of cable TV, TV, and radio stations to specific percentages in each region, and it sets pricing on some services.

IECs (the long-distance carriers) are now encouraged to interoperate. AT&T can use the lines of Level3, and Level3 can use the lines of AT&T. This is called *peering* in the industry and has led to the downfall of companies such as WorldCom and Global Crossing, because of improper accounting for peering charges.

The LECs were quick to react to the 1996 legislation. The FCC attempted to set prices at the local level as well as at the long-distance level. But the LECs won a court order that forced the FCC out of the local market. As a consequence, states can regulate prices, not the FCC. Today, the LECs are free to establish peering charges except where competitors cannot agree. In case of disagreement, states have the right to set pricing. So, today the FCC sets the wholesale price of long-distance service, but the peering fees charged by the local carrier are allowed to float within the limits of state regulation. This has resulted in instability in the industry not unlike the problem of wheeling in the electric power sector. Both sectors have become economically vulnerable because of this economic imbalance.

Because of the Telecommunications Act of 1996, the telecommunications sector is shaped by its transition from vertical monopoly to re-regulated competitive oligopoly, which has weakened the sector and made it economically vulnerable. Long-distance service has become a commodity and so lacks pricing power. There is a surplus of capacity because of extremes during the dot-com bubble of 1995–2000. On the other hand, local carriers still own the consumer, and their victory over the FCC has given them some pricing power. But LECs are undercapitalized and the transition from analog to digital is costing more than most can afford.

Currently, all players depend on long lines to make their service competitive, and local carriers "own" the consumer. But, local carriers must consolidate to gain scale, to support marginal operating costs that decline as the number of customers increase. The telecommunications industry is one of the original "increasing returns" industries; the larger they get, the more profitable they get, which means they get even larger. This leads to an urge to merge, which will once again result in a very small number of competitors.

The re-regulated telecommunications sector is yet another example of the Tragedy of the Commons, as described earlier. Recall the story: A community of herders sharing a grazing pasture for their herd of cattle soon depletes the pasture, because no single herder is responsible for the pasture. In fact, herders are encouraged to overpopulate the pasture because they pay for only a small increment of pasture, but benefit from the sale of a whole cow. The tragedy of telecommunications—and other re-regulated industries—is that many carriers in telecommunications depend on the old AT&T Long Lines (or pieces of it), but there is little incentive to maintain it. In other words, the national telecommunications sector is vulnerable to faults in the backbone of the network.

REGULATORY STRUCTURE

The first critical infrastructure legislation in U.S. history was prompted by the 1962 Cuban Missile Crisis that resulted in the creation of the NCS. Negotiations between President Kennedy and Premier Khrushchev were threatened because of telephone system "call completion" problems. It was not possible for the two leaders to simply pick up the telephone and place a call to anywhere in the world like it is today. In fact, Khrushchev was forced to use Radio Moscow to communicate, indirectly, with Kennedy.

> During this time, ineffective communications were hampering the efforts of the leaders to reach a compromise. Without the ability to share critical information with each other using fax, e-mail, or secure telephones such as we have today, Premier Khrushchev and President Kennedy negotiated through letters. Generally, Washington and Moscow cabled these letters via their embassies. As the crisis continued, hours passed between the time one world leader wrote a letter and the other received it. Tensions heightened. On October 27 and 28, when communications became urgent, Premier Khrushchev bypassed the standard communication channels and broadcast his letters over Radio Moscow.[5]

The so-called "hotline" established after the crisis was resolved was initially a Teletype set up in August 1963. Kennedy also established the NCS by executive order. Now part of the DHS, the NCS is one of several agencies overseeing the telecommunications infrastructure.

> Following the crisis, President Kennedy, acting on a National Security Council recommendation, signed a Presidential memorandum establishing the NCS. The new system's objective was "to provide necessary communications for the Federal Government under all conditions ranging from a normal situation to national emergencies and international crises, including nuclear attack.
>
> At its inception on August 21, 1963, the NCS was a planning forum composed of six Federal agencies. Thirty-five years later, it is a vital institution comprising 23 member organizations that ensure NS/EP [National Security/Emergency Preparedness] telecommunications across a wide spectrum of crises and emergencies. Today, the NCS structure includes the Secretary of Defense as the Executive Agent; the Manager, NCS, a Committee of Principals, composed of representatives from each of the 23

[5]http://www.ncs.gov/n5_hp/Customer_Service/Brochures/35Anniversary/begina.htm.

member organizations; and the Office of the Manager, NCS (OMNCS). The Deputy Manager oversees the daily operations of the OMNCS.[6]

In 1978 Presidential executive order EO 12046 consolidated two other telecommunications agencies into the NTIA. The Department of Commerce's NTIA was the result of combining the Whitehouse Office of Telecommunications Policy (OTP) with Commerce's Office of Telecommunications. The NTIA's principle role has been to sell spectrum to telephone, radio, and TV companies. But its involvement in telecommunications has sometimes extended beyond marketing of airwaves.

For example, in 1998–1999, it played a major role in commercialization of the Internet. A third big step in governmental oversight was taken in 1982 when President Reagan issued executive order EO 12382. This order established another watchdog organization that reported directly to the President, the NSTAC. NSTAC members are senior management (CEOs and Senior Vice Presidents) of telecom companies. Their job is to advise the President on matters of telecommunications security.

Senior executives are not the best people to get detailed work done, so the Committee of Principles (COP) and then the Committee of Representatives (COR) were formed to do detailed work. Today, much of responsibility of the NSTAC has been delegated to two subcommittees.

PDD 63 (1998) designated the U.S. Department of Commerce as the lead agency and the NTIA as the sector liaisons official for the information and telecommunications sector.

By 2000, the government had plenty of advice concerning telecommunications issues. The NCS was responsible for making sure the system worked, the NTIA regulated the airwaves, and the NSTAC advised the President. So before the consolidation of agencies that formed the DHS in February 2003, the regulatory structure of the telecommunications sector was distributed across at least three major influential organizations: the NCS, NTIA, and NSTAC (see Figure 11.1).

Meanwhile, the FCC was concerned with the Y2K problem (turnover of the millennium calendar that threatened to render computers and communications equipment inoperable). This prompted the FCC to temporarily create the NRIC in 1993. It was dismantled after the Y2K threat subsided, but then the FCC rechartered the NRIC in 2002, on the heels of 9/11. The NRIC seems to duplicate the roles of the NCS and the NCS-ISAC. In fact, a series of reports issued by the NRIC in 2002 remain the authoritative source of recommendations on how to secure the telecommunications infrastructure. Thus, after the formation of the DHS, the government regulatory structure seems to be even more distributed than before [see Figure 11.1(b)]. There are no less than four agencies overseeing telecommunications: FCC/NRIC, Commerce/NTIA, DHS/NCS, and NSTAC.

[6]Ibid.

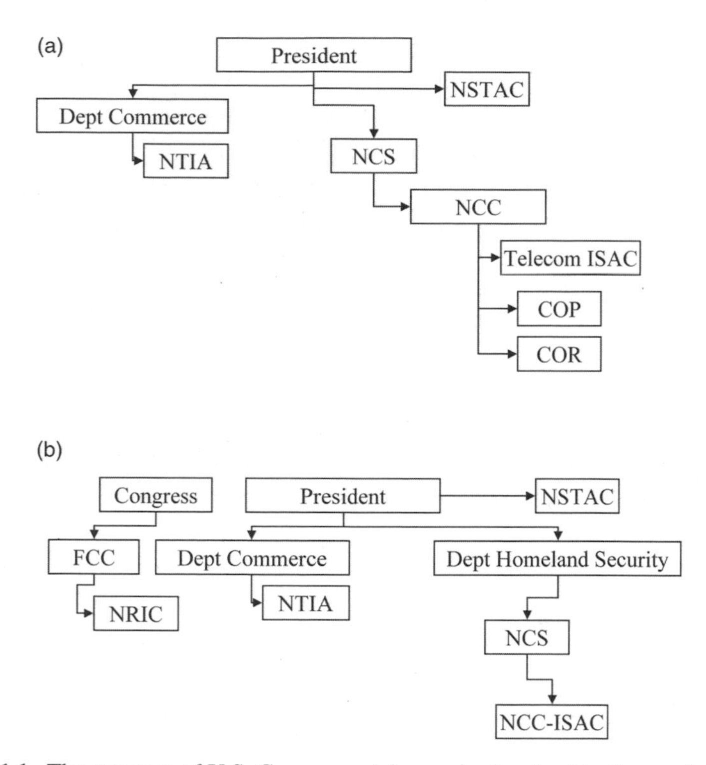

Figure 11.1. The structure of U.S. Governmental agencies involved in the regulation of the telecommunications and information sector: (a) before and (b) after the formation of the DHS in 2003.

DHS, NCS, NCC, AND NCC-ISAC

The NCS became part of the DHS in 2003. It is tucked under the Critical Infrastructure Protection Division. Within the NCS lies the National Coordinating Center (NCC), and within the NCC lies NCC-ISAC—the Information Sharing and Analysis Center for the telecommunications and information sector.

The NCS/NCC/NCC-ISAC membership consists of major telecommunications companies. In 2003, these were AT&T, Cisco Systems, Computer Sciences Corporation, COMSAT Corporation, EDS, ITT Industries, the National Telecommunications Alliance, Nortel Networks, Science Applications International Corporation, Sprint, United States Telecommunications Association, Verizon, and MCI/WorldCom. Contrast this list with the members of the NRIC: AT&T, Microsoft, Nokia, Nortel, Qwest, MCI/WorldCom, Motorola, Alcatel, Sprint, Verizon, Lockheed Martin, Boeing, AOL-Time Warner, Earthlink, Level3, Bellsouth, DHS, NCS, Hughes, Intelsat, Communication Workers of America, Comcast, Cox Communications, Cingular, and Cable & Wireless.

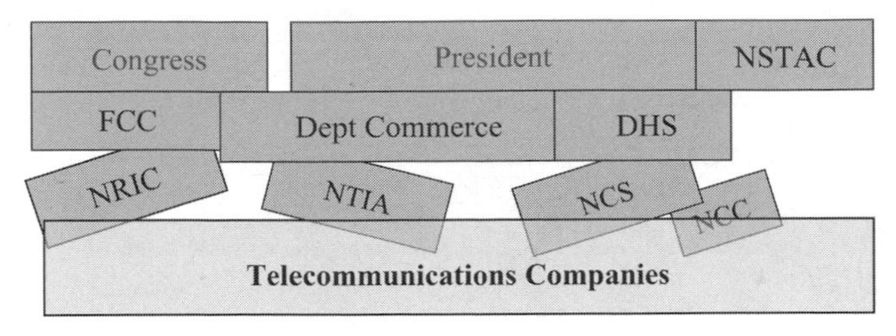

Figure 11.2. Pictorial illustration of the regulatory structure of the telecommunications and information sector. Command and control is diffused even after the consolidation of agencies following the creation of the DHS in 2003.

The major product of the DHS/NCS/NCC/NCC-ISAC seems to be a highly secretive Critical Infrastructure Warning Information Network (CWIN) that is designed for broadcasting alerts as well as information sharing:

> The mission of the CWIN is to facilitate immediate alert, notification, sharing and collaboration of critical infrastructure and cyber information within and between Government and industry partners.[7]

One interesting feature of CWIN's 24 × 7 alert and notification network is that it does *not* depend on existing telecommunications infrastructure. It has no logical dependency on the Internet or the Public Switched Network (PSN). It is "out-of-band," meaning it runs over a redundant network. Details of CWIN are classified, so little more is known about it.

Figure 11.2 represents the state of the regulatory structure of telecommunications in 2004. It is a picture of fuzzy command and control at best, requiring specialized knowledge of the industry to navigate. Although this is likely a transitory situation, the ability of such a structure to respond to changes in a timely fashion is perhaps limited and cumbersome at best. A more streamlined structure will be needed in the future.

The dot-com bubble, corporate malfeasance among the ranks of its executives, and re-regulation of its vertical monopolies have ravaged the telecommunications industry. Although the effects of 9/11 are still rippling through the country, the telecommunications industry has serious economic vulnerabilities of its own. Unconstrained retail pricing combined with constrained wholesale pricing results in lower pricing power. This in turn results in a negative incentive to improve the infrastructure that was protected for so many decades by a natural monopoly.

The Telecommunications Act of 1996 has created competitive local exchanges, but their ability to set prices is controlled by the incumbents, which inhibits true competition. In addition, establishment of a national footprint requires crossing boundaries—local-to-long-distance, and long-distance-to-local—which means local exchanges must pay to establish end-to-end connectivity to its customers.

[7]http://www.ncs.gov/cwin/.

Wireless adds to the list of competitors, because they threaten to disrupt the wired local loop business model, altogether.

The bottom line: The old monopolistic long-distance business has been commoditized, there is a surplus of capacity, and little profit to be had. At the same time, the long-distance "Bell system" is the backbone of the country's telecommunications network. The system will not work without it, and yet who will invest in its future? Economic vulnerability is the obvious consequence.

ARCHITECTURE OF THE NETWORK

Against this background lies a huge infrastructure undergoing massive technological change. How does it work, and how do we derive prevention policies from these mechanisms? First, we must understand the basic terminology and architecture of the telecommunications infrastructure.

Plain Old Telephone Service (POTS) is capable of transmitting 64 kbps of digital data over copper wires. This is the basic circuit of the telephone network, designated Digital Service Zero (DS0). A POTS call requires 8 kbps of control, so computer users get 56 kbps of data when they dial up a POTS line and use a modem to connect their computer to the Internet.

Circuits are combined to create more capacity. For example, a DS1, also known as a T1 line, is 24 DS0 lines working as one, and it yields 1.536 Mbps of data and 8 kbps of control. Therefore, a T1 line transmits 1.544 Mbps of data and control information.[8] Similarly, a T3 (also known as a DS3) line is 28 T1 circuits plus control bits, yielding 44.736 Mbps overall.

Capacity goes up by combining circuits or changing technology. An optical fiber cable (OC) transmits more information than a copper cable, so these lines are designated as OC-1 (51 Mbps), OC-3 (155 Mbps), OC-12 (622 Mbps), and so on. An ordinary cable TV coaxial cable can transmit from 3 to 10 Mbps.

Wireless transmission is governed by yet another technology—radio. Various bands (frequency ranges in the electromagnetic spectrum) have been set aside for cellular, satellite, and local area networks. For example, radios that connect earth orbiting satellites to ground stations operate in ranges similar to POTS. The so-called 3 G cellular wireless networks are expected to operate up to 2 Mbps. Wi-Fi networks that link together personal computers over short ranges are currently operating in the range of 54 Mbps. Each technology has its advantages and disadvantages, which is why they coexist in the marketplace.

Figure 11.3 illustrates how the three major technologies—landline, cellular, and extraterrestrial (communication satellite)—interoperate in one large network. For historical and regulatory reasons, these technologies have evolved along three subdivisions: LECs, IECs, and wireless carriers. LECs consist of the divested baby

[8]Here is how: $24 \times 64{,}000 = 1{,}536{,}000$, which is 1.536 Mbps. Add $8000 + 1{,}536{,}000$ to get a T1 line with capacity of 1.544 Mbps. A similar multiple is used to compute the capacity of a T3 line, as well: $28 \times 1{,}536{,}000 + 27 \times 64{,}000 = 43.008$ Mbps + 1.728 Mbps = 44.736 Mbps.

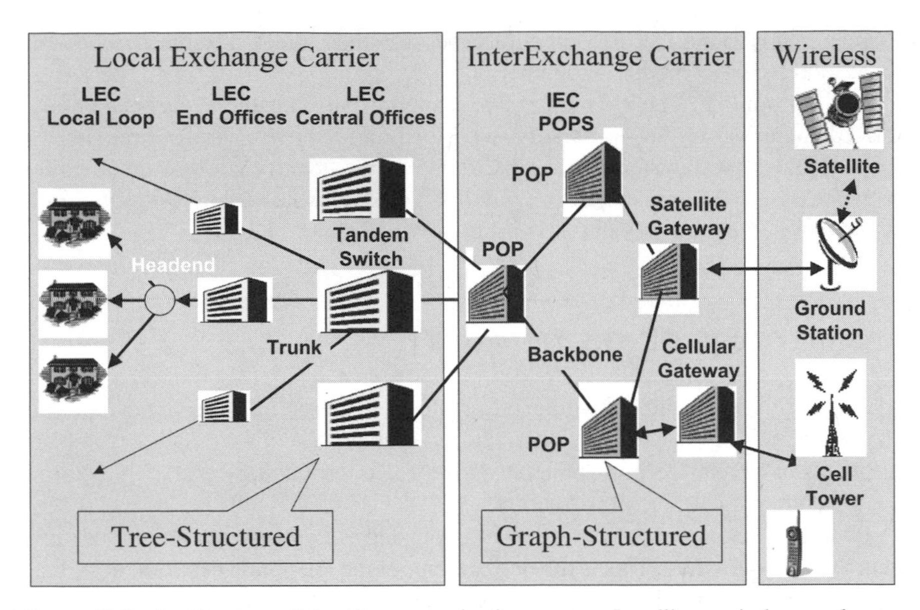

Figure 11.3. Architecture of the telecommunications sector: Landlines, wireless, and extraterrestrial transmissions interoperate across LEC, IEC, and wireless boundaries.

bells, often called CLECs, as well as new challengers. The IECs consist of the old Bell System plus new entrants into the long-distance market. The wireless component shown in Figure 11.3 consists of cellular carriers and communication satellite carriers.

Figure 11.3 is a simplification of the overall network architecture. It leaves out details such as microwave relay stations, laser links, and so forth. But it is sufficiently detailed for us to come to some conclusions about what leads to optimal security strategies. Once we have established a basic understanding of the telecommunications network, we will revisit the question of vulnerability analysis using Figure 11.3.

The major functional purpose of the IECs is to provide long-distance connectivity and to connect the LECs together into one national network. This is done by providing POPS, network access points (NAPS) for Internet users, and gateways for integrating satellite and cellular networks into the "backbone," as shown in Figure 11.3.

LECs provide local loop service. They connect to the backbone through POPS that switch calls to their central offices. In turn, central offices funnel calls to end offices, which in turn channel the call to the local loop. Local loops connect lines to consumers through a neighborhood switch known as a "headend." Headends handle approximately 1000 users at a time. Thus, LEC networks are shaped like a hierarchical tree, whereas the IEC networks are arbitrary graphs, not necessarily a tree or grid. We will come back to this question later in the section on vulnerability analysis and ask, "Is the IEC network scale-free?" "Is it another example of a small-world network?" The answer to these questions will reveal vulnerabilities in the sector.

WIRELESS NETWORKS

Wireless transmission encompasses three major technologies: communication satellites, cellular, and Wi-Fi, which is formally the IEEE 802.11 standard. Satellite communications has global coverage, but it is relatively low speed and expensive. Cellular is relatively low cost, but it lacks complete coverage. Wi-Fi is inexpensive, fast, and compatible with computers and the Internet, but it has a range of less than 100 m. No single technology serves all purposes. Instead, these technologies are integrated into the larger, IEC backbone as shown in Figure 11.3.

Extraterrestrial Communication

There are over 3500 communication satellites in use today. Although the public is mostly unaware of their presence in the national telecommunications infrastructure, they play a critical role in voice and data communications, broadcast television transmission, military surveillance and imaging, intelligence gathering, early warning systems, maritime and aeronautic navigation with GPS, weather forecasting, inspection of agricultural lands, rescue and disaster relief, oceanographic and natural resource observations, and so on.

Datacomm Research estimates that satellite operators earned revenues of \$2.3 billion—\$1.4 billion for moving data and \$900 million for providing voice services—in 2004.[9] Although this is small compared with the entire telecommunications industry, it suggests that extraterrestrial communications is a viable part of the overall sector. A brief timeline history of the industry is given in Table 11.1.

The idea of communication satellites circling the earth originated with science fiction writer Arthur C. Clarke in 1945.[10] Clarke was way ahead of his time. His article described how a rocket circling the earth at 22,300 miles above the equator would hover above the same land area, because such a rocket would circle the earth at the same speed as the earth rotates. He recommended three geosynchronous rockets be stationed above the earth at 120 degrees apart so together they would cover all of the earth's surface. Clarke invented the geosynchronous earth orbit (GEO) satellite, which was actually constructed and put into orbit 20 years later.

Today there are three kinds of communication satellites: LEO, MEO, and GEO. GEO is the oldest, followed by MEO and LEO networks. Each has its advantages and disadvantages in terms of latency (time delay due to the time it takes a radio signal to make a roundtrip from earth to satellite and back), bandwidth (the

[9]Datacomm Research Company, "Satellites in Cyberspace," http://www.researchandmarkets.com/index.asp.
[10]*Wireless World*, pp. 305–308, October 1945.

TABLE 11.1. A Selective Communications Satellite Chronology.[11]

1945 Arthur C. Clarke article: "Extra-Terrestrial Relays"
1955 John R. Pierce article: "Orbital Radio Relays"
1957 Sputnik: Russia launches the first earth satellite
1960 AT & T applies to the FCC for experimental satellite communications license
1961 Formal start of TELSTAR, RELAY, and SYNCOM Programs
1962 TELSTAR and RELAY launched (MEO)
1962 Communications Satellite Act (U.S.)
1963 SYNCOM launched
1964 INTELSAT formed
1965 COMSAT's EARLY BIRD: first commercial communications satellite
1969 INTELSAT-III series provides global coverage
1972 ANIK: first domestic communications satellite (Canada)
1974 WESTAR: first U.S. domestic communications satellite
1975 RCA SATCOM: first operational body-stabilized satellite
1976 MARISAT: first mobile communications satellite
1979 INMARSAT formed
1988 TAT-8: first fiber-optic trans-Atlantic telephone cable

transmission speed), coverage (how much of the earth's surface is served by one satellite), power (how much power it takes to send and receive the radio signal, and hence the size and weight of the handsets), and cost (how many satellites, how heavy, and how powerful). In simple terms, the further away a satellite is, the more surface it covers, but also the more power and size it takes to send and receive messages.

GEO satellites circle the earth at 22,300 miles, which exactly matches the rotational speed of the earth while giving the satellite enough centripetal force to offset gravity. Hence, they hover over the same location all the time, which also gives them large coverage. There are about 200 GEOs, which is the maximum amount possible, because they have to be separated by two degrees of arc to keep from interfering with one another. What happens when someone wants to launch another GEO satellite?

Reservations in space are made by the International Telecommunications Union (ITU), which regulates the GEO "spectrum." In 1967, the United Outer Space Treaty declared the geosynchronous orbit as the "common heritage of mankind." The ITU determined that slots in this orbit were up for grabs on a *first-come-first-serve* basis. A land rush ensued, and today the GEO orbit is full.

Inmarsat was the first and most successful GEO network system in the world. Started in 1979, its network consists of 5 satellites (4 older backups) linked to the global telecommunication network through 34 LESs, all run from a network operations center in London, U.K. Satellite coverage is 95% of the surface of the globe (North and South Pole have no coverage). Inmarsat currently provides

[11]http://www.hq.nasa.gov/office/pao/History/satcomhistory.html.

Products & Services

Inmarsat A
Inmarsat's original phone, fax and data system
Inmarsat B
Digital successor to Inmarsat-A
Inmarsat B HSD
64 kbit/s high speed data option
Inmarsat C
Store-and-forward data through briefcase terminals
Inmarsat D & D+
Global messaging and data broadcasts to pager-sized terminals
Inmarsat E
Global alerting services via Inmarsat
Inmarsat Mini M Phone
Inmarsat's smallest satellite phones for voice, fax and data

Figure 11.4. Inmarsat communication satellite services.

64-kbps voice and data, but in 2005, the system will provide a premium service that runs up to 432 kbps—comparable with DSL and digital cable services used by consumers with broadband access to the Internet.

Satellite services are key to many critical infrastructure sectors as well as homeland security operations (see Figure 11.4). For example, we discussed the use of communication satellites in SCADA pipeline monitoring and water sector monitoring in previous chapters. Inmarsat provides other kinds of monitoring as well: radiation leakage monitoring in power plants and oil refinery monitoring in the energy sector. Asset tracking is a major application of satellite communications because of global coverage: GPS container tracking by shippers, equipment tracking by large farms, train and car tracking by railway operators, and vessel tracking of fleets at sea.

MEO Satellites

New satellites must find other orbits, because of the limitations of the GEO orbit. Hence, the medium earth orbit (MEO), positioned from 6,000 to 13,000 above the earth is closer to earth than the GEOs, but it must be tracked as it moves across the surface of the earth. This requires more sophisticated telecommunications equipment. They also enjoy less coverage, and so more "birds" are needed to cover sufficiently large areas of the globe.

Telstar was the first MEO satellite launched by NASA for AT&T in 1962. Telstar II followed in 1963, and several ground stations were constructed to relay TV broadcasts across the Atlantic Ocean. But AT&T lost out to Comsat when President Kennedy decided to give the monopoly for international satellite communications to Comsat:

> Bell Telephone Laboratories designed and built the Telstar spacecraft with AT&T corporate funds. The first Telstar satellites were prototypes that would prove the concepts behind the large constellation system that was being planned. NASA's contribution to the project was to launch the satellites and provide some tracking and telemetry functions, but AT&T bore all the costs of the project reimbursing NASA $6 million. Although Telstar was not really a NASA project, NASA was able to negotiate an excellent deal with AT&T because NASA held a monopoly on launch services. NASA was able to claim Telstar as a NASA supported project and even publish the report on the results of experimentation as a NASA publication.
>
> Even as the success of Telstar I was becoming apparent, AT&T lost the chance to control commercial satellite communications. On August 31, 1962 President Kennedy signed the Communications Satellite Act, which gave a monopoly on international communications via satellite to a new corporation called Comsat. AT&T went ahead with Telstar II anyway to complete its experimental program. It was launched on May 7, 1963. The publicity from Telstar had been very positive for AT&T.[12]

LEO Satellites

Low earth orbit (LEO) satellites hover about 500 to 1500 miles up. They cover less surface of the earth; hence, it takes many LEO "birds" to span large enough portions of the earth to make the service commercially viable. But LEOs are closer to consumers, so its telephone devices require less power, and the message delay is shorter. LEOs may be short on coverage, but they are long on transmission capacity and low power consumption.

Even so, LEO companies have not been commercially successful, in general. Globalstar and Iridium are two spectacularly unsuccessful examples. Globalstar declared bankruptcy in November 2001. Thermo Capital Partners LLC bought the assets of Globalstar in 2004, keeping the company in business. Iridium declared bankruptcy in August 1999, and because of its value to the U.S. military, has been propped up by the Department of Defense.

Globalstar was started in 1991, but it did not launch its service until 1999. Its network consists of 48 satellites placed in LEO orbit 850 miles high. This "swarm" of satellites covers 80% of the globe. The Globalstar service was designed to support "CDMA in the sky"—essentially the same as third-generation (3G) cellular service here on earth. Handheld units communicate at POTS rates (56 kbps), and desktop units communicate up to 256 kbps.

[12]http://roland.lerc.nasa.gov/ ~ dglover/sat/telstar.html.

Satellites provide an alternative or redundant communications network. Because they work from outer space, they are available when landlines and cellular are not. Hence, they are especially important to emergency workers. For example, emergency satellite communication services (via Stratos, Inc.—a satellite service reseller) were employed after the 9/11 terrorist attacks on the Twin Towers:

> On Sept. 13, a Federal law enforcement agency contacted Stratos from the scene at Ground Zero in New York City, looking for a communications solution that *didn't require land-based facilities*. Stratos sent a shipment of Iridium phones to New York City, which arrived there hours after receiving the initial equipment request. After consulting with Federal officers at a command station a few blocks from the World Trade Center rubble in lower Manhattan, the Stratos team installed *two Iridium fixed-site terminals on a nearby roof and another in a mobile command station*. The equipment was used for emergency back up communications to help facilitate the agency's relief and damage containment efforts.[13]

Land Earth Stations

LESs handle bulk traffic between satellites and the terrestrial network. LESs are key assets in the IEC backbone because they handle large volumes of data and because they are easy targets for terrorists. One of the oldest LESs in the world is located at the southern tip of the British Isles—Goonhilly Station. Goonhilly is the largest satellite LES in the world (see Figure 11.5). It has 60 dishes spread across 140 acres of land in Cornwall. It transmits to every corner of the globe via space and through undersea fiber optic cable. Goonhilly is a critical node in the global telecommunications network because it handles millions of international phone calls, emails, and TV broadcasts:

> On 11 July 1962 this site transmitted the first live television signal across thc Atlantic from Europe to the USA, via TELSTAR. This Satellite Earth Station was designed and built by the British Post Office Engineering Department. Goonhilly-Downs covers 140 acres and is located at the westernmost end of the Cornwall coast in England. It was selected because of the topography of the land. The first satellite dish to be built on the site, Goonhilly 1, also known as Arthur, was a parabolic design and weighs 1118 tons and was 85 feet in diameter. It set a world standard for the open parabolic design of the dish.[14]

Large LESs also exist in the United States. For example, the Staten Island Teleport, owned by Teleport Communication Group (TCG), handles much of the broadcast telecommunications streaming in and out of media capital Manhattan. The 100-acre business park includes a 400-mile regional fiber optic network and an operations center linked to a satellite transmission facility.

[13]http://www.stratosglobal.com/caseStudies/usLawEnforcement.html.

[14]http://www.ieee.org/organizations/history_center/milestones_photos/telstar_ukri.html.

Figure 11.5. Historic dish at Goonhilly Station, circa 1962. Goonhilly remains the largest LES in the world. Photo courtesy of IEEE: http://www.ieee.org/organizations/history_center/milestones_photos/telstar_ukri.html.

CELLULAR NETWORKS

The cellular telephone wireless network also feeds into the IEC backbone as shown in Figure 11.3. Cellular telephones have become a pervasive commodity—nearly 1.5 billion handsets are in use in the world, and the number is rapidly growing. It is an alternative to landline telecommunications that cannot be ignored.

Cell phones operate on only one standard in Europe (GSM), which means networks interoperate across country borders. But in the United States, cellular networks have grown up somewhat like the landline LECs grew up in the 1890s—as sprawling competitors. The result: an overly complicated and confusing cellular network infrastructure. To understand this important infrastructure, we have to delve into the arcane world of cellular access methods and technology generations (see Table 11.2).

The cellular network derives its name from the fact that it is actually a honeycomb of 100-square-mile regions called cells—each cell acting like its own self-contained radio broadcast area. These cells communicate with a tower located in the middle of the region. The tower links each handset to a wired network that interfaces with a gateway to the IEC backbone through a POP. Cells divide a city into small areas about 10 miles in diameter.

Towers and their associated switching gear are called base stations. Each base station is connected to the mobile telephone switching office (MTSO), which ties

TABLE 11.2. Cellular Access Methods and Technology Generations.

Access Methods	Technology Generations
Analog (AMPS)	1G and CDPD
TDMA	2G, 2.5G, and GPRS
CDMA	3G
GSM	

into the wired phone system through a gateway (see Figure 11.3). As a cell phone moves around, the base station tracks it, and when the phone leaves one cell and enters another, the signal is handed off to the next tower. Thus, cell phones roam from one base station to the next, without interruption.

A cell phone needs three numbers to operate within its cell, cross over into another cell, and interoperate with the wired landline network. Each phone has a system identification number (SID), a unique five-digit number assigned by the FCC to each carrier; an electronic serial number (ESN), a unique 32-bit number programmed into the phone when manufactured; and a mobile identification number (MIN), a 10-digit number derived from your phone's dial-up number. The SID validates that your phone is "legal," the ESN validates that you have registered with a carrier such as Verizon, and the MIN identifies your phone uniquely.

Here is (roughly) how a cell phone works. When the handset is turned on, it is assigned 1 of 42 control channels to send its SID to the base station with the strongest signal. The MTSO switch monitors signal strength, and as you move from one cell to another, you are "handed off" to the cell with the strongest signal. If the SID does not match the SID of the base station, then the handset must be "roaming," which means the caller is outside of his or her "home" base station. The MTSO that is handling the call uses the SID, ESN, and MIN numbers to track the handset and pass its signal on to a gateway into the IEC backbone, and then to another MTSO. The receiving MTSO locates the destination handset and makes the connection. The switching equipment in each MTSO must be sophisticated enough to perform handoffs at both ends without the consumer realizing what is happening.

Access Methods

Most modern cell phones are *trimodal*, meaning they converse in three access methods: time-division multiple access (TDMA), frequency-division multiple access (FDMA), or code-division multiple access (CDMA) (see Table 2). The following is a greatly simplified explanation of each access method that serves only to argue why CDMA is the better method for homeland security applications.

TDMA divides time into slots that are assigned to different users. Thus, user A receives air time for a short period of time followed by user B, followed by user C, and then back to user A, again. Multiple users share the wireless communication channel, but because the switching is done so quickly (6.7 ms), the user does not notice the gaps in service.

FDMA is similar except the wireless channel is divided into (parallel) frequency bands; each band is capable of carrying a signal at the same time as another. Thus, user A is assigned one frequency band, user B another, and user C a third band. FDMA is as old as cell phones themselves.

CDMA is a clever mixture of TDMA and FDMA whereby signals are assigned a code—a unique identification number—and then transmitted through a kind of frequency-division wireless channel. At the other end, the signal is reconstructed according to its code and played back to the consumer—without the consumer noticing that his or her signal was shuffled in the process. CDMA signals hop from one frequency band to another, which makes it difficult to implement in an electronic device, but it also makes it difficult (perhaps impossible) to intercept and listen in on a conversation. This is why CDMA is the best cellular technology to use for homeland security applications.

CDMA is so clever and important that it was a closely guarded secret for many years. It uses spread-spectrum (frequency-hopping), which was invented by actress Hedy Lamarr and composer George Antheils in the early 1940s. This patented cryptology technology was used in military communications because it is difficult to crack:

> Many years ago, on the eve of World War II, a well-known actress of the day and my father, an avant-garde American composer, while at a dinner party, thought up an interesting scheme to control armed torpedoes over long distances without the enemy detecting them or jamming their transmissions. While they had the foresight to patent their invention, the term of the patent lapsed without either of them realizing any money from their invention, which formed the basis of what was to later become spread-spectrum communications.
>
> This invention becomes even more incredible when you consider that it came before the invention of digital electronics, however, it makes very substantial use of several key digital concepts.
>
> Yes, the term "ahead of it's time" would apply here, because over 50 years later, as high-speed microprocessors become inexpensive, spread-spectrum communications— Hedy Lamarr and my father, George Antheils' 'secret communications system'— adapted to use today's ultra fast microprocessors— is coming into it's own as a effective and inexpensive way to communicate over long distances, privately and efficiently.[15]

Generations

The shift that is rapidly advancing cell phone technology is known as generations—1G for first-generation phones, 2G for second-generation phones, and 3G for the latest generation of wireless phones. The following is also a simplification.

1G cellular networks ran on analog signals and are often called advanced mobile phone systems (AMPSs). 2G phones converted sound into digital signals of speech and data. So 2G cell phones introduced the first generation of *digital* cell phones. An interim generation called 2.5G runs on an all-digital network but is fast enough to support e-mail, Web browsing, video, and always-on connection with the Internet. 3G means more speed and therefore more functionality than 2.5G. The International

[15] http://www.ncafe.com/chris/pat2/.

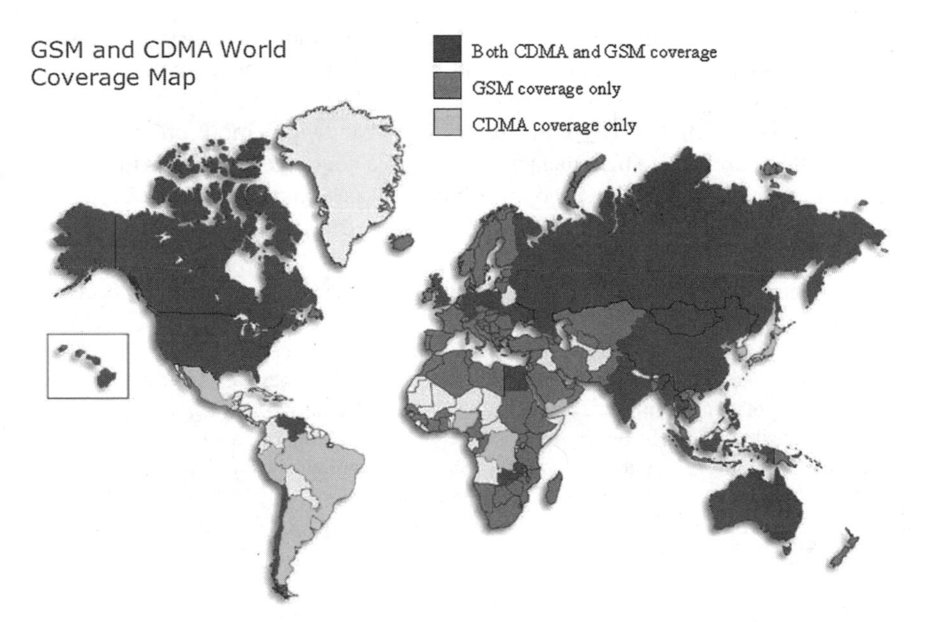

Figure 11.6. Global adoption of GSM and CDMA standards for mobile communication networks.[16]

Telecommunications Union (ITU) defines 3G networks as wireless digital networks supporting transmissions from 144 kbps to 2 mbps. In a sense, the 3G protocol and service is wireless DSL (broadband).

Along the way to 3G, several other interim generations were deployed. For example, cellular digital packet data (CDPD) are digital data transmitted over AMPS networks. CDPD are still used by police departments because CDPD covers larger areas than more modern networks.

Coincidentally, AMPS still has the largest overall coverage in the United States of any cellular network, which means that CDPD phones have the largest coverage of any wireless network in the United States. Estimates of coverage range upward of 80%, whereas 3G networks have less than 40% coverage. But AMPS with CDPD is not very secure, which suggests that the future of wireless networks for use in homeland security is CDMA and 3G. As 3G coverage expands, it will become more viable as a secure emergency network. It will support video as well as e-mail and voice communications.

Battle for Standards

There is a global battle between the European standard GSM and variants of CDMA (W-CDMA, CDMA-2000). GSM is the oldest and most widely adopted standard (see Figure 11.6), but CDMA is technologically more sophisticated and has advantages as far as network security is concerned.

[16]http://www.gsmcoverage.co.uk/maps/europe/gsm_cdma_map.jpg.

GSM uses a variation of TDMA. To overcome the deficiencies of TDMA, most GSM carriers add data compression and encryption to improve performance and enhance security.

On the other hand, CDMA inherently uses bandwidth more efficiently and is inherently secure. Other than its lack of wide area coverage, CDMA is the preferred technology for homeland security. In fact, it was used to encrypt transmissions during the 1962 Cuban Missile Crisis:

> After remaining classified for a long time, the CDMA technology was finally declassified in the mid-1980's. Only in 1995 CDMA was, for the first time, used for mobile communication in the U.S.A. Today, the CDMA customer base hovers around 80 million, concentrated mainly in South Korea, North America, Australia, Taiwan and parts of China. In fact the very entry of CDMA into non-U.S. countries is the direct result of politics by the U.S. giant, Qualcomm—the company that has put its weight behind CDMA technology. For example, Beijing clearly linked the entry of CDMA into the country to its United States WTO deal. A CDMA phone does not have a SIM card, and therefore you have to stick to the phone you have been provided with.[17]

To add to the confusion, GSM in the United States (1900 Mhz) is not compatible with European/Asian GSM (800/1800 Mhz) without a trimodal phone. Compatibility, corporate warfare, and international politics all seem to play a role in the GSM versus CDMA battle.

> Proponents of CDMA claim high communication security, high carrier efficiency meaning that the network can serve more subscribers at a time, smaller phones, low power requirement, ease of operation for the network operators, and extended reach beneficial to rural users. CDMA's detractors say that due to its proprietary nature, the engineering community does not yet know all of CDMA's flaws. Also, as CDMA is relatively new, the network is not set up to provide as many facilities as GSM (TDMA). Being the standard for mobile communication in very few countries, CDMA also cannot offer international roaming, a large disadvantage.[18]

The Wireless IEEE 802.11 Technology

The interesting thing is that GSM, CDMA, and the transition to 3G networks may be disrupted by yet another technology—Wi-Fi. Wi-Fi is the commercial name for a series of standards set by the IEEE: 802.11 b works on the same unlicensed band as a cordless phone and transmits 11 Mbps. 802.11a works on another unlicensed band but transmits much faster, 54 Mbps. IEEE 802.11 g is a compromise between the two earlier standards and remains compatible with 802.11 b. The 802.11 g standard transmits at 54 Mbps and interoperates with the original 802.11 b standard, but its utilization of the frequency band is less efficient. If you use a wireless laptop, it is most likely connected to the Internet by an 802.11 g link.

[17]http://www.mouthshut.com/readreview/34613-1.html.
[18]Ibid.

Wi-Fi has a very short range (100 m), but relatively high speed—20 times that of 3G networks. But it is unlicensed, which means access points can be installed anywhere by anyone. This has propelled Wi-Fi to mainstream use, not only in offices, but also in restaurants, libraries, shopping malls, and other public places.

The question is, "will 3G be snuffed out by Wi-Fi?" The answer is not known at this writing, but consider this: Wi-Fi is cheaper and faster than 3G, and it does not require an expensive license from the FCC. Andrei Jezierski and Sajai Krishnan, principals of LLC, express the challenge in succinct terms:

> Here's the problem—3G cellular data infrastructure is nearly unaffordable to the carriers for the foreseeable future, because of its massive capex [Capital Expense] requirements and, in Europe, the overhang from expensive spectrum licenses. Yet, ironically, and unlike current 2.5G deployments, 3G (once it finally gets deployed) offers the potential to be profitable by enabling affordable, rich media consumer services in a way that 2.5G cannot. In the meantime, we are 'stuck' with 2.5G deployments which promise half-a-loaf—services that, so far, seem less than compelling, and whose frequent use would be unaffordable by many consumers.
>
> Compounding the difficulty, "hotspot" Wi-Fi deployment (from Starbucks to airline clubs to corporate campuses) promises to capture business-based broadband data use with price-performance one to two orders of magnitude more attractive than as-yet-non-existent 3G services. Because Wi-Fi promises to skim the business cream off the wireless broadband market before 3G even makes its debut, it inconveniently disrupts the traditional carrier scenario, in which the mass market pays for scale deployment, while business customers deliver profitability.[19]

VULNERABILITY ANALYSIS

The redundancy provided by the three major telecommunication network infrastructure components—landlines, cellular, and extraterrestrial networks (communication satellites)—add strength to the sector because service can be switched from one to the other during an emergency. Landline and cellular service can be backed up by satellite communication services, for example.

But these three major networks also have vulnerabilities that are unique to each and are therefore subject to several threats. This complicates our analysis because we have to defend against an array of threats and weapons. First, the vulnerabilities (in order of criticality):

1. Telecom hotels—concentration of assets in the IEC network
2. Satellite LESs—critical link to communication satellites
3. IEC POPS—critical links between LECs and IECs
4. Cellular gateway POPS—critical link between cellular and land line networks

[19]http://www.bcr.com/bcrmag/2003/01/p11.asp.

5. LEC central offices—link between local and long distance services
6. Satellites, towers, cables, and fiber—less critical components
7. Network transport elements (Regenerators, Amplifiers)—long distance
8. Power outage—most protected by backup generators

Next, the threats—again in descending order of criticality:

1. Cyber-attack on all telecomm components—terrestrial and extraterrestrial
2. Physical attack on telecom hotel—destruction of concentrated assets
3. Physical attack on LES—damage to a critical link
4. High-power microwave (HPM) attack on telecom components
5. Physical attack on IEC POPS; gateways
6. Physical or HPM attack on satellite "bird"

Figure 11.7 assigns priorities and fault probabilities to each of these. We will use these estimates later on for resource allocation and policy decisions. Figure 11.8 puts it all together in the form of a fault tree. Notice that a complete telecommunications failure depends on all three failing: landlines, cellular, and extraterrestrial networks. Even so, the entire sector is highly vulnerable, with over 94% probability that it will fail! This is because the likelihood of a cyber-attack is rated HIGH in all four subcomponents.

We will evaluate the subtrees in Figure 11.8 in greater detail using the MBVA technique. At this point in the discussion, it is worthwhile noting some generalities. First, telecom hotels are the most important assets because all networks depend on them—wired as well as wireless. Second, the LESs are critical because they handle so much traffic, and third, the backbone carriers that make up the IEC POPS and gateways—for both satellite and cellular—are critical because they tie the entire

Component	Threat	Vulnerability	Cost	$Damage
Telecom Hotel	PHY	HIGH	$100	$250
	HPM	MEDIUM	$10	$100
	Cyber	HIGH	$10	$10
POPS	PHY	HIGH	$10	$25
	HPM	MEDIUM	$2	$10
	Cyber	HIGH	$1	$2
Gateway	PHY	HIGH	$10	$10
	HPM	MEDIUM	$2	$5
	Cyber	HIGH	$1	$1
LES	PHY	HIGH	$100	$250
	HPM	MEDIUM	$10	$100
	Cyber	HIGH	$10	$10

Figure 11.7. Vulnerabilities and threats to the telecommunications sector components.

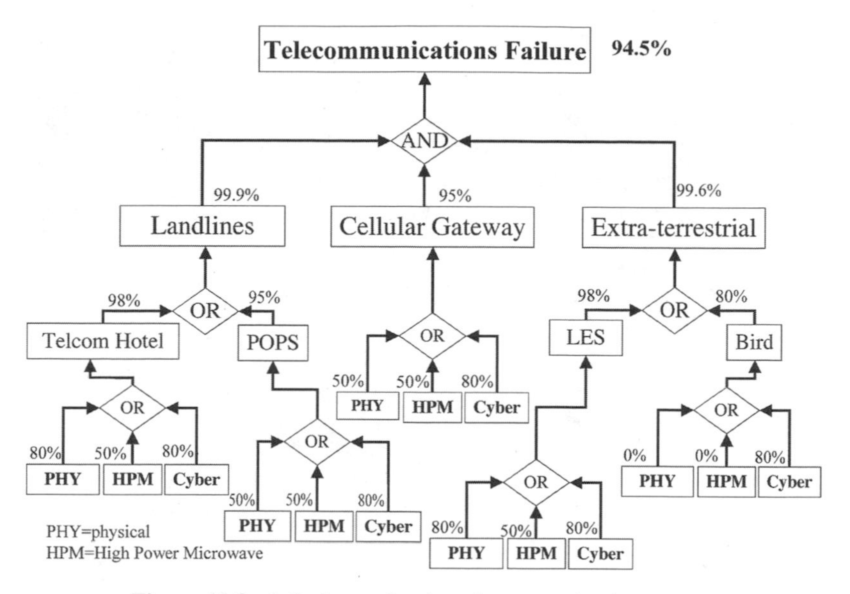

Figure 11.8. A fault tree for the telecommunications sector.

sector together. If a terrorist wants to unravel the landlines, he or she need only identify the busiest (highest degree) node in the backbone network.

A novel weapon will be introduced in the following threat analysis—HPMs. HPM machines are easy to assemble, easy to operate, and extremely damaging to electronic and communication equipment. Therefore, they have been given a MEDIUM rating in Figure 11.7.

SCALE-FREE NETWORK ANALYSIS

The telecommunications sector backbone is a vast network of *billions* of miles of copper and fiber optic cable, not to mention wireless links. Thus, it is impractical to protect every substation, power line, and tower in the system. Instead, our policy must focus on the critical nodes as identified by the following network analysis. The first question is, "does the telecommunications backbone network resemble a scale-free or small-world network?"

Figure 11.9(a) shows a map of the 30 busiest telecommunication backbone routes—in terms of traffic. Critical node analysis results obtained from the program *NetworkAnalysis* are shown in Figure 11.9(b). The power law that best fits this data yields an exponent of $p = 1.108$, which does not lie between 2 and 3, but is close enough for our purposes. Therefore, the major routes in the backbone form a structured network with a hub at Chicago.

(a)

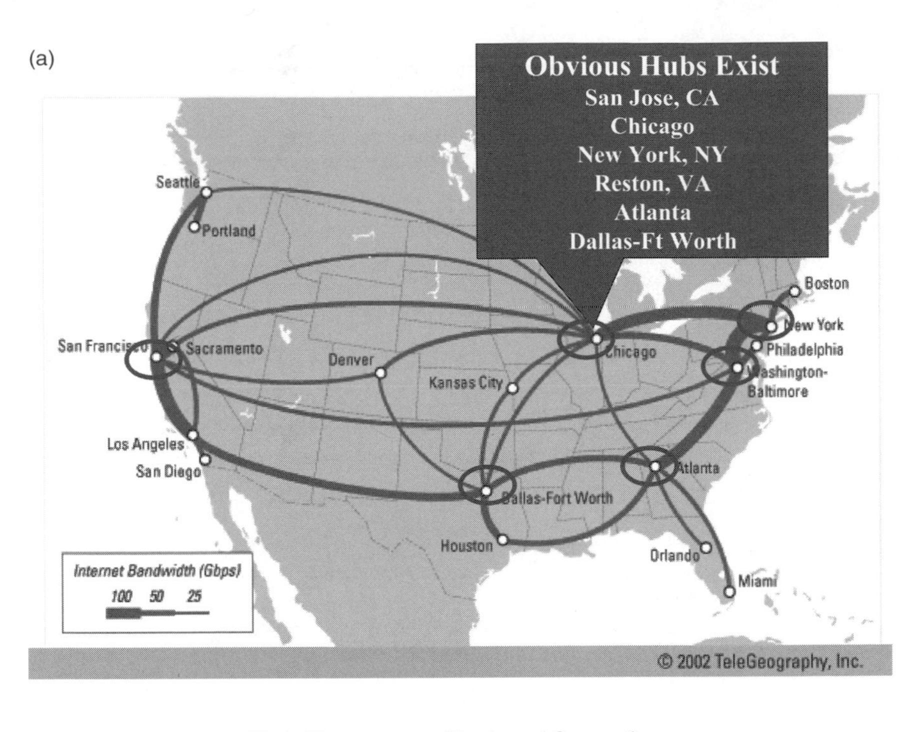

(b)

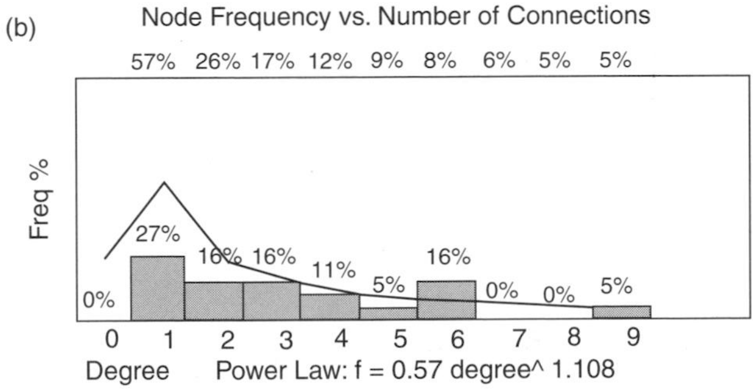

Figure 11.9. Network analysis of the top 30 telecommunications routes in the United States showing (a) the routes and (b) the network analysis of (a). The power law exponent is $p = 1.11$.

In fact, network-wide allocation of availability points confirms that the most vulnerable nodes in this network are also the hubs identified by scale-free analysis. This should not be too surprising because nodes of this network are assigned the same damage value. However, links are assigned values proportional to the capacity of the transmission lines connecting nodes. These estimates are based on the thickness of the links in Figure 11.9(a).

Top Node by *NetworkAnalysis*	Top Hub by Scale-Free Analysis
Chicago	Chicago
Atlanta	Atlanta
San Francisco	San Francisco
Dallas-Fort Worth	Dallas-Fort Worth
Washington-Baltimore	Washington-Baltimore

The forgoing comparison was calculated using information provided by Figure 11.9(a), which designates transmission line capacities of 100, 50, and 25 Gbps for each link. A 100-Gbps link is assigned a damage value of 100; a 50-Gbps link is assigned a damage of 50, and a 25-Gbps link a damage of 25. All nodes are assigned equal damage values of 100 each. Therefore, network-wide allocation of availability points will favor high-capacity links over lower capacity links, and so forth. Network-wide allocation of availability points tells the policy maker that the most vulnerable links in the backbone network are those emanating from San Francisco, especially the link between San Francisco and Sacramento, and the link between San Francisco and Los Angeles. After the critical nodes are protected, these critical links should be protected to minimize network-wide risk.

Telecom Hotels

Within these regions lie telecommunication facilities that over time have consolidated resources into critical nodes called *carrier hotels*, or *telecom hotels*. For economic reasons as well as the Telecommunications Act of 1996 that requires linking LECs and IECs, telecommunications companies, Internet ISPs, and businesses have tended to colocate their equipment and services in the same building. This also saves money, because infrastructure costs can be amortized over a large number of tenants. But telecom hotels are among the most vulnerable nodes in the telecommunications infrastructure. They are prime targets for asymmetric attacks.

Telecom hotels are attractive to carriers because they provide:[20]

- High-speed connections (fiber, satellite, microwave)
- Roof access for antennas
- Physical security:
  - Key card access
  - Video surveillance
  - Biometric scanners

[20]http://www.carrierhotels.com.

- Power and backup generators
- VESDA air sampling (imminent fire detection)
- Fire suppression—suppressors and sprinklers
- Redundant HVAC
- Seismic strength

Because these functions are expensive and bothersome for businesses to supply on their own, many telecom hotels also contain key assets outsourced by their clients. Computers, databases, and so on are often colocated in a telecom hotel. Telecom hotels contain switching equipment for LECs, Internet ISPs, IEC POPS, wireless gateways, storage and hosting servers for businesses, and application service providers (companies that run your applications for you). The vulnerability of a telecom hotel means businesses are vulnerable as well.

The original telcom hotels in the United States—60 Hudson St. in New York and 1 Wilshire Boulevard in Los Angeles—became telcom hotels in large part because they happened to sit on top of a big optical fiber intersection. One building, Number 1 Wilshire Boulevard, is home to nearly 100 telecommunication carriers alone, and is sometimes described as a direct jack to Europe and Japan.

Richard Clarke—the first cyber-security head at the DHS—recognized this national vulnerability:

> I'm told ... that although Transatlantic Fiber lands at about 10 different places in Massachusetts, Rhode Island, Long Island and New Jersey that, after having landed, it all goes to one of two facilities—60 Hudson Street or 111 Eighth Avenue in Lower Manhattan. If that's true, that would seem to be a problem. ... I suspect this statement ... is true, that if you blew up 60 Hudson Street and 111 Eighth Avenue, we could not communicate via fiber optic with Europe.[21]

Telecom hotel vulnerability was one of the immediate concerns expressed by the President via NSTAC—on the heels of 9/11. Although NSTAC's report to the President avoided an alarmist call to action, the report did identify telecom hotels as critical components of the telecommunications sector.

> Although no analyses performed to date have shown that the entire communications architecture would be adversely affected through the loss of a single telecom facility, according to JPO-STC, loss of specific telecommunications nodes can cause disruption to national missions under certain circumstances. As a result of these analyses, the JPO-STC not only has shown the dependencies of Department of Defense (DoD) missions on telecommunications, but also reports that there are further and more far-reaching implications to other national infrastructure sectors.[22]

[21] Richard Clarke, March 13, 2002.

[22] The president's National Security Telecommunications Advisory Committee, "*Vulnerabilities Task Force Report Concentration of Assets: Telecom Hotels,*" February 12, 2003.

The most critical nodes within the six critical geographical areas are telecom hotels containing IEC POPS and gateways and LESs that link communication satellites to terrestrial communication networks. We discuss the major LES locations in the foregoing (Staten Island Teleport), and one only need to go to http://www.carrierhotels.com to find the major telecom hotels in the six major metropolitan areas described above. For example, the largest telecom hotel in San Francisco is located at 200 Paul Street. The MAE-West network access point for the Internet is located in San Jose. Anyone can find these critical nodes.

Threats from HPM Attacks

Although these buildings are well protected against physical attack, they are susceptible to cyber- and HPM attacks as described below. HPM guns are low-cost weapons that cause havoc when unleashed on computers, telecommunications, radars, and other electronic devices. A burst of energy from an HPM weapon "fries" the circuits of most electronic machines.

HPM waves are created by discharging extremely short bursts of microwaves at high-energy levels—typically gigawatts of energy fired in nanosecond bursts. These waves are short, i.e., from a few meters to a few centimeters in length or from 100 Mhz to tens of Ghz in the frequency domain. This is the electronic equivalent of a sharp knife, cutting through walls and shielding to get to electronic circuits.

One way to think of HPM is to make an analogy with a high-heeled shoe. If the area of the heel is one square inch and the person wearing the shoe weighs 100 pounds, the heel presses against the floor with a force of 100 pounds per square inch. Now, if the area of the heel is reduced by half so it is now one half of a square inch, the 100-pound pressure is spread across one half as much area, so the downward pressure is 200 pounds per square inch. If we continue to reduce the size of the heel, say to one tenth of a square inch, the force against the floor is now 1000 pounds per square inch! If we apply the same weight to a smaller and smaller area, the force goes up and up. A person that weighs 100 pounds can apply a million pounds per square inch by simply wearing extremely pointed high heels. This is the idea behind HPM—energy is discharged over a very brief time interval, producing a large force—for a brief moment. But the force does not have to last very long to render damage.

Mathematically, power is defined as work divided by time. Work is defined as force multiplied by distance. Suppose we set the amount of work to a constant, say, 100. If this amount of work is done over 1 second, the amount of power is 100 (arbitrary units). If it is spread over one tenth of a second, the amount of power expended is ten times as much, or 1000 units. What happens if the work is confined to 1 ns? The power expended is 100 billion units! This is why HPM devices can be so destructive.

HPM devices are made from a variety of components, all of which can be purchased from almost any electronics store. In addition, there are a variety of methods for storing and discharging "work" in extremely short bursts, ranging from magnetic to electronic linear accelerometers. Figure 11.10 shows an

Figure 11.10. Photo of an ultra-wide-band HPM device constructed by students at the Naval Postgraduate School, December 2002.

ultra-wide-band HPM device constructed by students of Prof. Ryan Umstaddt of the Naval Postgraduate School. If students can build these devices with a limited budget, many others can also.

HPM weapons are ideal for asymmetric attacks on computer and electronic equipment, because:

1. They are silent and easy to conceal.
2. They can be easily transported by truck, van, or even briefcase.
3. They are difficult to locate and destroy.
4. They are effective against nearly any unshielded electronic device:
   a. Penetrate many materials
   b. Damage may not be apparent
   c. Not necessarily harmful to people

In summary, the most credible threats to highly secure facilities such as a telecom hotel are cyber-attacks and HPM attacks, followed by physical attacks. Cyber-attacks can travel into space and through cement walls. Weapons like HPM can penetrate the best physical defenses of most telecom hotels.

Cellular Network Threats

Next we turn to the analysis of cellular networks and cell phones. As cellular telephones become more and more like miniature computers, and the cellular network becomes more and more like a wireless Internet connection, the threats become more and more like cyber-security and Internet threats. This is the downside of *convergence*—the use of Internet protocol in all communication sectors including TV, radio, and cellular telephony.

As shown in Figure 11.8, the major threats to cellular networks fall into three categories: cyber, HPM-like, and physical (PHY). We examine the top contenders:

1. Cyber:
   Denial of service—flooding the airwaves with messages
   Cloning—intercepting the phone's SID and MIN
2. HPM-like:
   Radio-frequency (RF) jamming—blocking out the signal
3. PHY:
   Destruction of base stations—bombing, etc.
   Gateways and POPS—bombing of telecom hotels, etc.

Security experts call cyber-assaults *exploits*. One of the most common exploits is known as *denial of service* (DoS), because it renders the network useless by overloading the channel with meaningless messages. A DoS attack is like sending millions of automobiles onto a freeway to prevent ambulances and police cars from using the roadway. DoS in cellular networks works the same way—by overwhelming the network with calls, thus rendering the network useless during an emergency.

DoS attacks are not theoretical, they actually happen all the time. And because cell phones are becoming integrated with computer networks, a DoS attack can spill over into other parts of the telecommunications infrastructure, and vice versa. For example, Spain's Telefonica cellular network was attacked by "SMS bombing"—a short message system (SMS) DoS attack—in June 2000. Flooding of the Spanish cellular network was actually a side-effect of an e-mail virus—called Timofonica—that spread through computer networks, got into address books, and then dialed cell phone numbers at random. Timofonica-infected copies of Microsoft Outlook contained a *macro program* that randomly generated and dialed the phone numbers:

> Timofonica was marketed as a cell phone virus when in actual fact it was simply a clever variant of the good old email virus. Victims received an email with an exploitative attachment. When the attachment was executed an email was sent to every entry in the victim's address book and an SMS message was sent to random cell phones on the Telefonica network in Spain. The SMS message did not erase any critical information from the phone or cause any damage to the phone's operating system. It didn't spread from phone to phone. It was merely a variant of the spam we receive every day in our email inbox.[23]

Cloning—stealing phone identities and using them on unregistered handsets—is a far more insidious cyber-attack. Analog cell phone identities are snatched out of the air, as crooks use small electronic radio scanners to intercept cell phone

[23]C. McDonough, "Identifying the risk involved in allowing wireless, portable devices into your company," SANS Institute, 2003, p. 6.

transmissions. Later, they use the encoded information to "clone" a second phone, billing their calls to the account of the phone that was scanned. This exploit has diminished as analog phones are becoming replaced by digital handsets. But it is still a viable threat because there are still many analog phones in use by police and emergency personnel.

HPM-like *RF jamming* is the process of blocking wireless transmission by sending out an interfering signal that cancels the true signal.[24] These illegal devices can be easily purchased for less than $1000 from companies around the world: Special Electronic Security Products; U.K. Ltd. of Manchester, England; Intelligence Support Group, Ltd. based in China Lake, CA; an Israeli company called NetLine; and the manufacturer of the C-Guard; are only a few examples. A portable C-Guard sells for about $900. Another company offers the $890 M2 Jammer, which comes in a briefcase and can block phones within a radius of 50 feet. Hub-Giant of Taipei, Taiwan, sells its WAC1000 personal jammer, which has an operating radius of up to 30 feet, for $169. And Uptron of Lucknow, India, offers a full range of jammers with coverage ranging from 20 feet to over 1 mile.

Digital phones that use CDMA are more difficult to jam because a CDMA message "spectrum hops" to find a band that is free of interference. Frequency-hopping side steps the jammer, avoiding the interfering signal. This is another reason why emergency workers are advised to use CDMA instead of analog CDPD networks.

Manufacturers of jammers claim they are selling their devices to give anti-cell phone advocates a little peace and quiet from the ring of cell phones, especially in public places:

> Cell phone jammers are readily available on the Internet. Many can be battery-powered and fit in a pocket or briefcase for people who would like to enjoy a meal, movie or church service in peace.[25]

PHY threats (Figure 11.3) are the least sophisticated, and yet bombs are by far the most preferred weapon of terrorists. Gateways and POPS are typically concentrated in telecom hotels, so these become obvious targets. But other physical threats—even more asymmetric—may be employed. For example, chemical attacks against major telecom hotels are not out of the question. Similar threats confront large and unprotected LESs, such as those located at Staten Island, NY, and Niles Canyon, CA.

But PHY threats seem to have a limited payoff. HPM threats are more difficult to execute because of the sophistication of the devices, but the results are potentially more asymmetric. Cyber-attacks are the easiest to execute and have the highest payoff; hence, they are very asymmetric. Yet, the defender must protect against all serious threats.

[24]Jammers block access to a cell phone's radio tower by transmitting a high-powered signal on the same frequency, thus canceling the phone's signal. The result: a poor or erased signal.

[25]M. Wylie, "Cell Phone Jammers, Illegal in U.S., Can Create Silent Zones," 2000. http://www.newhouse.com/archive/story1a092200.html.

Communication satellites are too difficult to bring down, but they are as susceptible to cyber-attack as terrestrial networks. Cyber-worms and viruses can go into space at the speed of light.

Although LESs are lower valued targets, they are subject to the same threats as telecom hotels. The same goes for gateways and POPS. Hence, the fault tree of Figure 11.8 summarizes both the vulnerabilities and the threats for this sector. Where should we make investments to prevent a successful attack on telecommunications?

ANALYSIS

Redundant tandem switches and ring structures in local loops as well as some IEC loops provide a degree of security due to redundancy. In addition, the abundance of long line fiber across the country suggests that there is sufficient redundancy in the backbone. But the top 30 routes were shown to be vulnerable to disruption of network continuity simply because they carry such a large proportion of all traffic. Accordingly, we have argued against protecting the local loop and for protecting the most active metropolitan hubs, because this is where the assets are concentrated and this is where traffic levels are the highest. They are high-value targets. Indeed, network-wide analysis identified Chicago as the hub of the network formed by the top 30 routes.

The most active hubs are even more concentrated than they appear on a map, because profit-optimized carriers have chosen to house POPS and gateways in a select number of telcom hotels. This is the result of increasing returns economics. It does not take much research to locate the largest telecom hotels and to pinpoint the high-risk components of telecommunications.

Satellites in space may seem to be immune from attack, but they are as vulnerable as land-based facilities when it comes to cyber-attacks. A DoS attack on a satellite can render it as useless as if it was physically attacked. We have deferred detailed discussion of cyber-exploits for now—they will be covered in subsequent chapters. Suffice it to say that satellites may not be as immune to asymmetric cyber-attack as they may seem.

Asymmetric "energy weapons" such as HPM guns and RF jammers already exist and are proliferating. It would be a mistake to discount the threat of attack on communication and computer infrastructure from these weapons, simply because we know little about them. They already exist and are relatively inexpensive to acquire, hard to trace, and potentially very damaging.

The national strategy for protecting the telecommunications sector must focus resources on protecting telecom hotels, critical gateways and links, and address asymmetric threats such as those posed by cyber- and HPM attacks. The narrow concentration of key assets makes this sector vulnerable to attack by a relatively small force using inexpensive weapons.

A final note: Because of blurred roles and responsibilities in the regulatory agencies that oversee telecommunications, reorganization may be needed to implement an effective strategy. Currently there are too many agencies chasing too few industry-owned and operated assets. Does this sector have too much assistance and not enough help?

EXERCISES

1. The most vulnerable physical nodes in the telecommunications sector, from an asymmetric point of view, are (select only one):
   a. Headends
   b. Critical fiber routes
   c. Network transport elements
   d. Telecom hotels/multitenant locations
   e. Data centers
2. Which of the following can make cell phones unable to operate (select only one)?
   a. Jamming
   b. War driving
   c. Cloning
   d. Abuse of E-911 services
   e. The code red worm
3. The major failure in cell phones that occurred during the attack on the World Trade Center in 9/11/01 was (select only one):
   a. Base station overload
   b. Cyber-security breach
   c. Cloning
   d. Failure in the Verizon switch at 140 West St
   e. Running out of fuel for the backup generators
4. A major challenge facing protection, response policy, and unity of command within the telecom industry is (select only one):
   a. Cyber-intrusion
   b. GETS dispatch
   c. E-911 information sharing
   d. The large number of players in the industry
   e. Lack of an ISAC
5. Which of the following is an emergency telecom service?
   a. NSTAC
   b. GETS
   c. RSVP
   d. Telecommunications ISAC
   e. NRIA

6. Why should EMS personnel care about the differences between digital and analog telecom systems (select only one)?
   a. Digital packets are more efficient.
   b. Analog telecom is more vulnerable to cyber-attack.
   c. Digital telecom is more resilient to physical attack.
   d. Command and control is converging on digital telecom technology.
   e. CDMA networks are more secure than CDPD networks.
7. Cellular and other wireless networks depend on which of the following for call completion (select the best one)?
   a. Wired landlines
   b. Wi-Fi access points
   c. Towers with a range of 100 miles
   d. Telecom hotels
   e. Satellite ground stations
8. Frequency-hopping in a spread-spectrum communication system was patented by:
   a. Alexander Graham Bell
   b. Hedy Lamarr
   c. Qualcomm, Inc.
   d. Gardiner Hubbard and Thomas Sanders, founders of Bell Telephone
   e. Theodore Vail
9. The telecommunications Industry was deregulated (some say re-regulated) by which of the following?
   a. The breakup of AT&T in 1984
   b. The Telecommunications Act of 1935
   c. The Telecommunications Act of 1996
   d. The Tragedy of the Commons of 2003
   e. The creation of the NCS in 1963
10. Which of the following is legally responsible for the cyber-security of the telecommunications industry?
    a. NTIA
    b. NSTAC
    c. NCS
    d. NCC
    e. None of the above
11. The purpose of CWIN is to:
    a. Broadcasting alerts and information sharing.
    b. It is the SCADA of telecommunications.
    c. Tie together the NSTAC and NCS.

   d. Fund telecommunications improvements.
   e. Keep Congress informed.
12. 3G technology is:
   a. High-speed cellular network technology
   b. A competitor to GSM
   c. High-speed landline network technology
   d. The commercial name of IEEE 802.11 standard technology
   e. The protocol for satellite communications
13. Which of the following is the most critical component of telecommunications, from a vulnerability point of view?
   a. Headends
   b. Local loop service
   c. IEC network
   d. LEC network
   e. POPS
14. Communication satellites orbiting the earth were first envisioned by:
   a. Hedy Lamarr
   b. Alexander Graham Bell
   c. Theodore Vail
   d. President John F. Kennedy
   e. Arthur C. Clarke
15. There are currently three kinds of satellites in operation today: Which of the following describes these kinds of satellites?
   a. Wi-Fi, 802.11, and 3G
   b. LES, Goonhilly, and Staten Island Teleport
   c. LEO, MEO, and GEO
   d. Inmarsat, Marisat, and Westar
   e. Telestar, Intelsat, and Satcom
16. The ITU declared orbits in space, as:
   a. The common heritage of mankind
   b. The final frontier
   c. The property of the United Nations
   d. The property of property of Inmarsat
   e. There can be no more than GEO satellites
17. The largest LES in the world is:
   a. Goonhilly Station
   b. Staten Island Teleport
   c. NASA—Houston

d. NASA—Cape Kennedy
e. Arthur, named after Arthur C. Clarke

18. Which of the following wireless network has the highest bandwidth?
    a. Wi-Fi
    b. 3G
    c. CDPD
    d. AMPS
    e. CDMA
19. Which of the following wireless network has the best national coverage in the United States?
    a. Wi-Fi
    b. 3G
    c. CDPD
    d. AMPS
    e. CDMA
20. Which of the following wireless network has the best security?
    a. Wi-Fi
    b. 3G
    c. CDPD
    d. AMPS
    e. CDMA

Discussion Questions

1. For wired lines, explain why each is ranked accordingly:

Priority	Asset
High	Telecom Hotels
High	Satellite Gateways
Medium	IEC Points of Presence
Medium	LEC Central Offices
Low	Head ends
Low	Cables and Fiber
Low	Network Transport Elements (Regenerators, Amplifiers)
Low	Multitenant Locations (POP, Central Office, CLEC)
Low	Power Outage

2. For cellular, explain why each is ranked accordingly:

Priority	Threat
High	RF Jamming
High	Destruction of Base Stations
Medium	Denial of Service
Low	Cloning

3. Take stock of the critical telecommunication nodes in your state or region, and then produce a study, along with a presentation, on what to do about telecommunications services during an emergency similar to 9/11. How can your emergency services operation be guaranteed that services will be available when needed? Assume telecom critical nodes will be attacked during an incident.
4. Use the numbers in Figure 11.7 and the landlines subtree of the fault tree of Figure 11.8 to calculate the allocation of $50 million to protect landlines in San Lewis Rey. Calculate ranked order, apportioned, and minimal (optimal) risk reduction allocations, and compare. [Hint: Use the program *FTplus.*]
5. In #4, which strategy is best, in your opinion? Why?

CHAPTER 12

Internet

According to the Microsoft Encarta dictionary, the *Internet* is a network that links computer networks all over the world by satellite and telephone, connecting users with service networks such as e-mail and the World Wide Web.[1] A more specific definition defines the Internet as a global network that uses Transmission Control Protocol/Internet Protocol (TCP/IP). Although the Internet has been around for nearly 40 years, it began to spread like an epidemic only after TCP/IP was created and adopted by the U.S. Department of Advanced Research Projects Agency (ARPA and later DARPA). TCP/IP defines the Internet, World Wide Web, and many other forms of communication.

The Internet started as an idea on paper that grew into one of the largest manmade machines in the world. As of September 11, 2001, 174 million people or 65.6% of the U.S. population used computers from home or office, and 143 million people or 53.9% of the population used the Internet.[2] These numbers compared favorably with the number of cable TV subscribers, and the number continues to increase. More importantly, worldwide adoption of the Internet communication standard is many times greater than these numbers, easily doubling the total number of subscribers. Some estimates range as high as 665 million users.[3] The TCP/IP standard is one of the most ubiquitous standards in technology.

The global spread of TCP/IP and the confluence of computing and communicating is such an enormous topic that we devote an entire chapter to it. First, we review the basic components of the ubiquitous personal computer (PC), followed by a brief history of the Internet, so that we may understand network and computer security. One of the most significant aspects of the Internet is the way it came into existence and the culture that currently surrounds it. Curiously, the Internet has no centralized governing body. Instead, it is an open community of users—globally

[1]Encarta® World English Dictionary © 1999 Microsoft Corporation. All rights reserved. Developed for Microsoft by Bloomsbury Publishing Plc.

[2]http://www.ntia.doc.gov/ntiahome/dn/html/Chapter2.htm.

[3]According to Nielsen/NetRatings, there was a worldwide Internet population of 580 million users, as of 2002. The International Telecommunications Union estimates 665 million users.

Critical Infrastructure Protection in Homeland Security: Defending a Networked Nation,
edited by Ted G. Lewis

distributed—that govern themselves. This self-governance is socially as interesting as the technology.

It is necessary to understand the basic communication principles of the Internet before embarking on the subject of cyber-security. However, if the reader has already mastered these basics, he or she may skip this chapter.

In this chapter we survey the following topics:

1. Even though the Internet is much older than the PC, it was not commercialized until 1992. After that, it coevolved with the adoption of the consumer PC. Without the PC the Internet may not have been as significant as it is today, and conversely, without the Internet, the PC might not have become as ubiquitous as it has.
2. PCs and the Internet are inherently vulnerable: The hardware and software of PCs are the first link in information technology security. A breach of software security in one computer can spread, like an epidemic, to millions of other information systems—all through the global connectivity provided by the Internet. Although the Internet was designed to be resilient, it was not designed to be secure. And neither were the gateways to the Internet—the household PC!
3. The Internet is equivalent to the TCP/IP standard: Networks that communicate in TCP/IP are considered "the Internet," and conversely, the Internet is defined as any network that requires the use of TCP/IP. TCP/IP is rapidly becoming the universal protocol for electronic communication.
4. The Internet grew out of ARPANet, which was a product of the Cold War: In 1969 the ARPA began a project that created the first "Internet" called ARPANet; ARPANet begat NSFNet and then merged back with NSFNet. The National Science Foundation (NSF), which ran the NSFNet for a time, was directed by the U.S. Congress to commercialize the NSFNet, in 1992. "The Internet" has become the consumer name of the commercial NSFNet.
5. The biggest idea in the Internet is that data should be packet-switched rather than circuit-switched as it has been for over 100 years in the telephone network. Packets are blocks of data that contain their destination and return addresses so they can travel through the Internet on their own. Packet-switching is much more flexible and efficient than circuit-switching.
6. The Internet was invented by many people: Lickliter (the visionary); Taylor (the manager); Baran, Davies, and Kleinrock (packet-switching); Postel (names and addresses of users); Cerf and Kahn (TCP); Tomlinson and Roberts (e-mail); Crocker (governance); Metcalf (Ethernet IP); Postel, Mockapetris, and Partridge (DNS); and Berners-Lee (World Wide Web).
7. The Internet was designed to be resilient: Packets are verified for correctness; the global network rebuilds every day by updating a tree-structured network of DNS servers; TCP is a protocol that automatically routes packets around broken lines and reorders packets when necessary; and the Internet has its own built-in SCADA system called SNMP, for monitoring the devices on the Internet. However, the Internet was *not* designed to be secure.

8. The killer applications that ignited explosive growth of the Internet are e-mail and the World Wide Web (WWW). Marc Andreesen and Eric Bina created the first graphical user interface WWW browser in 1993, called MOSAIC, which set off consumer demand for WWW products and services throughout the world. By 2000 there were over 250 million WWW users—most using Andreesen and Bina's second-generation Netscape Navigator browser.
9. The Internet has unified the coding of all forms of digital information. E-mail follows the higher order rules of simple mail transport protocol (SMTP) and documents disseminated by the WWW follow the rules of Hyper Text Markup Language (HTML), hyper text transport protocol (http), and extensible Markup Language (XML)—universal standards for the encoding and transmission of text, pictures, sound, motion pictures, and animations.
10. The Internet is not owned by anyone or regulated by any single government. Rather, it is operated and governed by its users—volunteers who exert influence through an open process called the Request For Comment (RFC). Most decisions regarding Internet policies and standards are vetted through the Internet Society and its affiliated working groups such as the IETF and W3C.
11. The WWW is vulnerable to attacks on its hubs—primarily through the so-called Tier-1 internet service providers (ISPs), root servers, top-level domain servers (gTLDs), and highly connected e-commerce servers. In other words, even though there are more than 250 million nodes in the global Internet, it is vulnerable because of fewer than several hundred nodes.
12. The next step in Internet evolution is the "semantic web"—a WWW with meaning. XML will eventually displace HTML, the original language of the Web, because XML encodes meaning as well as syntax. The syntactic and semantic network, and its pervasive TCP/IP packet-switching protocol, is destined to unify all forms of communication including broadcast radio and TV, motion pictures, and telephony.

It should be noted that the Internet is the sum total of all TCP/IP networks that connect to one another. The WWW, on the other hand, is an application that runs on the Internet. These separate organisms coexist in a symbiotic relationship, but the two are not to be confused, one for the other. WWW and Web are synonyms for one another, but the Web and Internet are not synonyms.

This chapter drastically compresses the long and colorful history of the Internet and the WWW—perhaps too much. Several books and articles are available for the reader who wants a detailed history. The following discussion focuses on essentials needed for the policy maker to forge a strategy of prevention of damage to the Internet and WWW. The logic of this chapter goes as follows: The basis of the Internet is a protocol called TCP/IP. TCP/IP was never designed to be secure; hence, the first vulnerability of significance is in TCP/IP. Higher order structures like e-mail and the WWW are layered on top of TCP/IP. Each of these layers has its own weakness, which add to the list of Internet vulnerabilities. At the highest layer, the Internet is composed of major components called *autonomous system* or ISPs that carry most traffic and hence pose a high-level network-wide vulnerability.[4] Analysis of major autonomous systems reveals a hub-like asset concentration akin to other sectors. Add to this discovery the fact that the Internet is generally a scale-free network, and it becomes obvious that the 250 million nodes in the global Internet are susceptible to faults in fewer than 200 critical hubs.

Finally, the organizational and regulatory structure of the Internet and WWW borders on anarchy, which adds a third dimension to Internet and WWW vulnerability—the possibility of organizational instability in the organizations that run and govern the Internet. This weakness is not likely to be remedied any time soon.

COMPUTING 101

The rise of the Internet is a dramatic example of coevolution. *Coevolution* is the joint development of two or more interdependent species. In this case, one species is the PC and the other is the Internet. One would not exist without the other—or at least the success of each one is heavily dependent on the success of the other. Where would the public switched telephone network be without handsets and long-distance lines? Both are essential to the successful adoption of telephony throughout the world.

An increase in PC adoption feeds a corresponding increase in Internet use. And as Internet usage increases, so does the consumption of PCs. The Internet surfaced as a commercial success after many repetitions of *increasing returns* cycles alternating between PC and Internet adoption. Without the low-cost PC, the Internet would still be an exotic tool restricted to use by government and university laboratories. Without the Internet, the PC may never have reached the status of home appliance.

The PC adoption cycles (called *generational adoptions* by Bass and Bass) shown in Figure 12.1(a) highlight the importance of the Internet to the computer industry, and the PC in particular. Each successive adoption cycle propelled the PC into a larger and larger market, beginning with the IBM-compatible PC era, which commoditized the hardware and software; evolving through intermediate generations (32-bit processor, disk drive enhancement, Windows graphical user interface, and multimedia PC), ending with the Internet era. This took over 20 years, which is about the length of time most new technology products take to achieve mainstream status during the 1980s.

A model of coevolution is developed for the mathematically curious in **Derivation 12.1**. The results of this model are shown in Figure 12.1(b). The graph shows adoption as market share versus time. The PC began its ascent around 1980, and the Internet began its commercial ascent around 1992. Therefore, Internet adoptions feed off of PC adoptions, and the reverse. Together, they have pushed consumer adoption of PC and Internet to near saturation.

[4]An autonomous system (AS) is a collection of Internet routers and switches under a single administrative control, such as an ISP.

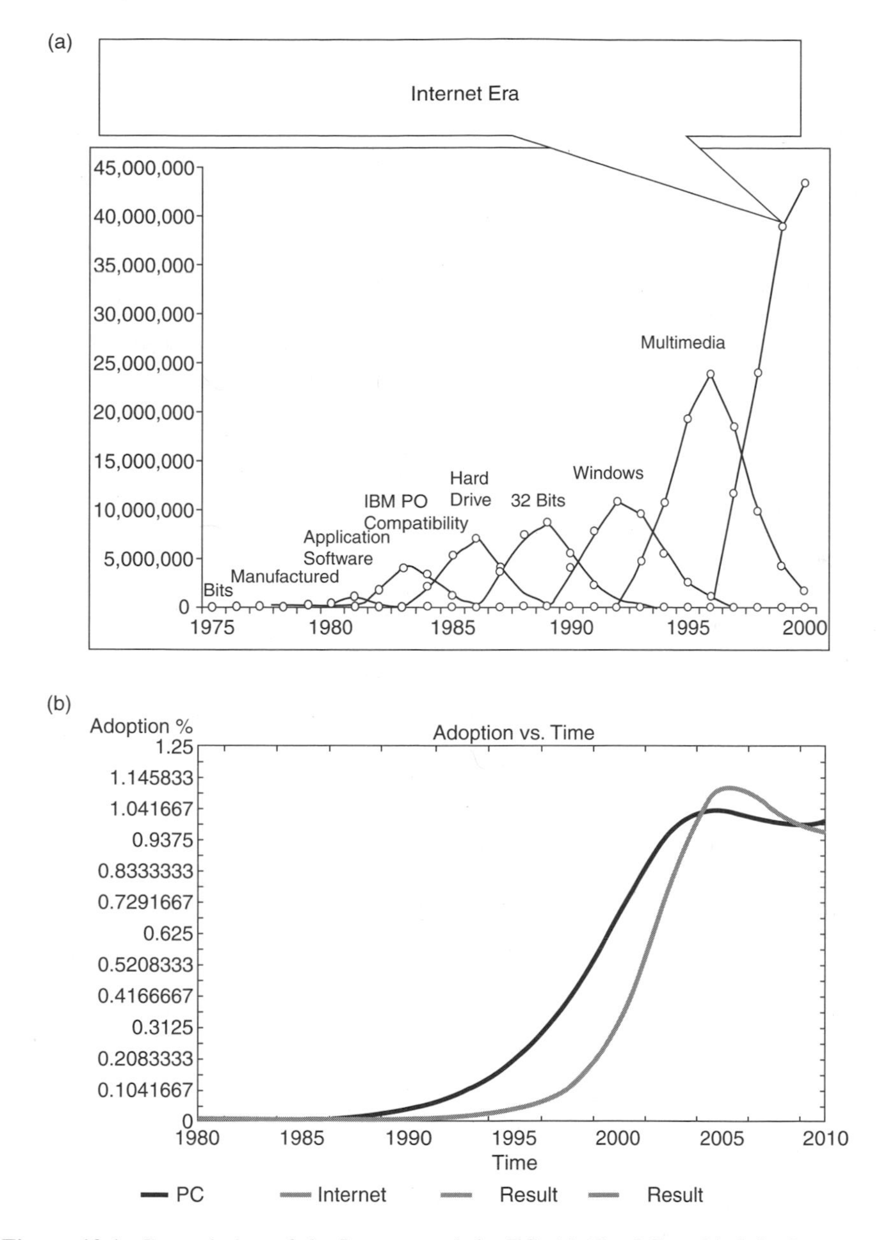

Figure 12.1. Co-evolution of the Internet and the PC. (a) The PC enabled the Internet to grow, and the Internet stimulated the adoption of PCs.[5] (b) Adoption rate model of the coevolution of PC and Internet.

[5]P. I. Bass and F. M. Bass, Modeling Diffusion of Technology Generations, Bass Economics Working Paper, 2001, http://www.basseconomics.com.

DERIVATION 12.1: A MODEL OF THE COEVOLUTION OF PC AND INTERNET

Figure 12.1(b) shows the output from a model of coevolution that is derived here. Let P(t) be the market share—expressed as a percentage—of the total market for PC purchases. Similarly, let I(t) be the market share of Internet users. Market share is a somewhat relative term, because the market is often expanding, which means the number of buyer's changes over time. Market share is a percentage of the expanding number of buyers at any instant in time.
Suppose the rate of change of PC market share, $\delta P/\delta t$, is proportional to the current market share P(t)—the increasing returns portions—and the "pent up demand" as represented by 100% − P(t). Thus, adoption of PCs is represented by the differential equation:

$$\delta P/\delta t = r_p \times P(t) \times [1 - P(t)], \text{ where } r_p \text{ is a constant of proportionality.}$$

Similarly, the rate of change of Internet adoption, $\delta I/\delta t$, is proportional to the current market share I(t)—the increasing returns portion—and the "pent up demand" for Internet products and services, as represented by P(t) − I(t). This expression says that Internet market share cannot exceed PC market share, because a user needs a PC to use the Internet. Internet adoption must coevolve with PC adoption. Thus, the adoption of Internet services is represented by the differential equation:

$$\delta I/\delta t = r_I \times I(t) \times [P(t) - I(t)], \text{ where } r_I \text{ is a constant of proportionality.}$$

Two differential equations in two unknowns can be solved simultaneously, given initial conditions for P and I, and the constants r_p and r_I. The solution given in Figure 12.1(b) assumes that adoption of the PC began in 1980 and adoption of the Internet began in 1990. The initial conditions and constants used in Figure 12.1(b) are

$$P(1980) = 1\%; \quad I(1990) = 1\%$$
$$r_p = 0.5, \quad \text{and } r_I = 0.5.$$

The oscillations around 2000 are caused by the nonlinearity of the differential equations and high adoption rates represented in the model by r_p and r_I. The equations become unstable for high values of these constants. They also remind us that the Internet Bubble burst around the year 2000!

Today, a PC is a very compact and sophisticated appliance consisting of a *processor* that does the textual, audio, video, and numerical processing; a *main memory* (RAM) that temporarily holds software and its immediate data; nonvolatile memory in the form of a hard disk and CD or DVD disk; and several peripherals such as printer, camera, music player, and so forth. In addition, every laptop and desktop PC has several *ports* or *sockets* for communication with the outside world. For example, Universal Serial Port (USB) ports are used to connect serial devices such as printers and cameras; Firewire (a.k.a. IEEE 1394 Multimedia Data Transfer) ports for connecting parallel data transfer devices such as audio/video recorders and players; Wi-Fi (a.k.a. IEEE 802.11) for short-range wireless communication; a 56Kbps analog modem for communication over a dialup telephone line; an *Ethernet* (local area network) port for local area communication; and an super video graphics adapter (SVGA) port for connecting the PC to a projection system or TV.

Each of these links with the outside world makes a PC more powerful than an isolated PC would be without them. But they also make a PC vulnerable, because these links can be unlocked like unattended doors when not properly protected. Whenever a piece of hardware can connect to the outside world through a 56-kbps modem, Ethernet port, or USB port, there is a possibility for cyber-attack. Each piece of hardware listed above is susceptible to security flaws, as we shall see, in Chapter 13.

Software turns PC hardware into a general-purpose machine. For example, a word-processing program turns PC hardware into a word-processing machine; a spreadsheet program turns a PC into a spreadsheet calculation machine; and a Web browser turns the same hardware into an Internet access machine. This is what separates a computer from other machines. Word processors, spreadsheets, and Web browsers are examples of *application software*. These applications are also vectors for the spread of malicious programs that become a cyber-threat. Application programs that look innocent but are actually malicious are called *Trojan horse* programs, because they are attack vehicles in disguise.

But two other major categories of software play an essential role in PCs: tools and operating systems. *Software tools* are the programs that programmers use to develop application programs. They are programs for developing other programs! Most people do not know that software tools exist, because only a handful of programmers need such software. This segment of the software industry is a multi-billion dollar business. It is also a vector for malicious software. At one time, Microsoft's Excel application program contained an entire computer game that could be launched from within Excel by using a combination of key presses. If Microsoft developers can insert a secret game into Excel, what other malicious software lurks inside of applications?

Operating systems are the underlying enablers of application software. An operating system provides services that link the application software program to the underlying hardware. Some of these services perform scheduling when each application is allowed to use the processor, manage how each application uses memory, provide the look and feel of the graphical user interface (a.k.a. windowing system), and enforce computer security. One service is of particular interest: the registry service within Microsoft's Windows operating system. This part of Windows stores the names of programs that are authorized to run on the PC. If a malicious program gets into the registry, it too can run on the PC. This is one of the principle ways that virus software takes control of a PC.

The entire software industry is the third largest industry in the United States, behind the automotive and electronics industries. In 2002, application software was approximately a $100 billion business, the tools segment was approximately $50 billion, and the operating systems segment was over $65 billion. Software is a major industry in the United States and throughout the world. Because it is so large and pervasive, the software industry and the products it produces are also vulnerable to purposeful attack.

When we consider cyber-security in Chapter 13, we will learn that software flaws are the primary source of vulnerability in all information systems. And because PCs are pervasive, they provide one of the best avenues of attack. The PC is a power tool of commerce, but it is also a powerful asset that can be used by terrorists to attack millions of other PCs and, hence, the entire information technology industry.

ORIGINS OF TCP/IP

TCP/IP is the Internet, and the Internet is TCP/IP. The story of how TCP/IP became the fundamental protocol of global information dissemination is a long and colorful one, but we will step through it lightly.[6] The purpose of this historical review is to prepare the reader for what comes later—a vulnerability analysis of the information sector.

The Internet is another consequence of the Cold War between the former USSR and the West. In 1957, the ARPA was formed by the U.S. government in response to the launch of Sputnik. The "missile gap" helped elect John F. Kennedy to the Presidency, and soon afterward, the United States launched the Space Program that put the first men on the moon. But there was one smaller step taken on the journey to the moon in 1969 that may have been just as important. What was to become the Internet was "invented" by employees of ARPA.

ARPA was, and still is, created for taking giant leaps forward to keep the United States technically ahead of its opponents. ARPA later become DARPA and initiated other forward-thinking ideas that would alter the world, but in its earliest days, it was focused on how to beat the Russians into space. The United States would need advanced computing capabilities—among other things—to make space exploration happen. The public relations similarity between the formation of ARPA and the formation in 2003 of the DHS is undeniable:

> All eyes were on ARPA when it opened its doors with a $520 million appropriation and a $2 billion budget plan. It was given direction over all US space programs and all advanced strategic missile research.[7]

In 1962, J. C. R. Lickliter moved from MIT to head the command and control program at ARPA. Lickliter surrounded himself with colleagues from Stanford University, MIT, UC-Berkeley, and UCLA—whom he dubbed as the "Intergalactic

[6]K. Hafner and M. Lyon, "Where Wizards Stay Up Late: The Origins of the Internet," Simon & Schuster, New York, p. 304, 1996.
[7]Ibid, p. 20.

Computer Network." In a memo to the Intergalactic Computer Network group 6 months after his arrival, Lickliter expressed frustration with the lack of interoperability and standards among computer centers:

> Consider the situation in which several centers are netted together, each center being highly individualistic and having its own special language and its own special way of doing things... is it not desirable or even necessary for all of the centers to agree upon some language, or at least, upon some conventions for asking questions as 'What language do you speak?'[8]

Thus was born the idea of networked computers. But it would be Lickliter's successor, Robert Taylor, who took the next important step. Taylor was frustrated with having to log onto three different computers from three different computer terminals—the so-called "terminal problem." Instead of using separate terminals for different computers, why not link all computers together through a network, and access each one from a single terminal? Computers should be just as easy to access as it is to call home through the telephone network.

Taylor convinced his ARPA boss to fund his project, arguing that his project would save money by solving the "terminal problem." A nationwide university network would make it possible for researchers all over the country to share expensive mainframe computers. In 1965, computers cost millions of dollars—a price barrier that prevented many academics from using them. But if a few expensive mainframes were made accessible via a network, then thousands of researchers could share the limited number of expensive machines. In 1968, ARPA contracted BBN (Bolt Beranek and Newman) to build ARPANet—the first version of what would become the Internet.

Meanwhile, others were thinking similar thoughts. One of the most profound ideas occurred to two people at about the same time. Paul Baran, and an Englishman named Donald Davies, both came up with the concept of a *packet*—"message blocks" of data that could travel through a network on their own rather than be harnessed to a single circuit. Telephone networks were *circuit-switched*, which meant that they communicated by connecting the sender and receiver together via an electronic circuit. The entire circuit had to be dedicated to the connection for the entire conversation. Only one pair of users can use a circuit-switched connection at a time. This is very inefficient.

When analog circuits are replaced by their digital equivalents, data can become "smart," so they no longer need to flow through a single circuit. Instead, a *packet-switched network* can share its wires or radio waves with packets from many users—all at the same time. Packets find their own way through a network of circuits. Packets are extremely efficient and flexible as compared with circuits, because multiple packets—all going to different destinations—can share a single circuit.

Packets are "smart," because they contain their own source and destination addresses, much like a letter that is sent through the U.S. Postal System. At each

[8]Ibid, p. 38.

juncture in the network, routing tables provide directions for where each packet should go next. Even if a portion of the physical network fails, an alternate path can be found and the packet re-routed. In this way, data communication becomes resilient—a failure of part of the network cannot disable the entire network.

While at UCLA working on an ARPA contract, Leonard Kleinrock not only conceived of packets, but he proved packet-switching is superior to circuit-switching. His theoretical analysis reinforced the intuition of Baran and Davies. Not only was packet-switching a good idea, it was now theoretically sound. The stage was set for a revolution in data communications. But change takes time, so the obscure ARPANet would take a few more decades to realize its potential.

By 1969 the ARPANet consisted of four computers located at UCLA, SRI (Palo Alto), UCSB, and Utah. Although extremely modest in terms of today's Internet, this was enough to get a small group of pioneers to start thinking about governance and a user's group! So, in 1969, Jon Postel started a list of ARPANet users, which eventually become the telephone directory of the Internet—the domain name server (DNS).[9] If you wanted to use the ARPANet, you had to ask Postel for a name and address in cyber-space. Once your name and address was entered into the DNS, you became "known" to everyone on the network.

DNS Basics

An Internet address—called a universal resource locator (URL)—is entered into the DNS system, where everyone on the network can find it. URLs have evolved into the familiar format of http://www.myname.tld, where http is shorthand for "hypertext transport protocol; www means "World Wide Web"; and tld means "top level domain." But this sophistication took decades to emerge. Modern URLs like http://www.amazon.com must be converted into an IP number such as 120.131.200.41.[10] The DNS converts every URL on the Internet into an IP number. For example, an e-mail message sent to myfriend@earthlink.com must be converted by a DNS into a number such as 120.131.200.41. Without the DNS, the Internet cannot function.[11]

Over time, the DNS organization has become almost identical to the structure of the Internet, because DNS servers are organized in a hierarchical (tree) structure as shown in Figure 12.2. In terms of *emergence theory*, the Internet has evolved into a hierarchical network. Initially somewhat random, the network is now structured as a tree. At the top of this tree are 13 root DNS servers. Hence, the Internet has 13 hubs—the most critical nodes in the network.

Three of the root DNS servers are physically located outside of the United States; one is at the University of Maryland, another one at NASA Internet Software Consortium, and one in Fairfax County, VA. Their exact locations are classified, because of their criticality. In Chapter 13, we will examine an attack on the DNS

[9]Paul Mockapetris[2] of USC/ISI invents DNS with the help of Jon Postel and Craig Partridge in 1983.
[10]An IP number corresponds to the network and specific machine that is connected to the network.
[11]DNS machine locations are secret because they are the anchor of the Internet.

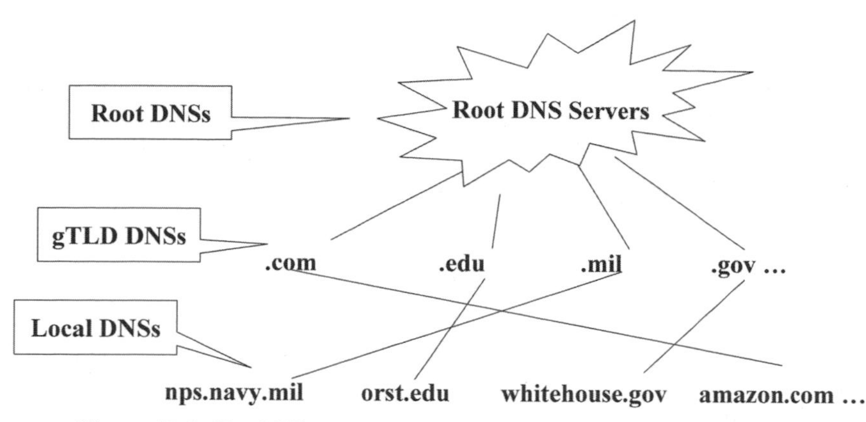

Figure 12.2. The DNS structure dictates the tree structure of the Internet.

root servers made in 2002, which led to their relocation. The perpetrators have not been found.

An additional 13 gTLD servers manage the "telephone directories" of the familiar. com, .gov, .edu, and other gTLD users of the Internet. Below these in the hierarchy are the millions of servers that connect desktop, laptop, cell phone, and PDA users to the Internet. Think of the DNS structure as a postal address where the zip code is equivalent to the gTLD and the street address is equivalent to the name of an organization's server. Therefore, myname@myserver.gtld is the address of a desktop computer connected to the Internet through a server named "myserver."

The structure of the DNS is dynamic. Names and addresses are constantly being added and changed. Thus, the DNS "telephone book" is published several times per day, keeping the Internet current. This is one of the strengths of the Internet; it essentially rebuilds itself every few hours! Because the hierarchical tree-structured collection of DNS machines that maintain the DNS system are constantly being updated, there is some redundancy in the network. Lower level DNS servers replicate the root server DNS information.

Registration of your name and address with Jon Postel become a tradition—and eventually a business—called Internet Assigned Names and Addresses (IANA). Postel was at the pinnacle of the IANA organization for 30 years until his death in 1998![12] He put the dot in "dot com" and contributed to many other standards between 1969 and 1998.

RFC

Internet governance processes started to appear during this period. Steve Crocker of UCLA created a public process called RFC, which became the major tool of Internet

[12]http://www.postel.org/postel.html.

development and decision making. RFC 1 was issued by Steve Crocker on April 7, 1969, and describes the first Internet switch—called the interface message processor (IMP). All modifications to the Internet are vetted through an RFC. For example, RFC 688 documented a new standard for e-mail in 1975. By 2004, there were over 3700 RFCs on record.[13] All major steps in the evolution of the Internet are documented and assigned an RFC number.

E-mail

In 1971–1972, Ray Tomlinson invented what we now know of as *e-mail*, and Larry Roberts quickly improved on it. Tomlinson started using the "@" character to separate user name from computer and domain name in his e-mail headers, e.g., name@machine.com. This convention soon became the international method of addressing e-mail. Also in 1972, ARPA was renamed DARPA. More researchers began using the growing network, because it now stretched from the East Coast to the West Coast.

TCP/IP

A seminal event took place in 1973 that marks the technical beginning of the modern Internet. Vinton Cerf of Stanford and Robert Kahn of DARPA invented TCP (Transmission Control Program)—the protocol that defines how messages are formatted and sent across the network. TCP standardizes packet-based communications and goes even further—it defined how messages are sent *reliably*. Today, any network that uses TCP is part of the Internet. But in the period of time, 1973–1976, the term "Internet" was just beginning to be used by advocates of TCP. By 1976, DARPA required the use of TCP in ARPANet.

Also during this time, Robert Metcalf was working on a protocol for connecting computers together over a local area network (LAN). While working at Xerox PARC, he invented ethernet for LANs. This "local protocol" would become the dominant LAN protocol and IEEE 802.3 standard. Ethernet was significant because it also solved a problem with TCP. When two or more computers attempt to send a message at the same time, an electronic *collision* occurs. Both messages are garbled because only one message can be transmitted at a time. Metcalf proposed an elegant solution. Whenever a collision occurs, both computers try transmitting again after waiting a random length of time. Because the probability of the two computers waiting the same random length of time approaches zero as the number of retry's increases, both computers eventually succeed in sending their messages.[14]

TCP was not perfect and still is not. But even as early as 1978, Vinton Cerf, Jon Postel, and Danny Cohen realized that TCP was trying to do too much. It was too big. So they decided to divide TCP into two parts: TCP and IP; thus, TCP/IP was born. As the protocol took on more functionality, it became necessary to further

[13]http://www.rfc-archive.org/.
[14]This is called CSMA/CD or Carrier Sense Multiple Access/Collision Detection.

divide it and put different functions into different layers. TCP and IP are two of the four layers that define the Internet today.

The first layer of the Internet (Layer 0: Physical) consists of a wire, optical cable, or some other physical device. The second layer (Layer 1: Data Link) defines the packet format and how to deal with collisions. Layer 1 is essentially ethernet. The third layer (Layer 2: Internet Protocol) is the "IP" part of TCP/IP and defines how packets are routed from sender to receiver. Inside of every Internet switch is a routing table that tells each IP packet where to go next. The next layer controls how IP packets are recovered at the other end. Layer 3: Transport can be implemented as one of two protocols: TCP or UDP. TCP guarantees delivery of all packets and reorders any packet that arrives out of order. UDP is faster but less reliable as it does not guarantee delivery and does not bother to reorder packets. TCP keeps track of the packet delivery order and the packets that must be resent. UDP can lose packets.

TCP is used for e-mail and most Internet transmissions, and UDP is used for streaming media such as video and audio, where a missing packet or two will not be noticed. UDP is fast because it does not have to reorder or retransmit packets.

The four-layer TCP/IP protocol described above defines the modern Internet. In 1988 the International Standards Organization released the Open Systems Interconnect (OSI) standard—a competitor to TCP/IP. OSI defines three more layers: layer 5: Session, Layer 6: Presentation, and Layer 7: Application. Figure 12.3 shows all layers of the ISO/OSI standard. Although ISO/OSI is the international standard, the popularity of PCs running TCP/IP, and the fact that most of the servers on the Internet run TCP/IP, means that the Internet is for all practical purposes, identical to the four-layer TCP/IP protocol.

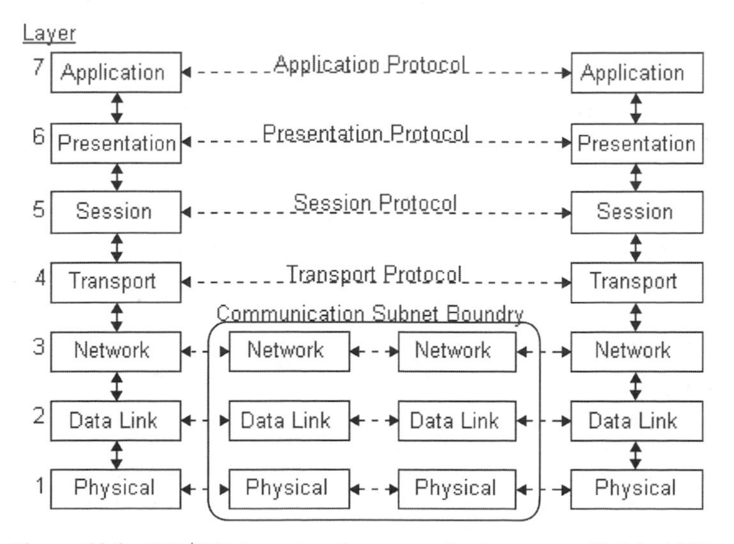

Figure 12.3. ISO/OSI for network communication as specified in 1988.

NSFNet

In 1981, the NSF established a research network based on TCP/IP called CSNet. It was aimed at serving the non-ARPANet users in the broader academic research community. Business was so good that CSNet was "outsourced" to MCI and re-named NSFNet. Then, in 1990, ARPANet merged back with NSFNet! Once again, the Internet was whole, and interoperable with anyone that adhered to the TCP/IP protocol.[15]

Meanwhile, the Internet was increasing in value and fashion appeal. Over one million users were paying subscribers by 1992. A National Research Council report chaired by Kleinrock suggested that the Internet be commercialized (at this time it was still the responsibility of the NSF). This report attracted the attention of Vice President Albert Gore, and in 1999, the Vice President of the United States claimed parentage of the Internet.[16] In 1992 the U.S. Congress gave the NSF permission to commercialize the Internet. This began a 5-year process of transition that ended with the privatization—indeed, the globalization—of the Internet.

A year later (1993), the number of subscribers had doubled to 2 million. The NSF created InterNIC to support the rapidly growing Internet and contracted with AT&T to maintain the DNS structure. In addition, the NSF awarded a 5-year contract to Network Solutions Inc to sell domain names for $50/year. During this period, millions of people became subscribers—fueling the Internet Bubble.[17]

After spending $200 million from 1986 to 1995, the NSF outsourced the Internet to four companies and turned the business of doing Internet business over to the U.S. Department of Commerce. In 1997, the Clinton administration directed the Secretary of Commerce to privatize the DNS, "in a manner that increases competition and International participation." True to the Internet culture as Steve Crocker defined it, a "white paper" was circulated like an RFC document, by the U.S. Department of Commerce. In 1998, the Internet was set free.

THE WORLD WIDE WEB

Progress continued at a rapid rate throughout the 1980s and 1990s as the Internet coevolved with the rise of the PC. In 1979 there were 100 users of ARPANet. In 1984 the number had grown by a factor of 10 to 1000 users, and another factor of 10 brought the total to 100,000 users by 1990! But the number of Internet users

[15]Cisco Systems, one of the most successful companies to commercialize TCP/IP equipment, was founded in 1984 by Leonard Bosack and Sandra Lerner, who later sold their interests in the company for $170 million.

[16]During a March 1999 CNN interview, while trying to differentiate himself from rival Bill Bradley, Gore boasted: "During my service in the United States Congress, I took the initiative in creating the Internet."

[17]The Internet Bubble (1995–2000) was a period of economic excess where billions of dollars were invested in "dot com" startups attempting to commercialize the Internet. A few of these startups survived, e.g., Amazon.com, Yahoo.com, and Ebay.com, but most of them went out of business, leaving many stock market speculators stunned.

would never rival that of radio or TV unless the Internet offered something more than connectivity. What the infant network needed was applications or, better yet, the *killer application!*

In 1982, when Jon Postel established SMTP as the standard for doing e-mail, the killer application of the Internet seemed to be e-mail. Most of the messages traveling over the Internet were e-mail messages. Even the defense, research, and university communities used the Internet mainly for e-mail. (This was a curious outcome, because the original purpose of the Internet was to share large centrally managed mainframes!)

The killer application for the Internet—the application that would ignite mainstream adoptions of networking—was invented by Tim Berners-Lee, while he was working for the world's largest particle physics research laboratory—the Center for European Nuclear Research (CERN). In 1989, Berners-Lee invented the *World Wide Web* (WWW)—a network of hyperlinked documents accessed via the Internet. Then, he built the first browser and invented HTML to support the sharing of hyperlinked documents across the Internet. His goal was to simplify the publication of research papers so that any physicist could disseminate his or her research electronically. What if an author could simply imbed a hypertext URL in any textual document, so that another document could be selected and retrieved merely by clicking on the embedded hyperlink? This would simplify the retrieval of referenced papers, regardless of where they were stored. One document could come from machine A, another document from machine B, and another document from machine C. Regardless of where the document lived, the collection of documents would pop up on the user's screen as if it was part of one large collection.

The hyperlinked document idea was not new, but Berners-Lee had to overcome the mindset of the Internet, which was that networking was designed to connect computers to users, and users to computers. The bigger the computer, the more users needed the network connection. But Berners-Lee had a better idea. Why not connect users to documents, regardless of where they were? Users wanted information, not connectivity, to hardware. This was an obvious notion in hindsight, but at the time, it was a contrarian's view of what the Internet was good for.

Berners-Lee called his software a *browser-editor*, because it combined a text editor with a web of hyperlinked documents. This provided a powerful tool for scientists, but it lacked the ease of use that consumers accustomed to a graphical user interface expected. What the WWW needed was a browser that worked like the graphical user interface on a PC. If an ordinary consumer can use a PC, he or she should be able to use the WWW, but this required a simpler browser.

Marc Andreesen and Eric Bina developed the first graphical browser for the WWW while students at the University of Illinois-Urbana. MOSAIC was a better mousetrap, because it simplified the user interface. When it became available for Macintosh and PC computers, it had the additional advantage that it ran on low-cost equipment and looked just like any other PC application. MOSAIC was more than an easier-to-use browser-editor. It enhanced the hypertext language invented by Berners-Lee. According to Andreesen:

> Especially important was the inclusion of the "image" tag which allowed to include images on web pages. Earlier browsers allowed the viewing of pictures, but only as separate files. Mosaic made it possible for images and text to appear on the same page. Mosaic also sported a graphical interface with clickable buttons that let users navigate easily and controls that let users scroll through text with ease. Another innovative feature was the hyper-link. In earlier browsers hypertext links had reference numbers that the user typed in to navigate to the linked document. Hyper-links allowed the user to simply click on a link to retrieve a document.[18]

Andreesen and Bina moved to California, co-founded Netscape Communication Corporation with money from Jim Clark, re-wrote MOSAIC and called in Netscape Navigator. The trio built the first WWW company—Netscape Communications Company. The highly successful enterprise was sold to AOL in 1999, but for a brief time, it was the fastest growing company in America.[19] Even more significant, Netscape ignited the commercial Internet. At the time of its public offering in 1995, Netscape claimed 35 million users. Five years later, the Internet had over 250 million users and 75% of them used Netscape's browser.

A large installed base of PC users, an easy-to-use graphical browser, and a cleverly designed WWW all came together in 1995 to propel the Internet into the mainstream. During the 5-year period from 1995 to 2000, adoption of the Internet far exceeded the 30-year adoption rate of cable TV, 20-year adoption rate of the home computer, and 15-year adoption rate of the VHS/VCR.[20] The Internet achieved 50% market penetration in 5 years—an adoption rate that has yet to be beat by any other technological product.

INTERNET GOVERNANCE

A question often asked is, "Who owns the Internet?" Other infrastructure sectors are owned by corporations or jointly by public–private partnerships. Public utilities (water and power) are often pseudo-private, meaning they are either heavily regulated monopolies or completely owned and operated by a municipality or metropolitan region. The Internet is different, because for one thing, it is a global organization. Its governance resembles the United Nations more than Microsoft Corporation or the Federal Government. The "UN of cyberspace" is actually a loose collection of societies mainly run by volunteers! Figure 12.4 lists some of the major groups that play a role in Internet standards, design, and ethics.

[18]http://www.ibiblio.org/pioneers/andreesen.html.

[19]In 1999, AOL paid $10 billion in stock for 5-year-old Netscape.

[20]It took 67 years for the public telephone to penetrate 50% of the homes in the United States.

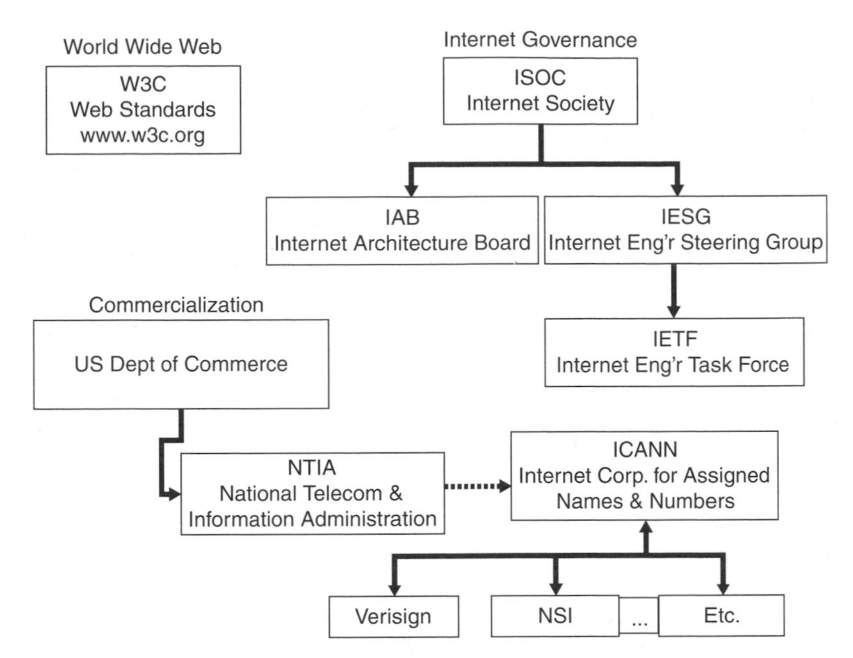

Figure 12.4. Governance of the Internet circa 2004.

IAB and IETF

The Internet is an open society of many volunteer organizations simultaneously contributing new ideas and technical recommendations for its evolution. It is a decentralized, freewheeling society that evolves standards rather than dictates them. One of the earliest of these voluntary organizations was the Internet Architectures Board (IAB) formed by Barry Leiner. According to RFC 1120, the IAB:

1) Sets Internet standards
2) Manages the RFC publication process
3) Reviews the operation of the IETF and IRTF[21]
4) Performs strategic planning for the Internet, identifying long-range problems and opportunities
5) Acts as a technical policy liaison and representative for the Internet community
6) Resolves technical issues that cannot be treated within the IETF or IRTF frameworks

Perhaps the most significant influence on the Internet has come from the activities of the IETF, formed in 1986. Since 1969, technical decisions regarding the Internet

[21]IETF = Internet Engineering Task Force; IRTF = Internet Research Task Force.

have been vetted by the user community using the RFC process established by Steve Crocker and the mediation powers of Jon Postel. Through the RFC process, any individual or group has a voice in Internet governance. Anyone can propose a modification, which is vetted by the IETF. This process has been formalized by a series of RFCs and is standard operating procedure for the ISOC, IAB, and IETF today.

This freewheeling approach to management should not work, but it does, perhaps because all decisions are documented and *Best Current Practices* (BCP) are catalogued and followed. Indeed, the open culture of the IETF has permeated the entire Internet culture and has had a profound impact on the way the information infrastructure sector has evolved. Excerpts from the RFC 3233 below underscore two key features of Internet governance: (1) its freewheeling—almost anarchical structure of governance, and (2) the culture of the Internet as it has evolved over 30 years.

According to RFC 3233:

> [BCP 9], the primary document that describes the Internet standards process, never defines the IETF. As described in BCP 11 ('The Organizations Involved in the IETF Standards Process') [BCP 11], the Internet Engineering Task Force (IETF) is an open global community of network designers, operators, vendors, and researchers producing technical specifications for the evolution of the Internet architecture and the smooth operation of the Internet. It is important to note that the IETF is not a corporation: it is an unincorporated, freestanding organization. The IETF is partially supported by the Internet Society (ISOC). ISOC is an international non-profit organization incorporated in the US with thousands of individual and corporate members throughout the world who pay membership fees to join. The Internet Society provides many services to the IETF, including insurance and some financial and logistical support. As described in BCP 11, Internet standardization is an organized activity of the ISOC, with the ISOC Board of Trustees being responsible for ratifying the procedures and rules of the Internet standards process. However, the IETF is not a formal subset of ISOC; for example, one does not have to join ISOC to be a member of the IETF. There is no board of directors for the IETF, no formally signed bylaws, no treasurer, and so on.[22]

As the number and scope of topics handled by the IETF broadened, the IESG (Internet Engineering Steering Group) was established by RFC 3710 to manage the expanded number of working groups:

> The Internet Engineering Steering Group (IESG) is the group responsible for the direct operation of the IETF and for ensuring the quality of work produced by the IETF. The IESG charters and terminates working groups, selects their chairs, monitors their progress and coordinates efforts between them. The IESG performs technical review and approval of working group documents and candidates for the IETF standards track, and reviews other candidates for publication in the RFC series. It also administers IETF

[22]http://www.faqs.org/rfcs/rfc3233.html.

logistics, including operation of the Internet-Draft document series and the IETF meeting event.[23]

Most decisions that deeply affect the technical evolution of the Internet come from the IETF, which are ratified by the ISOC and implemented by vendors. It is a remarkably decentralized and unfettered system that has reinforced the freewheeling culture of individuals, groups, and corporations that collectively comprise Internet governance. Most of the political and international governance of the Internet come from ICANN—the governing body set up by the U.S. Government when the Internet was spun out of the NSF in 1998.

ICANN Wars

The relatively self-governed Internet community is not without acrimony. In fact, there has been an abundance of disagreement over how the Internet should evolve, especially after it was set free by the U.S. Government. Most conspicuously was the so-called "ICANN Wars," which raged for years after the commercial Internet was born in 1998.

In June 1998, the U.S. National Telecommunications & Information Administration (NTIA) published the *White Paper* (Management of Internet Names & Addresses) in response to public comment on the *Green Paper*—an RFC-like proposal on how to commercialize the Internet. The NTIA proposed formation of a non-profit corporation—The Internet Corporation for Assigned Names & Numbers (ICANN), which subsequently assumed responsibility for management of the domain name system, allocation of IP address space, specification of protocols, and management of the root server system. ICANN does not register domain names itself. Instead, it delegates that responsibility to national registrars. However, this and other functions overlapped with much of what IANA was doing so Jon Postel revised the charter of IANA.

ICANN is governed by 19 directors who are broadly representative of the Internet community. Most members are appointed by their supporting organizations, but some are elected by members-at-large. For example, in 2003, the members of the ICANN Board were as follows:

Internet pioneer Vinton Cerf (Chair)
Mexican academic Alejandro Pisanty (Vice-Chair)
European lawyer Amadeu Abril i Abril
California lawyer Karl Auerbach
Brazilian businessman Dr. Ivan Moura
U.S. businessman Lyman Chapin
Canadian lawyer Jonathan Cohen
Mouhamet Diop

[23]http://www.faqs.org/rfcs/rfc3710.html.

Japanese businessman Masanobu Katoh
Netherlands businessman Hans Kraaijenbrink
Korean academic Dr. Sang-Hyon Kyong
Dr. M. Stuart Lynn (ICANN President & CEO)
German journalist Andy Mueller-Maguhn
Japanese academic Dr. Jun Murai
Dr. Nii Narku Quaynor
German businessman Helmut Schink
Francisco A. Jesus Silva
U.S. academic Dr. Linda Wilson

ICANN was envisioned to be more than an "FCC of the Internet" but fall short of "owning the Internet." But exactly what was the scope of ICANN's powers? In fact, several independent groups had other ideas about ICANN's power over the Internet. This difference of opinion evoked the ICANN Wars.

Dan Schiller—author of *Digital Capitalism*—called ICANN an "unelected parliament of the Web."[24] Karl Auerbach—ICANN board member in 2003—complained that ICANN was "essentially an organ of the trademark lobby."[25] Others accused ICANN of establishing policies that negatively impacted free expression and favored commercial interests over personal privacy. Milton Mueller lamented that the net's "role as a site of radical business and technology innovation, and its status as a revolutionary force that disrupts existing social and regulatory regimes, is coming to an end."[26] Criticism was not restricted completely to business. Network Solutions, Inc., the company that received a 5-year contract (1993–1998) to perform ICANN-like services on a temporary basis until the Internet was commercialized, complained in testimony to Congress that ICANN was out to destroy its business.

By the time you read this ICANN may have been replaced by another international body. More than 200 leaders from government and business attended the Global Forum on Internet Governance, held in 2004 by the United Nations Information and Communication Technologies (ICT) Task Force. The purpose of this meeting was "to contribute to worldwide consultations to prepare the ground to a future Working Group on Internet Governance to be established by Secretary-General Kofi Annan, which is to report to the second phase of the World Summit on the Information Society."[27] The United Nations, like so many other industrial age organizations, was slow to understand the significance of the Internet. But once they "got it," they began to organize their own brand of governance. The "UN of cyberspace" may turn out to be the UN after all.

[24]D. Schiller, *Digital Capitalism Networking the Global Market System*, MIT Press, Cambridge, 2000.
[25]http://www.icannwatch.org/.
[26]M. Mueller, *Ruling the Root: Internet Governance and the Taming of Cyberspace*, MIT Press, Cambridge, 2004.
[27]http://www.circleid.com/channel/index/C0_1_1/.

ISOC

In 1992, soon after Congress directed the NSF to commercialize NSFNet, Cerf and Kahn formed the Internet Society (ISOC), which has evolved into an umbrella organization, embracing social as well as technical issues.[28] Some topics of concern to ISOC:

- Censorship
- Copyright
- Digital divide
- Domain name systems
- E-Commerce
- Encryption
- Privacy
- Public policy
- Security
- Societal
- Spam

W3C

The startling success of the WWW and commercialization of the Internet prompted Berners-Lee and Al Vezza to form the World Wide Web Consortium (W3C) to create WWW technology and standards in 1994. According to the W3C, its charter is to formally nurture the Web as the Internet has traditionally been nurtured by volunteers. The Internet is a highway; the Web is a transportation system:

> The Web is an application built on top of the Internet and, as such, has inherited its fundamental design principles.
>
> 1. *Interoperability*: Specifications for the Web's languages and protocols must be compatible with one another and allow (any) hardware and software used to access the Web to work together.
> 2. *Evolution*: The Web must be able to accommodate future technologies. Design principles such as simplicity, modularity, and extensibility will increase the chances that the Web will work with emerging technologies such as mobile Web devices and digital television, as well as others to come.
> 3. *Decentralization*: Decentralization is without a doubt the newest principle and most difficult to apply. To allow the Web to "scale" to worldwide proportions while resisting errors and breakdowns, the architecture (like the Internet) must limit or eliminate dependencies on central registries.[29]

[28]http://www.isoc.org/.
[29]http://www.w3.org/.

The W3C has more profound objectives than making sure the Web is healthy. It seeks to take the Web to its next level. The WWW and its underlying HTML provided a standard *syntax* for information, but it did not define the *semantics* of the information. A "sentence" in HTML could be syntactically correct but meaningless. For example, the English sentence "The four sides of a square are circles" is syntactically correct, but meaningless. So, Berners-Lee set about to add meaning to the WWW. In 1996, W3C began working on XML and the "semantic network."

XML consists of three major parts: a language for encoding information—both as a document and as a message; eXtensible Style Language (XSL) software for rendering the information on a display (browser, printer); and Data Type Definition (DTD) a language for specifying the meaning of the information. Think of XML as a language (English, French, Italian), DTD as a dictionary, and XSL as an interpreter. Whenever an XML message is received, the receiving computer looks into a corresponding DTD to find the meaning of the tags in the message and then uses XSL to render the message on the user's screen. This is like an English-speaking person using an English-to-French dictionary to parse and understand French.

Today all browsers support XML and the "lingua franca" of XML. In fact, XML is the technology used to solve many homeland security problems such as interoperability between different computer systems, sharing of information among people with different levels of security, and data-mining to extract meaning out of databases.

IT-ISAC

The Internet is part of the information infrastructure as defined by the 2003 National Strategy for the protection of critical infrastructures and key assets. But it is not the entire sector. The nonprofit IT-ISAC is more comprehensive, covering all aspects of information technology. The IT-ISAC was formed in response to PDD-63 to share threat and vulnerability information among its members:

> The ISAC exists to enable members to share data. Using this shared data and other information, the ISAC operations staff gathers, analyzes, and disseminates to the members an integrated view of information system vulnerabilities, threats, and incidents that are relevant to the ISAC's sponsoring sector organization. The "other" information may be gathered from public and private sources, including semi-private organization like CERT or the publicly funded NIPC, or private organizations. The ultimate use of all of this data is to produce a coherent picture of the current state of the threat to the members. Likewise, the ISAC Operations also shares best security practices and solutions amongst its members.

The IT-ISAC was formed in 2001 and is funded by membership fees. Members are largely private sector corporations with an interest in protecting their industry. In 2004, the board members were Cisco, Computer Associates, BearingPoint, Verisign, Hewlett-Packard, International Business Machines, Microsoft Corporation, Oracle Corporation, and RSA Security Corporation.

INTERNET AND WWW TECHNOLOGY

The radical transformation of the telecommunications and information infrastructure sectors of the United States is a consequence of the bigger shift from analog to digital communications. Everything in communications is going digital—from computers to radio, television, music, and motion pictures. The Internet is evolving into the "digital backbone" of this transformation. Soon all communications on the planet will travel over a TCP/IP connection.

We can summarize this departure from analog as follows: Data are packaged and transmitted as *Packets of 1's and 0's*. That is, all information is encoded as a series of YES/NO, TRUE/FALSE, or 1/0 states. Whether we call these bits or states is immaterial. The important thing to understand is that all information can be reduced to fundamental pieces that take on only one of two states.

Rather than transmit single bits one at a time, it is more efficient to group them into blocks called *packets*. A packet is a collection of information bits, plus additional bits that tell switching machines where to send the packet, how to correct for errors, and what kind of information bits are contained in the packet. For example, a packet may contain bits that encode image, music, or numbers for a spreadsheet. Regardless of how the information in a packet is used, it all looks the same to a machine. But when it arrives at its destination, a packet must be *interpreted* as part of an image, sound, or numerical value.

The binary states of information are organized according to rules called protocols. *Protocols* are the rules that govern the transmission, security, and presentation of the packets. Without protocols, machines would not know what to do with packets. They control every aspect of data delivery.

Different protocols apply to different layers of communication. For example, the ISO/OSI standard as shown in Figure 12.2 defines seven layers or levels of rules that apply to a stream of packets. At each layer, a different protocol is used to manage the flow of information.

Some examples of protocols are as follows:

TCP/IP: Internet data communication.

DNS: Used in the IP addressing scheme.

HTML: Used to create Web pages.

XML: Used to add meaning to information.

HTTP: Hyper Text Transport Protocol: Rules of Web transmission.

SMTP: Simple Mail Transport Protocol: Rules for sending/receiving e-mail.

MIME: Multipurpose Internet Mail Extensions: Rules for e-mail attachments.

UDP: User Datagram Protocol used for streaming media like video.

SNMP: Simple Network Management Protocol used to manage a network.

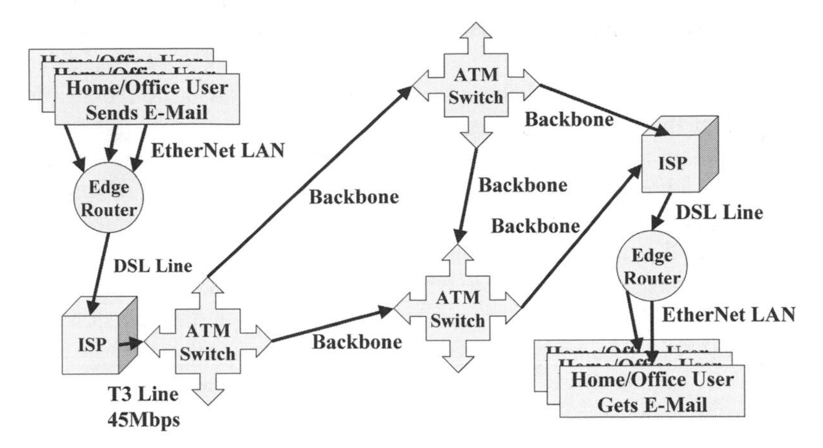

Figure 12.5. Example of an e-mail message as it travels through the Internet.

OSPF: Open Shortest Path First used to route packets through Internet switches.

BGP: Border Gateway Protocol used to establish routes through switches.[30] BGP governs the communication among autonomous systems (AS).

To illustrate the usefulness of these protocols, suppose we follow an e-mail message with an attachment, as it goes from the sender in the upper left-hand corner to a receiver in the lower right-hand corner of Figure 12.5.

Step 1: Encoding the message: The e-mail message is converted into plain text, HTML or XML on the desktop machine, and its attachment is "wrapped" inside of a MIME formatted message so that it is recognized as an attachment. The SMTP protocol defines how the e-mail is handled as it travels through the network. It is important to tag both message and attachment, because they may contain text, pictures, audio, and video information. Each of these types must be handled differently on the receiving end.

Step 2: The desktop sender's machine is connected to the Internet through an *edge router* and LAN that communicates in TCP/IP—or UDP if the data are streaming audio or video. So the e-mail and its attachment must be further broken down into packets—TCP/IP or UDP packets. These packets encapsulate the data as well as the source and destination addresses of sender and receiver. But these addresses are in the form of Myname@mymachine.mydomain, instead of an IP address such as 131.200.13.2. So the nearest DNS "telephone book" is consulted to translate the symbolic address from Myname@mymachine. mydomain into

[30]BGP first became an Internet standard in 1989 and was originally defined in RFC 1105. The current version, BGP-4, was adopted in 1995 and is defined in RFC 1771 and its companion document RFC 1772.

131.200.13.2. Now each packet can be given an address and sent out through the edge router to an ISP.

Step 3: The ISP provides an onramp to the faster backbone of the Internet. High-speed lines (45 Mbps) and fast switches like asynchronous transfer modes (ATMs) route the e-mail as individual packets along the backbone of the Internet. They use the OSPF and BGP protocols to select which routes and physical lines to use. The ISP manages a group of switches and routers; in other words, it is an autonomous system (AS). The e-mail must cross borders—from the AS managed by the sender's ISP to the AS managed by the receiver's ISP. The BGP rules govern this "border crossing" and routing of the e-mail. In addition, OSPF does exactly what its name implies: Using tables stored in the Internet's ATMs, it selects the shortest available path first. If the shortest route changes in mid-stream, one packet may take a different path than another. In fact, the packets from the sender's email message may arrive in different order because they took different paths through the network. If the TCP protocol was used, then TCP puts the out-of-order packets together again at the receiving end. If UDP was used, the out-of-order packets are discarded. In addition, if a packet is lost, TCP demands that it be sent again, which delays the message but assures that the entire e-mail message arrives intact. Thus, the routing information that essentially defines the graph-like structure of the Internet is established and updated using the BGP. BGP regulates communication among AS—collections of routers under a single administrative authority for routing packets. The OSPF protocol dictates how routes are selected.

Step 4: The packets travel across fast lines and ATM switches as they work their way across the network toward the recipient. Switches and transmission lines need to be maintained just like any other physical equipment. But the switches and routers are manufactured by different companies and may work in different ways. SNMP is an agreement among all manufacturers on how their devices will be managed. SNMP uses UDP to query and modify the behavior of every device in the Internet. SNMP is the Internet's "in-band" SCADA network. If something goes wrong, an SNMP agent signals this error condition so that the network operation center can take corrective action. Without SNMP, various devices from miscellaneous vendors would not work together, leading to interoperability chaos.

Step 5: The packets arrive at the recipient's desktop and are assembled into proper order according to the rules of TCP. Then the assembly process works its way up the ISO/OSI "stack," as shown in Figure 12.2. The SMTP and MIME protocols are worked in reverse order. The TCP packets are grouped into strings of HTML, XML, or pure text. Images and sound are tagged so that an application can recognize them as such. As the e-mail is reconstructed and tagged, it is stored on the recipient's disk drive as a formatted file. Clicking on it causes the appropriate application to open and read the message in the correct format. Note that the e-mail and its attachment can be anything—data, programs, and attachments containing audio, video, and pictures. In fact, the attachment can be a malicious program!

This example illustrates the use of routers and switches. Generally, switches (Layer 3: TCP) move packets between routers and switches, typically backbone networks. Routers (Layer 2: IP) move packets between local area routers and switches, typically within a LAN network. Switches are faster and more expensive, so they are used more in backbones. Routers are slower but cheaper, so they are used more in LANs. Routers and switches are managed from a distance, using the SNMP protocol and network operation center software.

VULNERABILITY ANALYSIS

Chapters 13 and 14 provide greater detail on information security. Before leaving this brief review of how the Internet works, it will be instructive to perform a risk analysis of the major autonomous systems of the Internet—the so-called *Tier-1 ISPs*.

The structure and activity of autonomous systems can be obtained freely from various "AS reports" obtainable from the WWW. Simply search on "AS Report." For example, the report for AS 1239 tells who owns it, and where it is located:

OrgName:	Sprint
OrgID:	SPRN
Address:	12502 Sunrise Valley Drive
City:	Reston
StateProv:	VA
PostalCode:	20196
Country:	US
ASNumber:	1239
ASName:	SPRINTLINK
ASHandle:	AS 1239

Every AS is given a number, which uniquely designates the administrative entity that controls the AS. Each AS report lists the current number of in-bound (upstream) and out-bound (downstream) connections, or adjacent ASs. For example, AS 1239 has 10 upstream and 902 downstream connections:

1239 SPRN Sprint Adjacency: 912 Upstream: 10 Downstream: 902
Upstream Adjacent AS list

AS 2764	UNSPECIFIED Connect Internet Solutions, Pty. Ltd.
AS 7474	OPTUSCOM-AS01-AU SingTel Optus Pty Ltd
AS 3356	LEVEL3 Level 3 Communications
AS 14742	PNAP Internap Network Services
AS 209	QWEST-4 Qwest
AS 3561	SAVVI-3 Savvis
AS 8057	VSNT Vision Net, Inc.

AS 19199	PAMAMS Pamamsat
AS 701	UU UUNET Technologies, Inc.
AS 4637	REACH Reach Network Border AS

Downstream Adjacent AS list

AS 17486	SWIFTEL1-AP Swiftel Communications
AS 17435	WXC-AS-NZ WorldxChange
AS 3643	SPRN Sprint
AS 7545	TPG-INTERNET-AP TPG Internet Pty Ltd

...

The adjacency list can be used to construct a network model of the entire Internet, where nodes are autonomous systems, and links are the upstream and downstream connections to other ASs. Indeed, this is how the network model of the top (Tier-1) ISPs was obtained as shown in Figure 12.6. However, only the

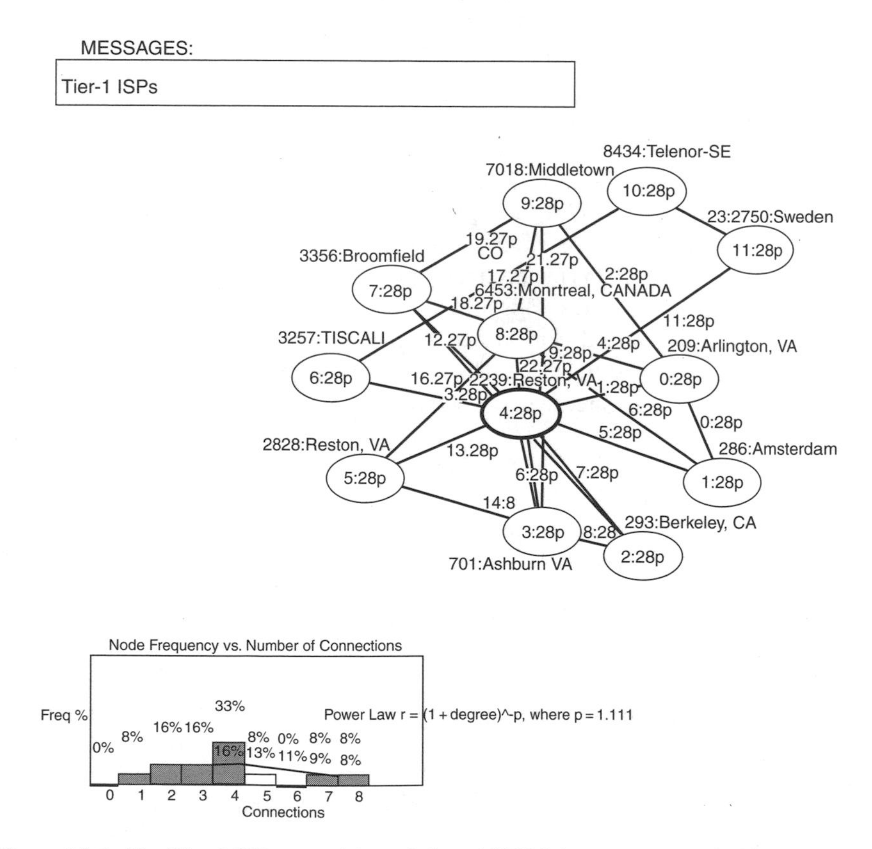

Figure 12.6. The Tier-1 ISP network has a hub at AS1239 that connects to the10 largest tier-1 ISPs as shown here.

top ISPs—in terms of their connectivity and number of customers—are analyzed in Figure 12.6.

Network-wide allocation of 1000 availability points to the 12 AS nodes in Figure 12.6 reveals the most critical nodes of the Internet:

AS#	Location or Name
209	Arlington, VA
286	Amsterdam
293	Berkeley, CA
701	Ashburn, VA
1239	Reston, VA
2828	Reston, VA
3257	TISCALI
3356	Broomfield, CO
6453	Montreal
7018	Middletown, NJ

Further detailed analysis ranks AS 1239 as the most critical node, followed by AS 701, AS 3356, AS 209, and AS 7018. Three of these are located near one another in Fairfax County, VA, which is already well known for its high concentration of Internet and WWW businesses. These results were obtained by assuming the damage values given by an independently derived AS analysis.[31] Damage is equated with number of customers, which roughly equates with the estimated value of each AS. These values were obtained for the month of June 2002.

Maps of the Internet, WWW, and Tier-1 ISPs are readily available on the Internet. These maps have been thoroughly studied, and it is a well-established fact that the Internet is scale free. This means it has hubs. In fact, its structure is hub-like from a variety of different measures. This means the Internet and WWW are vulnerable to asymmetric attack, but it also means it is much easier to protect the Internet and WWW than most policy makers realize. Careful application of network analysis suggests where the hubs are, and we already know that protection of hubs means protection of the entire sector.[32] Therefore, the best strategy for protecting the Internet is to protect the Tier-1 ISPs that we have identified here. Prevention of damage to the top dozen or so nodes and links connecting them is a much smaller problem that protecting 250 million nodes that comprise the global Internet.

ANALYSIS

By 1998 the Internet had matured to the point where it could be privatized. The NTIA produced a "Green Paper" describing how the Internet should be governed,

[31]J. I. Nieto-Hipolita, and J. M. Borcelo, "Analysis of Autonomous Systems," Technical Report: Computer Architecture Department, Polytechnic University of Catalonia, Barcelona, Spain, 2002.

[32]The Example in Figure 12.6 can be found in the program *NetworkAnalsys.html*, a.k.a. *NA.jar*.

how to transition the DNS to private ownership, how to add more gTLDs how trademarks should be honored as Internet names, how to reduce the $50 DNS registration fee to $20, and how to set aside 30% of the revenues from DNS registration for the Intellectual Infrastructure Fund (IIF). The Green Paper formalized Jon Postel's operation and created ICANN to sell blocks of names to several authorized resellers (see Figure 12.4).

The technical structure of the Internet has been shown to be similar to previously studied critical infrastructure sectors. Its most important components are highly concentrated hubs that are critical to the operation of the entire sector. Indeed, the Tier-1 ISPs are few in number and closely located, specifically around the Washington, D.C. "capitol area." Network-wide analysis reveals that the Internet is highly dependent on continuity of fewer than a dozen autonomous systems.

But understanding the technical structure of the Internet and WWW may be a small challenge compared with understanding the organizational and regulatory challenge posed by this vast infrastructure. The question of Internet ownership remains complex at the time of this writing. It does not belong to anyone or any company. Rather, it operates through a convoluted social network of volunteers, nonprofit organizations, government agencies, and for-profit corporations. It is a global social system—not controlled or augmented by any single government.

The Internet has been called the "information superhighway," but it is radically unlike the federally funded interstate highway system. It no longer receives subsidies, nor is it considered a natural monopoly, even though the entire U.S. society depends on the Internet as much as it does the interstate highway system. Destruction of the Internet would have severe consequences on the national economy.

The Internet has been called the most significant advancement in human communications during the past 500 years.[33] And yet the FCC does not regulate it as if it was radio or television, nor is it managed like roads, bridges, or power grids. Analog spectrum for radio and telephone broadcast (and cell phones) is sold to the highest bidder for billions of dollars by governments around the world. According to the FCC, the electronic spectrum belongs to the interstate public, and thus it is subject to federal oversight. The Internet, on the other hand, is not regulated by any federal agency. Anyone can buy broadcasting rights for $20/year! ICANN and its authorized resellers literally give away one of the most valuable rights in human history—the right to broadcast to everyone in the world without a license!

The Internet has been compared with a global publishing and printing machine and a global vending machine. It provides merchants a global distribution channel that will soon reach a billion consumers—all for minimal cost. This has enormous consequences for e-commerce. So far, no major government has imposed taxation on e-commerce, and only minimal restrictions have been placed on spam, freedom of speech, and pornography. Will the United Nations react to this

[33]One can easily argue that the printing press was the most significant advancement in communications before the Internet.

unprecedented freedom of expression? Will the Internet be banned in major parts of the world?

Like many other technological advances before it, the Internet and WWW have been exploited for both good and evil. The WWW supports human networks consisting of both terrorists and pen-pals. It has been a vehicle for positive social change as well as social unrest. The Internet is destined to have a major impact on critical infrastructure sectors ranging from Level 1 sectors (telecommunications, information, power/energy, and water) to Level 3 sectors such as public health and emergency services—and all the sectors in between. If the Internet is critical to the operation of other critical infrastructures in the United States, then what policies need to be enacted to prevent its destruction by terrorists? In Chapter 13, we address the question of the Internet's security.

EXERCISES

1. What is the Internet (select only one)?
   a. Any digital network
   b. Any packet-switching network
   c. Any TCP/IP network
   d. Any Ethernet network
   e. All of the above
2. What is an "Internet protocol"?
   a. Rules for communication between networked devices
   b. IEEE 802.11 standard
   c. ARPANet predecessor to the Internet
   d. A Microsoft product
   e. Rules proposed by Al Gore, former Vice President of the United States
3. Which of the following are Internet protocols (select all that apply)?
   a. TCP/IP
   b. http
   c. PDF
   d. JPEG
   e. DNS
4. Which of the following are examples of coevolution (select all that apply)?
   a. Scale-free emergence
   b. Formation of hubs and webs on the Internet
   c. Rapid adoption of the horseless carriage and parking garages
   d. Formation of the IAB and IETF
   e. Adoption of the Internet and PC

5. In terms of DNS structure, the Internet is shaped like a (select only one):
   a. Hierarchical tree
   b. Mesh or grid graph
   c. Random graph
   d. Hamiltonian graph
   e. Complete or full graph
6. Which of the following are Internet vulnerabilities (select only one)?
   a. PC hardware
   b. PC operating systems
   c. Internet switches
   d. UNIX servers
   e. All of the above
7. Packet-switching networks were studied and invented by (select only one):
   a. Kleinrock
   b. Baran
   c. Davies
   d. All of the above
   e. None of these people
8. Which of the following is an example of an Internet gTLD (select only one)?
   a. .com
   b. name@earthlink.com
   c. an e-mail attachment
   d. An e-mail format
   e. www.CHDS.us
9. What does a DNS server do (select only one)?
   a. Registers user names
   b. Runs the Internet
   c. Translates a URL into an IP address
   d. Rebuilds the Internet
   e. Implements TCP/IP
10. Who invented and sent the first e-mail (select all that apply)?
    a. Ray Bradbury
    b. Ray Tomlinson
    c. Larry Roberts
    d. Ray Robinson
    e. Jon Postel
11. [Discussion] The citizens of San Lewis Rey are upset about Internet spam—the unsolicited transmission of e-mail containing advertisements for

unwanted products. They want the owners of the Internet to do something about it. They are hesitant to write their representatives in Congress, and because they lean toward libertarianism, they do not want to take legal action. Besides, whom would they take legal action against? What should they do?

a. Take out an advertisement in the Wall Street Journal
b. Formulate and submit an RFC to the ISOC
c. Create a "Citizens Against Spam" nonprofit action committee
d. Join the IETF
e. Write to Tim Berners-Lee

Write an essay on the merits and demerits of each of these approaches, explaining why each approach may or may not succeed.

12. Which of the following are TRUE?
    a. Originally, TCP/IP was TCP.
    b. The ISO/OSI dictates what protocol is used by the Internet.
    c. TCP/IP is the same as Layer 2: Transport of the ISO/OSI stack.
    d. All Internet routers and switches use HTML.
    e. UNIX is the operating system of the Internet.
13. Which of the following protocols are authorized for use in Layer 3: Transport?
    a. TCP and UDP
    b. ISO/OSI
    c. MIME
    d. SMTP
    e. SNMP
14. How much did the U.S. Government spend on the Internet during the period: 1986–1995?
    a. $200 million
    b. $1 billion
    c. $1.5 billion
    d. $5 billion
    e. Nothing—it was commercialized by then
15. Which protocol guarantees delivery of packets over the Internet?
    a. UDP
    b. TCP
    c. IP
    d. TCP/IP
    e. DNS
    f. SNMP

16. Which of the following are major components of XML?
    a. DTD and XSL
    b. XML and HTML
    c. MIME and DNS
    d. SNMP and http
    e. HTTP and SSL
17. Which of the following government agencies commercialized the Internet?
    a. NTIA
    b. ICANN
    c. IANA
    d. IETF
    e. ISOC
18. Which of the following is TRUE:
    a. Internet governance is top-down, from the ISOC to the IETF.
    b. Internet governance is up to the U.S. Government.
    c. Internet governance is mainly through international volunteer organizations.
    d. Internet is owned by the IT-ISAC.
    e. Internet operation is regulated by the FCC.
19. Which of the following are TRUE (select all that apply)?
    a. The WWW and Internet are the same thing.
    b. The Internet is hardware; the WWW is software.
    c. W3C and ISOC have overlapping powers.
    d. XML is an extension of HTML.
    e. None of the above are true.
20. Two routes exist between the sender and the receiver of an e-mail message in Figure 12.5. What happens if parts of the e-mail are sent along one route and another part is sent along a second route (select only one)?
    a. The entire e-mail message is retransmitted.
    b. The switches and routers use OSPF to correctly route the pieces.
    c. TCP flags the error.
    d. IP flags the error.
    e. DNS translates myname@myserver.com into 131.200.13.4.
21. Write an essay on how the Internet compares with:
    a. Interstate highway system
    b. Broadcast networks like radio and TV
    c. Mail-order catalog commerce
    d. Electric power utilities and the four interconnection grids of the United States

CHAPTER 13

Cyber-Threats

A *cyber-threat* is a danger to a computer or network of computers. It is a potential attack that preys on weaknesses or flaws in hardware and software systems. An *exploit* is defined as an unauthorized action performed on an information system such as a corporate network, desktop PC, enterprise server, website, factory control system, supervisory control and data acquisition (SCADA) network, or home computer. A *remote exploit* is an unauthorized access to an information system from a distance—from across a network. This chapter is about exploits—the potential unauthorized acts against information systems for the purpose of gaining control, stealing information, destroying data, and denying service to the authorized users of the system.

We assume that the information systems of greatest interest are connected to one another via the Web—the predominant application running on the Internet. The Web is modeled as a *scale-free cascade network* where nodes are Internet infrastructure servers, e-commerce servers, desktops, and laptops. The links connecting these nodes are any TCP/IP connection, whether it is a wired or wireless communication link.

In this chapter, the following concepts are explained through a combination of theory and real-world example:

1. The cyber-threat is real: Estimates of the financial impact of computer hackers and crackers on the U.S. economy range from $200 million per year to hundreds of billions. Regardless of the wide range of estimates of the cost, information technology systems are under almost continual attack.
2. Cyber-thieves are generally divided into two major groups: script kiddies are inexperienced novices seeking fame; black-hats are knowledgeable hackers (fun and fame) and crackers (fame and fortune) with more serious damage in mind. Attacks are euphemistically called *exploits*, because hackers and crackers exploit weaknesses in information systems to gain access or control.
3. The tools of the exploit trade are viruses (malicious self-replicating programs) and worms (malicious self-replicating programs that spread via the network). These tools are used to render a website unusable by denying access (DoS or

Critical Infrastructure Protection in Homeland Security: Defending a Networked Nation,
edited by Ted G. Lewis

denial of service), infect files and databases, inflict loss of information on the operation, and stop or lower worker productivity. In some cases, an exploit can result in the hacker remotely taking control of a target computer.

4. Break-ins typically begin with a minor infraction such as an unauthorized login, and escalate to more serious infractions using *backdoor* programs (malicious programs stored on the victim's computer), *Trojan horses* (deceptive programs that look innocent but are actually malicious), and *zombies* (other people's computers that are used to launch an attack on even more computers).
5. Web vulnerability begins with the study of the structure of the Web. The most significant feature of the Web is its *giant strongly connected component* (GSCC), which is in the "center of the Web." The GSCC includes the most critical nodes of the Internet: the 13 DNS servers and 13 gTLD servers. Because the Web is a scale-free network, critical nodes are easily identified as hubs—both infrastructure and commercial concentrations—where most of the Web's assets are concentrated. Thus, the first order of business for cyber-security is to protect these obvious targets.
6. A model-based vulnerability analysis (MBVA) of the Web identifies the following threats: viruses and worms. Viruses infect disks and application software. Worms spread like an epidemic throughout the Web and exploit TCP/IP flaws, unattended ports, buffer overflow weaknesses in operating systems, e-mail weaknesses, and miscellaneous flaws in software at all levels.
7. The Web is an example of a *cascade network*, which means it has the same vulnerabilities (and strengths) as the cascade networks studied in Chapter 3: Worms spread like epidemics in human populations. Of particular concern is the recurrence of susceptible-infected-susceptible (SIS) contagions, where by a malicious program infects a large part of the Web, subsides, mutates, and re-infects the Web, repeatedly.
8. Vulnerability analysis of the Web suggests that the optimal strategy for preventing cyber-attacks is to protect the major hubs (root DNS servers, gTLD servers, major e-commerce sites) and emphasize defenses against worms, especially DoS exploits. Because the Web is also a cascade network, Web epidemics spread and subside according to the SIS epidemic model. The more we protect hubs, the less pervasive is the epidemic, and the less likely that a certain infection will persist. From the policy maker's point of view, this means that resources should be focused on protection of nodes in the GSCC and the major hubs of the Web.

SCRIPT KIDDIES AND BLACK-HATS

According to the Computer Security Institute's 2002 CSI/FBI Computer Crime Survey, 98% of respondents experienced some form of cyber-break-in.[1] A total of 44% acknowledged financial losses of over $2 million, 70% were attacked from

[1] http://www.gocsi.com.

the Internet; 40% experienced a DoS attack, and nearly all (94%) reported they had detected the presence of at least one computer virus. Estimates of the financial loss due to exploits such as these range from $200 million to hundreds of billions, depending on how financial loss is defined. Regardless of the exact numbers, cyber-threats are real, and they cost real money.

A century ago the U.S. national economy depended on railroads and heavy industries to create wealth. Today, the U.S. economy is heavily dependent on information, and information is captured, stored, moved, processed, and delivered by information systems. So the train robber has been replaced by the cyber-thief—the so-called *script kiddies* and *black-hats* that prey on vulnerabilities in information systems.[2] Curiosity motivates script kiddies who use automated tools that are readily available over the Web to probe other people's computer systems. More pernicious are the black-hats—people that are driven by more serious motivations. *Hackers* typically break in because they can, whereas *crackers* break in to destroy or steal information. Both are knowledgeable experts that often develop their own malicious programs—mainly *worms*.[3] Cyber-threats are not acts of nature, but instead they are *manufactured* by hackers and crackers.

According to some experts, a cyber-"Pearl Harbor" is unlikely, because such an operation would be highly complex, require extreme coordination effort, and result in dubious damage.[4] A more likely scenario is that future black-hats will use cyber-attacks asymmetrically as a force multiplier in concert with a physical attack. For example, a cyber-attack might be used to interrupt emergency services, manipulate traffic control signals, hinder disaster recovery, and so forth, in concert with a bomb, biological, chemical, or other physical assault.

As we shall see once again, the Web is a scale-free network, which means it can survive random attacks or failures instigated by a cyber-attack. However, the Web is highly susceptible to attacks on its hubs. And there are many obvious "super hubs" in sectors like banking where the entire sector depends on a scale-free financial network. It turns out that these electronic networks are also *cascade networks*, and most exploits are aimed at disabling them through malicious programs that spread like an epidemic. We show that proper hardening of these hubs can prevent exploits from turning into successful attacks, but this requires constant updating of counter-measures. Cyber-threats are constantly evolving and adapting to measures used to block them.[5]

The most common exploits prey on flaws in software programs and hardware. Due to the complexity and size of most software in use today, many flaws or "holes" exist in the operating systems, application programs, and hardware that they run on. Software flaws—called *defects*—are often discovered years after consumers have deployed the software. Once a defect is discovered, the software

[2]Script kiddies are amateurs out for fun and glory. Black-hats are serious professionals working for financial gain.

[3]A worm is a self-activating program that spreads through a computer network like an epidemic.

[4]C. J. Dunlevy, "Protection of Critical Infrastructures: A New Perspective," CERT Analysis Center, 2004, http://www.cert.org.

[5]Use the keyword "exploits" in a search engine such as http://www.google.com and a very large number of cyber-threats will be retrieved.

manufacturer may offer a repair—a *patch*—that fixes the problem. Unfortunately, many of these patches are never installed, leaving the information system vulnerable. One of the most effective counter-measures to combat cyber-attack—the patch—is often overlooked.

Hackers and crackers use a variety of methods for penetrating corporate and home systems. One of the oldest is called *war dialing*, where a hacker programs his or her computer to dial all the telephone numbers listed in a telephone book, until a modem is sensed. Once the war dialing computer senses a modem tone, it repeatedly sends login and password combinations—words taken from the English dictionary. If the password is a proper English word, the war dialer will eventually discover it.

War dialing is a tedious brute-force method of breaking into someone's computer, but as it is done by another computer, it is easy and inexpensive for the hacker. Today the large number of open wireless access points encourage a variation of war dialing called war driving. A mobile hacker equipped with a laptop computer simply drives around a neighborhood until an 802.11 Wi-Fi signal is detected and then begins exhaustively sending a series of user name and password codes (generated from a dictionary), until the login is successful.

Once in, a hacker or cracker will attempt to escalate his or her access privileges. Can the invader open up password files to get more user names and passwords? Can he or she intercept another user's e-mail? Is the corporate database accessible from the login?

In some cases, the professional cyber-thief can store a program on the cracked system for use at a later time. This is called a *backdoor*—a program that the hacker activates from outside of the security zone of the cracked system. A backdoor program may lie dormant for a long period of time before it is activated or activate itself periodically. It can look like an authorized part of the system but instead become destructive. A *Trojan horse* program is a deception, it looks valid, but it is not.

War dialing and war driving are not the only means of hacking into a system. A large number of exploits come from employees or trusted associates—the so-called *insiders*. But perhaps the most disturbing exploits come from outside the organization. If information systems can be attacked from anywhere in the world, no infrastructure sector is secure. Anyone in the world can attack any Web-connected system located anywhere else in the world with inexpensive equipment and knowledge of how computers and networks operate. In fact, the construction of malicious programs for the purpose of carrying out *remote exploits* has become an entire cottage industry of virus, worm, and Trojan horse software developers.

Attacks from inside the organization and its information system perimeter are called *insider attacks*. Whether an attack comes from the inside or outside, exploits are not difficult to initiate. The tools already exist, and for the most part, they can be acquired at little expense. Many of these tools are available from the Web. They fall into the following general categories:

- Virus programs (user-activating software that spreads via files, etc.)
- Backdoor programs (black-hat takes remote control)

- Trojan horse programs (deceptive software)
- Worm programs (self-activating software that spreads via a network)

These tools are used for a variety of nefarious activities, including, but not limited to:

- Stealing passwords or credit card information
- Taking control of a remote computer or network
- Destroying or corrupting files and databases
- Using a remote computer to spread viruses and worms to others
- Turning a remote computer into a *zombie*—a computer that launches a subsequent DoS attack on a website or corporate network

MODEL-BASED VULNERABILITY ANALYSIS OF THE WEB

We apply the five-step MBVA method to the Web and show that the most critical nodes are located in the GSCC "neighborhood" of the Internet, and the most serious cyber-threats come from hackers and crackers who are continually devising new exploits that take advantage of networks, computers, hardware, and software that make up the information technology sector.

Recall the steps in MBVA are as follows:

1. Take inventory: What assets do you have?
2. Reveal the sector's architecture—are there hubs?
3. Build fault tree model of vulnerabilities.
4. Build event tree model of outcomes.
5. Decide resource allocation strategy.

Take Inventory

The first step in MBVA is to identify the components of the infrastructure and try to understand their relationship to one another. In the case of the Web, components are the key pieces of hardware and software that make up the computers and network equipment of the Web. The links are telecommunications hardware and software that comprise the Internet. Therefore, nodes are primarily computers, and links are communication lines.

It has long been established that the Web is a scale-free network. Indeed, it is scale free in several ways (see Table 13.1). Broder et al. discovered a deeper secret of the Web: It is made up of components with distinct properties.[6] One component consists of sparsely connected input nodes, designated as IN (see Figure 13.1). Another

[6]A. Broder, R. Kumar, F. Maghoul, P. Raghavan, S. Rajagopalan, R. Stata, A. Tomkins, and J. Wiener, *Computer Networks*, 33, p. 309, 2000.

TABLE 13.1. Properties of the Web and Corresponding Scale-Free Exponents.

Property of the Web	Exponent in Power Law
Number of in-links	2.1
Number of out-links	2.72
Strongly connected components	2.54
Number of web pages in a site	2.2
Number of visitors to a site during a day	2.07
Number of links clicked by Web surfers	1.5
Page rank	2.1

component consists predominantly of output nodes, designated as OUT. One large component of 56 million nodes—called the GSCC—consists of a strongly connected core. Recall that a graph is strongly connected if every node is connected to every other node through a series of one or more links. Hence the GSCC is a graph of densely linked nodes. Figure 13.1 shows other nodes: 8.3% are relatively disconnected or isolated, and 21% are "feeder" nodes called *tendrils*.

Most of the traffic on the Web goes through the GSCC. Only a relatively few number of packets go through express links, shown as *tubes* in Figure 13.1. Thus, for the most part, the GSCC is critical to the Web because without it, the Web would become fragmented and break into islands of isolated components.[7]

Figure 13.1 lists the percentage of Web nodes belonging to each component. Most nodes are nearly evenly distributed among IN, OUT, TENDRIL, and GSCC components. The even distribution of nodes to each type of component suggests a prevention strategy that evenly amortizes target hardening across every component. But this conclusion would be wrong. The criticality of each component depends on which Web application is used most by consumers. For example, if e-mail traffic is traced through the bow-tie structure of Figure 13.1, the most critical components are GSCC and OUT (see Table 13.2). Most e-mail exploits traverse the GSCC component on their way to the OUT nodes. Hence, GSCC is more critical than the other components.

Newman et al. showed that nodes in the DISCONECTED and OUT components account for most e-mail activity.[8] The DISCONNECTED component contains 41% of the nodes, but they are disconnected, which means an e-mail exploit will have little effect on the Web. On the other hand, 54% of the Web's nodes can be reached via the OUT and GSCC components. Therefore, e-mail exploits greatly impact the Web because they reach nodes in the OUT and GSCC components. This is why e-mail exploits are so virulent.

The nodes of the GSCC also form a *small world*. The average path length is 16 (directed path exists) and 7 (undirected) hops, respectively. The shortest directed

[7]The Web is somewhat like most of the power grid in this respect.

[8]M. E. J. Newman, S. Forrest, and J. Balthrop, "Email Networks and the Spread of Computer Viruses," *Physical Review E*, 66, 035101-(R), 2002.

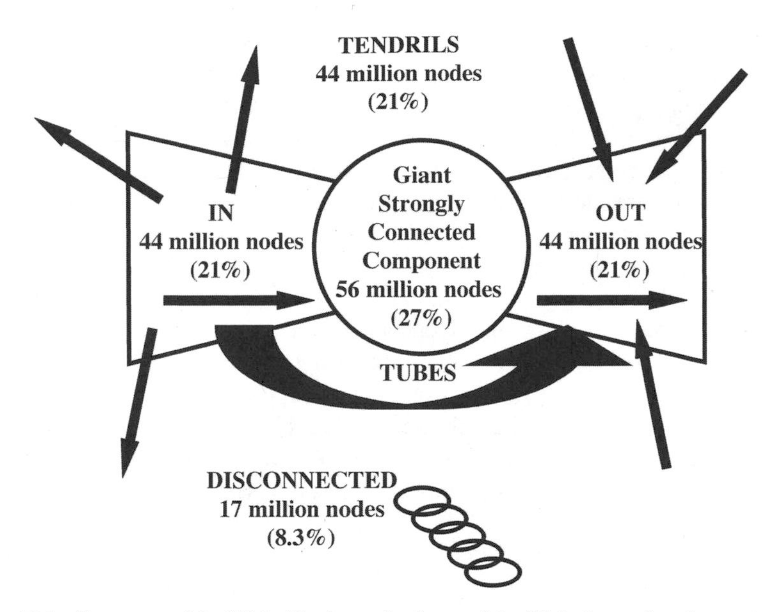

Figure 13.1. Structure of the Web: The bow-tie shape of the Web shows that the center of the Web is a GSCC.

path between two nodes in GSCC are 16–20 hops on average, and the power law distribution has an exponent of approximately 1.6 to 1.9, versus approximately 2.1 for the Web as a whole. These observations reinforce the notion that the GSCC is a vulnerable component. Inside GSCC are several scale-free network hubs that are even more vulnerable, because without them, the GSCC will fall apart. And the Web strongly depends on GSCC.

Reveal the Sector's Architecture

What is inside of GSCC? Perhaps the most critical of all Web nodes are the 13 DNS root servers at the top of the tree-structured Web. Recall that DNS servers are the Web's telephone books. They translate symbolic URLs into numerical IP addresses.

TABLE 13.2. Demographics of e-mail Traffic on the Web Relative to the Bow-Tie Structure Depicted in Figure 13.1.

Web Component	Percentage of all Web Nodes in the Component
Disconnected nodes	41%
OUT	34%
GSCC	20%
Tendrils	3%
IN	2%

The DNS root servers, enumerated by the letters A through M, are located in cyber-space at a.root-servers.net, b.root-servers.net, through m.root-servers.net. Until attacked by a cyber-exploit in 2002, they were located in Herndon, VA; Marina del Rey, CA; College Park, MD; Mountain View, CA; Palo Alto, CA; Vienna, VA; Aberdeen, MD; Stockholm, Sweden; London; and Tokyo, Japan. Subsequently, these most critical of all Web nodes have gone into hiding.

The next-most critical nodes in the Web are the 13 gTLD servers. Recall that gTLD servers manage the top-level domains designated as .com, .edu, .net, and so on. Of course, the "telephone book" that converts www.name.com into an IP number is distributed to lower level servers, increasing the redundancy of information. Thus, removal of all 13 gTLD servers would not completely ruin the Web. However, the gTLD servers are the final word on domain name translation. And they cannot be quickly and easily replaced. Therefore, they are critical nodes in the GSCC.

Like the DNS root servers, the gTLD servers are also enumerated from A to M and physically distributed to locations around the world:

1. a.gtld-servers.net Herndon, VA
2. b.gtld-servers.net Mt. View, CA
3. c.gtld-servers.net Dulles, VA
4. d.gtld-servers.net Herndon, VA
5. e.gtld-servers.net Los Angeles, CA
6. f.gtld-servers.net Seattle, WA
7. g.gtld-servers.net Mt. View, CA
8. h.gtld-servers.net Amsterdam, N.L.
9. i.gtld-servers.net Stockholm, S.E.
10. j.gtld-servers.net Tokyo, J.P.
11. k.gtld-servers.net London, U.K.
12. l.gtld-servers.net Atlanta, GA
13. m.gtld-servers.net Hong Kong, C.N.

In addition to this infrastructure, the Web stretches across the United States through a network of metropolitan area exchanges (MAEs) and network access points (NAPs). The principle MAE and NAP facilities are located in the following major metropolitan areas:

San Francisco NAP
San Jose MAE
Santa Clara NAP
Los Angeles MAE
Dallas MAE
Houston MAE
Chicago NAP

Fairfax County, VA MAE
College Park, MD NAP
New York/New Jersey NAP

Many of these are colocated with telecom hotels, which adds to their attractiveness to attackers. Other obvious targets are the e-commerce sites such as Amazon.com, Buy.com, CNN.com, eBay.com, E-Trade.com, Yahoo.com, and ZDNet.com. In fact, these e-commerce hubs were attacked by a 15-year-old Canadian teenager in February 2000.[9]

The Web has plenty of hubs—websites with a high concentration of connectivity and high economic value. These hubs are the most-likely targets of terrorism and deserve special attention when deciding how to protect the Web. Before we build a fault tree and complete the final steps of the MBVA process, we examine the weapons of choice of the black-hats.

TOOLS OF THE BLACK-HAT TRADE

A *virus* is a malicious self-replicating program. A *worm* is a malicious self-replicating program that spreads through a network. A *Trojan horse* is a data file or program containing a malicious program. It is a computer program that seems to be harmless but it actually does damage. For our purposes, viruses, worms, and Trojan horses are all the same—malicious programs used by hackers and crackers to do damage or take control of Web infrastructure. In the following description of exploits and how they work, we will treat viruses, worms, and Trojan horses as tools of the black-hat trade.

In 1988, Robert Tappan Morris—a 23-year-old Ph.D. student at Cornell University in Syracuse, NY—remotely launched the first cyber-worm aimed at a MIT machine located miles away in Cambridge, MA. The worm quickly infected and disrupted 6000 computer systems across the United States and their users. In some cases, the worm forced users to disconnect from the Internet to stop the contagion.

How did this happen? Morris had discovered two flaws in the operating system of the Internet's servers that allowed him to gain unauthorized access to machines all over the Internet. Working alone for several months, Morris' worm exploited operating system flaws to gain access to each target machine. Once inside the target machine, the worm used the target machine's routing tables and user names/passwords to find new victims. The worm copied itself onto other machines, where the process was repeated.

The worm attempted three exploits on target machines:

1. Execute a remote command that gives the attacker access to the target machine.

[9]MafiaBoy caused and estimated $1.7 billion in lost revenues, but its perpetrator was fined only $250 and sentenced to 8 months in jail.

2. Force a so-called *buffer overflow* exploit on the target machine, which inadvertently relinquishes control to the attacker.
3. Access the e-mail program on the target machine, and command it to download the virus onto the target machine.

Morris was caught, charged with computer fraud and abuse, and found guilty on 4 May 1990 in Syracuse, NY. He was sentenced to 3 years probation, levied a $10,000 fine, and required to contribute 400 hours of his time to performing community services. Estimates of financial loss range from $100,000 to $10,000,000.

The Morris exploit illustrates several features of cyber-threats. First, flaws in the software of networked computers make it possible for hackers and crackers to gain access to remote machines. Cyber-threats depend on these flaws to gain a foothold. Second, the malicious program replicates by copying itself onto other vulnerable machines. In other words, a computer virus or worm works much the same way that a biological virus or worm does—it reproduces itself and travels to a new host, where the process is repeated. Malicious programs can infect many remote systems at the speed of the Internet. Third, this historical example illustrates what is still true: Hackers may cause millions of dollars of loss, but they typically get modest sentences. Historically, cyber-crime pays!

Viruses existed long before worms. In the early days of the PC, viruses traveled by infecting floppy disks, document files, and application programs. They did not depend on the Internet, but rather, they spread through physical contact. One of the oldest exploits used special tracks on software distribution disks, called the *boot record*. When the infected disk is inserted into a PC, the boot record is copied into the main memory of the PC. Once inside, the infected boot record made copies of itself on every disk inserted into the PC. The virus spread to new target machines whenever a human computer user shared the disk with another user. Today a computer user inadvertently activates viruses, whereas a worm spreads on its own.

Other viruses work through other vectors. A virus might attach itself to a document file, such as an Excel or Word file. Wherever the file goes, the virus goes also. The user activates the virus when it is loaded into his or her computer. Trojan horse viruses frequently traveled this way before the Internet became widely used.

Microsoft Office products are designed to allow a programmer to imbed a program inside of a Word or Excel document. These programs are called *macros*.[10] When activated, macros perform routine tasks or add additional capability to the Office application. But macros are vulnerabilities in Microsoft products, because a macro can be a Trojan horse.

Worms can be thought of as mobile viruses, because they copy themselves onto target computers and do not require a user to initiate them. They are a favorite of black-hats because they can infect the entire Web with little time, effort, or expense on the part of the attacker. Hence, worms pose an asymmetric threat to

[10]Macros are written in Visual Basic and activated by the user whenever the document is loaded into an Office application.

the Internet, SCADA networks, financial networks, power grids, and telecommunication networks.

Worms have become notorious because of the far-flung Web. A worm can be launched from a hacker's living room and sent to all parts of the world within hours. Because the network does the work of spreading the contagion, the hacker merely launches the attack and then waits. This is one of the motivations of hacking: An obscure outlaw can gain the attention—and sometimes adulation—of the entire world with little more than a PC and telephone.

Worms can do anything any other software can do. Worms have been known to e-mail the entire contents of a victim's hard disk to others, install a backdoor Trojan horse on the victim's computer, observe and record a user's keyboard key strokes (key-logger), launch DoS attacks, disable anti-virus software, and steal or destroy a victim's files. Worm exploits were the most frequent type of exploit in 2004.

How do worms travel? There are currently five fundamental ways that worms propagate from computer to computer:

1. Fundamental flaws in TCP/IP
2. Unprotected or open input/output ports on target machines
3. Operating system flaws: buffer overflow exploits
4. E-mail protocols and attachments: SMTP and MIME
5. Flawed applications and system software

TCP/IP Flaws

In Chapter 12, we surveyed TCP/IP's historical past and observed that it was designed over 30 years ago to enable the United States to regain leadership in missile technology, save money by sharing expensive computers among university and research laboratories, and withstand thermo-nuclear attack from the former Soviet Union. But it was not designed to withstand cyber-attacks. Unfortunately, the many known flaws in TCP/IP make it extremely easy to exploit.

Each TCP/IP packet contains both source and destination address. The source address identifies the server that sent the packet, and the destination address identifies the intended recipient of the packet. These IP addresses are clear; they can be read and changed by anyone clever enough to intercept and modify them.

TCP/IP's packet address vulnerability was exploited by an unknown black-hat in 2003, stalling the largest banking network in the country. Inserting of source and destination addresses into millions of randomly generated packets denied service to 400,000–700,000 computers worldwide (200,000 in North America) in what was named the SQL Slammer. One of the infected computers happened to also be connected to the Bank of America ATM network. When this computer stalled, it also stalled the ATM network. SQL Slammer was launched on the weekend of 24–26 January 2003. It caused systems running Microsoft SQL Server to generate a massive number of messages with random source and destination addresses. This

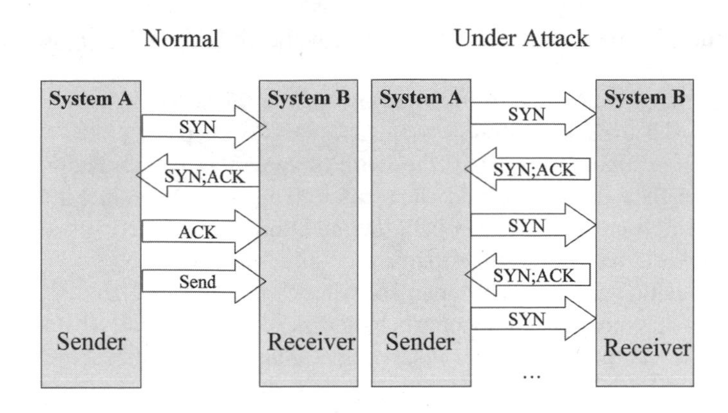

Figure 13.2. SYN Flooding exploit uses the TCP/IP protocol to flood a server with SYN messages. Normal operation results in a successful SEND. Flooding never sends a message.

generated a flood of traffic between pairs of Microsoft SQL Servers, crowding out all other traffic on the network.

Coincidentally, the Bank of America ATM network was affected, because the financial network and Web access were both hosted on the same machine. The worm got past this machine's firewall, and it became flooded with millions of short messages. Normally, this machine exchanged information between ATM and non-ATM networks, but on Monday morning, the server was so loaded down with messages generated by the SQL Slammer that it was useless. All 13,000 Bank of America ATMs became unusable until port 1434 (used by Slammer) was filtered.

Another flaw in TCP/IP is responsible for another type of DoS attack. The SYN Flooding exploit works because TCP/IP was designed to be simple, not necessarily flawless. SYN Flooding is traced to the three-way handshake used by TCP to establish a connection between two computers A and B. Figure 13.2 shows what is supposed to happen, and what can happen when SYN Flooding is used to overload a server with ceaseless unresolved SYN messages.

In Figure 13.2, system A (sender) initiates a connection to system B (receiver) by sending a SYN message.[11] System B responds to the SYN request by returning a SYN followed by an ACK within a reasonable time interval. Sending a confirming ACK from sender to receiver confirms the three-way handshake. Once the ACK is received, system A sends the message to system B.

But what if the three-way handshake never completes? An exploitation of this initiation protocol occurs when system A *never* returns the expected ACK corresponding with its initial SYN. System A (sender) and system B (receiver) shake hands by exchanging a SYN and ACK as before. But the receiver never gets an ACK. Instead, system B (receiver) gets a stream of more SYNs. This ceaseless

[11]SYN is short for synchronize and ACK is short for acknowledge—two hold-over signals from the days of teletypes and Western Union "e-mail" delivery.

stream keeps the receiver busy doing nothing but expecting ACKs, which never arrive.

SYN Flooding is an elementary DoS exploit that burdens the network (and system B) with useless unfulfilled messages. If millions of SYNs are sent to a single receiver, both network and receiving system get bogged down with handshaking, which leaves little time to process valid messages. The remedy is to close the port or filter the stream of SYNs coming from the attacker.

DoS attacks using TCP/IP flaws are commonplace, although they are not as frequent as other exploits. But they can be dramatic. Code Red was launched against the White House website on October 21, 2002. Within hours it was detected on millions of computers around the globe. Here is how Code Red works:

- The worm enters the target computer through its port 80.
- Then it finds and infects the Microsoft Internet Information server software.
- It copies itself onto other targets generated at random for 20 days.
- Then it goes dormant until a certain date, when all copies are activated.
- Millions of distributed copies flood the White House server with messages.

Code Red inundated the http://www.whitehouse.gov server with messages generated by *zombie* computers selected at random when the worm spread. The number of zombies numbered in the millions, because the worm replicated itself for 20 days before launching the denial of service exploit against http://www.whitehouse.gov.

Fortunately, Code Red used the numerical IP address of the White House server, instead of the symbolic http://www.whitehouse.gov. The DoS attack was diverted by simply changing the DNS address book, pointing http://www.whitehouse.gov to a different server. The White House administrators essentially *spoofed* the attacker.[12]

Code Red infected servers on all continents including Europe, Africa, Russia, China, and North and South America. It illustrated several vulnerabilities in theWeb:

1. Port 80 (the port used by all Web browsers) was used, showing that a worm can travel through ports.
2. Code Red showed that a simple worm could cause widespread damage.
3. DoS attacks are simple but effective.

DoS attacks do not destroy information or cause physical damage. They simply render the website they attack useless. In emergency or national security crises, information and information technology systems are essential to the operation of police, fire, military, medical, power, energy, transportation, and logistical systems. In the information age, denial of access to information can be damaging to critical financial and command-and-control systems. Without our

[12]Spoofing means that source address of packets returned from DNS are changed to something else, which changes the identity of the sender.

information technology systems, information societies are crippled, if not permanently damaged.

Open Ports

Code Red used port 80 to travel through cyber-space. Every computer has ports—doors through which information enters and leaves a system. Ports are numbered from 1 to 65535, but only a few are actually used in a given computer. Some well-known ports are as follows:

Port	How It Is Used
25	TELNET
80	HTTP
443	HTTPS
21	FTP
110	POP3
25	SMTP
1433	SQL Server
53	DNS

For example, port 21 is the preferred doorway to a commonly used data transfer program called File Transport Protocol (FTP). FTP provides a fast way to transfer large files. It also provides a fast way for worms to spread through exploitation of ports. Virus mutations called Sasser.C and Sasser.D swept through Windows XP and Windows 2000 systems using FTP, mostly infecting home computers (500,000 to 1 million) in 2004. The Sasser worm is not a single worm, but it is a series or strain of worms that have mutated over time.

The worm scans random IP addresses for exploitable systems. When one is found, the worm exploits the vulnerable system, through a buffer overflow exploit. Here is how a typical Sasser worm works:

- The worm initiates an infection by creating a remote program via port 9995.
- The remote program creates an FTP program on the target computer, which downloads the remainder of the malicious program onto the target computer, thus completing the infection.
- Now the infected target accepts any FTP traffic on port 5554, which gives the attacker access to the target computer.

Note that Sasser uses a combination of open ports and buffer overflow. What is buffer overflow?

Buffer Overflow Exploits

One of the oldest, and still most difficult, exploits to prevent is called a *buffer overflow* exploit. Essentially, this exploit uses the fact that a computer does not know the difference between data and program code. All information looks the same to a computer. In a buffer overflow exploit, a virus, disguised as data, is sent from the attacker to the victim, but once it arrives at the target computer, it turns into a malicious program! How is this possible?

Figure 13.3(a) shows what is supposed to happen when data enter a computer operating system from an open port. Normally, the operating system acts as an intermediary between the outside world and the application (user) program. Input data are temporarily stored in a storage area called a *buffer*, along with a return address that tells the operating system where to return control once the data have been transferred. After the buffer fills up, the return address is used to return control back to the user program. The user program then transfers the input data into their own processing area.

Figure 13.3(b) shows how a buffer overflow can be exploited to wrest control away from the operating system (and the user program) and turn control over to a

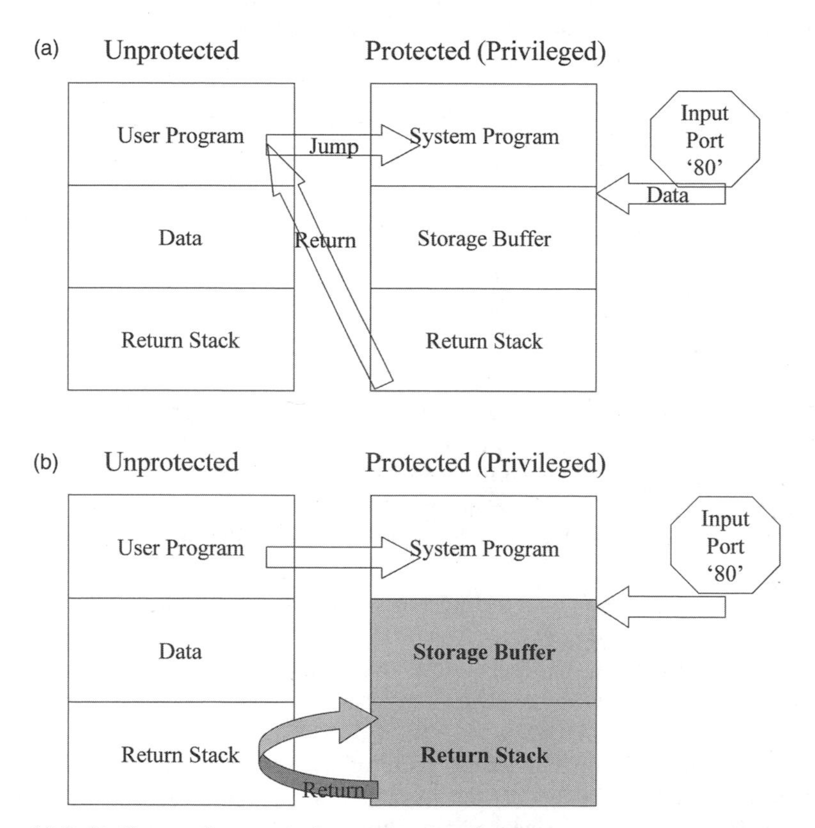

Figure 13.3. Buffer overflow works by sending a malicious program to the target computer's input buffer as if it was pure data. (a) Normally, data arrives via a port and is stored in a temporary storage buffer, after which control is returned to the user program. (b) A buffer overflow exploit over writes the return address, passing control to the data itself, which turns out to be a malicious program.

malicious program. Data enter the target computer as before, but this time it overflows the storage buffer. In fact, it writes over the return address stack and inserts a new return address that returns control to the buffer and stack. The operating system uses the hacked return address to pass control to the malicious program, which now resides in the storage buffer or stack. What was thought to be data actually turns out to be a malicious program.

Perpetrators of buffer overflow attacks must discover the size of the storage buffer and return stack of each system by trial-and-error. That is, they have to guess where to place the malicious return address and viral code. This is done by launching thousands of buffer overflow attacks containing one, two, three,... hundreds of different trial return addresses, until one works.

In July 2003, the Win32:Blaster worm (a.k.a. msblast.exe) used port 135 and a buffer in Windows to spread throughout the Web. RPC is the remote procedure call mechanism that allows two computers to communicate with one anther. TFTP is trivial FTP, and the Windows registry is a table inside of Windows that holds the names of programs that are allowed to run on a specific machine. Here is how the buffer overflow exploit worked:

- Exploits buffer overflow in the Microsoft Windows RPC interface.
- It scans 20 hosts at a time, trying to connect to port 135.
- When an open port is found, the worm copies itself to the host using TFTP.
- Activated whenever Windows is started (via Windows registry).
- Can force Windows to restart.

DDoS Attacks

Open ports, buffer overflows, and various flaws in software provide contamination vectors for the spread of cyber-viruses and worms. These vectors can be exploited in thousands of computers at once to turn innocent victims into collaborators in DoS attacks. These collaborators are called *zombies*, and the attack is called a *distributed denial of service* attack (DDoS). The DDoS is one of the most effective exploits known.

A DDoS exploit starts by infecting a large number of zombies with a virus that lies in wait until a certain date arrives or signal occurs. Then, the zombies simultaneously flood a single target computer with meaningless data. The objective is to overload the target with messages, rendering it useless for ordinary processing. Figure 13.4 shows how DDoS works.

A 15-year-old Canadian teenager calling himself and his exploit MafiaBoy launched a DDoS strike against the most popular e-commerce sites in February 2000. This worm flooded Amazon.com, Buy.com, CNN.com, eBay.com, E-Trade.com, Yahoo.com, and ZDNet.com with millions of messages, resulting in an estimated loss of $1.7 billion in revenue. The MafiaBoy worm electronically recruited an army of zombie computers around the world, which in turn flooded the e-commerce servers with thousands of simultaneous requests for service,

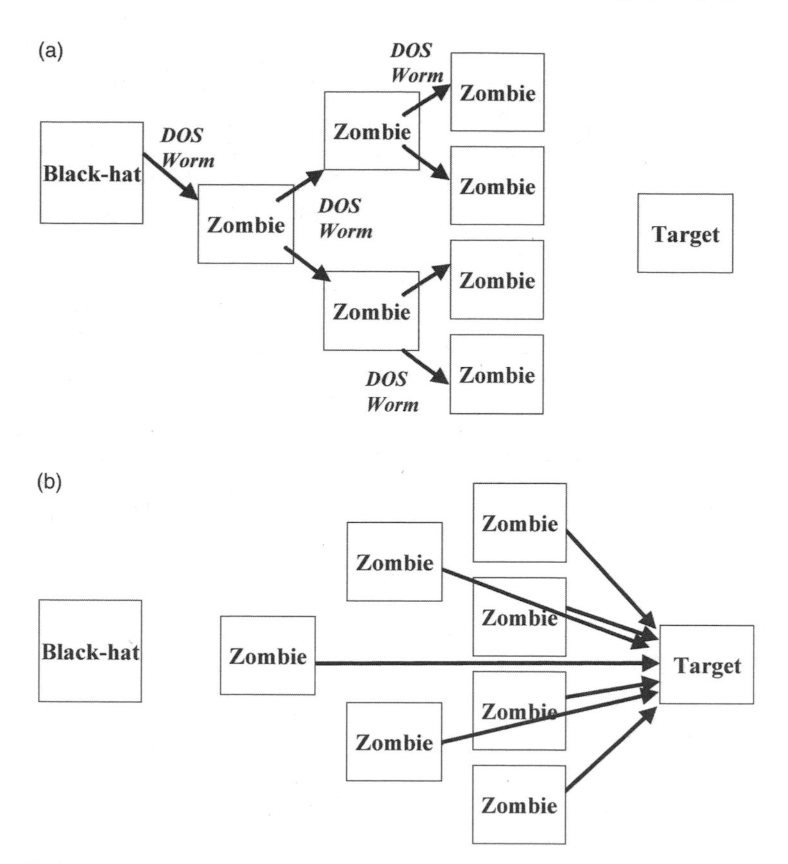

Figure 13.4. DDoS recruits innocent zombies to participate in a DoS attack against a single target computer. (a) Phase 1: Spread the worm to thousands of zombies. (b) Phase 2: Zombies flood the target with messages.

forcing them to shut down for several hours. The teenager was fined $250 and given an 8-month jail sentence.

MafiaBoy illustrates how DDoS attacks work, but it also illustrates another unfortunate fact: Billions of dollars of damage can be done with very inexpensive weapons. Even more disturbing is the social consequences of hacking: An underage offender can render billions of dollars of damage but suffer almost no punishment, or punishment that is extremely disproportionate to the amount of damage caused.

E-mail Exploits

E-mail exploits are carried out by hackers using malicious programs that predominately spread by e-mail attachment and a user's address book. They depend on the victim to run the worm when received, and typically they modify the Microsoft

Windows registry.[13] E-mail worms can do all the things that other worms can, such as installing backdoors, key-loggers, and compromising security and data integrity.

One of the most virulent e-mail exploits in 2002 was the w32.klez.e@mm worm, also known simply as the Klez virus. It traveled via entries on the affected computer's Microsoft Outlook address book. Klez tried to disable the user's anti-virus programs, copy itself to network disks, and mass mail itself to all entries in the user's Outlook address book.

In 2003, Bugbear used e-mail attachments and MIME vulnerability to spread and install a backdoor, key-logger, and its own SMTP engine that sent spoofed e-mail using the victim's address book. Bugbear could do serious damage to the target computer, including deletion of the user's files.

Flawed Application and System Software

The number of exploitable flaws in application and system software is legion. Software flaws are used by cyber-outlaws in thousands of ways. We spare the reader a detailed explanation of them all and merely describe a few examples from the following categories:

- HTML and XML as clandestine message software.
- HTTP may leave open access doors.
- ActiveX: Code from the Web can access your computer.
- SMTP and POP3: E-mail can give away your password.
- SNMP: Manages networks, but also opens it to the outside.
- SOAP/XML: RPCs can be used against you.

HTML and XML uses tags to tell a Web browser such as Microsoft Internet Explorer what each line of data means. But if Internet Explorer encounters an unknown tag, it simply skips over the line of data and continues looking for meaningful tags. What if the unknown tag is actually a malicious program? This is called steganography, which has been used by spies for thousands of years to conceal messages.[14] It is also a tool of hackers.

HTTP is the protocol that dictates how a Web browser communicates with a Web server. There are several vulnerabilities in this fundamental software. First, it communicates in the clear—transmissions can be intercepted and substituted by unscrupulous people looking for credit card names and numbers. Second, version HTTP 1.1 leaves sessions open, because repeated opening and closing of sessions is inefficient. But open sessions can be used like open ports. Hackers can exploit

[13]The registry is where Windows keeps the names and authorizations for every program that is allowed to run on a Windows machine.

[14]Steganography is the art and science of hiding information by embedding messages within other, seemingly harmless messages. It is used for both good and evil. Steganography is used to electronically protect intellectual property by embedding watermarks in digital documents as well as to conceal secret messages.

port 80, which is the port used by HTTP. Once port 80 is hacked, the currently running session can be hacked too.[15]

HTTP Secure server (HTTPS) and Secure Socket Layer (SSL) should be used instead of HTTP, when security is important. HTTPS/SSL encrypts transmissions between server and Web browser. Most secure e-commerce applications, such as credit card buying over the Internet, are run on HTTPS/SSL encrypted sessions.

HTTPS/SSL transmissions may still be vulnerable to ActiveX programs that are transmitted between server and desktop. ActiveX is a Microsoft system for downloading programs and running them on the user's computer. Secure ActiveX programs ask for the user's permission and require a security certificate. But most users grant access to every ActiveX program without knowing what each program does! The ActiveX program may be a virus that destroys information or installs other malicious programs on the user's machine. How can a user know?

ActiveX programs should not be allowed to write to a user's local disk drive or alter a Windows registry file. Without prior knowledge of what the ActiveX program will do, granting access is like inviting a stranger to take over your house for the weekend!

ActiveX software has been employed by unscrupulous merchants and advertisers to promote their products and services. Called *spyware* for good reason, these ActiveX programs collect information about the user so that the unscrupulous merchant can target him or her for copious amounts of advertising. For example, file-sharing music pirates used Kazaa in 2003 to download spyware to home computer users and then subsequently spammed the unsuspecting users with popup ads.

An EarthLink.com study found more than 29 million spyware-related files on 1 million of their subscriber's computers. Dell Computer customer support reported 12% of their support calls were complaints about spyware. In 2004, Microsoft attributed 50% of reported crashes of Windows XP to spyware.

Professional black-hats possess even deeper knowledge of how computers and the Web operate. Because of this knowledge, they have devised complex exploits that go far beyond the scope of this book. Exploits involving the SNMP and POP3 e-mail servers are known for exposing passwords; the network management protocol, SNMP, used to maintain the hardware of the Internet is also vulnerable to hacks; and the SOAP/XML protocols used by e-commerce companies are vulnerable to knowledgeable hackers and crackers. The list continues to get longer, and the exploits continue to get more sophisticated.

InternetVirus SIMULATOR

Web worms and viruses spread like an epidemic through human social networks. So why not model Internet epidemics as if they are human epidemics? We did this in Chapter 4 and noted that epidemics behave like cascade failures and the

[15]A session and an application is almost identical, so hacking an open session is tantamount to hacking a running application.

reverse—cascade networks model epidemics. In this section we show that cascade networks also model the Web and the exploits that infect the Web. But there is one major difference: Internet epidemics, unlike human epidemics, often reverberate in waves of "susceptible, infected, susceptible" cycles. These are modeled as SIS cascade networks here.

In this section we examine the results obtained from the program *InternetVirus*, which simulates worm exploits by treating the Web as a scale-free SIS cascade network. Then we ask questions like, "Is it inevitable that a worm will spread to all corners of the Web, or do worms quickly die out?"

Figure 13.5(a) shows the interface to *InternetVirus*. The screen displays a network with blue nodes and links. These nodes turn red when they are infected and yellow when they have recovered from being infected. The program is controlled by several buttons along the bottom of the screen.

The simulation is started by pressing GO and restarted by pressing RESET. When the VirusON button is pressed, a node is selected at random and infected. The infection propagates along links with a probability set by the user. Each node stays infected for a length of time also set by the user. A graph of percentage of nodes infected versus time (for the last 50 time units) is generated as the infection spreads.

Figure 13.5(b) illustrates the results from one of several modes of simulation. Pressing the ScaleFreeOn button causes a scale-free network to emerge by repeatedly applying the increasing returns organizing principle to each link. After a short period of time, a scale-free network and its hub emerge.

In addition to the formation of a scale-free hub, the simulation run of Figure 13.5(b) shows the results of pressing SISon. This mode causes nodes to cycle among infected, recovered, and infected. In a SIS network, the pathogen spreads through the network, infecting nodes with a probability of say 5% and then subsides because the number of infected nodes grows, thus diminishing the likelihood that more nodes will become infected. After a period of time, called DURATION (set by the user), each infected node recovers and becomes susceptible to the virus again. Thus, each node passes through three states: susceptible, infected, and then recovered (but susceptible again).

As shown in Figure 13.5(b), the number of infected nodes never returns to 0%. Instead, a wave of new infections is followed by a wave of recoveries, which is followed by another wave of infections, over and over again. The percentage of infected nodes versus time hovers between 70% and 80%. The network does not become 100% infected, nor does it completely recover.

Figure 13.5(c) shows the results of pressing the PROTECT button, which immunizes the hub node so it cannot be infected. In this case, the infection rate hovers between 60% and 70%. In other words, the average number of infected nodes declines when the hub is protected. In general, the more we protect hubs, the lower is the percentage of infected nodes.

InternetVirus can be used to learn many things about epidemics. This introduction reconfirms the results obtained in Chapter 4 for cascade networks. The Web is a cascade network when it comes to worms and viruses. The more random a section of the Web is, the more the epidemic spreads. But viruses and worms

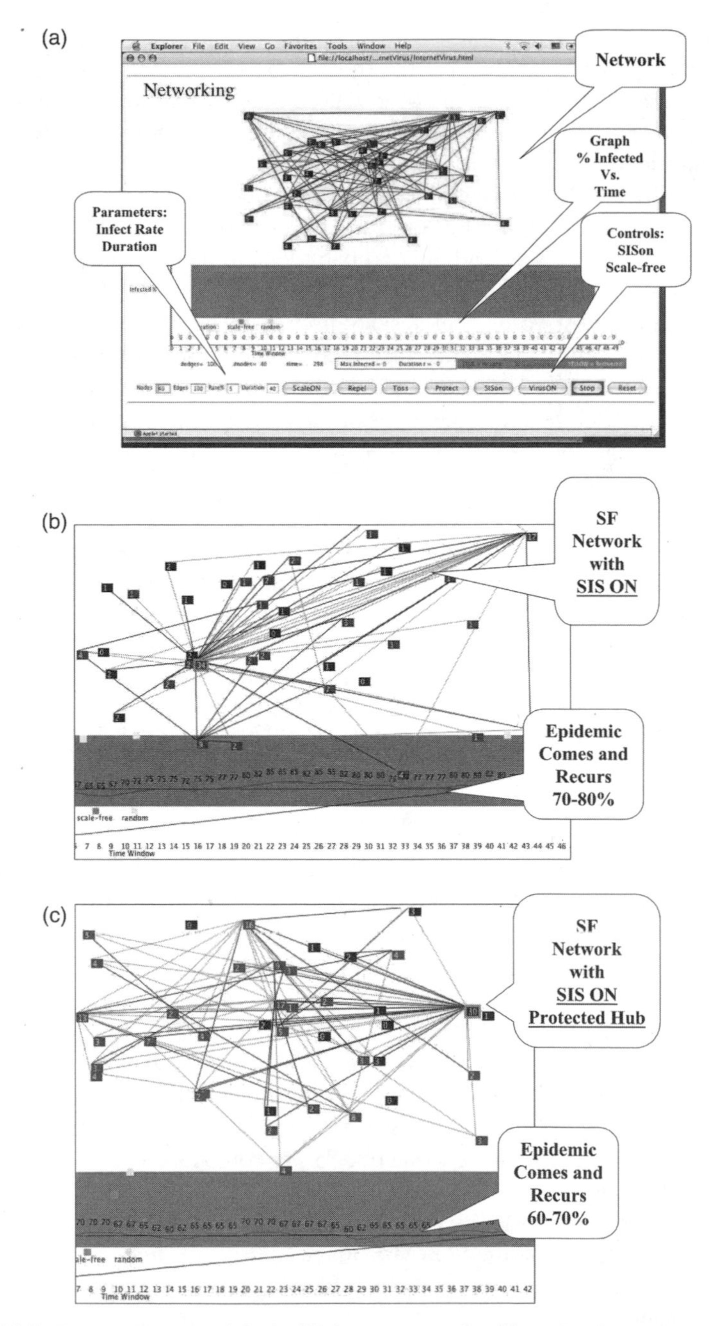

Figure 13.5. *InternetVirus* models the Web as a network with nodes that start out healthy or susceptible, then become infected, and then recover. (a) Display showing network, graph of percent infected versus time, and program controls. (b) Virus epidemic in a scale-free network. (c) Virus epidemic in a scale-free network with protected hub.

spread faster in scale-free networks than random networks. But protecting hubs decreases the spread of epidemics; the more hubs we protect, the fewer infected nodes there are. Protecting the hub of a scale-free network reduces the percentage of infected nodes even more than protecting a hub in a random network.

The major lesson we can learn from *InternetVirus* is that the Web has hubs, and protecting these hubs from infection is the most effective strategy for preventing the spread of worms and viruses. In 2004, the number of exploits declined over the number in 2003. If this is a trend, it is probably due to the increased cyber-security being deployed at critical nodes such as the DNS root servers, gTLD servers, and busy e-commerce hubs. Millions of desktop users have been advised to take preventive anti-virus measures, but the most effective counter-measure is to harden the extremely small number of hubs that dominate the Internet. Protecting the most critical hubs in the GSCC should be the focus of the national strategy for cyber-security.

VULNERABILITY ANALYSIS CONTINUED

Now that we are equipped with a fundamental understanding of cyber-threats, it is possible to take the next step in MBVA—building a fault tree model of the vulnerabilities confronting the Web. This fault tree should represent the most critical nodes of the Web, which are e-commerce hubs, root servers, tier-1 ISPs, and other hubs whose loss would dramatically impact the Web's connectivity. This involves perhaps fewer than 200 hubs (servers) throughout the world.

Build Fault Tree Model of Vulnerabilities

Figure 13.6 shows a simple fault tree for the threats noted in the foregoing sections. On the left branch are the older exploits properly designated as viruses. On the right branch are the more recent threats, which are properly designated as worms. All we need to complete the MBVA process is to find realistic estimates of fault probabilities, fault reduction costs, and financial damage estimates. These estimates can be obtained by studying known data on cyber-exploits over the past few years.

According to a 2003 study, 43% of system administrators surveyed did not know how their systems got infected.[16] But they estimated that exploits had a minor financial consequence: 75% of the exploits detected caused less than $100 damage. Multiplying by 75% yields a damage value of $75 per computer. This may seem small, but worms affect millions of computers once they spread. Therefore, an exploit that infects a million computers really cost $75 million per million victims. We will use $75 million as the estimate of WORM damage in the fault tree of Figure 13.6 and divide this estimate evenly across the five threats. Therefore, each worm exploit causes a damage of $15.

The 2003 survey also concluded that 30% of the exploits caused some loss of data, typically due to a virus. So once again, we can assume approximately

[16]http://www.avast.com.

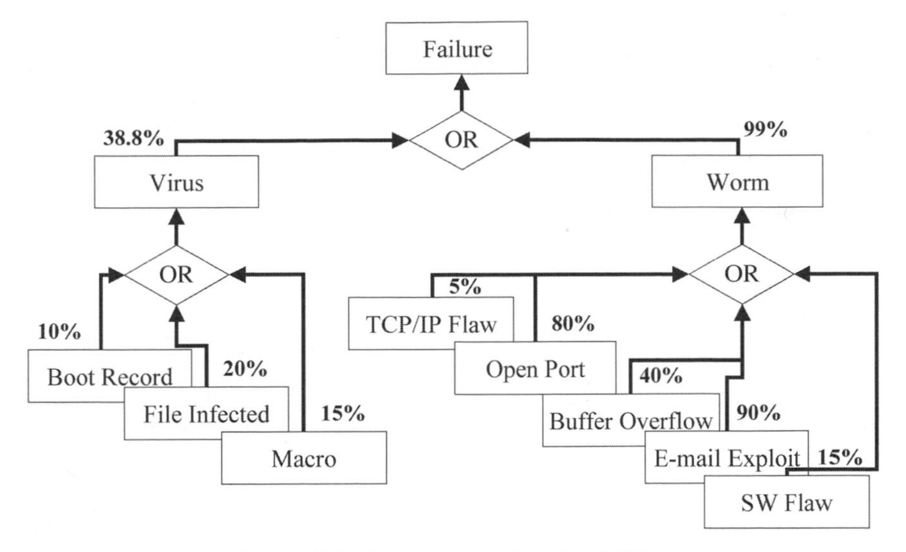

Figure 13.6. Fault tree of web vulnerabilities.

300,000 per million are victims of data loss. If we also assume each data loss incident cost \$100, then such exploits cost \$30 million per million victims. We will use \$30 million as the VIRUS damage estimate in Figure 13.6 and divide this estimate evenly across the three threats. Therefore, each virus exploit causes a damage of \$10.

Finally, assume a minimal cost to prevent virus or worm exploits; e.g., an investment of \$1/machine will reduce its vulnerability by 1%. Hence, the cost per percentage point is assumed to be \$1 million per million exploits. Multiplying this by the fault probability of each threat, we obtain the necessary data to run *FTplus* on the fault tree.

In Figure 13.6, we have used similar arguments for estimating the fault probabilities shown as labels on the boxes of the fault tree. Combining fault probability estimates with cost and damage estimates yields the following for each threat:

- Boot file exploits are going out of style: v = 10%, c = \$10, d = \$10.
- File infection loses data: v = 20%, c = \$20. d = \$10.
- Macro infection is easy: v = 15%, c = \$15, d = \$10.
- Unless you are a hub, TCP/IP (DoS) exploits are rare: v = 5%, c = \$5, d = \$15.
- Buffer overflow attacks on databases: v = 40%, c = \$40, d = \$15.
- Ports are almost always being scanned: v = 80%, c = \$80, d = \$15.
- E-mail exploits were the most common in 2003: v = 90%, v = \$90, d = \$15.
- Software flaws are similar to Macro exploits: v = 15%, c = \$15, d = \$15.

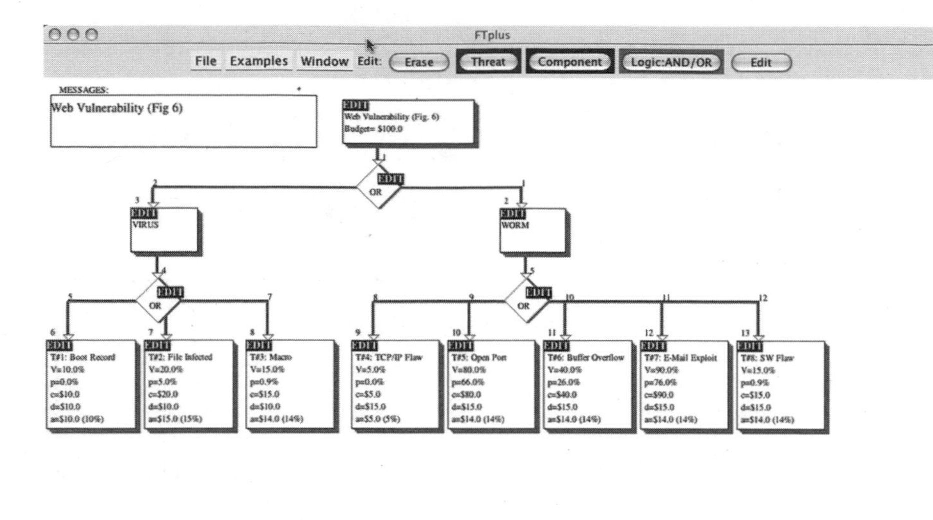

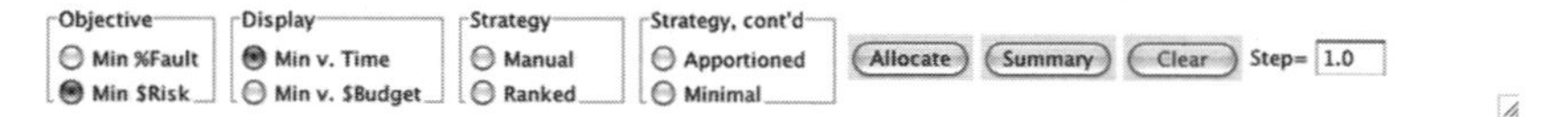

Figure 13.7. Results of resource allocation obtained from *FTplus* as it is applied to the fault tree of Figure 13.6.

The fault tree of Figure 13.6 and the estimates above yield an allocation of \$100 million (see Figure 13.7).

Resource Allocation

The fault tree of Figure 13.6 has eight threats, which means there are 128 events in the full event tree! Figure 13.7 and Table 13.1 show the results of allocation of \$100 million to the fault tree using ranked, apportioned, and minimal risk reduction allocation strategies. Initially, the sector is 99% vulnerable to some kind of exploit. After allocation of \$100 million, sector vulnerability drops to 87% using fault minimization. Instead of 128 combination events in the resulting event tree, only 23 correspond with nonzero fault probabilities. In fact, the most likely event after allocation is threat #7 (E-mail exploit).

Table 13.3 says to ignore virus exploits and focus on worms. In general, the investment strategies focus on reducing OPEN PORT and E-MAIL exploits, when the ranked allocation strategy is used, and OPEN PORT, BUFFER OVERFLOW, and E-MAIL exploits for apportioned and minimal allocations. Only the optimal allocation strategy allocates funding to TCP/IP and Software Flaw exploits. MBVA results mainly point to a strategy focused on OPEN PORT and E-MAIL exploits, perhaps the most frequent types of cyber risks.

TABLE 13.3. Summary of Results from Resource Allocation of Figure 13.7.

	Risk Reduction			Fault Reduction		
Threat	Ranked	Apportioned	Minimal	Ranked	Apportioned	Minimal
Boot	$0	$0	$0	$00	$00	$0
Fil	$0	$0	$0	$0	$0	$0
Macro	$0	$0	$0	$0	$0	$0
TCP/IP	$0	$0	$4	$0	$0	$0
Open Port	$10	$40	$28	$10	$40	$43
Buffer Over	$0	$14	$32	$0	$13	$4
Email	$90	$46	$22	$90	$47	$54
SW Flaw	$0	$0	$14	$0	$0	$0

ANALYSIS

Cyber-exploits that threaten Web security are currently divided into two broad categories: viruses and worms. Worms are essentially viruses that infect computers connected to the Web using a variety of flaws in the network and computer systems. Attackers exploit flaws in various information systems to disrupt or deny service to e-commerce, SCADA, banking, airline reservation systems, and many other modern organizations. These attacks are called *exploits*, hacks, or cyber-threats.

Amateurs seeking notoriety and professionals seeking a hacker's reputation and financial reward carry out exploits. The most serious hackers and crackers are called black-hats. They are responsible for major financial losses due to the decline or loss of productivity that results when information systems are rendered useless while worms and viruses are removed.

How likely is it that cyber-exploits will be used by terrorists? In a report released immediately after 9/11, Michael Vatis, Director of the Institute for Security Technology Studies at Dartmouth University, claimed that cyber-attacks are highly correlated with physical attacks and terrorism:

> In the Israel/Palestinian conflict, following events such as car bombings and mortar shellings, there were increases in the number of Cyber-attacks. Subsequent to the April 1, 2001 mid-air collision between an American surveillance plane and a Chinese fighter aircraft, Chinese hacker groups immediately organized a massive and sustained week-long campaign of Cyber-threats against American targets.[17]

Vatis argues that cyber-attacks immediately accompany physical attacks, and they increase in volume, sophistication, and coordination. He also correlates these attacks with high-value targets.

[17]M. Vatis, "Cyber Attacks During the War on Terrorism: A Predictive Analysis," Institute for Security Technology Studies, Dartmouth College, September 22, 2001, 45 Lyme Road, Hanover, NH 03755, http://www.ists.dartmouth.edu/ISTS.

Thus far, nobody has died from a cyber-attack. In fact, James Lewis argues that the threat of cyber-attacks from terrorists has been exaggerated:

> Digital Pearl Harbors are unlikely. Infrastructure systems, because they have to deal with failure on a routine basis, are also more flexible and responsive in restoring service than early analysts realized. Cyber attacks, unless accompanied by a simultaneous physical attack that achieves physical damage, are short lived and ineffective. However, if the risks of cyber-terrorism and cyber-war are overstated, the risk of espionage and cyber crime may be not be fully appreciated by many observers. This is not a static situation, and the vulnerability of critical infrastructure to Cyber-threats could change if three things occur. Vulnerability could increase as societies move to a ubiquitous computing environment when more daily activities have become automated and rely on remote computer networks. The second is that vulnerability could increase as more industrial and infrastructure applications, especially those used for SCADA (Supervisory Control and Data Acquisition), move from relying on dedicated, proprietary networks to using the Internet and Internet protocols for their operations. This move to greater reliance on networks seems guaranteed given the cost advantage of Internet communications protocols (Transmission Control Protocol/Internet Protocol), but it also creates new avenues of access. These changes will lead to increased vulnerabilities if countries do not balance the move to become more networked and more dependent on Internet protocols with efforts to improve network security, make law enforcement more effective, and ensure that critical infrastructures are robust and resilient.[18]

Cyber-attacks seem to be declining since 9/11. According to the 2003 Computer Crime and Security Survey:[19]

- Overall financial losses from 530 survey respondents totaled $201,797,340. This is down significantly from 503 respondents reporting $455,848,000 last year. (75 percent of organizations acknowledged financial loss, though only 47 percent could quantify them.)
- The overall number of significant incidents remained roughly the same as last year, despite the drop in financial losses.
- Losses reported for financial fraud were drastically lower, at $9,171,400. This compares to nearly $116 million reported last year.
- As in prior years, theft of proprietary information caused the greatest financial loss ($70,195,900 was lost, with the average reported loss being approximately $2.7 million).
- In a shift from previous years, the second-most expensive computer crime among survey respondents was denial of service (DoS), with a cost of $65,643,300 - up 250 percent from last year's losses of $18,370,500.
- But for the fourth year in a row, more respondents (78 percent) cited their Internet connection as a frequent point of attack than cited their internal systems as a frequent point of attack (36 percent).
- Forty-five percent of respondents detected unauthorized access by insiders.

[18]J. A. Lewis, "Assessing the Risks of Cyber Terrorism, Cyber War and Other Cyber Threats," Center for Strategic and International Studies, Washington, D.C., December 2002. http://www.csis.org.
[19]http://www.gocsi.com.

Cyber-threats still exist and are responsible for major financial losses. They will continue to be a threat for as long as computer systems have flaws. And flaws are expected to remain a part of this sector for a long time, because fallible humans build information systems.

EXERCISES

1. What is a virus (select only one)?
   a. A malicious self-replicating program.
   b. An e-mail attachment.
   c. A malicious program disguised as a safe program.
   d. A malicious program that travels via the Internet.
   e. A flaw in TCP/IP.
2. What is a worm (select only one)?
   a. A malicious self-replicating program.
   b. An e-mail attachment.
   c. A malicious program disguised as a safe program.
   d. A malicious program that travels via the Internet.
   e. A flaw in TCP/IP.
3. What is a Trojan horse (select only one)?
   a. A malicious self-replicating program.
   b. An e-mail attachment.
   c. A malicious program disguised as a safe program.
   d. A malicious program that travels via the Internet.
   e. A flaw in TCP/IP.
4. Software patches are (select one):
   a. Often not installed on vulnerable systems.
   b. A defect that opens a computer to attack.
   c. A Microsoft repair kit.
   d. A kind of virus.
   e. A way to repair an open port.
5. Which one of the following is an old method of breaking into a computer:
   a. War dialing
   b. War driving
   c. SIS
   d. SOS
   e. SOB
6. A zombie is a:
   a. Defective computer

   b. Network of dead computers
   c. Innocent participant in a DoS attack
   d. Computer cyber-thief
   e. Black-hat cyber-criminal

7. The GSCC is:
   a. Giant secure connected component
   b. Grand secure connected component
   c. Giant strongly connected component
   d. Where most of the Web's nodes are located
   e. Where most of the Web's vulnerabilities are located
8. E-mail exploits are particularly virulent because:
   a. Almost everyone uses e-mail.
   b. E-mail attachments are easy to infect.
   c. SMTP is flawed.
   d. E-mail uses the OUT and GSCC components.
   e. Users fail to install patches.
9. Which of the following is NOT a critical node in the Web:
   a. The e-mail server at the police department.
   b. The NAP in College Park, MD.
   c. The MAE-West in San Jose, CA.
   d. The Chicago NAP.
   e. The server at a.root-servers.net.
10. As far as we know, the first cyber-worm was launched in:
    a. 1998
    b. 2001
    c. 9/11/01
    d. 1988
    e. 1984
11. A malicious program can be a macro that travels by:
    a. A Word or Excel document.
    b. Downloading itself through a buffer overflow attack.
    c. Downloading itself through spyware.
    d. Attaching itself to an e-mail attachment.
    e. Embedding themselves in Trojan horses.
12. A DoS attack:
    a. Floods a victim computer with a huge number of messages.
    b. Uses e-mail to send fake messages to users listed in address books.
    c. Is a special kind of worm.

   d. Uses macros to travel.
   e. Blocks ports.
13. The Bank of America ATM network was stalled by:
   a. A SYN flooding DoS attack
   b. SQL Slammer
   c. MS-Blaster
   d. Klez
   e. The Morris worm
14. The Whitehouse of the United States was attacked in 2002 by:
   a. Bugbear
   b. MafiaBoy
   c. Code Red
   d. Microsoft IIS
   e. Changing the DNS server address
15. What are ports?
   a. Input/output channels through which Web information flows
   b. Vulnerable flaws in the Web
   c. 65,535 doors
   d. TELNET input
   e. FTP input/output
16. In a buffer overflow attack:
   a. A program enters a computer as if it was data.
   b. A malicious program travels through ports.
   c. A worm exploits FTP.
   d. A worm exploits port 21.
   e. An operating system is exploited.
17. A DDoS attack uses:
   a. SYN flooding
   b. Zombies
   c. Web servers
   d. Microsoft Windows flaws
   e. Routing tables
18. MafiaBoy caused $1.7 billion in financial loss. How much was the fine?
   a. $10,000,000
   b. $1,000,000
   c. $10,000
   d. $250
   e. None

19. Steganography is defined as:
    a. Hiding information
    b. Transmitting viruses via software flaws
    c. Placing copyright in digital music
    d. Hiding malicious software in HTML
    e. All of the above
20. From Figures 13.6 and 13.7, what is the reduced vulnerability of the sector using the optimal (minimal) vulnerability reduction allocation—for an investment of $100 million?
    a. V = 99%
    b. V = 91%
    c. V = 87%
    d. V = 12%
    e. V = 0%
21. The virtual city of San Lewis Rey (SLR) has an active Internet that reaches out and connects with the global as well as local economy, public services, and three GSCC connected servers. This network is shown in Figure 13.8. The SLR counter-terrorism committee wants to know how to protect the SLR Internet, and hence the following study was commissioned. Your job is to determine where to invest limited municipal funds to reduce network-wide risk. The damages corresponding to nodes and links in Figure 13.8 are given below.

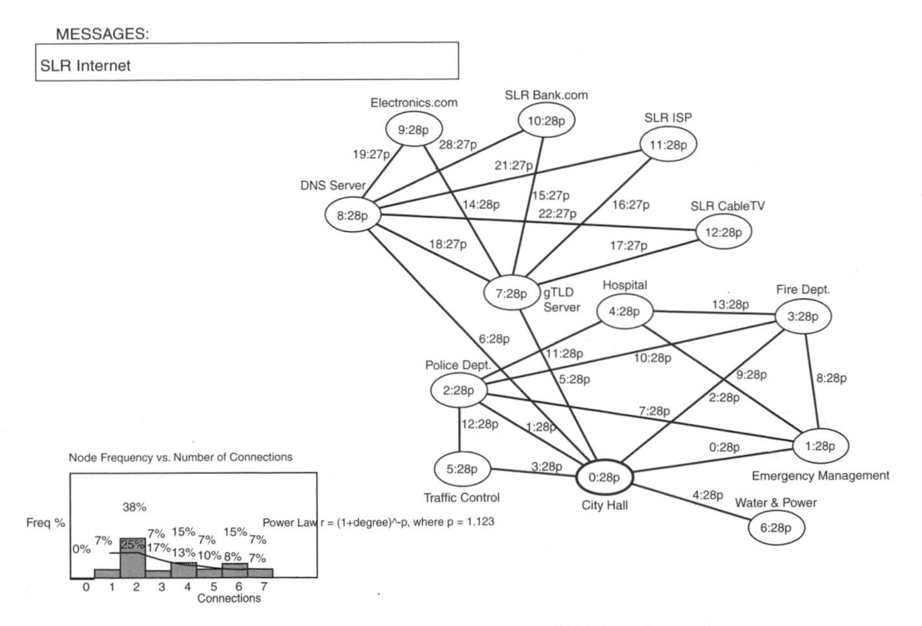

Figure 13.8. SLR Internet network [*NA.jar* display].

Damages are expressed in (000)'s, so for the purposes of network analysis, $150,000 corresponds with 150 in *NetworkAnalsyis*.

a. Police Department Computers: $100,000
b. Fire Department Computers: $150,000
c. Emergency Management Computers: $75,000
d. DNS root server located somewhere in SLR: $120,000
e. gTLD server located somewhere in county: $120,000
f. E-commerce site**: www.slr-electronics.com** with annual revenues of $2 billion: $250,000
g. Government website: www.slr-cityhall.gov with 20,000 hits per month: $150,000
h. Medical website: www.slr-hospital.org: $150,000
i. Regional Bank: www.slr-bank.com: $250,000
j. The major ISP in SLR serving 200,000 home computers: $200,000
k. SLR Water and Power network: $80,000
l. LR Cable TV network: $70,000
m. SLR Freeway Traffic Control network: $100,000
n. SLR Telecommunications network: $230,000 ($10,000 per each of 23 links)

Q1: Assuming the definition of scale free means the power law exponent, $p > 1.0$, is the SLR Internet network scale free?

a. Yes
b. No

Q2: Can it be argued that the SLR Internet network is small world?

a. Yes
b. No

Q3: Which node is the hub of the SLR network?

a. SLR Police
b. SLR ISP
c. SLR City Hall
d. gTLD server
e. DNS server

Q4: How many nodes and links does the SLR Internet network contain?

a. 14
b. 23 nodes and 13 links
c. 23 links and 13 nodes
d. 23 nodes and links
e. 13 nodes and links

Q5: What is the sum total damage value of all links and nodes in the SLR Internet?

a. 23 links + 13 nodes = 39
b. 3900 points
c. $1 million
d. $1.925 million
e. $1.900 million

Q6: What is the total number of points needed to reduce vulnerability to zero?

a. 3600
b. 3500
c. 2300
d. 1000
e. 1925

Q7: How many availability points correspond to 25% funding?

a. 900
b. 960
c. 1000
d. 250
e. 2500

Q8: Assuming 25% funding (one quarter the funding needed to reduce network-wide risk to zero), what is the allocation of points to links and nodes that minimizes risk (according to *NetworkAnalsyis*)?

a. All nodes and links get 1/36th of the total budget.
b. All nodes get 100 points; all links get 0 points.
c. All but Traffic, Water and Power, and SLR Bank get 100 points; all links get 0 points.
d. All nodes get 100 points; 4 links get the remaining points.
e. 25% of nodes get 100 points; everything else gets 0 points.

CHAPTER 14

Cyber-Security

Cyber-security is defined by the U.S. Congress as "the vulnerability of any computing system, software program, or critical infrastructure, or their ability to resist, intentional interference, compromise, or incapacitation through the misuse of, or by unauthorized means of, the Internet, public or private telecommunications systems or other similar conduct that violates Federal, State, or international law, that harms interstate commerce of the United States, or that threatens public health or safety."[1] For our purposes, cyber-security is the study and practice of securing assets in cyber-space—the world of computers and computer networks. Cyber-security is more than defending against viruses and worms, as described in Chapter 13. It encompasses *information assurance* in *enterprise computing*.

This chapter surveys the policies and technologies of securing information and the information technology (IT) systems. Specifically, this chapter will discuss the following concepts:

1. Cyber-security involves a wide range of information assurance policies, and practices, including, but not limited to, loss of access to information, loss of data, and loss of security associated with IT and human information-handling processes.
2. Secure IT systems are founded on a *trusted computing base* (TCB) and employ *trusted paths* (TP) through a network of IT components and human users to ensure the security of information stored and processed by the IT system.[2] To be secure, all IT processes must run within a TCB and communicate via a TP.
3. The major components of a TCB and TP are firewalls, proxies, intrusion detection systems, encryption, public key encryption (PKI), and policies that enforce a certain level of security. Security is not an absolute "secure" or "not secure" decision, but rather a tradeoff with other factors.

[1] HR 4246 introduced into the 106th Congress, 2nd session, April 2000.

[2] A trusted path is a mechanism by which a person using a terminal can communicate directly with the TCB. The TP can only be activated by the person or the TCB and cannot be imitated by untrusted software.

Critical Infrastructure Protection in Homeland Security: Defending a Networked Nation, edited by Ted G. Lewis

4. There are two basic types of encryption: *symmetric* and *asymmetric*. Symmetric encryption is used to secure information between trusted parties; asymmetric is used to secure information between anonymous parties. Both kinds of encryption have political implications because ciphers have historically been viewed as munitions.
5. Symmetric ciphers such as the data encryption standard (DES), triple-DES, and advanced encryption standard (AES) have evolved out of the Lucifer project started in the 1960s by IBM, but the ideas go further back—perhaps as far back as the classified work of the British during World War II. As computers get faster, symmetric codes are broken, requiring longer and longer keys. One major disadvantage of symmetric codes is that they are symmetric, which leads to a vulnerability.
6. AES is the latest symmetric code to be standardized by the U.S. National Institute of Standards and Technology (NIST) (2002), and besides being strong (256-bit keys), it is suitable for small computers such as those used in SCADA systems. 3DES and AES have been adopted by the U.S. Federal Government and are required for an IT system to be FIPS compliant.[3]
7. Asymmetric ciphers rediscovered by Diffie and Hellman in 1976 use a public key to encode and a private key to decode. The private key is not shared; hence, it is less vulnerable to cracking. The RSA Algorithm implements the ideas of Diffie–Hellman and makes it possible to authenticate users (digital signatures) as well as to protect the privacy of both sender and receiver.
8. PKI authenticates the identity of users by assuring that the sender is who he or she claims; guarantees the integrity and security of the message by assuring that it has not been modified by an intermediary; assures privacy by making sure the message is decodable only by the intended recipient; guarantees authentication, security, and privacy is enforceable by assuring that the message is signed by the verified parties; and guarantees nonrepudiation by assuring that both parties cannot disavow or deny involvement with the transaction.
9. Cyber-security will improve when the following information infrastructure improves: TCP/IP encryption of source/destination addresses; vendors remove software flaws; software application defaults are configured for the highest level of security; users are better informed and trained to prevent security breaches; organizations adopt stronger standard operating procedures; consumers demand better IT security; and vulnerability and risk analysis are standardized and used routinely.
10. More research needs to be done to make software virus-proof, reduce software errors that hackers can exploit, standardize vulnerability analysis including the use of quantitative techniques, and develop new methods to analyze cascade effects, predictive methods, and recovery.

[3]FIPS is the Federal Information Processing Standard.

ENTERPRISE SYSTEM VULNERABILITIES

An enterprise system is an IT system that is used by an enterprise—corporation, government agency, school, military command, etc.—regardless of size. Because an entire organization depends on it, an enterprise system demands high availability, integrity, reliability, and security. A desktop computer may be a member of an enterprise system, but it is not an enterprise system on its own. Thus, enterprise systems span entire organizations and provide a stable core of hardware and software components that support the mission of an organization. Indeed, enterprise systems are the IT infrastructure of most modern organizations. They consist of computers of all sizes, networks for connecting them, and software for making them useful. The enterprise system consists of e-mail programs, word-processing software, payroll applications, database storage, network operation centers, wired and wireless networks, and other hardware and software.

A secure enterprise system is essential to continuity of operations. Unfortunately, it is theoretically impossible to determine whether a system is secure.[4] The best we can do is institute policies that diminish the likelihood that an enterprise system is compromised either maliciously or inadvertently. Therefore, cyber-security is largely a practice rather than an exact science.

Generally, the goal of cyber-security is to protect an enterprise system from loss of service, loss of data, and loss of security. *Loss of service* typically means the system is down, slow, or otherwise unable to respond to its users. *Loss of data* means that information is lost, and *loss of security* means the system has been compromised, either by a break-in or lack of proper controls such as access rights (failed password, user privileges, etc.).

Figure 14.1(a) shows a fault tree for these three major categories, and Figures 14.1(b)–(d) show details of each category. Lower levels of the fault tree of Figure 14.1(a) are shown in Figures 14.1(b)–(d).

Loss of Service

Loss of service can occur in at least three ways: power failure, telecommunications failure, and a denial of service (DoS) attack. Power and telecommunication failures may be accidental or perpetrated incidents. A DoS attack is an exploit perpetrated by an attacker. Figure 14.1(b) shows one way in which a DoS might be successful: if a SYN-flooding attack is perpetrated AND the enterprise system fails to note the SYN-flooding attack because it does not contain an intrusion detection system (IDS) for detecting the attack.

Of course, there are other ways for loss of service to occur, such as a malfunction of equipment and software defects that cause the enterprise system to stop. Each analysis must be tailored to the enterprise system under investigation.

[4]Deciding whether a computer system is secure has been shown to be impossible, by mathematical logic. Consider the following paradox: Tom says, "Sally always tells the truth," and Sally says, "Tom always lies." Is Sally lying now? It is impossible to decide. In a similar fashion, system security can be shown to be undecidable.

Loss of Data

Loss of data can occur for several reasons: A file might be inadvertently deleted, a virus might be responsible for file deletions, or the deletion might be the result of an exploit that uses a flaw in an application such as Microsoft Excel, Oracle database, or human relations management software.

File deletion may not result in a loss of data if the enterprise system maintains up-to-date backups. Therefore, a break-in made possible by a clear password file may result in a malicious act such as an important file being deleted, but if there is a backup, then the file can be restored. Thus, a backup policy can assure the security of information even when files are deleted. Figure 14.1(c) contains an AND logic connector under "File Deleted," because all three faults must be true for loss of data to occur.

Data can be lost due to a virus or worm attack as shown in Figure 14.1(c). In this case, the lack of a firewall for filtering out the virus, plus the lack of anti-virus

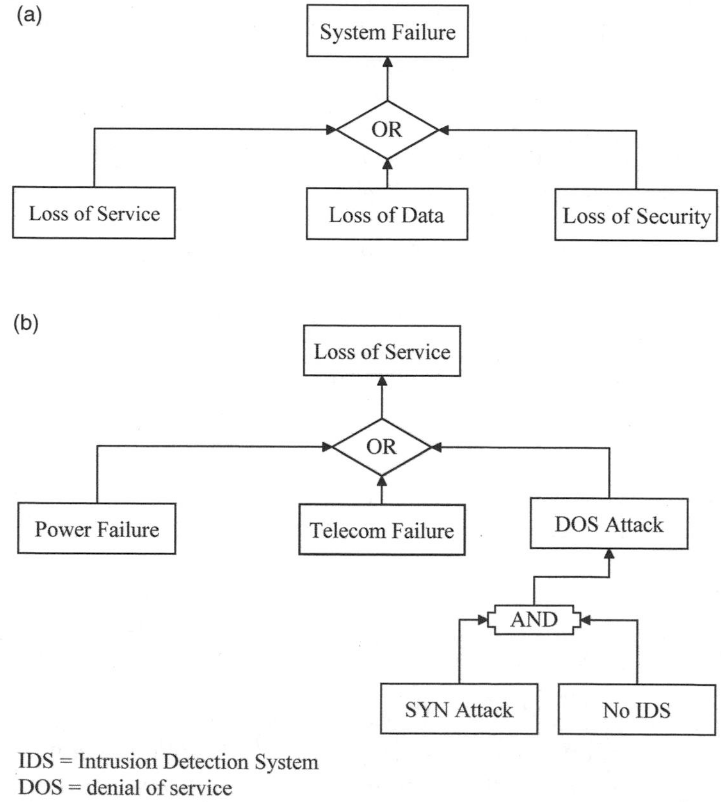

Figure 14.1. A fault tree containing cyber-security faults in a typical enterprise system. (a) Top-level fault tree. (b) Loss of service. (c) Loss of data. (d) Loss of security.

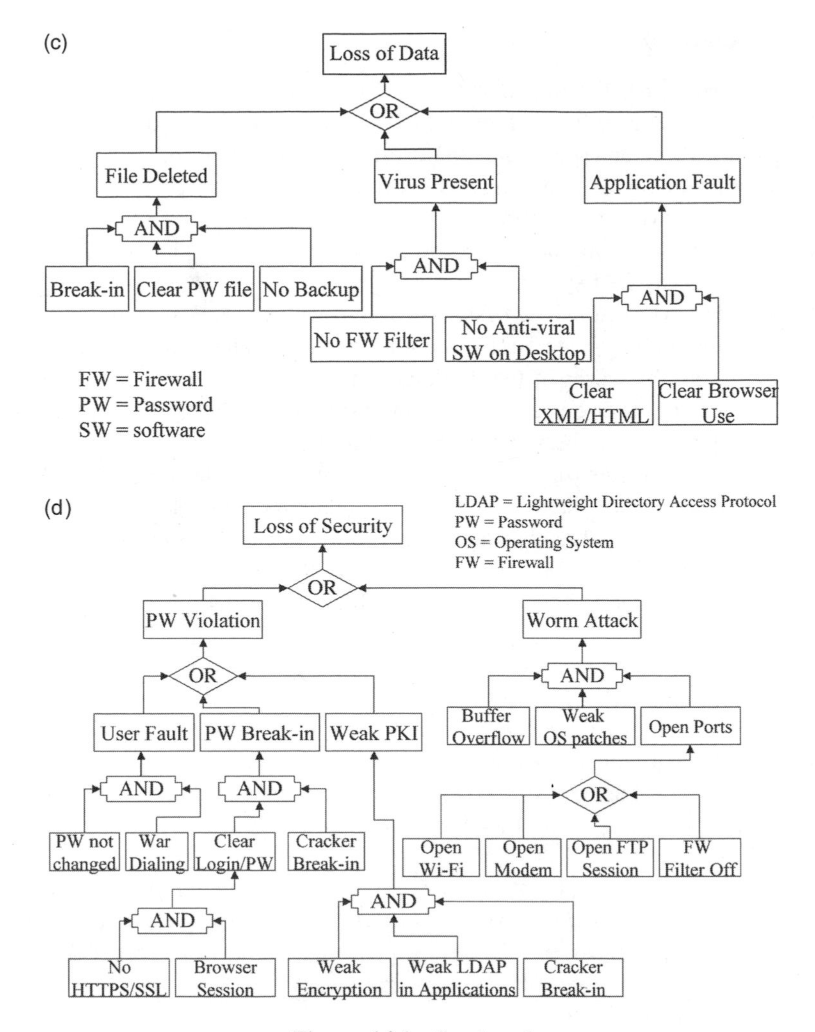

Figure 14.1. *Continued.*

software on the victim's computer, leads to a successful attack. If the virus is not filtered by a firewall AND the desktop computer containing the information is not protected by anti-viral software, loss of data occurs.

Finally, an application running on the victim's machine may be vulnerable to an attack or inadvertent loss of data because of a flaw in the application software. For example, an ActiveX control may be downloaded through a Microsoft Internet Explorer browser that operates in the clear—nonencrypted. The control is allowed to delete the user's files, because of inadequate security settings in the browser.

Loss of Security

Loss of security is what most people think of when they think of cyber-security. This category includes many faults. Figure 14.1(d) shows only two: password violations leading to break-in, and worm attacks leading to loss of control.

Password violations are among the most prevalent faults in enterprise systems. They occur because the user fails to protect his or her password, the enterprise system fails to protect the password file by encrypting it, or the enterprise system implements a weak PKI system.

Passwords can be discovered or obtained in several ways. War dialing is the act of systematically dialing phone numbers until a computer answers, and then systematically trying all the words in the dictionary until one of them works. This is why passwords should never be real words. Instead they should be nonsense strings of characters including numbers and special symbols.

Login and password strings can also be obtained by an attacker that breaks into a system and obtains the system's password file, records keystrokes of the user (keylogging), or observes traffic over the network connecting the user to the Internet. If the user's login and password are transmitted in the clear, instead of encrypted, a "man-in-the-middle attack" may succeed in obtaining the user's password and username. Hence, a browser should always connect with the enterprise system through a hyper text transport protocol-secure (HTTPS) server that uses secure socket layer (SSL) encryption.

PKI is a technology for managing passwords, encryption, and authentication of users through the use of *certificates*. Thus, PKI is an essential ingredient in the maintenance of TPs—secure links between the system and the user. We will devote a major part of this chapter to PKI later on. A weak PKI is an enterprise system vulnerability.

An encryption algorithm that can be cracked,[5] a flaw in the lightweight directory access protocol (LDAP) *directory* system that stores and manages passwords and access rights, and a break-in all add up to a breach due to a weak PKI. Thus, a final example of a password violation is an incident that exploits the very core of security—the PKI subsystem.

Loss of security can also occur because of a worm attack that succeeds in entering a victim's enterprise computer and then spreading to users connected to the enterprise computer. As described in Chapter 13, one such exploit begins with a buffer overflow, which succeeds because the victim has not installed the latest patch and the attacker has found an open port. If all of these are TRUE, a worm can take control of the system.

Why would anyone leave a port open? This is like leaving the doors to your house open and leaving for the weekend. But open ports are necessary, because they are the doorways through which the Internet is accessed and information is shared among authorized users. Ports must be open for wireless Wi-Fi communications, dialup modems, FTP file transfers, and all other forms of input/output to occur. In addition,

[5]Cracking is the process of illegally decoding ciphertext into plaintext.

the enterprise system's firewall must allow data to flow into and out of the enterprise through an open port. Thus, the paradox of cyber-security is that ports must be opened to make the system useful, and yet open ports are open to attacks..

DEFENSE

Cyber-defense is a matter of policy. On the one hand, security is costly and inconvenient for users. On the other hand, defense is necessary, at some level, to secure the information managed by an enterprise system. Therefore, cyber-security policy will end up reflecting many compromises between assurance and convenience. This leads to a question, "what is the minimum security, possible?" In this section we describe a set of minimum policies for ensuring a basic or foundational level of cyber-security. Such a cyber-security foundation is called a TCB.

Definitions of TCB vary, but for our purposes, a TCB is the totality of protection mechanisms within a computer system, including hardware, firmware, and software, that is responsible for enforcing a security policy. It creates the most fundamental protection environment possible along with some additional user services required for a trusted computer system. The ability of a trusted computing base to correctly enforce a security policy depends solely on the mechanisms within the TCB and on the correct input of parameters by system administrative personnel (e.g., a user's clearance) related to the security policy.

Figure 14.2 shows a TCB made up of a TCB system and a TP between the computing system and the user. The user will typically be a person sitting at a desktop or laptop computer connected to an enterprise system through the Internet. The trusted path will typically be a secured Internet connection. The core component of the trusted computing base will typically be the enterprise servers running behind a firewall.

To make the TCB less abstract we define TCB in more practical terms using the architecture of a typical TCB. The core of a typical TCB shown in Figure 14.3 is the *de-militarized zone* (DMZ), shown as a shell surrounding the enterprise system. Inside of this core are the components necessary for enforcing security policies such as user authentication, encrypted data, and access privileges.

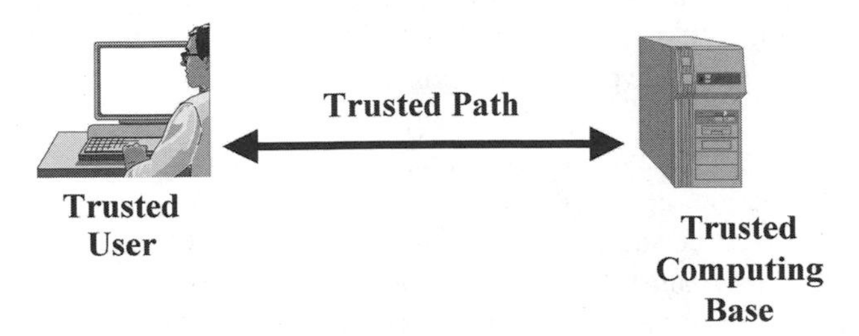

Figure 14.2. Simplified architecture of a TCB.

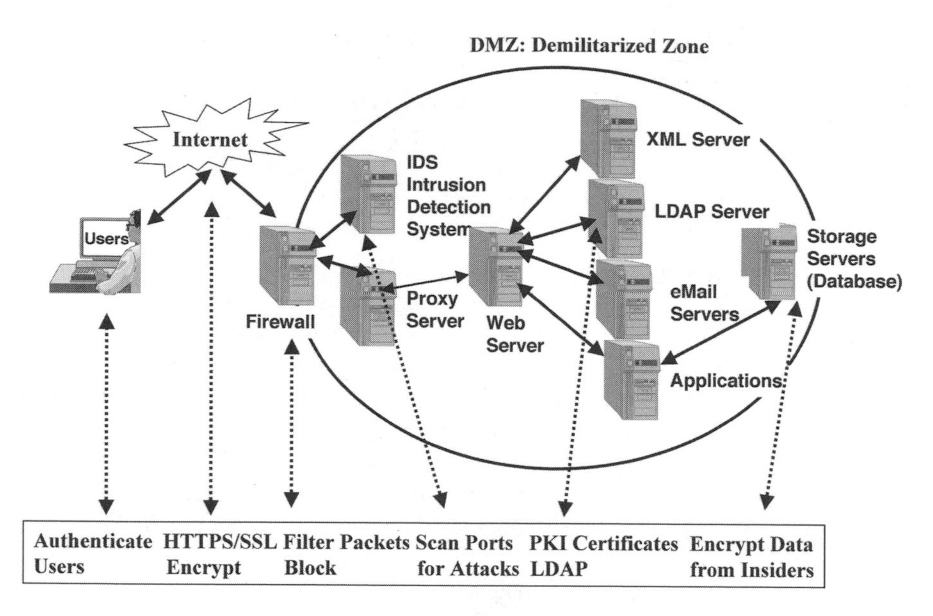

Figure 14.3. The architecture of a typical TCB.

Outside of the DMZ core is a TP connecting users to the DMZ. Although a TP can be implemented by any kind of network, our example uses an encrypted Internet connection. In addition, a TCB must ensure that the users are who they say they are. The identity of the system's end-users must be authenticated, usually by employing a user login and password.

The goal of this architecture is to establish a TCB made up of the components protected within the DMZ, a TP between user and DMZ, and a trusted user. The list of mechanisms at the bottom of Figure 14.3 suggests ways to ensure the correctness of the TCB, TP, and authenticated users.

Authenticate Users

Starting from the user's perspective, the first step in establishing a TP is to verify the authenticity of the users. The login/password mechanism currently used by most enterprise systems is perhaps the simplest. But more sophisticated biometric techniques should be used if a higher level of trusted computing is desired. This is an example of a policy decision that will affect the implementation of cyber-security.

Suppose, for example, a certain enterprise is required to guarantee the security of classified information. In this case, it is likely that a login/password mechanism will be inadequate for user authentication. Instead, users might be authenticated using some biometric mechanism such as a voiceprint, fingerprint, or both. Retinal scanning, synchronized timers, and smart cards are other authentication mechanisms used in high assurance systems.

Trusted path

Next, the TCB needs a TP that connects the users to the information they need to do their jobs. The purpose of a TP is to guard against man-in-the-middle attacks or impersonations of the users. Using a naval metaphor, we not only want to protect the ports, but also ships on the high seas. Similarly, cyber-security aims to protect information stored in the core of the TCB as well as information that is in transit between users and the TCB.

The most elementary means of protecting communication links is to use a browser that supports SSL encryption. SSL requires HTTPS running on the enterprise server. SSL encrypts each session so that an intercepted communication cannot be hacked. E-commerce sites use the SSL/HTTPS combination to provide a TP for their online customers. Credit card information is encrypted by SSL, for example, and delivered by HTTPS to the e-commerce store.

SSL implements a modest level of cyber-security. A stronger form is called a virtual private network (VPN).[6] Recall that each TCP/IP packet is transmitted in the clear, meaning that source and destination addresses can be seen by a man-in-the-middle attacker. Although the data may be encrypted, the remainder of the packet is not. IP version 6 (Ipv6) supports encryption of the TCP/IP packets. This conceals not only the contents but the sender and receiver's identity as well. The "V" in VPN stands for "virtual," which means that virtual source and destination addresses are used in place of real addresses. To get through a firewall, these virtual addresses must be recognized and translated back into their real address equivalents. This is called IP *tunneling*, or VPN tunneling, because the VPN establishes a "tunnel" through the firewall. Tunneling involves establishing and maintaining a logical network connection with possibly intermediate hops. A VPN allows corporations to establish a proprietary network on an open public network such as the Internet.

The bottom line is this: A VPN can be constructed on top of Ipv6, making the TP even more secure—even though data travels over the open Internet. This can be costly in terms of hardware and software, and it can slow down a network because of the translation between virtual and real addresses. So the tradeoff is cost, speed, and convenience versus enhanced security. Thus, the decision to use a VPN must be a policy decision.

Inside the DMZ

Once inside the DMZ of Figure 14.3, implementation of cyber-security becomes a more complex and sophisticated challenge, because a successful hack into the DMZ can have disastrous repercussions. The DMZ cannot be compromised. The question is, "what are the minimum mechanisms for achieving a minimally secure DMZ?"

[6]Virtual Private Network. A network that uses the Internet rather than leased lines for connections. Security is guaranteed by means of a *tunnel* connection in which the entire TCP/IP packet (content and addresses) is encrypted and encapsulated.

A typical minimum set of mechanisms for assuring the security of a DMZ are the following:

- Firewalls
- Proxies
- IDS
- Web servers
- XML servers
- LDAP servers
- PKI software and policies for enforcing the TCB

The first line of cyber-defense is the firewall. A *firewall* is a special-purpose computer that manages ports, inspects and filters network packets, and determines whether to allow packets into the DMZ. Firewalls come in two varieties: *static packet filtering firewalls* that block packets based on the source and destination addresses in each packet, and *stateful packet filtering firewalls* that block packets based on content, level of protocol, and history of packets. Stateful firewalls are sometimes called *dynamic filtering* firewalls.

Firewalls are not perfect. In fact, they are far from perfect, because they cannot block many malicious programs. Simply stated, a firewall is mainly used to manage ports and VPNs. They are not sufficient to detect Trojan horses, prevent DoS attacks, and thwart e-mail viruses. Therefore, they should not give the administrator a false sense of security, but instead they should constitute the first step in establishing a TP between users and information.

A *proxy server* is a special-purpose computer that sits between a user and a real server. It intercepts all requests to the real server to see if it can fulfill the requests itself. If not, it forwards the request to the real server. The purpose of a proxy server is twofold: to improve security, and to enhance performance. It improves performance and security by caching incoming requests on behalf of an external website or user. In this way, unauthorized requests can be thwarted by the proxy and never reach the inner components of the TCB. The enterprise system does not expose all of its information to the outside world, only the portions that are public.

A proxy server can also perform the functions of a *gateway* by accessing external pages on behalf of an internal user. Each time a user requests a page from a remote website, the proxy server is consulted, and if the page is already inside of the DMZ, the proxy server supplies the page instead. This avoids delays and enhances security because the entire transaction is performed within the DMZ. It is not necessary to venture beyond the firewall. Gateway proxies are often used to prevent employees from viewing unauthorized websites.

Every good TCB needs an intrusion detection system (IDS). An IDS is a special-purpose computer that inspects all inbound and outbound network activity and identifies suspicious patterns that may indicate that a network or system attack is underway. It uses a variety of algorithms to detect when someone is attempting to break into or compromise the DMZ. For example, it may employ *misuse detection* or the process of comparing "signatures" against a database of attack signatures to determine if an attack is underway. Or the IDS may employ *anomaly detection* by comparing the state of the network against a "normal" baseline.

An IDS can be network-based or host-based. A *network-based IDS* protects an entire network, whereas a *host-based IDS* protects a single computer such as a home PC. It can also be passive or reactive. A *passive IDS* simply logs network traffic status and only signals a human operator when an unusual pattern is observed. A *reactive IDS* automatically logs off a user or blocks network traffic from the suspected source when it records a suspicious pattern.

A *Web server* is a computer with special software to host Web pages and Web applications. It is the component that hosts HTTP/HTTPS and delivers HTML/XML pages to users. Figure 14.3 shows how a Web server acts like a traffic cop, handing off actual processing to other computers. For example, e-mail messages are handed off to an e-mail server, XML messages are handed off to the XML server for parsing, database queries are handed off to a database application server, and security functions are handed off to the LDAP directory server.

Some well-known examples of Web servers are as follows:

- Apache
- MS Internet Information Server (IIS)
- Netscape Enterprise Server
- Sun Microcomputer Java System Application Server

An LDAP directory server is an essential part of any TCB.[7] Its function is twofold: to participate in the authentication of users through password storage and verification, and to hand out permissions—called *privileges*—to running applications.[8]

Access privileges are transferred among users of TCBs through a ticketing system called X.509 certificates. A *certificate* is a digitally signed message that transfers privileges from the sender to the recipient. X.509 is a recommended standard as defined by the International Telecommunications Union (ITU).[9] Most computer users see certificates as dialog boxes that pop up in the middle of a Web browsing session. The dialog asks the user to allow the foreign request access to

[7]LDAP—defined by the IETF—defines a relatively simple protocol for updating and searching directories running over TCP/IP.

[8]Additions to version 3 of LDAP rectified many of the shortcomings of the original LDAP and allowed LDAP servers to correctly store and retrieve X.509 attributes, but searching for them was still impossible. This is because the protocol fields, i.e., the X.509 attributes, are simply transferred and stored as binary blobs by LDAPv3, with the server having no knowledge about their structure and contents. D. W. Chadwick, E. Ball, and M.V. Sahalayev, "Modifying LDAP to Support X.509-basedPKIs," University of Salford, Salford, M5 4WT.

[9]X.509 is actually an ITU recommendation, which means that it has not yet been officially defined or approved for standardized usage. Nonetheless it has been widely adopted by the Internet community.

his or her computer. If the user agrees, the certificate transfers permission from the user to the foreign requestor. More on this later.

How do users get privileges? They do so by providing a user login and password to an authentication program to verify the authenticity of the login. The authentication program obtains each user's access privileges from a (LDAP) directory that is safely stored within the DMZ. As the user moves from one application to another, each application consults the list of user privileges to determine if he or she has the necessary access rights. For example, one user may have the right to read a database record and another user may have the right to change the record. In this way the user does not have to login repeatedly to different applications, and the entire system is protected from unauthorized internal access. Certificates and access privileges form the basis of a security infrastructure called PKI, which we describe in more detail later in this chapter. But first, we need to understand the basics of encryption.

BASICS OF ENCRYPTION

Encryption, turning messages into secret codes, is at least 4000 years old. Encryption converts *plaintext* words into *ciphertext* codewords using a key and some encoding algorithm. The result is called a *cipher*. The reverse process—converting ciphertext into plaintext—is called *decryption*. The *key* is a special word that enables encoding. If the same key is used to decode the secret message, we say the encryption is *symmetric*. If a different key is used, we say the encryption is *asymmetric*. Cryptology is the study of ciphers, keys, and encryption algorithms.[10]

During most of its history, cryptography did not change much. Find a way to translate plaintext into ciphertext and then transfer the ciphertext to a recipient, who reverses the process using the secret *key*. The cipher is symmetric, because both parties use the same key to encode and decode the secret message. Thus, the key must be protected, because anyone with the key can unravel the cipher.

Perhaps the best-known symmetric cipher is the logical EXCLUSIVE-OR cipher—widely known because of its simplicity. The EXCLUSIVE-OR cipher performs the logical EXCLUSIVE-OR operation on each bit of the binary representation of plaintext (see Table 14.1). It works bit-by-bit across the plaintext by taking one bit from the plaintext word and another bit from the key, and performing the EXCLUSIVE-OR operation shown in Table 14.1. To reverse the process—from ciphertext to plaintext—do the same thing over again: EXCLUSIVE-OR the key with the ciphertext. This is a symmetric encryption algorithm because the same key is used for both encoding and decoding.

For example, suppose the secret key is 1101, and the sender wants to encrypt the plaintext 1001 and send it to the receiver, who also knows the key. Encoding is done

[10]A simple definition of cryptology is the study of secret messages.

TABLE 14.1. EXCLUSIVE-OR Logic: A EXCLUSIVE-OR B: Only One of the Two Operands Can Be 1 to Produce a 1, and the Operands Must Be Identical to Produce a 0.

EXCLUSIVE-OR	B = 0	B = 1
A = 0	0	1
A = 1	1	0

by EXCLUSIVE-ORing each bit of the message 1001 with the each corresponding bit in the key. The same process is repeated to recover the plaintext from the ciphertext.

Sender Encodes 1001 using key *1101*, as follows:

1101 EXCLUSIVE-OR 1001 = 0100 = ciphertext.

Receiver decodes 0100 as follows:

1101 EXCLUSIVE-OR 0100 = 1001 = plaintext.

Keys in the EXCLUSIVE-OR cipher are limited to no more than 2^k possible values for a key with k bits. That is, the time it takes to enumerate all possible keys is proportional to 2^k. For example, a 20-bit key can have no more than 2^{20}—approximately 4 million—distinct values. This may seem like a lot, but even a key with 128-bits is not too large for a modern computer to run through in a relatively short period of time. The EXCLUSIVE-OR cipher can be cracked by simply trying every key value from 0 to $2^k - 1$. But a key with 256 bits would take a computer 2^{256} units of time to crack—many times more than the time to crack a cipher with half as many bits. In other words, the strength of a cipher is *exponentially* related to the number of bits in the key. Key length determines encryption strength. Cyber-security needs *strong encryption*, and this means ciphers with large keys.[11]

DES

In the 1960s, a team of IBM researchers designed a symmetric cipher for commercial applications that they called the *Lucifer Algorithm*. Lucifer was not unique, but it was destined to become the first standard encryption technique for the U.S. Federal Government. Indeed, Lucifer was adopted by the NIST for use by non-military customers in 1977 and revised in 1994. Simply called the DES, the Lucifer algorithm has been widely used by banks, insurance companies, and handheld Blackberry e-mail machines. It is also known as 56-bit encryption, because it uses a 56-bit key.

[11]The strength of a cipher is measured by how long it takes for a computer to break it. Today, strong encryption means a computer the size of the universe would need all of recorded time to break the cipher.

In the late 1960's, IBM's chairman Tomas Watson, Jr., set up a cryptography research group at his company's Yorktown Heights research laboratory in New York. The group, led by Horst Feistel, developed a private key encryption system called "Lucifer." IBM's first customer for Lucifer was Lloyd's of London, which bought the code in 1971 to protect a cash-dispensing system that IBM had developed for the insurance conglomerate.

In 1968, the National Bureau of Standards (NBS, since renamed National Institute of Standards and Technology, or NIST) began a series of studies aimed at determining the U.S. civilian and government needs for computer security. One of the results indicated that there was a strong need for a single, interoperable standard for data encryption that could be used for both storage and transmission of unclassified data (classified stuff was still the domain of the NSA).[12]

DES uses 64 bits: 56 for data and 8 for parity. It is also called a *block cipher* because it breaks the message into 64-bit blocks and encodes each block separately. There are actually four variants of DES:

1. ECB = Electronic codebook (standard DES algorithm)
2. CBC = Cipher block chaining
3. CFB = Cipher feedback
4. OFB = Output feedback mode

The DES algorithm is described in more detail in **Algorithm 14.1**.

ALGORITHM 14.1: DES ENCRYPTION

Encode:

Use permutation tables to scramble the plaintext to be encoded:
The 56-bit key + tables produce 16 48-bit subkeys: K1, K2 ... K16.
Do this 16 times:

Split 64-bit input data into two halves, L and R of 32-bits each.
Expand and permute R into 48-bits and XOR with Ki, i = 1 ... 16.
Further scramble with a table that produces eight 4-bit blocks.
Permute the result again, then XOR with Li, and swap L and R.
L16 and R16 are joined back together to form the 64-bit pre-output.
Use a table to permute the pre-output one last time.

Decode: Apply subkeys in reverse order: K16, K15, ... K1 using the Encode Algorithm.

Note: XOR is the EXCLUSIVE-OR operation.

[12]http://library.thinkquest.org/27158/concept2_1.html.

Unfortunately, DES was cracked in 3 days in 1998 using a special-purpose computer. In 1999, it was cracked in 22 hours using 100,000 personal computers working together over the Internet. Today, cracking the 56-bit DES cipher is child's play for most home computers. But DES can be strengthened by using longer keys.

3DES

The easiest way to make DES stronger is to make the keys longer—three times longer, in fact. Triple-DES, or 3DES, simply applies DES three times with three keys: Key1, Key2, and Key3. This effectively increases key length threefold, from 56 to 168 bits. It also increases the difficulty of breaking the code by a factor of 2^{112}, or about 168 years of doubling in computer processing speed.[13] 3DES is strong but somewhat cumbersome.

AES

Modern symmetric encryption uses the AES, adopted by NIST and officially standardized by the U.S. Government in 2002. It replaces DES and 3DES, because it uses even longer keys: 128-, 192-, or 256-bit keys. In May 2002, the NIST adopted the Rijndael (Daemen–Rijmen) Algorithm as the basis of AES.[14] A 256-bit Rijndael cipher is 2^{200} times stronger than DES and 2^{88} times stronger than 3DES. In other words, it will take 120 years of progress in computing to achieve the necessary speeds to crack AES the way that DES was cracked in 1999.

One major advantage of AES, in addition to its strength, is that Rijndael works on small machines, which means AES is suitable for SCADA applications. But it takes 10, 12, or 14 rounds, depending on key size, to encode and then to decode messages. This is slower than other symmetric codes, but not too much of a burden for modern processors, even the commodity processors used in SCADA. The future of symmetric encryption is AES.

Asymmetric Encryption

Codebreakers have learned to use exhaustive brute-force methods to defeat symmetric key ciphers. In fact, high-powered computers have become very good at finding keys for symmetric encryption. Although the future of symmetric encryption is AES, even AES has faults. One of these is the exposure that comes with sharing the secret key among many users. In symmetric encryption, both the sender and the receiver use the same key, which exposes the most critical piece of the cipher—the key.

[13]Moore's Law says processing speed doubles every 1.5 years. So, $1.5 \times 112 = 168$.

[14]J. Daemen and V. Rijmen, "AES Proposal: Rijndael, AES Algorithm Submission," September 3, 1999, http://www.nist.gov/CryptoToolkit.

Fortunately, the cryptographic loophole caused by key sharing was overcome in 1976 by two clever mathematicians named Diffie and Hellman.[15] They found an elegant way for two parties to share a secret message without sharing the same secret key. Instead, each party shares *public keys* but conceals his or her own *private keys*. Public keys are used to encode, and private keys are used to decode the secret message. In this way, two parties can keep both the message and their private keys a secret. Public key encryption is *asymmetric*, because a different key is used to encode than to decode.

The Diffie–Hellman invention of asymmetric cryptography using a public key was profound and disconcerting to intelligence agencies of the U.S. Federal Government, because it allowed anyone to build ciphers that nobody could crack, not even the powerful National Security Agency. Before Diffie–Hellman, ciphers were routinely cracked by the NSA. Afterward, criminals and law-abiding citizens alike were able to keep their messages completely secure and completely unbreakable by even the NSA.

The history of cryptology is long and colorful—too long and colorful to do it justice in this chapter. But one of the most interesting events of recent history is the peculiar case of Phil Zimmermann and the U.S. Customs. Zimmerman was accused of trafficking in munitions simply because he wrote a computer program that implemented the Diffie–Hellman asymmetric algorithm! In 1995, Charles Gimon described the essence of asymmetric encryption and Phil Zimmermann's program called *Pretty Good Privacy* (PGP):

> In 1976, a completely new way to encrypt messages was published by Whitfield Diffie and Martin Hellman. This new method was called *public key encryption*. In this system, each person has two keys, a public key and a private key. The public key is broadcast around so anyone can use it, the private key is known only to the owner. You can encode a message with the recipient's public key so that only they can decode it with their private key. This public key encryption not only provides privacy, it also makes it possible to be certain that only the sender wrote the secret message you received. It ensures both privacy and identity.
>
> Public key encryption is fantastically difficult for even computers to break. The longer you make the keys, the more difficult it is to break. You can make the keys long enough so that, using today's technology, anyone's best guess is that it would take so-and-so many billions of years to break the code. One cute phrase you hear to describe this situation is "acres of Crays." There's even wild talk of making keys so long that using the codebreaking methods we have right now, you'd need a computer with more circuits than there are atomic particles in the known universe working for a longer period of

[15]W. Diffie and M. Hellman, "New Directions In Cryptography," *IEEE Transactions on Information Theory*, vol. IT-22(6), pp. 644–654, November 1976. Actually, three British Security Service researchers, Ellis, Cocks, and Williamson, discovered public key encryption in 1968/69, but because their work was classified, they could not publish their results. Diffie and Hellman discovered public key encryption, independent of the British Security Service researchers.

time than has passed since the Big Bang to break it. In other words, a metaphysically unbreakable code - talk about tough math homework.

Many companies, including AT&T, SCO and Sun Microsystems, have used public key encryption in their products. In order to give the power of public key encryption to folks like you and me, a programmer in Boulder, Colorado named Phil Zimmermann put together a shareware program called PGP—"Pretty Good Privacy"—which lets anyone with a PC use public key cryptography.

Governments like ours have a healthy respect for cryptography; it's sometimes said that the U.S. and Britain won the Second World War by breaking German and Japanese codes. In the United States, strong, "unbreakable" encryption is considered a weapon for export purposes, just like hand grenades or fighter planes are. In theory, it's illegal to export public key cryptography, on paper or as a computer program.

In 1991, right after the Gulf War, there was a bill before the U.S. Senate (S.266) that would have had the effect of banning public key encryption altogether. Faced with this situation, some activists in the [San Francisco] Bay Area decided that if they could spread public key encryption around widely enough, the genie would be out of the bottle and there'd be no way for Uncle Sam to get it back in again. They took a copy of Zimmermann's program and uploaded it to as many bulletin boards and Internet sites as they could.

It took the Feds two years to react. In February 1993, Mr. Zimmermann received a visit from Customs. Even though he didn't do the uploading himself, the Feds say that Zimmermann allowed his program to be uploaded to Internet sites that can be reached from anywhere in the world, and therefore he has supposedly exported a munition without a license. It sounds like something an oily guy in Miami or Beirut would be involved in—but a computer geek in Boulder, Colorado?

As of this writing, any indictments from the grand jury are still pending—hanging over Phil Zimmermann's head. Businesses that have been using strong encryption (but have not been visited by customs agents) are peeved and worried, and some of them have been pressuring the Government to lay off the heat, at least. Thousands of regular Net users have come to Zimmermann's defense, some of them sending e-mail to the appropriate politicians, some of them donating to Zimmermann's legal defense fund.

David-and-Goliath aspects aside, the case is important for two reasons. The obvious one is the First Amendment one—computer software ought to be considered speech, something that Congress is not supposed to pass any law abridging the freedom of. Anyway, encryption is just math, and restricting or banning it is not that much different than banning the knowledge that two plus two equals four.

The other thing about the [Zimmermann incident] is its impact on America's software industry. Restricting the export of strong encryption is a joke—you can buyit shrink-wrapped in Moscow. The restrictions are an outdated, artificial leg-iron on American companies, and if they were enforced on everybody, it would make American encryption software a second-rate choice in every other part of the world. Public key encryption lets you do secure transactions on the Internet. That means buying and selling and free enterprise—all the things that we won the Cold War for—with little risk of theft or fraud. It's a shame that exporting what could be a great crime-fighting device could end up being a crime itself.[16]

[16]C. Gimon, "The Phil Zimmerman Case," http://www.skypoint.com/members/gimonca/philzima.html.

The socio-political implications of encryption mathematics is obvious from this description. Encryption is critical to secure operation of IT infrastructure. But it is not just a topic for computer experts, because it affects everyone. However, the remainder of this chapter will be devoted to the discussion of encryption's role in establishing a PKI. A thorough discussion of the politics of encryption will be left to another author!

Public Key Encryption

The nontechnical reader may want to skip the following section, which describes, by example, how public key encryption works. It is the backbone of trusted computing. Without public key encryption, privacy would not be possible in the Internet Age.

The Diffie–Hellman paper described the concept of public key encryption, but it did not describe how to actually do it. The problem was that translation from plaintext to ciphertext had to be one-way. That is, the process had to be irreversible. Otherwise, the receiver of a secret message could work backward and discover the sender's key. Most mathematical operations are two-way: $3 + 2 = 5$ is reversible to $3 = 5 - 2$, and division, $6/3 = 2$, can be reversed by multiplication $2 \times 3 = 6$. Asymmetric encryption needed a mathematical operation that worked one way but not the other.

In 1977, three mathematicians, Ronald Rivest, Adi Shamir, and Leonard Adleman (RSA), started their journey into the annals of encryption history by attempting to prove Diffie and Hellman wrong. Instead, they showed how to implement the Diffie–Hellman ideas, which led to the famous RSA cipher in 1977.[17] Today, RSA is the most common form of encryption used in PKI. **Algorithm 14.2** shows how to do public key encryption.

Public key encryption is an extremely clever application of big numbers—really big numbers![18] RSA is based on prime numbers raised to large powers, which results in extremely large numbers.[19] The numbers are so large that it takes a computer to add, subtract, multiply, and divide them. In fact, the larger the number, the better, because a codebreaker must be able to find extremely large prime numbers just to start the process of cracking an RSA cipher. Prime numbers containing hundreds of digits are common, and primes with millions of digits are well known to the intelligence community.

RSA translates a series of plaintext words into a series of codewords that look random (see Figure 14.4). This makes it difficult for codebreakers to analyze long sequences of codewords using pattern-matching software to unravel the key. Instead of producing an intelligible pattern, pattern analysis produces random noise.

[17]R. L. Rivest, A. Shamir, and L. Adelman, "On Digital Signatures and Public Key Cryptosystems," MIT Laboratory for Computer Science Technical Memorandum 82, April 1977.

[18]Public key encryption uses numbers in excess of 200 digits long!

[19]A prime number is a positive number that is divisible by one and itself only. Prime numbers can be found by computerized mathematical sieve techniques that take time proportional to the size of the prime number.

ALGORITHM 14.2: RSA ENCRYPTION

Let a public key P be a pair of integers (n, e) and a private key V be a pair (n, d). The public and private keys share, $n = p \times q$, where p and q are randomly chosen primes.

Make sure that n is larger than the largest plaintext character you want to encode.

To encrypt a plaintext character m:

$$\text{Encode}(m) = m^e \bmod n, \text{ where mod is the modulo function.}$$

To decrypt a ciphertext character c:

$$\text{Decode}(c) = c^d \bmod n.$$

How are the numbers n, e, and d in P:(n, e) and V:(n, d) found?

1. Select large prime numbers p and q at random.
2. Calculate the product $n = p \times q$.
3. Choose a number e such that:

   e is less than n.

   e has no factors in common with either $(p - 1)$ or $(q - 1)$.

4. Find d, such that $e \times d \bmod (p - 1) \times (q - 1) = 1$. One way to find d is to search for values of k and d that make this true: $e \times d = 1 + k (p - 1) \times (q - 1)$, for some $k > 1$.

   The mod operation is simply the remainder of a/b after division. For example, if $a = 8$ and $b = 5$, $a/b = 8/5 = 1$ with a remainder of 3. Therefore, $8 \bmod 5 = 3$.

Here is an example. We want to send a secret message containing the date, December 7, 1941—the three plaintext words {12, 7, 41}—from Honolulu to Washington, D.C. using $p = 5$ and $q = 11$, $n = 55$, which is large enough to encrypt plaintext words ranging from 0 to 54. Using the algorithm above, we select e less than 55 and make sure it has no factors in common with either $(p - 1) = 4$ or $(q - 1) = 10$. Note that $(p - 1) \times (q - 1) = 4 \times 10 = 40$. The number e must be prime relative to 40. Suppose $e = 3$, which satisfies this requirement (as does many other numbers such as 7 and 11). Because $p \times q = 5 \times 11 = 55$, the public key is $P = (55, 3)$, which the sender in Honolulu uses to encrypt plaintext {12, 7, 41} into {23, 13, 6} as follows:

Ciphertext word 1 = $12^3 \bmod 55 = 1728 \bmod 55 = 31$ with remainder 23.
Ciphertext word 2 = $7^3 \bmod 55 = 343 \bmod 55 = 6$ with remainder 13.
Ciphertext word 3 = $41^3 \bmod 55 = 68921 \bmod 55 = 1253$ with remainder 6.

Now we need a public key V = (55, d), where d satisfies the requirement (e × d) mod 40, which is the same as saying $e \times d = 1 + 40 \times k$ for some k. We have already chosen e = 3, so we want to find a, d, and k such that $3 \times d = 1 + 40 \times k$. The smallest value is d = 27, for k = 2. [Check: $3 \times 27 = 1 + 40 \times 2 = 81$.] Thus, Washington's private key is V = (55, 27).

Washington, D.C. receives the cipher containing codewords {23, 13, 6} and uses its private key V = (55, 27) to reverse the encryption, transforming each codeword back into plaintext as follows:

$$\text{Plaintext word } 1 = 23^{27} \bmod 55 = 12.$$
$$\text{Plaintext word } 2 = 13^{27} \bmod 55 = 7.$$
$$\text{Plaintext word } 3 = 6^{27} \bmod 55 = 41.$$

Computing large numbers such as 23^{27} can tax even the most capable computer, so we take advantage of the fact that $27 = 3 \times 3 \times 3$ and $23^{27} = ((23^3)^3)^3$. At each step in the calculation, we can apply the mod function to reduce the size of the number. Therefore, $23^{27} \bmod 55 = ((23^3)^3)^3 \bmod 55 = (12{,}167 \bmod 55)^3)^3 \bmod 55 = ((12)^3)^3 \bmod 55 = (1{,}728 \bmod 55)^3 \bmod 55 = (23)^3 \bmod 55 = 121{,}677 \bmod 55 = 12$. If we keep reducing the number modulo 55 after each exponentiation, the intermediate result never gets too large.

Note that the choice of private key exponent d is arbitrary, except that it must be relatively prime to $(p - 1) \times (q - 1)$. We used d = 27, but d = 67 is also relatively prime to 40, because there are no factors of 67 that are also factors of 40. (Actually, 67 is prime.) If we used the private key V = (55, 67), we would get the same result: {12, 7, 41}. Many private keys decrypt messages produced by P = (55, 3). Does this weaken the cipher?

Algorithm 14.2 illustrates the RSA technique using December 7, 1941 (12/2/41) as an example of a message to be sent from Honolulu to Washington, D.C. Honolulu uses Washington's public key to encode, and Washington uses its own private key to decode. So 12/2/41 is encoded as {23, 13, 6} using Washington's public key P = (55, 3). When Washington receives the ciphertext, it decodes {23, 13, 6} into (12, 7, 41) using its private key, V = (55, 27).

Honolulu does not know V, and so only Washington can decode the message. However, other private keys can also decode the message. For example, V = (55,67) also unscrambles the cipher. But nobody knows the exact values used in these other keys.

Public key encryption cleverly uses the one-way property of modulo arithmetic. Its strength is based on the (large) size of keys, which are large prime numbers. Although these are not difficult to compute, there are so many of them with hundreds of digits that it takes a long time to crack.

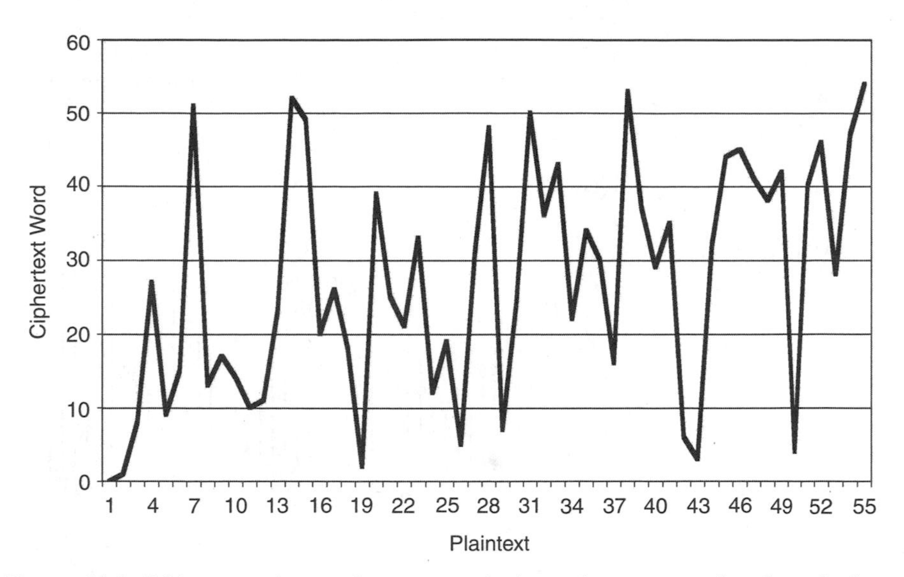

Figure 14.4. RSA encryption produces a seemingly random stream of codewords from plaintext. The codewords were produced from the public key P = (55, 3).

PROGRAM RSA

A program called *RSA.jar* performs RSA encryption and allows the user to experiment with different public and private keys (see Figure 14.5).[20] Given two prime numbers, p, q, and a plaintext message such as, "Now is the time," RSA calculates a public and private key, encodes the plaintext message using the public key, and then decodes the cipher using the private key. All inputs are keyboard characters, which are converted into numerical equivalents, and all outputs are alphanumeric characters.

RSA uses Algorithm 2. To encrypt a plaintext character m:

$$\text{Encode}(m) = m^e \bmod n,$$

where mod is the modulo function, and $n = p^*$.

To decrypt a ciphertext character c:

$$\text{Decode}(c) = c^d \bmod n,$$

where d satisfies $e \times d = 1 + k \times (p - 1) \times (q - 1)$ for some k.

Program RSA finds the smallest k and d that produce a satisfactory d. If the user

[20]*RSA.java* can be found on the Web or the disk accompanying this book.

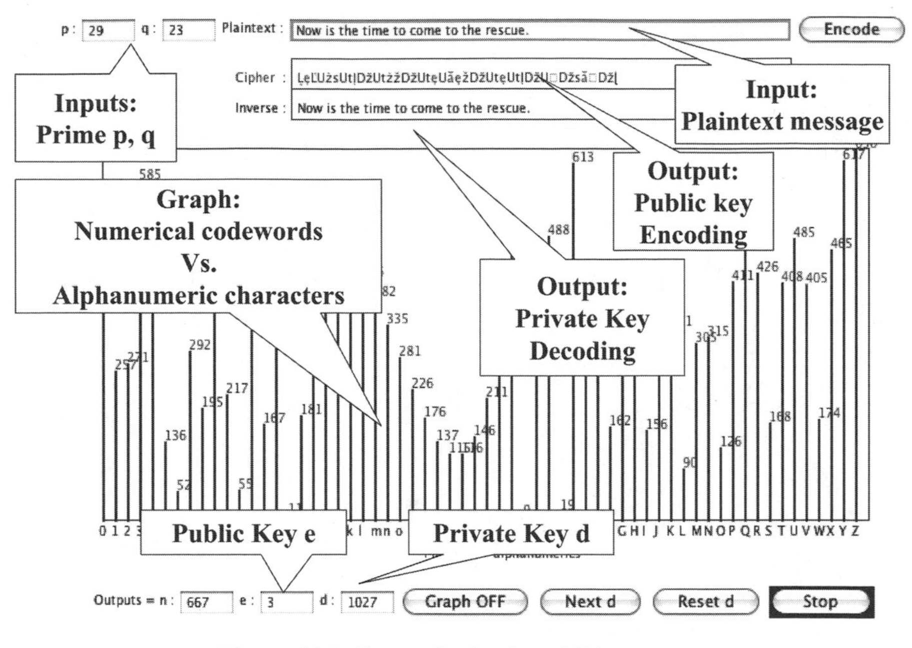

Figure 14.5. Screen display from RSA program.

wants to increase k and d, press NEXT-d. Pressing Reset-d returns the private key values to the smallest possible k and d that works.

Program RSA also displays the numerical value of codewords (encrypted plaintext characters) versus the alphanumeric plaintext characters "0" through "9," "A" through "Z," and "a" through "z." Note that each public key (p, q, e) produces a different graphical display, but they all look random. Also note that each private key (p, q, d) produces the *same* result; they all correctly decode the cipher. Only the public key determines the cipher. The private key simply decodes the cipher, returning it back to its original plaintext.

PKI

The topic of PKI is vast and complex, so we will attempt to simplify. As a result of simplification, some accuracy will be sacrificed. The foundation of TPs in cybersecurity is public key encryption, which is only as good as the public and private keys used to encrypt and decrypt messages. Thus, secure key management becomes critical. Hackers and crackers will try to break into a system and steal encrypted passwords, for example. If the attacker unravels the encryption key, all passwords will be exposed. Cracked password files give attackers access to bank accounts and criminal databases, for example.

PKI combines encryption, key management, and user authentication into a comprehensive system for implementing TPs and TCBs. It enables users who do not know each other, and perhaps may never meet in reality, to trust one another in cyber-space. That is, PKI defines the way users exchange trust in cyber-space.

PKI has to manage authentication, privileges, keys, and secrecy. In addition, PKI has to be standardized so that authentication, privileges, keys, and secrecy can be exchanged among different systems. Thus, standardization, which is still emerging, is a critical element of PKI.

Two IETF working groups—PKIX (PKI X.509) and SPKI (Simple PKI)—are developing PKI standards. Some of the more important RFCs related to PKI are as follows:

RFC 2401 (Security Architecture for the Internet Protocol, November 1998)

RFC 2437 (PKCS #1: RSA Cryptography Specifications Version 2.0, October 1998)

RFC 2527 (Internet X.509 Public-Key Infrastructure Certificate Policy and Certification Practices Framework, March 1999)

RFC 2692 (SPKI Requirements, September 1999)

RFC 2693 (SPKI Certificate Theory, September 1999)

RFC 2898 (PKCS #5: Password-Based Cryptography Specification, Version 2.0, September 2000)

Definition of PKI

PKI has been defined in several ways, but the following definition was selected because of its simplicity:

> A public-key infrastructure (PKI) is a full system for creating and managing public keys used for encrypting data and exchanging those keys among users. A PKI is a complete system for managing keys that includes policies and working procedures. PKI is about distributing keys in a secure way. Whitfield Diffie and Martin Hellman developed the concept of asymmetric public-key cryptography in 1976, but it was RSA (Rivest, Shamir, Adleman) Data Systems that turned it into a workable and commercial system. Today, RSA is the most popular public-key scheme.[21]

The PKI system described here includes the management of certificates—permissions or privileges shared by a sender and receiver of documents—the management of encryption, and the management of authentication. Therefore, it is more comprehensive than simple encryption or simple authentication. When combined with XML, PKI is called XKI, and it usually incorporates login and password authentication services as well as certificate services.[22]

[21]http://www.linktionary.com/p/pki.html.

[22]Defining boundaries among LDAP directories, authentication, and PKI is a moving target because these technologies seem to be merging into a comprehensive system of security.

The goals of PKI are as follows:

- *Authentication*: PKI assures that the sender is whom he or she claims. This is done by a combination of public key encryption and the use of certificates.
- *Security*: PKI guarantees the integrity of the message; e.g., it has not been modified by an intermediary. This is enforced by encryption.
- *Privacy*: PKI assures the message is decodable only by the intended recipient. This is achieved by encrypting the message and authenticating the recipient.
- *Enforceable*: PKI assures that the message is signed by verified parties. This is implemented in PKI by authenticating the users and trusting the certificate authorities.
- *Nonrepudiation*: PKI guarantees that both parties cannot disavow or deny involvement with the transaction. This is achieved by attaching the private keys of the sender (receiver) to the message.

In the following, we illustrate how a typical PKI system works, and how each of these goals is met by public key encryption.

Certificates

Certificates are tickets for communicating trust. Like a passport or birth certificate, X.509 certificates have become the ad hoc standard for exchanging trust over the Internet. The assumption underlying certificates is that they emanate from a trusted source—the so-called *certificate authorities* (CAs). Thus, trust is based on a root authority that says you are who you say you are, and that you have the privileges stated on your certificate.

X.509 certificates are created by a CA, i.e., a trusted LDAP server or trusted third party. Minimally, they contain the identities and keys of the parties that want to enter into a trusted relationship. For example, the following certificate contains the identity and public keys of Alice and Bob, two users who want to enter into a trusted relationship:

Real Name	Username	Public Key
Alice	A	Public (3, 3233)
Bob	B	Public (17, 6345)

For simplicity, assume Alice's username is A and Bob's username is B. The certificate contains other information, but this simple example will be sufficient to illustrate how PKI works.

Figure 14.6 illustrates how PKI works. Alice, working within her TCB, sends a message to Bob, working within his TCB, through a TP established by encryption,

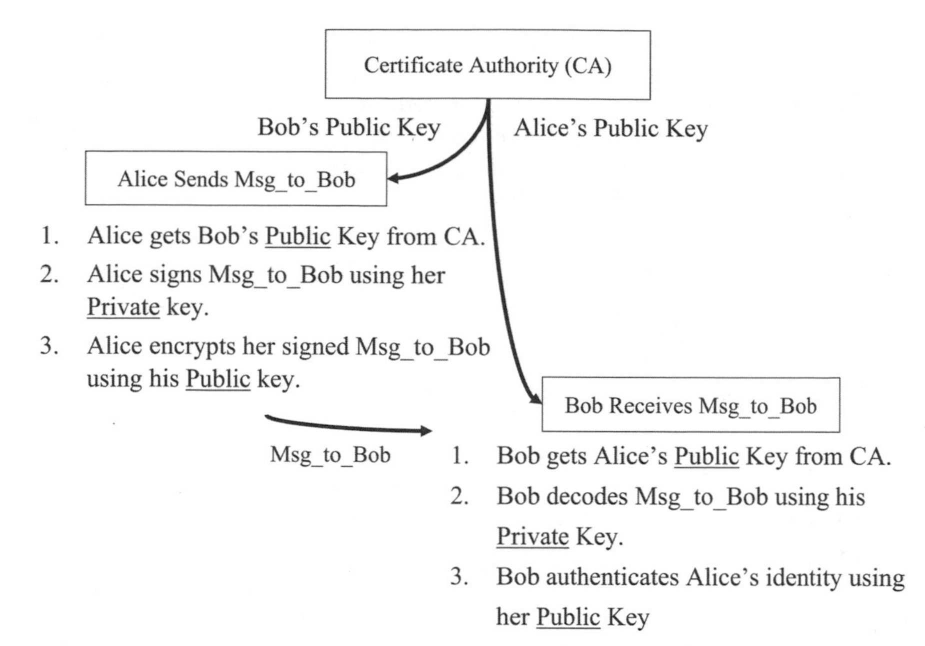

Figure 14.6. An example of PKI: Alice sends Msg_to_Bob to Bob. A TP is established by issuing certificates and encrypting the message using the RSA Algorithm.

authentication, and certificates. The process is started by Alice, who obtains Bob's public key from her CA, signs the message labeled Msg_to_Bob, encrypts it, and sends it to Bob. The CA can be an LDAP server where certificates like the one above are stored and served up whenever the TCB needs to transfer trust from one user to another.

Next, Alice digitally signs her message and encodes her signature using her private key, so only she can unlock her signature. This guarantees that only Alice has signed the message. She then appends her public key to the encoded message so that Bob can verify that she sent the message. A digital signature is an electronic signature that can be used to authenticate the identity of the sender of a message or the signer of a document, and possibly to ensure that the original content of the message or document is unchanged. By using her private key, Alice can determine that the message was actually created and sent by her, because only she knows her private key.

In the final step Alice uses Bob's public key to encrypt Msg_to_Bob, and then she sends the encrypted, signed, message to Bob. This message contains the original message content, Alice's digital signature, and Alice's public key, all encrypted using Bob's public key. By using Bob's public key to encrypt the whole package, only Bob can decode it, and therefore, only Bob and Alice know what was sent. This assures privacy. By including Alice's digital signature, only Alice could have sent the message, because only Alice knows her private key. And by including

Alice's public key, Bob can verify that Alice is the sender. Bob knows the message came from Alice because she is an authenticated user.

At the other end, Bob receives the encrypted Msg_to_Bob from Alice. He looks up her identity from the CA, and it returns a certificate containing Alice's public key. Bob uses his private key to decode the message, which also contains Alice's public key. He verifies that this message actually came from Alice by comparing the public key obtained from the CA with the public key decoded from Msg_to_Bob. If Alice tries to repudiate that she sent the message, she will have a difficult time, because her private key was used to encrypt the signature. Only she could have done this.

The only way someone besides Bob and Alice could have sent or received the message is if someone stole their private keys. Certificates guarantee that Bob and Alice are who they say they are, RSA encryption guarantees security, privacy and enforceability is assured by certificates and CAs, and nonrepudiation is assured by digital signatures.

PKI establishes TPs. If implemented correctly, PKI assures a TCB. But if the CA is cracked, PKI cannot guarantee security, privacy, enforceability, or nonrepudiation. Therefore, it is critically important that keys be protected, certificate authorities be secure, and that the RSA Algorithm never be cracked.

Certificate authorities are hierarchical directories that vouch for one another. At the highest level within the hierarchy—the *root CA*—certificates are signed by the root CA. That is, the root CA vouches for itself. The top-level CAs use digital signatures and certificates to vouch for sublevel CAs. Thus, trust is passed down from a root CA to sublevel CAs. Certificates are signed by trusted CAs using the trusted CAs private key and authenticated using the recipient's public key, just like any other message. Remember, public keys are used to encrypt and private keys to decrypt. Thus, a CA encrypts each certificate using the user's public key. The user decrypts the certificate using his or her private key. The certificate can be verified just like any other message.

COUNTER-MEASURES

The reader is referred back to Figure 14.1 and Table 14.2 in the following. What counter-measures should an IT system use to assure cyber-security? Table 14.2 contains a list of typical counter-measures for the vulnerabilities modeled by the fault tree of Figure 14.1. In general, counter-measures consist of the following:

- Providing backup to power and telecommunications services.
- Installing and operating at least one IDS.
- Installing and operating at least one firewall.
- Installing and updating vender-released software patches.
- Encrypting password files and periodically updating passwords.
- Performing frequent backups.

TABLE 14.2. Sample Counter-Measures Corresponding to Figure 14.1 Vulnerabilities.

Vulnerability	Counter-Measure
Power failure	Install backup power supply
Telecom failure	Buy redundant telecom service
SYN attack	Install IDS
	Install firewall: filter ports
No IDS	Install IDS
Break-in	Install IDS
	Install firewall: filter ports
	Install latest patches
Clear password file	Encrypt password files
No backup	Do periodic backups
No firewall filter	Install firewall: filter ports
No anti-viral SW on desktop	Install patches
	Install anti-virus SW
Clear XML/HTML	Install HTTPS/SSL
	Install PKI/VPN
Clear browser use	Time-out inactive sessions
Password not changed	Change password periodically
War dialing	Close modem ports
No HTTPS/SSL	Install HTTPS/SSL
Browser session open	Time-out inactive sessions
Weak encryption	Install 3DES or AES
	Install PKI
Weak LDAP in applications	Install LDAP directory
	Modify applications
Buffer overflow	Install patches
	Update patches
Weak OS patches	Update patches
	Install IDS
	Install firewall: filter ports
Open Wi-Fi ports	Install IDS
	Install firewall: filter ports
	Encrypt Wi-Fi sessions
	Authenticate Wi-Fi users
Open modem	Close dialup modems or use VPN
Open FTP ports	Close FTP or filter ports
Firewall filter off	Turn on firewall filtering

- Managing ports, especially dialup modem ports.
- Using symmetric and asymmetric encryption to achieve desired level of security.
- Security can be achieved in layers: HTTPS/SSL at the low end, and full PKI at the high end. 3DES/AES can be used where appropriate.

Cyber-security is a tradeoff among expense, effort, inconvenience, privacy, security, and target hardening. Strong encryption protects the Internet from attack, but it also protects the terrorist. Surveillance infringes on privacy, but it is also a weapon in the global war on terrorism. High-assurance systems may be secure, but users are inconvenienced and productivity suffers. Cyber-security is a balancing act.

Richard Pethia, Director of CERT, gave the following testimony before the a subcommittee of the U.S. House in 2003:

> The current state of Internet security is cause for concern. Vulnerabilities associated with the Internet put users at risk. Security measures that were appropriate for mainframe computers and small, well-defined networks inside an organization are not effective for the Internet, a complex, dynamic world of interconnected networks with no clear boundaries and no central control. Security issues are often not well understood and are rarely given high priority by many software developers, vendors, network managers, or consumers.[23]

Pethia goes on to list the following general vulnerabilities of cyber-space:

1. Other critical infrastructures are becoming increasingly dependent on the Internet and are vulnerable to Internet-based attacks.
2. Cyber-space and physical space are becoming one. The growing links between cyber-space and physical space are being exploited by persons bent on causing massive disruption and physical damage.
3. System administration and management is often being performed by people who do not have the training, skill, resources, or interest needed to operate their systems securely.
4. Users often lack adequate knowledge about their network and security. Thus, misconfigured or outdated operating systems, mail programs, and websites result in vulnerabilities that intruders can exploit. A single naive user with an easy-to-guess password can put an entire organization at risk.
5. Product security is not getting better: Developers are not devoting sufficient effort to apply lessons learned about the sources of vulnerabilities. In 1995, CERT received an average of 35 new reports each quarter, 140 for the year. By 2002, the number of annual reports received had skyrocketed to over 4000. Vendors concentrate on time to market, often minimizing that time by placing a low priority on security features.
6. It is often difficult to configure and operate many products securely.
7. There is increased reliance on "silver bullet" solutions, such as firewalls and encryption, lulling organizations into a false sense of security. The security situation must be constantly monitored as technology changes and new exploitation techniques are discovered.

[23] R. D. Pethia, "Cyber security—Growing Risk from Growing Vulnerability," CERT, Software Engineering Institute, Carnegie Mellon University, Pittsburgh, PA. Testimony given before the House Select Committee on Homeland Security Subcommittee on Cybersecurity, Science, and Research and Development, June 25, 2003. http://hsc.house.gov/files/Testimony.pethia.pdf.

8. Compared with other critical infrastructures, the Internet seems to be a virtual breeding ground for attackers. Unfortunately, Internet attacks in general, and DoS attacks in particular, remain easy to accomplish, hard to trace, and a low risk to the attacker. Technically competent intruders duplicate and share their programs and information at little cost, thus enabling novice intruders to do the same damage as the experts. In addition to being easy and cheap, Internet attacks can be quick. In a matter of seconds, intruders can break into a system; hide evidence of the break-in; install their programs, leaving a "back door" so they can easily return to the now-compromised system; and begin launching attacks at other sites.
9. Attackers can lie about their identity and location on the network. Senders provide their return address, but they can lie about it. Most of the Internet is designed merely to forward packets one step closer to their destination with no attempt to make a record of their source. There is not even a "postmark" to indicate generally where a packet originated. It requires close cooperation among sites and up-to-date equipment to trace malicious packets during an attack. Moreover, the Internet is designed to allow packets to flow easily across geographical, administrative, and political boundaries. Consequently, cooperation in tracing a single attack may involve multiple organizations and jurisdictions, most of which are not directly affected by the attack and may have little incentive to invest time and resources in the effort. This means that it is easy for an adversary to use a foreign site to launch attacks at U.S. systems. The attacker enjoys the added safety of the need for international cooperation to trace the attack, compounded by impediments to legal investigations. We have seen U.S.-based attacks on U.S. sites gain this safety by first breaking into one or more non-U.S. sites before coming back to attack the desired target in the United States.
10. There is often a lack of unambiguous or firmly enforced organizational security policies and regulations.
11. There is a lack of well-defined security roles and responsibilities or enforcement of accountability in many organizations, including failure to account for security when outsourcing IT services, failure to provide security awareness training for all levels of staff, nonexistent or weak password management, and poor physical security leading to open access to important computers and network devices.
12. Other practices lead to:
    a. Weak configuration management that leads to vulnerable configuration
    b. Weak authentication practices that allow attackers to masquerade as valid system users.
    c. Lack of vulnerability management practices that require system administrators to quickly correct important vulnerabilities.

d. Failure to use strong encryption when transmitting sensitive information over the network.
e. Lack of monitoring and auditing practices that can detect attacker behavior before damage is done.

Finally, Pethia recommends the following remedies and actions:

1. Incentives for vendors to produce higher quality information technology products with security mechanisms that are better matched to the knowledge, skills, and abilities of today's system managers, administrators, and users. For example:
   a. Vendors should ship their products with "out of the box" configurations that have security options turned on rather than require users to turn them on.
   b. The government should use its buying power to demand higher quality software. The government should consider upgrading its contracting processes to include "code integrity" clauses, clauses that hold vendors more accountable for defects in released products.
2. Wider adoption of risk analysis and risk management policies and practices that help organizations identify their critical security needs, assess their operations and systems against those needs, and implement security improvements identified through the assessment process. What is often missing today is management commitment: senior management's visible endorsement of security improvement efforts and the provision of the resources needed to implement the required improvements.
3. Expanded research programs that lead to fundamental advances in computer security. For example:
   a. Make software virus-resistant/virus-proof.
   b. Reduce implementation errors by at least two orders of magnitude.
   c. Develop a unified and integrated framework for all information assurance analysis and design.
   d. Invent rigorous methods to assess and manage the risks imposed by threats to information assets.
   e. Develop quantitative techniques to determine the cost/benefit of risk mitigation strategies.
   f. Develop methods and simulation tools to analyze cascade effects of attacks, accidents, and failures across interdependent systems.
   g. Develop new technologies for resisting attacks and for recognizing and recovering from attacks, accidents, and failures.
4. Increase the number of technical specialists who have the skills needed to secure large, complex systems.

5. Increase awareness and understanding of cyber-security issues, vulnerabilities, and threats by all stakeholders in cyber-space. For example, children should learn early about acceptable and unacceptable behavior when they begin using computers just as they are taught about acceptable and unacceptable behavior when they begin using libraries.

EXERCISES

1. An open Wi-Fi port can result in (select only one)?
   a. A loss of security
   b. A loss of data
   c. A loss of service
   d. A loss of control
   e. A loss of continuity of operations
2. A secure link between user and system is defined as (select only one)?
   a. VPN
   b. PKI
   c. TCB
   d. TP
   e. Certificate
3. A TCB is defined as (select only one)?
   a. The country's best yogurt
   b. An example of an Internet threat
   c. A mechanism for enforcing minimal security
   d. A malicious program that travels via the Internet
   e. A protocol for ensuring security
4. The DMZ enforces (select one):
   a. Cyber-security policies
   b. PKI standards
   c. X.509 standards
   d. Complete security
   e. Security standards
5. Which of the following are methods of authenticating users (select all that apply)?:
   a. Passwords
   b. DMZ
   c. Biometrics
   d. PKI
   e. X.509 certificates

6. HTTPS/SLL is a protocol for (select the best one):
   a. Serving X.509 certificates to users
   b. Authenticating users
   c. Encrypting communication between user and Web server
   d. Encrypting credit card numbers
   e. Catching man-in-the-middle thieves
7. Tunneling is used in (select only one):
   a. 3DES
   b. RSA
   c. PKI
   d. DMZ
   e. VPN
8. A firewall is a special-purpose computer for guarding (select one):
   a. Ports
   b. PKI
   c. Password files
   d. User logins
   e. Packets
9. An IDS is a special-purpose computer for (select one):
   a. Checking passwords
   b. Preventing break-ins
   c. Nonrepudiation detection
   d. Information assurance
   e. Detecting suspicious patterns
10. San Lewis Rey is home to one of the most successful dot coms in the state called WestBay.com. WestBay does $4 billion per year in e-commerce and has over 5 million repeat customers. It enjoys high customer loyalty because it has never been successfully hacked and, even more important, none of its customers have ever had their identity stolen or been incorrectly charged for purchases made from WestBay. The company wants to keep its reputation intact, so it has allocated $250 million to bolster cyber-security over the next 12 months.

    The fault tree model of WestBay.com vulnerabilities shown in Figure 14.7 was developed by the vice president of cyber-security and used to convince the Board to spend $250 million to bulletproof the servers and network at the heart of the company. A total of $220 million will be spent on training and equipment upgrades, but the remaining $30 million will be invested to reduce the vulnerabilities given by Figure 14.7.

    Assume the following cost and damage estimates, and use the fault probabilities in Figure 14.7 to allocate $30 million in such a way as to minimize

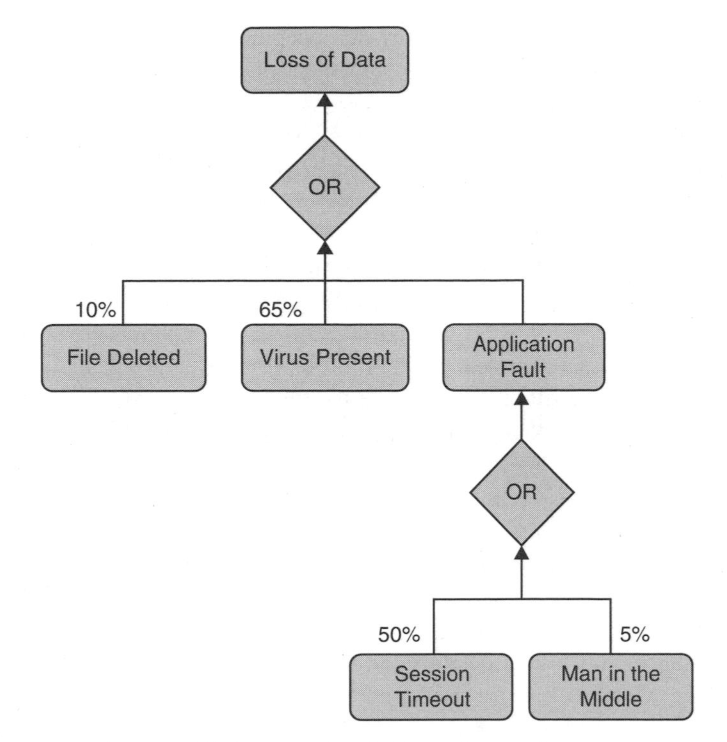

Figure 14.7. Loss of data fault tree for WestBay.com.

financial risk. How should WestBay.com spend its $30 million? How does the ranked allocation differ from the minimal (optimal) allocation strategy?

File Deleted
Cost: $50 million
Damages: $250 million

Virus Present
Cost: $65 million

Damages: $50 million
Session Timeout
Cost: $10 million

Damages: $1 million
Man-in-the-Middle Attack
Cost: $20 million
Damages: $400 million (mostly in consumer confidence).

11. Algorithm 14.2 illustrates public key encoding and decoding for the message {12, 07, 41}, which represents the date December 7, 1941. The public key uses n = 55, but this is too small to encode all years from 0 to 99 (1900 to

1999). Suggest a new public key that accommodates numbers from 0 to 99, and give a private key that decodes the cipher.

12. WestBay.com, San Lewis Rey's largest e-commerce company, uses RSA in its PKI software. Recall that it has over 5 million regular customers. The company wants to guarantee the security of its customers; hence, it uses https/SSL. Which encryption algorithm should WestBay.com use for its 5 million consumers, and why?
13. WestBay.com of San Lewis Rey has an auction department that has to guarantee the security and authenticity of its buyers and sellers. Each time money is exchanged via WestBay.com, the company must know who the buyer is so that it can assure the seller that credit card information is authentic. Which encryption algorithm should it use for this purpose, and why?
14. WestBay.com of San Lewis Rey wants to issue RSA keys to its 1 million customers who regularly use its auction services; how many public and private keys must it issue? Be sure your answer considers the e-commerce transactions that may take place among all 1 million buyers and sellers. If each buyer/seller is allowed to buy and sell to any other buyer/seller, there are $1{,}000{,}000 \times (1{,}000{,}000 - 1)/2 = 500{,}000 \times 499{,}999 =$ nearly 250 billion possible interactions among buyers and sellers.
15. In Question 14, assume Bob and Alice are two WestBay.com customers. Alice wants to send her street address to Bob so that Bob can send her a used computer he recently sold to her via WestBay. But Alice does not want her address known to everyone on the Web, so she elects to use the WestBay.com PKI system that uses RSA public and private keys.
    a. What is the process that takes place between Alice and Bob (see Figure 14.6).
    b. Use program RSA to find public and private keys for Alice and Bob.
    c. Explain how nonrepudiation is implemented in this case: If Alice disavows that she sent her street address to Bob, how can it be shown that she is the only person that could have sent her address to Bob?
    d. Suppose, for some unknown reason, Alice's message went to Albert by mistake. Explain why it is impossible for Albert to decode her message.

INDEX

Critical Infrastructure Protection in Homeland Security: Defending a Networked Nation, edited by Ted G. Lewis

CUSTOMER NOTE: IF THIS BOOK IS ACCOMPANIED BY SOFTWARE, PLEASE READ THE FOLLOWING BEFORE OPENING THE PACKAGE.

This software contains files to help you utilize the models described in the accompanying book. By opening the package, you are agreeing to be bound by the following agreement:

WILEY